(LXII) LAGRANGIAN, HAMILTONIAN

Instructor's Manual to Accompany

Part 1

PHYSICS

For Science and Engineering

Jerry B. Marion
William F. Hornyak

University of Maryland

Prepared and Edited by: Karen Waldo

SAUNDERS GOLDEN SUNBURST SERIES
SAUNDERS COLLEGE PUBLISHING
Philadelphia New York Chicago
San Francisco Montreal Toronto
London Sydney Tokyo Mexico City
Rio de Janeiro Madrid

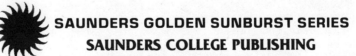

Address orders to:
383 Madison Avenue
New York, NY 10017

Address editorial correspondence to:
West Washington Square
Philadelphia, PA 19105

Instructor's Manual to accompany
PHYSICS FOR SCIENCE AND ENGINEERING, Part 1 ISBN 0-03-058361-6

Library of Congress catalog card number 81-53070.

2345 149 987654321

CBS COLLEGE PUBLISHING
Saunders College Publishing
Holt, Rinehart and Winston
The Dryden Press

TABLE OF CONTENTS

PREFACE

Primarily this Instructor's Manual is intended to give the completely worked out solutions to all the problems in part 1 of Physics for Science and Engineering.

In addition two suggested reading assignment and lecture schedules are given. Various surveys indicate that these two choices may satisfy the course objectives commonly desired. The two volume text is designed to accomodate many choices for coverage of the general material and it goes without saying that many variations from these two suggestions are possible.

Syllabus A: intended for a three semester (or equivalent) course scheduling a total of 135 possible lecture hours. It was assumed that for each semester only 40 hours of actual lectures were possible with 5 hours set aside each time for exams and other uses. It was assumed that one semester of calculus was a prerequisite and that a modest coverage of modern physics was desired. On the average each assignment requires reading less than 9 pages of text.

Syllabus B: intended for a two semester (or equivalent) course scheduling a total of 90 possible lecture hours. It was assumed that for each semester only 40 hours of actual lectures were possible with 5 hours set aside each time for exams and other uses. It was assumed that one semester of calculus was a prerequisite and that no coverage of modern physics was desired. On the average each assignment requires reading less than 9 pages of text.

Teaching experience suggests that an average assignment of four to five problems per lecture hour offers adequate coverage (either syllabus) without requiring excessive work on the part of the students.

FIRST SEMESTER OF THREE SEMESTER TRACK

(40 LECTURE HOURS)*

CHAPTERS		LECTURE HOURS	COMMENTS
1	Introduction	1	
Calculus I		½	** Review
Calculus II		½	
2	Kinematics of Linear Motion	3	-
3	Vectors and Coordinate Frames	2	-
4	Kinematics in Two and Three Dimensions	3	-
5	Dynamics and Newton's Laws	2	omit section 5
6	Applications of Newton's Laws	2	omit section 3
Calculus III		1	** Review
7	Mechanical Encrgy and Work	3	-
Calculus IV		1	-
8	Forces and Potential Energy	1(2)	perhaps include section 3
9	Linear Momentum	3	-
10	Angular Momentum, Dynamics of Systems	2	brief attention to section 3
Calculus V		1	-
11	Rotation of Rigid Bodies	3	brief attention to section 4
12	Dynamics of Rigid Bodies	2	omit section 3
13	Forces in Equilibrium	1	-
14	Gravitation and Planetary Motion	3	-
Calculus VI		½	combine Calculus V and section 1,
15	Oscillatory Motion	2½	omit sections 5, 6, and 7
16	Deformations of Solids	2(1)	perhaps omit sections 3 and 4
	TOTAL	40 Lecture hours	

* 5 Lecture hours set aside for tests and other uses

** Assuming one semester of calculus as prerequisite

SECOND SEMESTER OF THREE SEMESTER TRACK

(40 LECTURE HOURS)*

CHAPTERS		LECTURE HOURS	COMMENTS
17	Static Fluids	2	omit section 5
18	Fluid Dynamics	2	omit sections 5, 6, and 7
19	Mechanical Waves	4	omit sections 8 and 9
20	Sound	2	omit sections 6 and 7
21	Special Relativity	4**	brief attention to section 9
22	Temperature and Heat	2	-
23	The Behavior of Gases	2	omit section 5
24	Kinetic Theory	3	brief attention to section 4
25	Thermodynamics I	2	-
26	Thermodynamics II	3	include much of section 6
27	Electric Charge	1	-
28	The Electric Field	2	-
29	Electric Potential	2	omit section 5
30	Capacitance and Dielectrics	3	brief attention to sections 5 and 6
31	Current and Resistance	3	include much of section 6
32	DC Circuits	3	include much of sections 4 and 5
	TOTAL	40 Lecture hours	

* 5 Lecture hours set aside for tests and other uses

** Special Relativity is sometimes postponed to the third semester. In order to be grouped with modern physics topics it may exchange with Chapters 33 and 34.

THIRD SEMESTER OF THREE SEMESTER TRACK

(40 LECTURE HOURS)*

CHAPTERS		LECTURE HOURS	COMMENTS
33	The Magnetic Field	2	-
34	Sources of Magnetic Fields	2	-
35	Electromagnetic Induction	3	brief attention sections 5 and 6
36	Inductance and Magnetic Materials	4(3)	brief attention sections 4 and 5
Calculus VII		½	combine Calculus VII with section 3,
37	Circuits with Time-Varying Currents	3½	omit section 6
38	Electromagnetic Waves	2(3)	-
39	Radiating Systems	2(1)	give chapter brief attention
40	Light	2	omit section 6
41	Geometrical Optics	3	-
42	Interference	2	omit section 5
43	Diffraction	2	brief attention sections 3 and 4
44	Polarized Light	2	omit section 5
45	Optical Instruments	2(3)	omit section 6
46	Quanta, Waves, and Particles **	4	-
47	Atoms	4	-
	TOTAL	40 Lecture hours	

* 5 Lecture hours set aside for tests and other uses

** Special Relativity is sometimes scheduled to just preceed this Chapter; may exchange with Chapters 33 and 34.

FIRST SEMESTER OF TWO SEMESTER TRACK

(40 LECTURE HOURS)*

CHAPTERS		LECTURE HOURS	COMMENTS
1	Introduction	1	-
Calculus I		½	brief review
Calculus II		½	
2	Kinematics of Linear Motion	2	omit section 6
3	Vectors and Coordinate Frames	2	-
4	Kinematics in Two and Three Dimensions	2	brief attention to section 6
5	Dynamics and Newton's Laws	2	omit sections 4 and 5
6	Applications of Newton's Laws	2	omit section 3
Calculus III		½	combine as needed
7	Mechanical Energy and Work	2½	
Calculus IV		-	omit chapter
8	Forces and Potential Energy	-	omit chapter
9	Linear Momentum	2	brief attention to section 5
10	Angular Momentum, Dynamics of Systems	2	brief attention to section 3
Calculus V		-	combine as needed and give brief attention to section 11-4.
11	Rotation of Rigid Bodies	2	
12	Dynamics of Rigid Bodies	2	omit section 3
13	Forces in Equilibrium	1	-
14	Gravitation and Planetary Motion	2	omit section 5, cover 6 briefly
Calculus VI		-	combine as needed and omit sections 15-5, 6, and 7
15	Oscillatory Motion	2	
16	Deformations of Solids	-	omit chapter
17	Static Fluids	-	omit chapter
18	Fluid Dynamics	-	omit chapter
19	Mechanical Waves	3	omit sections 8 and 9
20	Sound	2	omit sections 6 and 7
21	Special Relativity	-	omit chapter
22	Temperature and Heat	2	-
23	The Behavior of Gases	1	brief attention to sections 3,4, and 5
24	Kinetic Theory	2	omit section 4
25	Thermodynamics I	1	-
26	Thermodynamics II	1	omit sections 5 and 6
TOTAL		40	Lecture hours

* 5 Lecture hours set aside for tests and other uses.

SECOND SEMESTER OF TWO SEMESTER TRACK

(40 LECTURE HOURS)*

	CHAPTERS	LECTURE HOURS	COMMENTS
27	Electric Charge	1	-
28	The Electric Field	2	-
29	Electric Potential	3	omit section 5
30	Capacitance and Dielectrics	2	brief attention to section 5, omit 6
31	Current and Resistance	2	perhaps brief attention to section 6
32	DC Circuits	2	omit sections 4 and 5
33	The Magnetic Field	2	brief attention to section 5
34	Sources of Magnetic Fields	2	-
35	Electromagnetic Induction	3	brief attention to sections 5 and 6
36	Inductance and Magnetic Materials	3	-
Calculus VII		½	combine as needed, and omit sections 5 and 6
37	Circuits with Time-Varying Currents	2½	
38	Electromagnetic Waves	2	-
39	Radiating Systems	1	brief attention to chapter
40	Light	2	omit section 6
41	Geometrical Optics	2	omit section 5
42	Interference	2	omit section 5
43	Diffraction	2	omit sections 3 and 4
44	Polarized Light	2	omit section 5
45	Optical Instruments	2	omit sections 5 and 6
46	Quanta, Waves, and Particles	-	omit chapter
47	Atoms	-	omit chapter
	TOTAL	40	Lecture hours

* 5 Lecture hours set aside for tests and other uses

Chapter 1

1-1 From the text, the speed is 40 mi/h.

$$40 \text{ mi/h} = \left(40\ \frac{\text{mi}}{\text{h}}\right)\left(5280\ \frac{\text{ft}}{\text{mi}}\right)\left(12\ \frac{\text{in.}}{\text{ft}}\right)\left(2.54\ \frac{\text{cm}}{\text{in.}}\right)\left(\frac{1}{100}\ \frac{\text{m}}{\text{cm}}\right)\left(\frac{1}{3600}\ \frac{\text{h}}{\text{s}}\right)$$

$$= 17.88 \text{ m/s}$$

1-2 The signs will read in km/h; thus, 55 mi/h $= \left(55\ \frac{\text{mi}}{\text{h}}\right)\left(\frac{1.609 \text{ km}}{1 \text{ mi}}\right) = 88.5$ km/h

1-3 $1 \text{ y} = (1 \text{ y}) \times \left(365\ \frac{\text{d}}{\text{y}}\right)\left(24\ \frac{\text{h}}{\text{d}}\right)\left(3600\ \frac{\text{s}}{\text{h}}\right) = 3.15 \times 10^{7} \text{ s}$

1-4 $\frac{1500 \text{ m}}{1 \text{ mi}} = \frac{1.5 \text{ km}}{1.609 \text{ km}} = 0.932$

1-5 The circumference of the Earth is $C = 4.01 \times 10^{7}$ m.

To find the diameter,

$$d = \frac{C}{\pi} = 1.28 \times 10^{7} \text{ m} = 12{,}800 \text{ km}$$

$$= (12{,}800 \text{ km})\left(\frac{1}{1.609}\ \frac{\text{mi}}{\text{km}}\right) = 7955 \text{ mi}$$

1-6 1 in. = 2.54 cm

$(1 \text{ in.})^3 = (2.54 \text{ cm})^3$

$1 \text{ in.}^3 = 16.39 \text{ cm}^3$

1-7 Speed $= \left(\frac{100 \text{ yd}}{9.52 \text{ s}}\right)\left(36\ \frac{\text{in.}}{\text{yd}}\right)\left(0.0254\ \frac{\text{m}}{\text{in.}}\right) = 9.61$ m/s

$100 \text{ m}\left(\frac{1 \text{ s}}{9.61 \text{ m}}\right) = 10.4 \text{ s}$

10.4 s to complete the 100 m distance.

1-8 We have,

1 in. = 2.54 cm = 0.0254 m

$1 \text{ mil} = 10^{-3} \text{ in.} = (0.0254 \times 10^{-3}) \text{ m}$

$= 2.54 \times 10^{-5} \text{ m}$

$1 \text{ microinch} = 10^{-6} \text{ in.} = (0.0254 \times 10^{-6}) \text{ m}$

$= 2.54 \times 10^{-8} \text{ m}$

1-9 $1 \text{ acre} = \frac{1}{640} \text{ mi}^2$

$$= \left(\frac{1}{640} \text{ mi}^2\right) \times \left(\left(1.609\ \frac{\text{km}}{\text{mi}}\right)^2\right) \times \left(\left(1000\ \frac{\text{m}}{\text{km}}\right)^2\right)$$

$$= 4045 \text{ m}^2$$

1-10 $1\ yd = 36\ in. = (36)(2.54)\ cm = 91.44\ cm$

$(1\ yd)^2 = (91.44\ cm)^2$

$1\ yd^2 = 8360\ cm^2$

$3.65\ yd^2 = (3.65)(8360)\ cm^2 = 3.05 \times 10^4\ cm^2$

1-11 Let n be the number of years required for an accumulated error of one day.

$n[365.25\ d - (365\ d, 5\ h, 48\ min., 45.6\ s)] = 1\ d$

$n\ (365.25\ d - 365.2422\ d) = 1\ d$

so that

$n = 128.2\ y$

To correct this error, century years (1700, 1800, 1900, 2100, 2200, ...) are not to be counted as leap years, unless the first two digits are divisible by 4; thus, 2000, 2400, etc., are leap years. With this scheme, how long is required to accumulate an error of one day?

1-12 We have

$x = v_o t + \frac{1}{2} at^2$

We know that the dimensions of x and $\frac{1}{2} at^2$ must be the same, that is,

$[x] = [\frac{1}{2} at^2] = L$

And,

$[t] = T$

Thus,

$[a]\ [t]^2 = L$

$[a] = L/T^2 = LT^{-2}$

1-13 We have

$[h] = L$

$[g] = LT^{-2}$

And,

$[v] = LT^{-1}$

Thus,

$[v]^2 = [g][h]$

$v^2 = K(gh)$

where K is the constant of proportionality.

1-14 $\rho_s = 7.86 \times 10^3\ kg/m^3, \quad M_s = 4.86\ kg$

$$V = \frac{M_s}{\rho_s} = 6.18 \times 10^{-4}\ m^3$$

$$M_m = V \rho_m = 1.58\ kg$$

1-15 $V = \frac{M}{\rho} = \frac{62\ kg}{8.93 \times 10^3\ kg/m^3} = 6.94 \times 10^{-3}\ m^3$

$$V = T(\pi r^2)$$

where T is the thickness of the plate.

$$T = \frac{V}{\pi r^2} = \frac{6.94 \times 10^{-3} m^3}{\pi (0.243\ m)^2} = 3.74 \times 10^{-2}\ m$$

1-16 $\rho_a = 1.293\ kg/m^3$

$$M = V \rho_a = (8\ m)(6\ m)(3.2\ m)\ (1.293\ kg/m^3)$$

$$M = 198.6\ kg$$

1-17 $\rho_w = 1.00 \times 10^3\ kg/m^3$

$$= \left(1 \times 10^3 \frac{kg}{m^3}\right) \left(2.20 \frac{lb}{kg}\right) \left[(0.305 \frac{m}{ft})^3\right]$$

$$= 62.4\ lb/ft^3$$

1-18 a) $\rho_E = \frac{M_E}{V_E} = \frac{M_E}{\frac{4}{3} \pi R_E^3}$

From the table,

$$R_E = 6.38 \times 10^6\ m$$

$$M_E = 5.97 \times 10^{24}\ kg$$

Thus,

$$\rho_E = 5.49 \times 10^3\ kg/m^3$$

b) $\rho_S = \frac{M_S}{V_S} = \frac{M_S}{\frac{4}{3} \pi R_S^3}$

From the table,

$$R_S = 6.96 \times 10^8\ m$$

$$M_S = 1.99 \times 10^{30}\ kg$$

Thus,

$$\rho_S = 1.41 \times 10^3\ kg/m^3$$

I.

I-1 $G(v) = \frac{1}{3} v^2 + \sin(\frac{\pi}{2} v)$

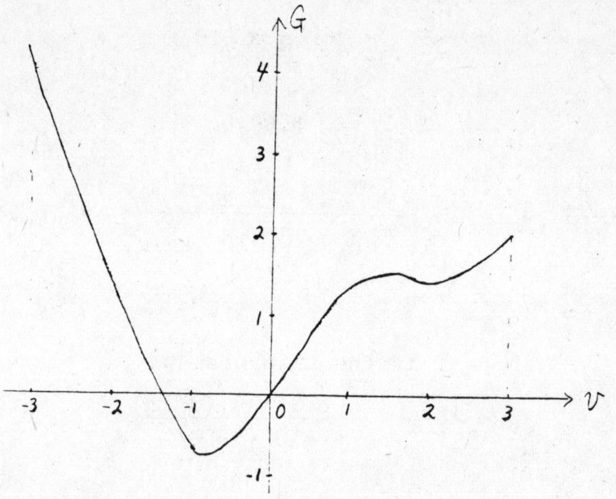

I-2 By a method of successive approximations, we want to solve $\frac{1}{3} v^2 = 3 - \sin(\frac{\pi}{2} v)$

1st: $\frac{1}{3} v_1^2 = 3 \qquad v_1 = 3$

2nd: $\frac{1}{3} v_2^2 = 3 - \sin(\frac{\pi}{2} \times 3) = 3 + 1 = 4 \qquad v_2 = \sqrt{12} = 3.46410$

3rd: $\frac{1}{3} v_3^2 = 3 - \sin(\frac{\pi}{2} \times 3.46410) = 3.74583 \qquad v_3 = \sqrt{3 \times 3.74583} = 3.35224$

4th: $\frac{1}{3} v_4^2 = 3 - \sin(\frac{\pi}{2} \times 3.35224) = 3.85080 \qquad v_4 = \sqrt{3 \times 3.85080} = 3.39888$

5th: $\frac{1}{3} v_5^2 = 3 - \sin(\frac{\pi}{2} \times 3.39888) = 3.81005 \qquad v_5 = \sqrt{3 \times 3.81005} = 3.38085$

6th: $\frac{1}{3} v_6^2 = 3 - \sin(\frac{\pi}{2} \times 3.38085) = 3.82633 \qquad v_6 = \sqrt{3 \times 3.82633} = 3.38807$

$v = (v_5 + v_6)/2 = 3.384$

I-3 $y = v_o t - \frac{1}{2} g t^2$

(a) $0 \le y \le \frac{v_o^2}{2g}$ Domain

$0 \le t \le \frac{2v_o}{g}$ Range

(b) $t^2 - \frac{2v_o}{g} t + \frac{2}{g} y = 0$

$$t = \tfrac{1}{2}\left(\frac{2v_o}{g} \pm \sqrt{\left(\frac{2v_o}{g}\right)^2 - 4\left(\frac{2}{g} y\right)}\right) = \frac{v_o}{g}\left(1 \pm \sqrt{1 - \frac{2gy}{v_o^2}}\right)$$

The branch from the positive sign is the solid line.

The branch from the negative sign is the broken line.

This is a multiple-valued function so each branch must be considered separately or one may get the inverse of the original function by turning f(t) 90° clockwise and inverting the vertical axis.

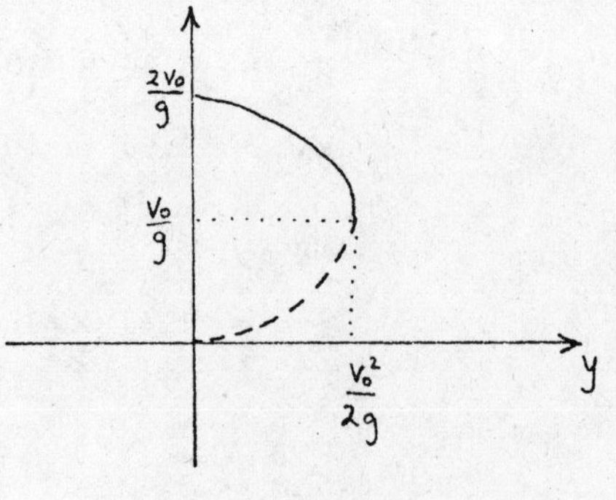

(c) $0 \le t \le \dfrac{2v_o}{g}$ Domain

$0 \le y \le \dfrac{v_o^2}{2g}$ Range

I-4 $y = \dfrac{1}{4}x^4 - x^2$

$x^4 - 4x^2 - 4y = 0 \qquad x^2 = 2 \pm 2\sqrt{1+y}$

So $x = \pm\sqrt{2 \pm 2\sqrt{1+y}}$

1st branch: $x = \pm\sqrt{2 - 2\sqrt{1+y}}$ where $-1 \le y \le 0$

(Dotted line on the graph)

2nd branch: $x = \pm\sqrt{2 + 2\sqrt{1+y}}$ where $-1 \le y \le \infty$

(Solid line on the graph)

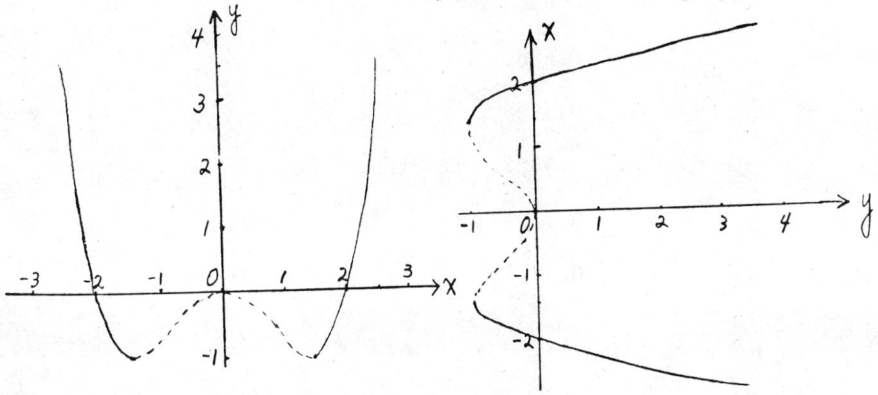

To get a graph of $x = g(y)$, turn $y = f(x)$ counterclockwise (clockwise) by 90° around the origin and invert the y (x-axis).

I-5 (a) $\Delta y = (m(x + \Delta x) + b) - (mx + b) = m\Delta x$

(b) $\Delta y = e^{x+\Delta x} - e^{x} = e^{x}(e^{\Delta x} - 1)$

(c) $\Delta y = \ell n\,(x + \Delta x) - \ell n\, x = \ell n\,\dfrac{x+\Delta x}{x} = \ell n(1 + \dfrac{\Delta x}{x})$

I-6 $n = 0 \quad e \cong 1.0$

$n = 1 \quad e \cong 2.0$

$n = 2 \quad e \cong 2.50$

$n = 3 \quad e \cong 2.666$

$n = 4 \quad e \cong 2.7083$

$n = 5 \quad e \cong 2.7166$

$n = 6 \quad e \cong 2.71805$

$n = 7 \quad e \cong 2.718254$

$n = 8 \quad e \cong 2.7182788$

$n = 9 \quad e \cong 2.7182815$

$e = 2.7182818$ to eight digits

I-7 Euler's constant: $\gamma = \lim_{s\to\infty} [1 + \frac{1}{2} + \frac{1}{3} + \cdots \frac{1}{s} - \ln s] = 0.5772157$

Let $\gamma_n = 1 + \frac{1}{2} + \frac{1}{3} + \cdots \frac{1}{n} - \ln n$

n	γ_n
2	0.8069
4	0.6970
8	0.6384
16	0.6081
32	0.5928
100	0.5822 (% diff. ~ 0.86%)
1000	0.57772 (% diff. ~ 0.087%)
10000	0.5777268 (% diff. ~ 0.0091%)

I-8 $\lim_{x\to 0} \frac{10^x - 1}{x} = \ln 10 = 2.302585093$

x	$10^x - 1$	$(10^x - 1)/x$
1/2	2.16227766017	4.32455532034
1/8	0.33352143216	2.66817145728
1/64	0.03663292844	2.34450742016
1/128	0.01815172172	2.32342038016
1/256	0.00903504484	2.31297147904
1/512	0.00450736425	2.30777049600
1/2048	0.0011249414	2.3038799872
1/16384	0.00014054852	2.30274695168
1/13072	0.00001756748	2.30260473894

I-9 To calculate $\sqrt{10}$ (N = 10) by an iterative process:

(a) Guess A = 3 $\quad$ ($\sqrt{10} = 3.1622777...$)

$B = \frac{1}{2}[A + (N/A)] = \frac{1}{2}[3 + (10/3)] = 3.1666667$

(% diff. ~ 0.1388%)

$$C = \frac{1}{2}[B + (N/B)] = \frac{1}{2}[3.1666667 + (10/3.1666667)] = 3.1622807$$

(% diff. ~ 0.0000949%)

$$D = \frac{1}{2}[C + N/C)] = \frac{1}{2}[3.1622807 + (10/3.1622807)] = 3.1622777$$

(% diff. ~ 0%)

(b) Guess A = 5

$B = \frac{1}{2}[A + (N/A)] = 3.50$ (% diff. ~ 10.680%)

$C = \frac{1}{2}[B + (N/B)] = 3.1785714$ (% diff. ~ 0.5153%)

$D = \frac{1}{2}[C + (N/C)] = 3.1623196$ (% diff. ~ 0.0013%)

$E = \frac{1}{2}[D + (N/D)] = 3.1622777$ (% diff. ~ 0%)

Try N = 19 with guess A = 4

$B = \frac{1}{2}[A + (N/A)] = 4.375$ (% diff. ~ 0.4267%)

$C = \frac{1}{2}[B + (N/B)] = 4.3589286$ (% diff. ~ 0.00068%)

$D = \frac{1}{2}[C + (N/C)] = 4.3588989$ (% diff. ~ 0%)

($\sqrt{19} = 4.3588989...$)

I-10 (a) $y = \frac{1}{x}$

$$\frac{\Delta y}{\Delta x} = \frac{\frac{1}{\Delta x + x} - \frac{1}{x}}{\Delta x} = \frac{\frac{x - (x + \Delta x)}{x(\Delta x + x)}}{\Delta x} = \frac{1}{x^2 + x\Delta x}$$

$$\frac{dy}{dx} = \lim_{\Delta x \to 0} \frac{\Delta y}{\Delta x} = -\frac{1}{x^2}$$

(b) $y = \cos\theta$

$$\frac{\Delta y}{\Delta \theta} = \frac{\cos(\theta + \Delta\theta) - \cos\theta}{\Delta\theta} = \frac{-2\sin(\theta + \frac{1}{2}\Delta\theta)\sin(\frac{1}{2}\Delta\theta)}{\Delta\theta}$$

$$= -\sin(\theta + \tfrac{1}{2}\Delta\theta)\frac{\sin(\frac{1}{2}\Delta\theta)}{\Delta\theta}$$

$$\frac{dy}{d\theta} = \lim_{\Delta\theta \to 0} \frac{\Delta y}{\Delta\theta} = (-\sin\theta)(1) = -\sin\theta$$

(c) $y = \tan\phi$

$$\frac{\Delta y}{\Delta\phi} = \frac{\tan(\phi + \Delta\phi) - \tan\phi}{\Delta\phi} = \frac{1}{\Delta\phi}\left[\frac{\tan\phi + \tan\Delta\phi}{1 - \tan\phi\tan\Delta\phi} - \tan\phi\right]$$

$$= \frac{\tan\Delta\phi}{\Delta\phi}\left(\frac{1 + \tan^2\phi}{1 - \tan\phi\tan\Delta\phi}\right) = \left(\frac{\sin\Delta\phi}{\Delta\phi}\right)\left(\frac{1}{\cos\Delta\phi}\right)\left(\frac{1 + \tan^2\phi}{1 - \tan\phi\tan\Delta\phi}\right)$$

$$\frac{dy}{d\phi} = \lim_{\Delta\phi \to 0} \frac{\Delta y}{\Delta\phi} = (1)(1)\left(\frac{1 + \tan^2\phi}{1 - 0}\right) = \sec^2\phi$$

I-11 (a) $\frac{d}{dx}(mx + b) = \frac{d}{dx}(mx) + \frac{db}{dx} = m\frac{dx}{dx} + 0 = m \cdot 1 = m$

(b) $\frac{d}{dt}(\frac{1}{2}at^2 + v\,t_0 + s_0) = \frac{1}{2}a\frac{dt^2}{dt} + v_0\frac{dt}{dt} + \frac{ds_0}{dt} = \frac{1}{2}a \cdot 2t + v_0 \cdot 1 + 0 = at + v_0$

(c) Set $y = a^2 - x^2$.

$$\frac{d}{dx}\left(\sqrt{a^2 - x^2}\right) = \frac{dy}{dx}\frac{d}{dy}(\sqrt{y}) = (-2x)(\tfrac{1}{2}y^{-\frac{1}{2}}) = \frac{-x}{\sqrt{y}} = \frac{-x}{\sqrt{a^2 - x^2}}$$

I-12 (a) Set $y = 2ax$.

$$\frac{d}{dx}(2ax)^{\frac{1}{2}} = \frac{d}{dx}(y)^{\frac{1}{2}} = \frac{dy}{dx}\frac{d}{dy}(y)^{\frac{1}{2}} = 2a \cdot \tfrac{1}{2}\, y^{-\frac{1}{2}} = \frac{a}{\sqrt{2ax}} = \sqrt{\frac{a}{2x}}, \quad (x > 0)$$

(b) Set $y = 2x$. $\dfrac{d}{dx}(\sin 2x) = \dfrac{dy}{dx}\dfrac{d}{dy}(\sin y) = 2 \cos y = 2 \cos 2x$

(c) Set $u = \cos\theta$ and $v = \theta$.

$$\frac{d}{d\theta}\frac{\cos\theta}{\theta} = \frac{d}{d\theta}\left(\frac{u}{v}\right) = \frac{v\frac{du}{d\theta} - u\frac{dv}{d\theta}}{v^2} = \frac{\theta \cdot (-\sin\theta) - \cos\theta \cdot 1}{\theta^2}$$

$$= -\frac{\theta \sin\theta + \cos\theta}{\theta^2} \quad \text{for } \theta \neq 0$$

I-13 (a) $y = x \sin x$

$\dfrac{dy}{dx} = x \cos x + \sin x$. At $x = \dfrac{\pi}{2}$, $\dfrac{dy}{dx} = \dfrac{\pi}{2} \cdot 0 + 1 = 1$

$\dfrac{d^2y}{dx^2} = -x \sin x + 2 \cos x$. At $x = \dfrac{\pi}{2}$, $\dfrac{d^2y}{dx^2} = -\dfrac{\pi}{2} \cdot 1 + 2 \cdot 0 = -\dfrac{\pi}{2}$

(b) $s = \tfrac{1}{2} at^2 + v_o t + s_o$, $\dfrac{ds}{dt} = at + v_o$, $\dfrac{d^2s}{dt^2} = a$

At $t = 0$, $\dfrac{ds}{dt} = a \cdot 0 + v_o = v_o$, $\dfrac{d^2s}{dt^2} = a$

(c) $u = e^{-\alpha t} \cos \omega t$, $\dfrac{du}{dt} = e^{-\alpha t}(-\omega \sin \omega t - \alpha \cos \omega t)$

$$\frac{d^2u}{dt^2} = e^{-\alpha t}(-\omega^2 \cos \omega t + 2\,\omega\,\alpha \sin \omega t + \alpha^2 \cos \omega t)$$

$$\text{At } t = 0: \begin{cases} \dfrac{du}{dt} = e^{-0}(-\omega \sin 0 - \alpha \cos 0) \\ \quad = 1\,(-\omega \cdot 0 - \alpha \cdot 1) = -\alpha \\ \dfrac{d^2u}{dt^2} = e^{-0}(-\omega^2 \cdot 1 + 2\,\omega\alpha \sin 0 + \alpha^2 \cdot 1) = \alpha^2 - \omega^2 \end{cases}$$

$$\text{At } t = \infty \quad \begin{cases} \dfrac{du}{dt} = e^{-\infty}(-\omega \sin\infty - \alpha\cos\infty) = 0 \\[2ex] \dfrac{d^2u}{dt^2} = e^{-\infty}(-\omega^2 \cos\infty + 2\,\omega\alpha \sin\infty + \alpha^2\cos\infty) = 0 \\[2ex] (\text{Note: } e^{-\infty} = 0,\ \sin\infty \text{ and } \cos\infty = \text{finite number.}) \end{cases}$$

I-14 $y = x^n = uv$ with $u = x^{n-m}$, $v = x^m$

$$\frac{dy}{dx} = \frac{dx^n}{dx} = \frac{d}{dx}(uv) = u\frac{dv}{dx} + v\frac{du}{dx}$$

$$= x^{n-m}\, mx^{m-1} + x^m (n-m)x^{n-m-1} = mx^{n-1} + (n-m)x^{n-1} = nx^{n-1}$$

I-15 $\ell n(1 + x) = x - \dfrac{x^2}{2} + \dfrac{x^3}{3} - \dfrac{x^4}{4} + \cdots + (-1)^{r-1}\dfrac{x^r}{r} + \cdots$

Differentiating both sides,

$$\frac{1}{1+x} = 1 - x + x^2 - x^3 + \cdots + (-1)^{r-1} x^{r-1} + \cdots$$
$$= 1 - x + x^2 - x^3 + \cdots + (-1)^r x^r + \cdots$$

From Eq. A-13 with a = 1, b = x, and n = -1,

$$\frac{1}{1+x} = 1 + (-1)x + \frac{(-1)(-2)}{2!}x^2 + \frac{(-1)(-2)(-3)}{3!}x^3 + \cdots$$
$$\cdots + \frac{(-1)(-2)(-3)\cdots(-r)}{r!}x^r + \cdots$$

Now we know $\dfrac{(-1)(-2)(-3)\cdots(-r)}{r!} = (-1)^r \dfrac{r!}{r!} = (-1)^r$. So,

$$\frac{1}{1+x} = 1 - x + x^2 - x^3 + \cdots + (-1)^r x^r + \cdots,$$

which agrees with Eq. A-16

I-16 Eq. A-12: $e^x = 1 + \dfrac{x}{1!} + \dfrac{x^2}{2!} + \dfrac{x^3}{3!} + \cdots + \dfrac{x^n}{n!} + \cdots$

$$\frac{d}{dx}(e^x) = e^x = \frac{d}{dx}\left(1 + \frac{x}{1!} + \frac{x^2}{2!} + \frac{x^3}{3!} + \cdots + \frac{x^n}{n!} + \cdots\right)$$
$$= 0 + 1 + \frac{2}{2!}x + \frac{3}{3!}x^2 + \cdots \frac{n}{n!}x^{n-1} + \cdots$$
$$= 0 + 1 + \frac{x}{1!} + \frac{x^2}{2!} + \cdots + \frac{1}{(n-1)!}x^{n-1} + \cdots$$

$$e^x = 1 + \frac{x}{1!} + \frac{x}{2!} + \frac{x}{3!} + \cdots + \frac{x^n}{n!} + \cdots,$$

which is Eq. A-12

I-17 (a) $y = \sin x, \frac{dy}{dx} = \cos x, \tan \alpha = \left[\frac{dy}{dx}\right]_{x=\frac{\pi}{4}} = \cos \frac{\pi}{4} = \frac{\sqrt{2}}{2}$

(b) $y = e^{-ax}, \frac{dy}{dx} = -a\, e^{-ax}, \tan \alpha = \left[\frac{dy}{dx}\right]_{x=0} = -a\, e^{-0} = -a$

(c) $y = xe^{-x}, \frac{dy}{dx} = e^{-x}(1 - x), \tan \alpha = \left[\frac{dy}{dx}\right]_{x=1} = e^{-0}(1 - 1) = 0$

I-18 (a) $y = \frac{1}{2} x^2 + 1$

$\frac{dy}{dx} = x \qquad \tan \alpha = \left[\frac{dy}{dx}\right]_{x=2} = 2$

tangential line: $y = mx + b \quad (m = \tan \alpha)$

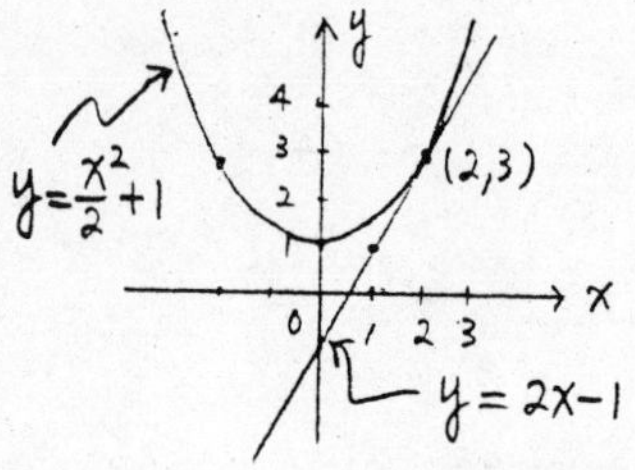

$$\left.\begin{aligned} y &= 2x + b \\ [y]_{x=2} &= \tfrac{1}{2} \cdot 2^2 + 1 = 3 \end{aligned}\right\} \qquad 3 = 2 \cdot 2 + b \qquad \text{So} \quad b = -1$$

Thus, the equation of the tangent line is

$$y = 2x - 1$$

(b) $y = (x - 1)^3, \frac{dy}{dx} = 3(x - 1)^2, \; y = \begin{cases} -1 & \text{at } x = 0 \\ 0 & \text{at } x = 1, \end{cases} \frac{dy}{dx} = \begin{cases} 3 & \text{at } x = 0 \\ 0 & \text{at } x = 1 \end{cases}$

Tangential line at $x = 0$: $y = 3x + b;\ -1 = 3 \cdot 0 + b;\ b = -1$
So $y = 3x - 1$ is the equation of the tangential line at $x = 0$

Tangential line at $x = 1$: $y = 0x + b;\ 0 = 0 \cdot 1 + b;\ b = 0$
So $y = 0$ is the equation of the tangential line at $x = 1$

(b)

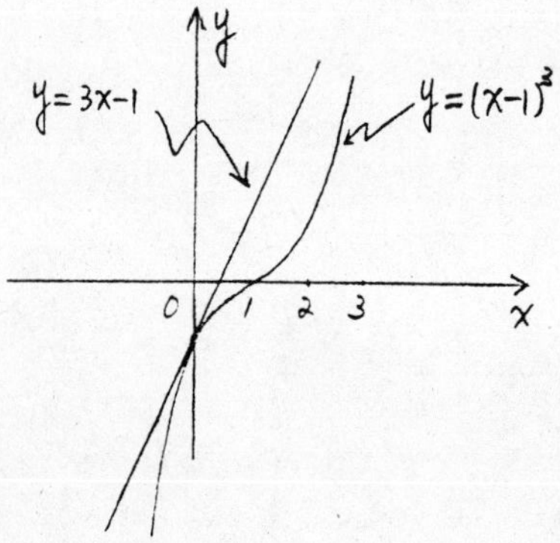

(c)

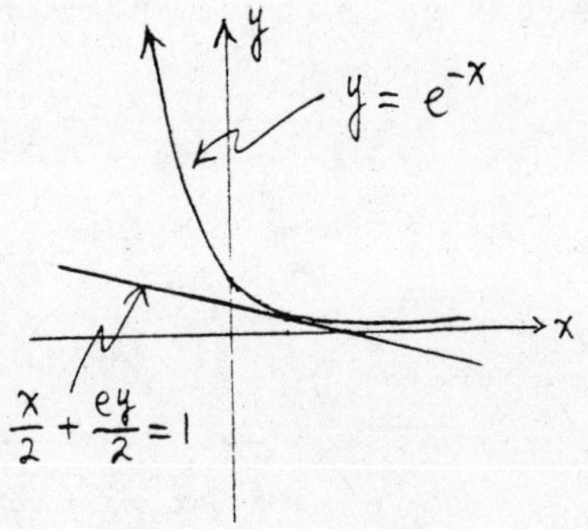

(c) $y = e^{-x}, \frac{dy}{dx} = -e^{-x}, \quad \left[\frac{dy}{dx}\right]_{x=1} = -e^{-1}$

Tangential line at $x = 1$:

$y = -e^{-1} x + b;\ e^{-1} = -e^{-1} \cdot 1 + b;\ b = 2e^{-1}$

$y = -e^{-1} x + 2e^{-1}$ or $\frac{x}{2} + \frac{ye}{2} = 1$

I-19 (a) $y = \dfrac{ax}{x^2 + 1}$; $\dfrac{dy}{dx} = \dfrac{a(x^2 + 1) - ax \cdot 2x}{(x^2 + 1)^2} = \dfrac{a(1 - x^2)}{(x^2 + 1)^2}$

$$\frac{d^2y}{dx^2} = \frac{(x^2 + 1)^2 (-2ax) - a(1 - x^2)\, 2\,(x^2 + 1) \cdot 2x}{(x^2 + 1)^4}$$

$$= \frac{(x^2 + 1)x[-2ax^2 - 2a - 4a + 4ax^2]}{(x^2 + 1)^4} = \frac{2ax(x^2 - 3)}{(x^2 + 1)^3}$$

Extremum: $\dfrac{dy}{dx} = \dfrac{a(1 - x^2)}{(1 + x^2)^2} = 0$; $x^2 = 1$; $x = \pm 1$

For x = 1: $\left[\dfrac{d^2y}{dx^2}\right]_{x=1} = \dfrac{2a \cdot 1(1-3)}{(1 + 1)^3} = \dfrac{-4a}{8} = -\dfrac{a}{2} < 0$

y attains a maximum value of $\frac{a}{2}$ at $x = 1$

For x = -1: $\left[\dfrac{d^2y}{dx^2}\right]_{x=-1} = \dfrac{2a\,(-1)(1-3)}{(1 + 1)^3} = \dfrac{4a}{8} = \dfrac{a}{2} > 0$

y attains a minimum value of $\frac{-a}{2}$ at $x = -1$.

Points of inflection:

$\dfrac{d^2y}{dx^2} = 0$; $x(x^2 - 3) = 0$; $x = 0, \pm\sqrt{3}$.

$(0, 0)$, $(\sqrt{3}, \frac{\sqrt{3}a}{4})$ and $(-\sqrt{3}, \frac{-\sqrt{3}a}{4})$ are points of inflection.

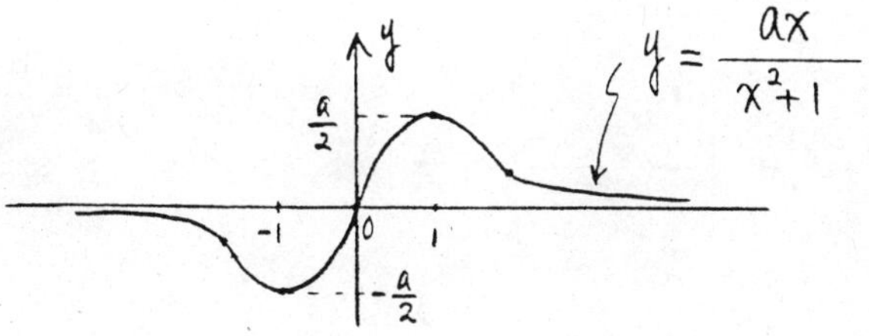

(b) $y = (x - a)^3$, $a > 0$

Extremum: $\dfrac{dy}{dx} = 3(x-a)^2 = 0$; $x = a$

$\dfrac{d^2y}{dx^2} = 6(x-a) = 0$ at $x = a$

(no information for minimum or maximum)

So we need to check the sign of $\frac{dy}{dx}$ for $x \gtrless a$.

$$\frac{dy}{dx} = 3(x-a)^2 > 0 \quad \text{for } x \gtrless a.$$

So (a,0) is a point of inflection; the function $y = (x - a)^3$ has neither a maximum nor a minimum.

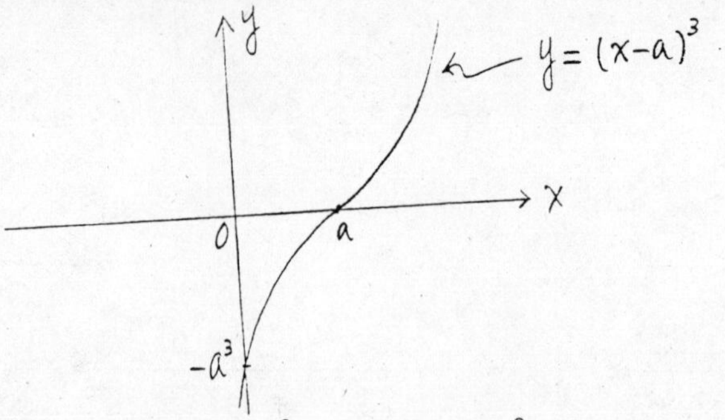

(c) $y = \sin\frac{1}{2}\pi x, \quad \frac{dy}{dx} = \frac{\pi}{2}\cos\frac{1}{2}\pi x, \quad \frac{d^2y}{dx^2} = -\left(\frac{\pi}{2}\right)^2 \sin\frac{1}{2}\pi x$

Extremum: $\cos\frac{1}{2}\pi x = 0; \quad \frac{\pi}{2}x = 2n\pi \pm \frac{\pi}{2}; \quad x = 4n \pm 1$

$(n = 0, \pm1, \pm2, \cdots)$

$$\frac{d^2y}{dx^2} = -\left(\frac{\pi}{2}\right)^2 \sin\frac{\pi}{2}x \begin{cases} > 0 \text{ for } x = 4n - 1 \\ < 0 \text{ for } x = 4n + 1 \end{cases}$$

So y attains a maximum value, +1, at $x = 4n + 1$ and a minimum value, -1, at $x = 4n - 1$, where $n = 0, \pm1, \pm2, \ldots$.

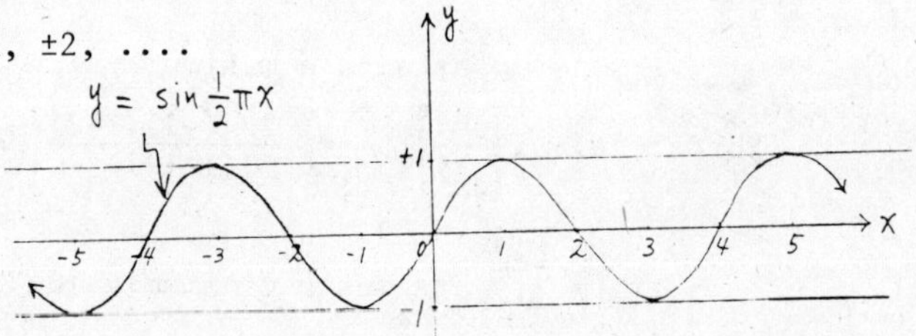

I-20

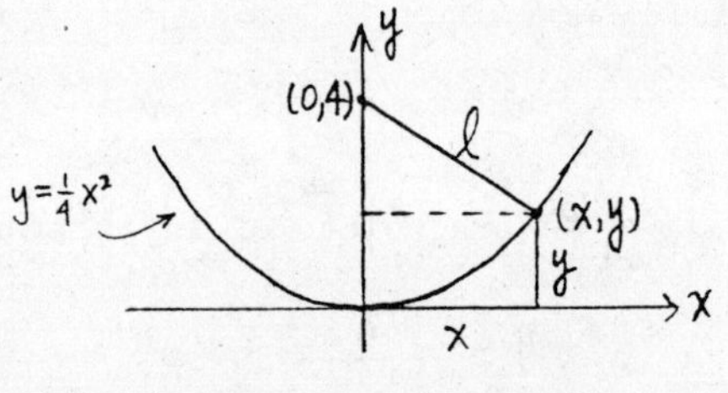

$$\ell^2 = x^2 + (4-y)^2$$

$$= x^2 + \left(4 - \frac{x^2}{4}\right)^2$$

$$\frac{d\ell^2}{dx^2} = 2x + 2\left(4 - \frac{x^2}{4}\right)\left(-\frac{x}{2}\right)$$

$$= 2x - 4x + \frac{x^3}{4}$$

$$\frac{d\ell^2}{dx} = -2x + \frac{x^3}{4} = x\left(\frac{x^2}{4} - 2\right) = 0; \quad x = 0, \ x = \pm 2\sqrt{2}$$

$\frac{d^2\ell^2}{dx^2} = -2 + \frac{3}{4}x^2 = -2 < 0$ for $x = 0$. So $x = 0$ gives a maximum ℓ^2 value.

$\frac{d^2\ell^2}{dx^2} = -2 + \frac{3}{4} \cdot 8 = -2 + 6 = 4 > 0$ for $x = \pm 2\sqrt{2}$. So $x = \pm 2\sqrt{2}$ gives a minimum ℓ^2 value.

Two points, $(\pm 2\sqrt{2}, 2)$ on the curve $y = \frac{1}{4}x^2$ are nearest to the fixed point (0,4).

I-21

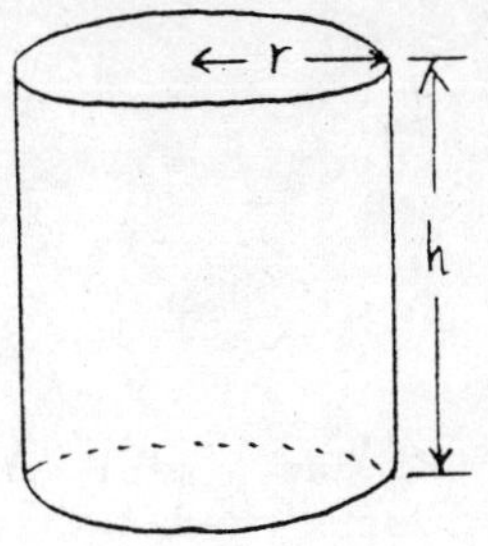

$$V = \pi r^2 h$$

(a) Surface area of tin:

$$A = 2\pi rh + \pi r^2$$

$$= 2\pi r \frac{V}{\pi r^2} + \pi r^2 = \frac{2V}{r} + \pi r^2$$

$$\frac{dA}{dr} = 0; \quad -\frac{2V}{r^2} + 2\pi r = 0$$

$$r^3 = \frac{V}{\pi} \quad r = \left(\frac{V}{\pi}\right)^{1/3}$$

$$\text{diameter} = 2r = 2\left(\frac{V}{\pi}\right)^{1/3}$$

(b) $A = 2\pi rh + 2\pi r^2$

$$= 2\pi r \frac{V}{\pi r^2} + 2\pi r^2 = \frac{2V}{r} + 2\pi r^2$$

$$\frac{dA}{dr} = 0 = -\frac{2V}{r^2} + 4\pi r; \quad r^3 = \frac{V}{2\pi}; \quad r = \left(\frac{V}{2\pi}\right)^{1/3}$$

$$\text{diameter} = 2r = 2\left(\frac{V}{2\pi}\right)^{1/3} = \left(\frac{4V}{\pi}\right)^{1/3}$$

I-22

$$\text{Area} = A = 2\cdot\frac{1}{2}xh + ah$$

$$h = (a^2 - x^2)^{\frac{1}{2}}$$

$$A = (a+x)(a^2 - x^2)^{\frac{1}{2}}$$

$$\frac{dA}{dx} = (a+x)\frac{1}{2}(a^2 - x^2)^{-\frac{1}{2}}(-2x) + (a^2 - x^2)^{\frac{1}{2}} = 0$$

$$\frac{(a^2 - x^2) - x(a+x)}{(a^2 - x^2)^{\frac{1}{2}}} = 0 \qquad a^2 - x^2 - ax - x^2 = 0$$

$$2x^2 + ax - a^2 = 0 \qquad (2x-a)(x+a) = 0$$

$$x = \frac{a}{2} \quad (x = -a \text{ is discarded.})$$

So the width of the opening across the top for maximum carrying capacity is $a + 2x = a + 2\cdot\frac{a}{2} = 2a$. Note that the angle α is 60°. ($\cos\alpha = \frac{1}{2}$)

I-23 Assume a, b > 0

Recall $|A| = \begin{cases} A \text{ if } A \geq 0 \\ -A \text{ of } A \leq 0 \end{cases} = \sqrt{A^2}$.

$$y = |a + bx| = \begin{cases} a + bx & \text{for } x > -a/b \\ -(a + bx) & \text{for } x < -a/b \end{cases}$$

$$y = \sqrt{(a + bx)^2 + \varepsilon^2} \to |\varepsilon| \quad \text{as } x \to -a/b$$

$$\frac{dy}{dx} = \frac{2(a+bx)}{2\sqrt{(a+bx)^2 + \varepsilon^2}} \begin{cases} > 0 & \text{for } x > -a/b \\ = 0 & \text{at } x = -a/b \\ < 0 & \text{for } x < -a/b \end{cases} \quad \begin{array}{l} y \text{ has a min.} \\ \text{at } x = -a/b \end{array}$$

When $(a+bx)^2 >> \varepsilon^2$, $y \cong \sqrt{(a+bx)^2} = a+bx$

When $(a+bx)^2 << \varepsilon^2$ (i.e. near a minimum),

$$y = |\varepsilon| \sqrt{1 + \frac{(a+bx)^2}{\varepsilon^2}} \cong |\varepsilon| \left[1 + \frac{1}{2}\frac{(a+bx)^2}{\varepsilon^2} + \cdots\right]$$

Thus, for $\varepsilon^2 << 1$, the two functions are almost the same except

$y = \sqrt{(a+bx)^2 + \varepsilon^2}$ has a continuous derivative at $x = -a/b$.

That is, $y = \sqrt{(a+bx)^2 + \varepsilon^2}$ can correspond to a real physical situation.

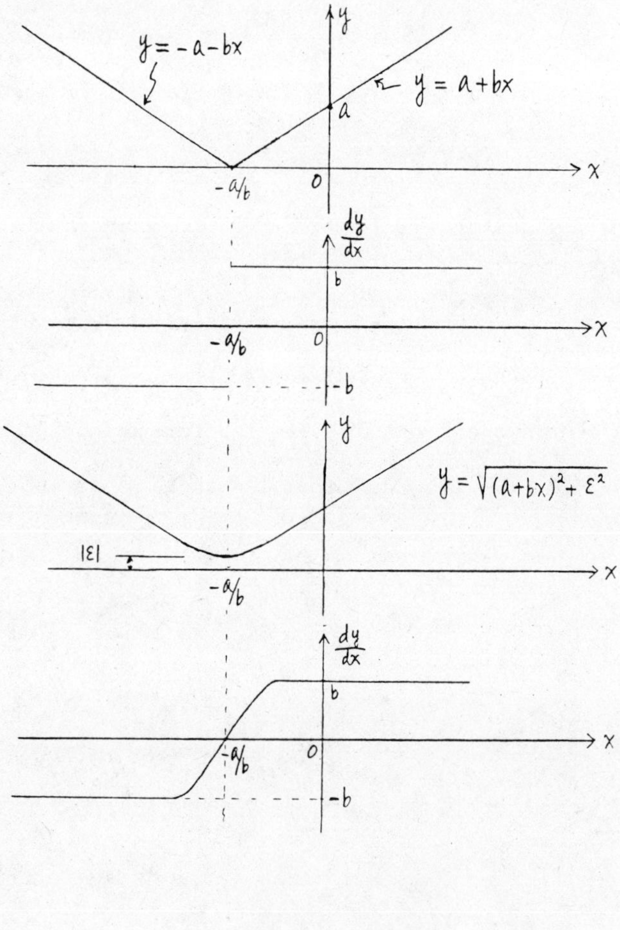

I-24 (a) $y = 3x^3 + 7x^2 - 2x + 1$

$$dy = (9x^2 + 14x - 2)dx$$

(b) $y = \sin(2\pi x^2)$

$$dy = \cos(2\pi x^2) \cdot 4\pi x\, dx$$

(c) $y = e^{-ax} \sin(\frac{\pi}{2} x)$

$$dy = [(-ae^{-ax}) \sin\left(\frac{\pi}{2} x\right) + e^{-ax} \frac{\pi}{2} \cos\left(\frac{\pi}{2} x\right)]dx$$

$$= e^{-ax} [-a \sin\left(\frac{\pi}{2} x\right) + \frac{\pi}{2} \cos\left(\frac{\pi}{2} x\right)]dx$$

I-25 Let ℓ be the length of circular band, and r its radius. Then $\ell = 2\pi r$.

By use of differentials: $\frac{d\ell}{dr} = 2\pi$, $dr = \frac{d\ell}{2\pi} = \frac{1}{2\pi}$ m

By exact solution: $(\ell + \Delta\ell) = 2\pi(r + \Delta r)$

$$\ell + 1\ \text{m} = 2\pi r + 2\pi\, \Delta r$$

$$\Delta r = \frac{1}{2\pi}\ \text{m} = dr$$

There is no difference between using differentials and an exact solution. Since Δr does not depend on ℓ or r, this solution is applicable to the Moon also.

I-26 $V = \frac{4\pi}{3} r^3$; $\frac{dV}{dr} = 4\pi r^2$

By an approximation, Eq. 2-33, $\Delta V \cong \frac{dV}{dr} \Delta r$

$$\Delta V = 4\pi r^2\ \Delta r \qquad \Delta V = \text{surface area} \times \Delta r$$

This implies that the increment of volume is approximately equal to the product of surface area and increment of radius.

I-27 $V = x^3$, $\frac{dV}{dx} = 3x^2$

$\Delta V \cong 3x^2\ \Delta x$

Now $\Delta x = 2t$, $t = 0.2$ cm

$x = 10$ cm

$\Delta V \cong 6x^2 t = 6(10^2\ \text{cm}^2)(0.2\ \text{cm}) = 120\ \text{cm}^3$

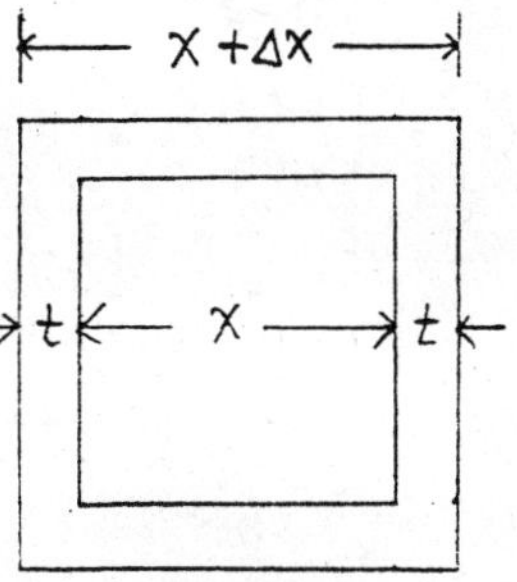

Here $\Delta V \cong 6x^2 t$ means that the increment of volume, ΔV, is equal to the surface area of cube, $6x^2$ multiplied by the increment of length of each side.

II-1 (a) $\int(2x^2 - 7)dx = 2\int x^2dx - 7\int dx = \frac{2}{3}x^3 - 7x + C$

(b) $\int \cos\frac{2\pi}{\tau}t\ dt = \frac{\tau}{2\pi}\int \cos(\frac{2\pi}{\tau}t)\frac{2\pi}{\tau}dt$

Let $u = \frac{2\pi}{\tau}t,\ du = \frac{2\pi}{\tau}dt$

$\int \cos\frac{2\pi}{\tau}t\ dt = \frac{\tau}{2\pi}\int \cos u\ du = \frac{\tau}{2\pi}\sin u + C = \frac{\tau}{2\pi}\sin(\frac{2\pi}{\tau}t) + C$

(c) $\int e^{-at}dt = -\frac{1}{a}\int -ae^{-at}dt = -\frac{1}{a}e^{-at} + C$

II-2 $df(x) = \sin x \cos x\ dx$

$f(x) = \int \sin x \cos x\ dx = \frac{1}{2}\sin^2 x + C$

II-3 Let $u = 2x + 1,\ du = 2\ dx,\ x = \frac{1}{2}(u - 1)$

$\int x\sqrt{2x+1}\ dx = \int[\frac{1}{2}(u - 1)u^{½}]\frac{1}{2}du = \frac{1}{4}\int(u^{3/2} - u^{½})du$

$= \frac{1}{4}[\frac{2}{5}u^{5/2} - \frac{2}{3}u^{3/2}] + C$

$= \frac{1}{10}(2x + 1)^{5/2} - \frac{1}{6}(2x + 1)^{3/2}$

II-4 Let $v = u^3 + 1,\ dv = 3u^2\ du$

$\int\frac{u^2}{u^3 + 1}du = \frac{1}{3}\int\frac{dv}{v} = \frac{1}{3}\ln v + C = \frac{1}{3}\ln(u^3 + 1) + C$

II-5 (a) $\frac{d}{dx}\frac{df(x)}{dx} = 2x + 1$

$\frac{df(x)}{dx} = \int(2x + 1)dx = x^2 + x + C_1$

$f(x) = \int(x^2 + x + C_1)dx = \frac{1}{3}x^3 + \frac{1}{2}x^2 + C_1x + C_2$

(b) $\frac{d}{dx}\frac{df(x)}{dx} = \cos(\frac{1}{2}\pi x)$

$\frac{df(x)}{dx} = \int \cos(\frac{1}{2}\pi x)dx = \frac{2}{\pi}\sin(\frac{1}{2}\pi x) + C_1$

$f(x) = \int(\frac{2}{\pi}\sin(\frac{1}{2}\pi x) + C_1)dx = \frac{-4}{\pi^2}\cos(\frac{1}{2}\pi x) + C_1x + C_2$

(c) $\frac{d}{dx}(\frac{d^2f(x)}{dx^2}) = \frac{1}{2}(x + a)^2$

$\frac{d^2f(x)}{dx^2} = \int\frac{1}{2}(x + a)^2\ dx = \frac{1}{6}(x + a)^3 + C_1$

$\frac{df(x)}{dx} = \int(\frac{1}{6}(x+a)^3 + C_1)dx = \frac{1}{24}(x + a)^4 + C_1x + C_2$

$f(x) = \int(\frac{1}{24}(x + a)^4 + C_1'x + C_2)dx = \frac{1}{120}(x + a)^5 + C_1x^2 + C_2x + C_3$

II-6 $f(x) = \int(ax + b)dx = \frac{a}{2}x^2 + bx + C$

$f(0) = 0 = \frac{a}{2}\cdot 0 + b\cdot 0 + C = C$

$f(x) = \frac{a}{2}x^2 + bx$

II-7 $\frac{df(t)}{dt} = \int a\,dt = at + C_1$

At $t = 0$, $\frac{df(t)}{dt} = v_o$.

$v_o = a \cdot 0 + C_1 = C_1$

$f(t) = \int (at + v_o)dt = \frac{1}{2} at^2 + v_o t + C_2$

$f(0) = s_o = \frac{1}{2} a \cdot 0^2 + v_o \cdot 0 + C_2 = C_2$

$f(t) = \frac{1}{2} at^2 + v_o t + s_o$

II-8 From the above problem,

$f(t) = \int (at + C_1)dt = \frac{1}{2} at^2 + C_1 t + C_2$

$f(0) = s_o = \frac{1}{2} a \cdot 0^2 + C_1 \cdot 0 + C_2 = C_2$

$f(1) = s_1 = \frac{1}{2} a \cdot 1^2 + C_1 \cdot 1 + s_o = \frac{1}{2} a + C_1 + s_o$

So,

$C_1 = s_1 - s_o - \frac{1}{2} a$

$f(t) = \frac{1}{2} at^2 + (s_1 - s_o - \frac{1}{2} a)t + s_o$

II-9 $\frac{df(t)}{dt} = \int e^{-t}dt = -e^{-t} + C_1$

As $t \to \infty$, $\frac{df(t)}{dt} \to 0$.

$0 = \lim_{t\to\infty} (-e^{-t} + C_1) = 0 + C_1 = C_1$

$f(t) = \int -e^{-t}dt = e^{-t} + C_2$

$f(0) = 0 = e^{-0} + C_2 = 1 + C_2$

So,

$C_2 = -1$

$f(t) = e^{-t} - 1$

II-10 $\frac{df(x)}{dx} = \int \sin(\frac{1}{2} \pi x)dx = \frac{-2}{\pi} \cos(\frac{1}{2} \pi x) + C_1$

At $x = 1$,

$\frac{df(x)}{dx} = 0 = -\frac{2}{\pi} \cos(\frac{1}{2} \pi) + C_1 = 0 + C_1 = C_1$

$f(x) = \int \frac{-2}{\pi} \cos(\frac{1}{2} \pi x)dx = -\frac{4}{\pi^2} \sin(\frac{1}{2} \pi x) + C_2$

$f(1) = (\frac{2}{\pi})^2 = -(\frac{2}{\pi})^2 \sin(\frac{\pi}{2}) + C_2 = -(\frac{2}{\pi})^2 + C_2$

$C_2 = 2(\frac{2}{\pi})^2$

$f(x) = -\frac{4}{\pi^2} \sin(\frac{1}{2} \pi x) + \frac{8}{\pi^2}$

II-11 $\frac{dN(t)}{dt} = kN(t)$

$dt = \frac{1}{k} \frac{dN(t)}{N(t)}$

$$t = \frac{1}{k}\int \frac{dN(t)}{N(t)} = \frac{1}{k}\ln N(t) + C$$

$$\ln N(t) = kt + C'$$

$$N(t) = e^{(kt+C')} = C''e^{kt}$$

$$N(1970) = C''e^{k(1970)}$$

$$N(1970 + 35) = N(2005) = C''e^{k(2005)} = 2C''e^{k(1970)}$$

Solving the above equation for k,

$$e^{k(2005)} = 2e^{k(1970)}$$

$$k = \frac{\ln 2}{(2005-1970)} = 0.02$$

To find C'',

$$N(1970) = C''e^{0.02(1970)} = 3.63 \times 10^9$$

$$C'' = \frac{3.63 \times 10^9}{e^{0.02(1970)}} = 2.8 \times 10^{-8}$$

$$N(2500) = (2.8 \times 10^{-8})e^{0.02(2500)} = 1.45 \times 10^{14}$$

The area each person will have is

$$\frac{1.36 \times 10^8 \text{ km}^2}{1.45 \times 10^{14}} = 9.38 \times 10^{-7} \text{ km}^2 = 0.938 \text{ m}^2$$

2-1 $s(t) = \frac{1}{2} at^4 - bt^2$

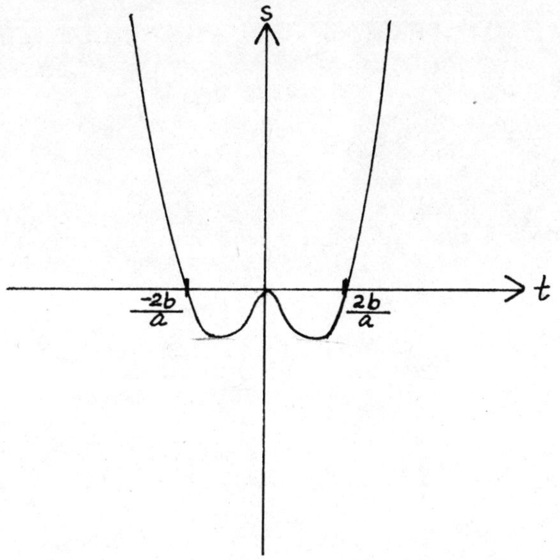

The particle starts ($t = -\infty$) at $s = +\infty$, reaches a minimum value at $t = -\frac{b}{a}$, and again at $t = \frac{b}{a}$; it then returns ($t = +\infty$) to $s = +\infty$.

The value of t is not restricted for this function.

2-2 $s(t) = \pm s_o \sqrt{1 + \alpha t}$

where $s_o = 1$ m and $\alpha = \pm 1\ s^{-1}$

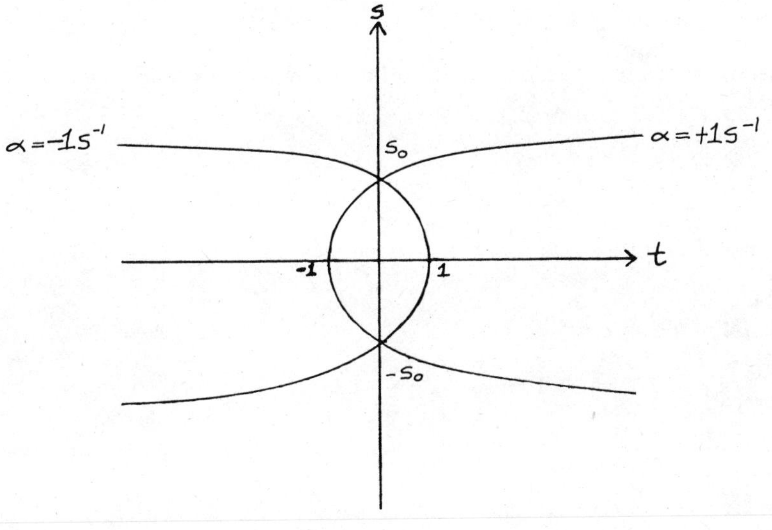

For $\alpha = +1\ s^{-1}$, $t \geq -1$;

For $\alpha = -1\ s^{-1}$, $t \leq 1$.

2-3 $s(t) = \alpha t^3 + \beta t$

Depending on the sign of α, the particle will either come from positive or negative infinity. It will always cross the origin with a velocity $+ \beta$. If $\beta = 0$, the particle will momentarily stop at $t = 0$.

2-4 a) $v_T = \overline{v}_H = \dfrac{300 \text{ m}}{7300 \text{ s}}$

$= 0.0411$ m/s

b) During the first 40 s,

$$\overline{v}_H = \frac{100 \text{ m}}{40 \text{ s}} = 2.5 \text{ m/s}$$

During the last 60 s,

$$\overline{v}_H = \frac{200 \text{ m}}{60 \text{ s}} = 3.33 \text{ m/s}$$

2-5 $$\overline{v} = \frac{X_{AB}}{t_1 + t_2} = \frac{X_{AB}}{\frac{1}{2}\frac{X_{AB}}{v_1} + \frac{1}{2}\frac{X_{AB}}{v_2}}$$

$$\overline{v} = 2\left(\frac{1}{\frac{1}{v_1} + \frac{1}{v_2}}\right) = 2\left(\frac{v_1 v_2}{v_1 + v_2}\right)$$

Notice that $\overline{v} \neq \frac{1}{2}(v_1 + v_2)$.

2-6 a) $\overline{v} = \dfrac{1500 \text{ m}}{230 \text{ s}} = 6.52$ m/s

b) $v_f = \overline{v} + (0.20)\overline{v} = 7.83$ m/s

The time difference is the time required for each runner to move 3.0 cm.

$$t = \frac{0.03 \text{ m}}{7.83 \text{ m/s}}$$

$$= 3.83 \times 10^{-3} \text{ s}$$

To measure this time difference an accuracy of a millisecond is needed.

2-7 $s(t) = \frac{1}{2} at^2 + bt + c$

$a = 1 \text{ m/s}^2, \quad b = -1 \text{ m/s}, \quad c = 2 \text{ m}$

a)

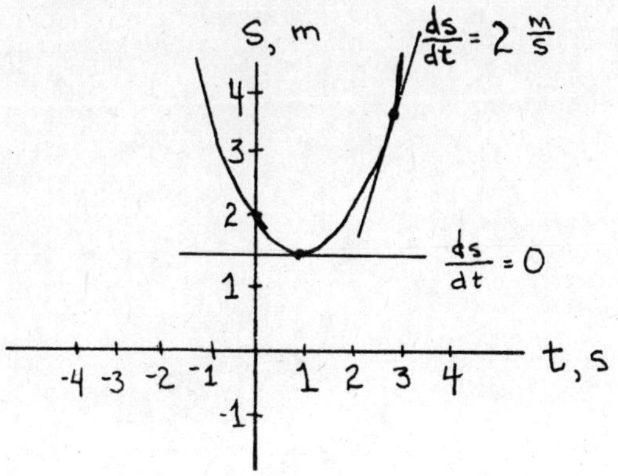

b) At $s = \frac{3}{2}$ m,

$$\frac{3}{2}\text{ m} = \tfrac{1}{2}\, a\, t^2 + bt + c$$

Solving for t, we find

$$t_1 = 1\text{ s}$$

At $s = \frac{7}{2}$ m,

$$\frac{7}{2}\text{ m} = \tfrac{1}{2}\, a\, t^2 + bt + c$$

$$t_2 = 3\text{ s}$$

$$\bar{v} = \frac{s_2 - s_1}{t_2 - t_1} = \frac{2\text{ m}}{2\text{ s}} = 1\text{ m/s}$$

c) $v(t) = \frac{ds}{dt} = at + b$

At $s = \frac{7}{2}$ m, we had t = 3 s; thus,

$$v(3\text{ s}) = 2\text{ m/s}$$

d) At $s = \frac{3}{2}$ m, we had t = 1 s; thus,

$$v(1\text{ s}) = 0$$

2-8 $s(t) = \alpha t^3 + \beta t^2 + \delta t$

$\alpha = 1\text{ m/s}^3$, $\beta = -9\text{ m/s}^2$, $\delta = 24\text{ m/s}$

a)

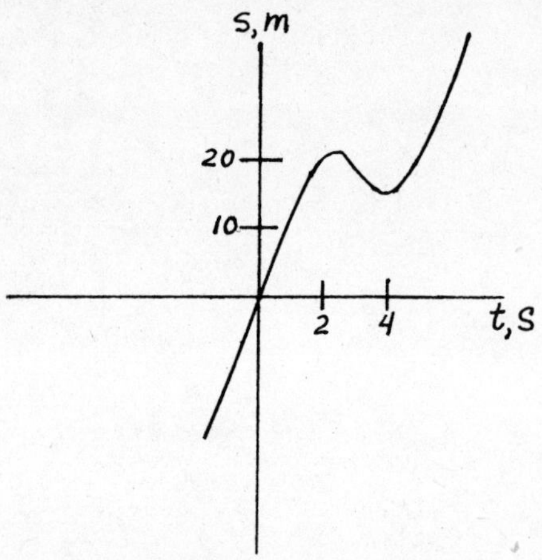

b) $v(t) = \frac{ds}{dt} = 3\alpha t^2 + 2\beta t + \delta$

$v(2\ s) = 3(1\ m/s^3)\ (2\ s)^2 + 2(-9\ m/s^2)\ (2\ s) + 24\ m/s = 0$

$v(4\ s) = 3(1\ m/s^3)(4\ s)^2 + 2(-9\ m/s^2)\ (4\ s) + 24\ m/s = 0$

$v(4.5\ s) = 3(1\ m/s^3)(4.5\ s)^2 + 2(-9\ m/s^2)\ (4.5\ s) + 24\ m/s = 3.75\ m/s$

c) For the interval 0 - 2 s,

$\overline{v} = \frac{s(2\ s) - s(0)}{2\ s - 0} = \frac{20 - 0}{2 - 0} = 10\ m/s$

$\frac{\Delta \ell}{\Delta t} = \frac{s(2\ s) - s(0)}{2\ s - 0} = 10\ m/s$

For the interval 0 - 4 s,

$\overline{v} = \frac{s(4\ s) - s(0)}{4\ s - 0} = \frac{16 - 0}{4 - 0} = 4\ m/s$

$\frac{\Delta \ell}{\Delta t} = \frac{[s(2\ s) - s(0)] + [s(2\ s) - s(4\ s)]}{4\ s - 0} = \frac{20 + 20 - 16}{4} = 6\ m/s$

For the interval 0 - 4.5 s,

$\overline{v} = \frac{s(4.5\ s) - s(0)}{4.5\ s - 0} = \frac{16.9 - 0}{4.5} = 3.8\ m/s$

$\frac{\Delta \ell}{\Delta t} = \frac{[s(2\ s) - s(0)] + [s(2\ s) - s(4\ s)] + [s(4.5\ s) - s(4\ s)]}{4.5\ s - 0} = 5.5\ m/s$

2-9 $\frac{\ell}{x} = \frac{L}{x + y}$ or $\ell y = x(L - \ell)$

Differentiating,

$\frac{dy}{dt} = \frac{(L - \ell)}{\ell}\frac{dx}{dt}$

$\frac{dx}{dt} = \frac{\ell}{L - \ell} \cdot \frac{dy}{dt} = \frac{\ell}{L - \ell}v$

$v_s = \left(\frac{1.8\ m}{5.0\ m - 1.8\ m}\right)2\ m/s = 1.125\ m/s$

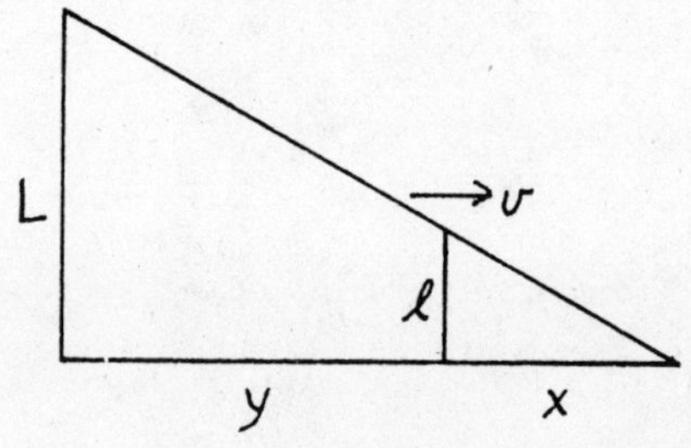

2-10 $\overline{a} = \frac{v_f - v_o}{t_f - t_o} = \frac{0\text{-}60\ m/s}{15\ s - 0} = -4\ m/s^2$

The negative sign in the result shows that the velocity decreases with time.

2-11 $v(t) = 3(t - t_o) + v_o$

$v_o = 10\ m/s,\ t_o = 5\ s$

$v(15\ s) = 40\ m/s$

$$\bar{a} = \frac{v(15\ s) - v(0)}{15\ s - 0} = 2\ m/s^2$$

2-12 a) $$\bar{a} = \frac{v_f - v_o}{t_f - t_o} = \frac{100\ km/h - 0}{8\ s - 0}$$

$$\bar{a} = \frac{27.8\ m/s}{8\ s} = 3.47\ m/s^2$$

b) At t = 4 s and t = 8 s and all other values in the interval 0 - 8 s, the instantaneous acceleration, $\underline{a}$, equals $\bar{a}$ because the speed increase is linear and the acceleration is constant.

2-13 $x(t) = A \sin \omega t$

$A = 5\ cm\ ,\ \omega = 3\ rad/s$

$$v(t) = \frac{dx}{dt} = A\omega \cos \omega t$$

The maximum value of v(t) occurs when

$\cos \omega t = 1$. Thus,

$v_m = A\omega = 15\ cm/s$

$$a(t) = \frac{dv}{dt} = -A\omega^2 \sin \omega t$$

The maximum value of a(t) occurs when

$\sin \omega t = -1$. Thus,

$a_m = A\omega^2 = 45\ cm/s^2$

2-14 $s(t) = at^2 + bt + c$

$a = 5\ m/s^2\ ,\ b = 2\ m/s\ ,\ c = -1\ m$

a) $s(1\ s) = 6\ m$

$s(3\ s) = 50\ m$

distance traveled $= 50\ m - 6\ m = 44\ m$

b) $$\bar{v} = \frac{50\ m - 6\ m}{3\ s - 1\ s} = 22\ m/s$$

c) $$v(t) = \frac{ds}{dt} = 2at + b$$

$v(1\ s) = 12\ m/s$

$v(3\ s) = 32\ m/s$

$$\overline{a} = \frac{32\ m/s - 12\ m/s}{3\ s - 1\ s} = 10\ m/s^2$$

$$\frac{\overline{v}}{\Delta t} = 11\ m/s^2$$

Thus, $\frac{\overline{v}}{\Delta t} \neq \overline{a}$

2-15 $x(t) = At \sin \omega t$

$A = 1\ m/s,\ \omega = 2\pi\ s^{-1}$

a) $v(t) = \frac{dx}{dt} = A \sin \omega t + \omega At \cos \omega t$

$a(t) = \frac{dv}{dt} = \omega A \cos \omega t + \omega A \cos \omega t - \omega^2 At \sin \omega t$

$a(t) = 2 \omega A \cos \omega t - \omega^2 At \sin \omega t$

$a\left(\frac{5}{4} s\right) = -5 \pi^2\ m/s^2$

b) $a(2\ s) = (4 \pi)\ m/s^2$

2-16 a) Using Eq. 2-11,

$v^2 - v_o^2 = 2a(s - s_o)$

$(125\ km/h)^2 = 2a(250\ m) = (34.7\ m/s)^2$

$a = 2.41\ m/s^2$

b) Using Eq. 2-9,

$v = a t + v_o$

$34.7\ m/s = (2.41\ m/s^2)\ t$

$t = 14.4\ s$

c) Again, using Eq. 2-9,

$v = (2.41\ m/s^2)\ (25\ s)$

$= 60.25\ m/s$

2-17 Using Eqs. 2-9 and 2-10,

$v = a t + v_o = (60\ m/s^2)(10\ s) = 600\ m/s$

$s = \frac{1}{2} a t^2 + v_o t + s_o = \frac{1}{2}(60\ m/s^2)(10\ s)^2 = 3000\ m$

2-18 Using Eqs. 2-9 and 2-10,

$v = a t + v_o$

$$a = \frac{v - v_o}{t} = \frac{85\ m/s - 60\ m/s}{12\ s} = 2.08\ m/s^2$$

$s = \frac{1}{2} at^2 + v_o t + s_o = \frac{1}{2}(2.08 \text{ m/s}^2)(12 \text{ s}) + (60 \text{ m/s})(12 \text{ s})$

$= 732.5 \text{ m}$

2-19 Using Eq. 2-11,

$$v^2 - v_o^2 = 2 a (s - s_o)$$

$$a = \frac{v^2 - v_o^2}{2(s - s_o)} = \frac{(40 \text{ m/s})^2 - (60 \text{ m/s})^2}{2(250 \text{ m})}$$

$$= -4 \text{ m/s}^2$$

2-20 a) Using Eq. 2-9,

$$v = at + v_o$$

$$t = \frac{v - v_o}{a}$$

Adding the driver's reaction time,

$$t = \frac{0 - 90 \text{ km/h}}{-2.5 \text{ m/s}^2} + 0.2 \text{ s} = \frac{0 - 25 \text{ m/s}}{-2.5 \text{ m/s}^2} + 0.2 \text{ s}$$

$$= 10 \text{ s} + 0.2 \text{ s} = 10.2 \text{ s}$$

b) Using Eq. 2-10 with reaction distance added to s,

$$s = (\tfrac{1}{2} a t_d^2 + v_o t_d + s_o) + v_o t_r$$

where t_d = deceleration time = 10 s

and t_r = reaction time = 0.25

$$s = \tfrac{1}{2}(-2.5 \text{ m/s}^2)(10 \text{ s})^2 + (25 \text{ m/s})(10 \text{ s}) + (25 \text{ m/s})(0.2 \text{ s})$$

$$= 130 \text{ m}$$

2-21 $v_a = 100 \text{ km/h} = 27.8 \text{ m/s}$

$v_p(10 \text{ s}) = 60 \text{ km/h} = 16.7 \text{ m/s}$

$$a_p = \frac{v_p}{t} = \frac{16.7 \text{ m/s}}{10 \text{ s}} = 1.67 \text{ m/s}^2$$

a) The chase ends when $s_p = s_a$ Thus,

$$\tfrac{1}{2} a_p t^2 + v_{op} t + s_o = v_a t$$

$$\tfrac{1}{2} a_p t^2 - v_a t = 0$$

Eliminating the trivial solution, t = 0,

$$\tfrac{1}{2} a_p t - v_a = 0$$

$$t = \frac{2 v_a}{a_p} = \frac{2(27.8 \text{ m/s})}{1.67 \text{ m/s}^2} = 33.3 \text{ s}$$

b) $s = v_a t = (27.8 \text{ m/s})(33.3 \text{ s})$

$s = 925.7 \text{ m}$

c) $v_p = a_p t + v_o = (1.67 \text{ m/s}^2)(33.3 \text{ s})$

$= 55.6 \text{ m/s}$

2-22 a) Using Eq. 2-14,

$y = -\frac{1}{2} g t^2 + v_o t + y_o$

$0 = -(4.9 \text{ m/s}^2)t^2 + (10 \text{ m/s})t + 80 \text{ m}$

Solving for t, we have

$$t = \frac{-10 \pm \sqrt{10^2 - 4(-4.9)(80)}}{-9.8}$$

Using only the positive value for t, we have

$t = 5.19 \text{ s}$

b) $v = -gt + v_o$

$= -(9.8 \text{ m/s}^2)(5.19 \text{ s}) + 10 \text{ m/s}$

$= -40.9 \text{ m/s}$

2-23 From Eq. 2-11, we can write

$v^2 - v_o^2 = -2g(y - y_o)$

$v_o = \sqrt{v^2 + 2g\,(y - y_o)}$

$= \sqrt{(7 \text{ m/s})^2 + 2(9.8 \text{ m/s}^2)(10 \text{ m})}$

$= 15.6 \text{ m/s}$

2-24 a) Using Eq. 2-13, where $v_m = v$ at time $t = t_f - 1$ s,

$v_f - v_m = -g(1 \text{ s})$

Using Eq. 2-14, we have

$0 - \frac{1}{3} x_o = -\frac{1}{2}g\,(1 \text{ s})^2 + v_m\,(1 \text{ s})$

Substituting

$-\frac{1}{3} x_o + \frac{1}{2} g = (v_f + g)$

$-\frac{1}{3} x_o - \frac{1}{2} g = v_f$

Using Eq. 2-15,

$-\frac{1}{3} x_o - \frac{1}{2} g = \sqrt{2g\,x_o}$

Squaring,

$\frac{1}{9} x_o^2 + \frac{1}{3} g\,x_o + \frac{1}{4} g^2 = 2g\,x_o$

2-24 cont'd $\frac{1}{9}\ x_o^2 - \frac{5}{3}\ g x_o\ + \frac{1}{4}\ g^2\ =\ 0$

Solving for x_o,

$$x_o\ =\ \frac{\frac{5}{3}\ g \pm \sqrt{\frac{25}{9}\ g^2\ -\ \frac{1}{9}\ g^2}}{\frac{2}{9}}$$

$$=\ 145.5\ \text{m},\ 1.485\ \text{m}$$

For x_o = 1.485 m, v_m > 0; this could not be the case, so

$$x_o\ =\ 145.5\ \text{m}$$

b) We have from Eq. 2-14,

$$0 - x_o\ =\ -\ \tfrac{1}{2}\ g t_f^2$$

$$t_f\ = \sqrt{\frac{2x_o}{g}}$$

$$t_f\ =\ 5.45\ \text{s}$$

2-25 We have

$$y_1\ =\ y_2\ =\ -200\ \text{m}$$

$$y_1\ =\ -\tfrac{1}{2}\ g t_1^2\ =\ -200\ \text{m}$$

$$t_1\ =\ 6.39\ \text{s}$$

We are given that

$$t_2\ =\ t_1\ -\ 1.50\ \text{s}\ =\ 4.89\ \text{s}$$

$$y_2\ =\ -\tfrac{1}{2}g\ t_2^2 + v_o\ t_2$$

$$v_o\ =\ \frac{y_2\ +\ \tfrac{1}{2}g\ t_2^2}{t_2}\ =\ -\ 16.9\ \text{m/s}$$

2-26 Using Eq. 2-10,

$$s\ =\ \tfrac{1}{2}\ at^2\ +\ v_o\ t\ +\ s_o$$

$$1.4\ \times\ 10^3\ \text{m}\ =\ \tfrac{1}{2}(980\ \text{m/s}^2)\ t^2$$

$$t\ =\ \sqrt{2.86}\ \text{s}\ =\ 1.69\ \text{s}$$

Using Eq. 2-9,

$$v\ =\ at\ +\ v_o\ =\ 1656\ \text{m/s}$$

2-27 We have

$$y\ =\ -\tfrac{1}{2}\ g t^2\ +\ v_o\ t\ +\ y_o$$

$$0\ =\ -(4.9\ \text{m/s}^2)t^2\ -\ (8\ \text{m/s})\ t\ +\ 30\ \text{m}$$

Solving for t,

$$t = \frac{8 \pm \sqrt{64 + 588}}{-9.8}$$

Using only the positive value for t,

$$t = 1.79 \text{ s}$$

2-28 Clearly, $v_1 = v_2 = v$. Therefore,

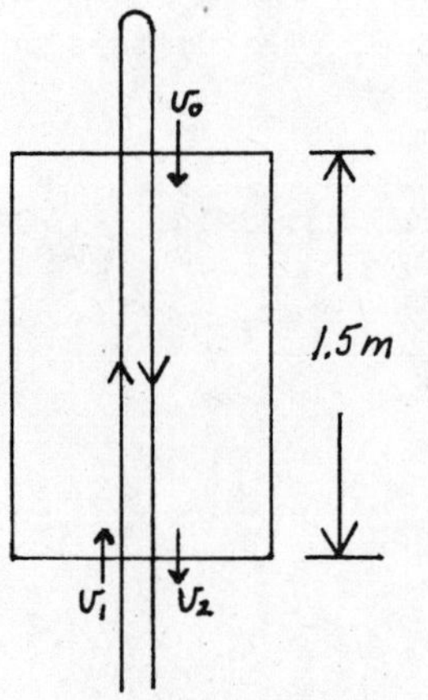

$$t_1 = t_2 = \frac{0.6\text{ s}}{2} = 0.3 \text{ s}$$

Now,

$$v = v_o + gt$$

$$= v_o + 2.94 \text{ m/s}$$

And, $$\overline{v} = \frac{v + v_o}{2}$$

$$s = \overline{v}\, t$$

$$1.5 \text{ m} = \frac{v + v_o}{2} \times 0.3 \text{ s}$$

$$v + v_o = 10.0 \text{ m/s}$$

Substituting,

$$v = (10.0 \text{ m/s} - v) + 2.94 \text{ m/s} = 6.47 \text{ m/s}$$

To find the total height above the ledge, we use

$$v^2 = 2g\, s$$

$$s = \frac{v^2}{2g} = 2.14 \text{ m}$$

2-29 $a(t) = \beta t, \quad x(0) = 0, \qquad x(\tau) = \gamma$

$$v(t) = \int a(t)dt = \int (\beta t)dt$$

$$v(t) = \tfrac{1}{2}\beta t^2 + C_1$$

$$x(t) = \int v(t)dt = \int (\tfrac{1}{2}\beta t^2 + C_1)dt$$

$$s(t) = \frac{1}{6}\beta t^3 + C_1 t + C_2$$

Because $x(0) = 0$,

$$C_2 = 0$$

And because $x(\tau) = \gamma$,

$$\frac{1}{6}\beta\tau^3 + C_1 \tau = \gamma$$

$$C_1 = \frac{\gamma - \frac{1}{6}\beta\tau^3}{\tau} = \frac{\gamma}{\tau} - \frac{1}{6}\beta\tau^2$$

Thus,

$$x(t) = \frac{1}{6}\beta t^3 + \left(\frac{\gamma}{\tau} - \frac{1}{6}\beta\tau^2\right) t$$

2-30 $v(t) = \alpha(1 - e^{-\beta t})$

a) $a(t) = \frac{dv}{dt} = \alpha\beta e^{-\beta t}$

b) $x(t) = \int v(t)dt = \int \alpha(1 - e^{-\beta t})dt$

$$x(t) = \alpha t + \frac{\alpha}{\beta} e^{-\beta t} + C$$

We have

$$x(0) = 0 = \frac{\alpha}{\beta} + C$$

$$C = -\frac{\alpha}{\beta}$$

$$x(t) = \alpha t + \frac{\alpha}{\beta}e^{-\beta t} - \frac{\alpha}{\beta}$$

c) For $\beta > 0$,

$$\lim_{t\to\infty} e^{-\beta t} = 0$$

Thus,

$$\lim_{t\to\infty} v(t) = \alpha$$

d) $v(0) = \alpha(1 - e^{0}) = \alpha(1 - 1) = 0$

e) Since $\lim_{t\to\infty} e^{-\beta t} = 0$,

$$\lim_{t\to\infty} a(t) = 0$$

2-31 We have

$$a(t) = kt$$

where k is a constant of proportionality.

$$v(t) = \int a(t)dt = \int (kt)dt$$

$$v(t) = \tfrac{1}{2} kt^2 + C$$

To find C,

$$v(0) = 0 = 0 + C$$

$$C = 0$$

To find k,

$$v(20\ s) = 20\ km/h = 5.56\ m/s = \tfrac{1}{2} k(20\ s)^2$$

$$k = 0.0278\ m/s^3$$

$$x = \int v(t)dt = \int dt \ \tfrac{1}{2} kt^2 = \frac{1}{6} kt^3 + C$$

Assuming $x(0) = 0$, $C = 0$.

$$x = \frac{1}{6}(0.0278 \text{ m/s}^3)(60 \text{ s})^3 = 1000 \text{ m}$$

2-32 $x(t) = at^3 + bt^2 + ct$

$$a = 2 \text{ m/s}^3,\ b = -5 \text{ m/s}^2,\ c = 2 \text{ m/s}$$

$$v(t) = \frac{dx}{dt} = 3\,at^2 + 2bt + c$$

$$a(t) = \frac{dv}{dt} = 6\,at + 2b$$

a) For $x(t) = 0$,

$$2t^3 - 5t^2 + 2t = 0$$

$$t(2t^2 - 5t + 2) = 0$$

$t = 0$ is one solution; also,

$$t = \frac{5 \pm \sqrt{25 - 16}}{4} = \tfrac{1}{2} \text{ s},\ 2 \text{ s}$$

Thus, $t = 0,\ \tfrac{1}{2} \text{ s},\ 2 \text{ s}$

Now,

$$a(0) = 0 - 10 \text{ m/s}^2 = -10 \text{ m/s}^2$$

$$a(\tfrac{1}{2} \text{ s}) = 6 \text{ m/s}^2 - 10 \text{ m/s}^2 = -4 \text{ m/s}^2$$

$$a(2 \text{ s}) = 24 \text{ m/s}^2 - 10 \text{ m/s}^2 = 14 \text{ m/s}^2$$

b) For $v(t) = 0$,

$$6t^2 - 10t + 2 = 0$$

$$t = \frac{10 \pm \sqrt{100 - 48}}{12} = 0.2 \text{ s},\ 1.4 \text{ s}$$

Now,

$$a(0.2 \text{ s}) = 2.4 \text{ m/s}^2 - 10 \text{ m/s}^2 = 7.6 \text{ m/s}^2$$

$$a(1.4 \text{ s}) = 16.8 \text{ m/s}^2 - 10 \text{ m/s}^2 = 6.8 \text{ m/s}^2$$

2-33 $a(t) = (3 \text{ m/s})t + (2 \text{ m/s}^2)$

$$v(t) = \int a(t)dt = \int (3t + 2)dt$$

$$= \frac{3}{2} t^2 + 2t + C_1$$

$$v(1) = 4 = \frac{3}{2} + 2 + C_1$$

$$C_1 = \tfrac{1}{2}$$

$$v(t) = \frac{3}{2}t^2 + 2t + \tfrac{1}{2}$$

$$s(t) = \int v(t)dt = \int (\frac{3}{2} t^2 + 2t + \tfrac{1}{2})dt = \tfrac{1}{2} t^3 + t^2 + \tfrac{1}{2}t + C_2$$

$$s(1) = 3 = \tfrac{1}{2} + 1 + \tfrac{1}{2} + C_2$$

$$C_2 = 1$$

$$s(t) = \tfrac{1}{2}t^3 + t^2 + \tfrac{1}{2}t + 1$$

2-34 $s(t) = \pm A\sqrt{\dfrac{t}{t - t_o}}$

$A = 1 \text{ m}, \; t_o = 1 \text{ s}$

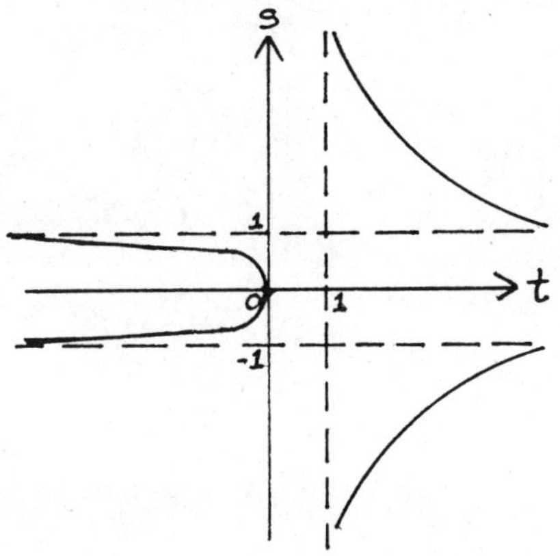

t must have values such that $\dfrac{t}{t - (1 \text{ s})}$ is non-negative, and t - (1 s) is not zero.

Therefore,

$t \leq 0$ or $t > 1$ s

2-35 $t_1 = \dfrac{d}{v_1} = \dfrac{200 \text{ km}}{40 \text{ km/h}} = 5 \text{ h}$

We have

$$t_2 = t_1 - 1 = 4 \text{ h}$$

$$\bar{v}_2 = \frac{d}{t_2} = \frac{200 \text{ km}}{4 \text{ h}}$$

$$\bar{v}_2 = 50 \text{ km/h}$$

2-36 Let v_g = glider velocity, and v_s = shadow velocity

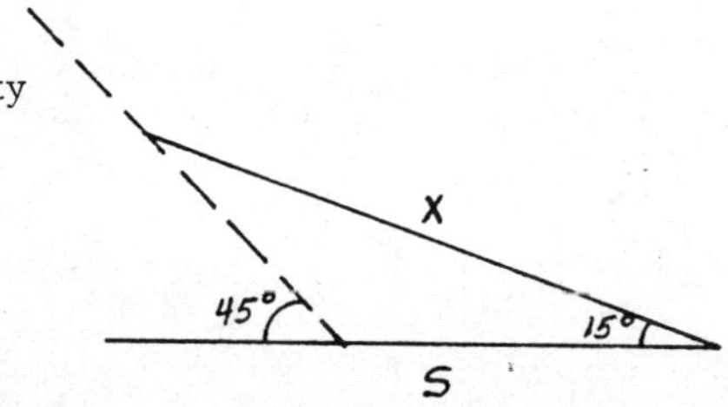

$$t_s = t_g = t$$

$$x = v_g t$$

$$s = v_s t$$

Using the law of sines,

$$\frac{x}{\sin 135°} = \frac{s}{\sin 30°}$$

$$s = (0.707)x$$

Now,

$$v_s = \frac{s}{t} = \frac{(0.707)x}{t} = \frac{(0.707)v_g t}{t} = (0.707)v_g$$

2-37 $\bar{a} = \frac{\Delta v}{\Delta t} = \frac{(20 \text{ m/s}) - (-14 \text{ m/s})}{0.01 \text{ s}}$

$$\bar{a} = \frac{34 \text{ m/s}}{0.01 \text{ s}}$$

$$\bar{a} = 3400 \text{ m/s}^2$$

2-38 $s(t) = -(\frac{1}{10} \text{ m/s}^3)t^3 + (6 \text{ m/s})t$

$$v(t) = \frac{ds}{dt} = -\frac{3}{10}t^2 + 6$$

$$a(t) = \frac{dv}{dt} = -\frac{6}{10}t$$

a) A maximum excursion will occur if $\frac{ds}{dt} = 0$.

$$-\frac{3}{10}t^2 + 6 = 0$$

$$t = \pm\sqrt{20} = \pm 2\sqrt{5}$$

Using only the positive value of t,

$$s(2\sqrt{5} \text{ s}) = 17.9 \text{ m}$$

b) The particle returns to the origin when

$$s(t) = -\frac{1}{10}t^3 + 6t = 0$$

Ignoring the trivial solution, $t = 0$, we have

$$-\frac{1}{10}t^2 + 6 = 0$$

$$t = \pm\sqrt{60}$$

Using only the positive value of t, we have

$$v(\sqrt{60} \text{ s}) = -12 \text{ m/s}$$

c) $a(10 \text{ s}) = -\frac{6}{10}(10 \text{ s})$

$$a(10 \text{ s}) = -6 \text{ m/s}^2 = (-6 \text{ m/s}^2)(g/9.8 \text{ m/s}^2) = 0.61g$$

2-39 $a = \frac{\Delta v}{\Delta t}$

$$a = \frac{-285 \text{ m/s}}{1.5 \text{ s}} = -190 \text{ m/s}^2 = -19.4g$$

2-40 $a = \frac{\Delta v}{\Delta t} = \frac{80 \text{ km/h}}{5 \text{ s}} = \frac{22.2 \text{ m/s}}{5 \text{ s}}$

$$a = 4.44 \text{ m/s}^2$$

The velocity after 4 s is

$$v_o = at = 17.78 \text{ m/s}$$

Now, using Eq. 2-10,

$$s = \tfrac{1}{2} at^2 + v_o t + s_o$$

$$= \tfrac{1}{2}(4.44 \text{ m/s}^2)(1 \text{ s}) + (17.78 \text{ m/s})(1 \text{ s}) = 20 \text{ m}$$

2-41 d = distance to range of radar

$$= 40 \text{ km} - 38 \text{ km} - 0.8 \text{ km} = 1200 \text{ m}$$

$$v_o = 120 \text{ km/h} = 33.3 \text{ m/s}$$

$$v_f = 90 \text{ km/h} = 25 \text{ m/s}$$

To avoid a ticket, the motorist must decelerate from v_o to v_f in a distance d. Thus,

$$a = \frac{v_f^2 - v_o^2}{2d} = -0.20 \text{ m/s}^2$$

2-42 We have

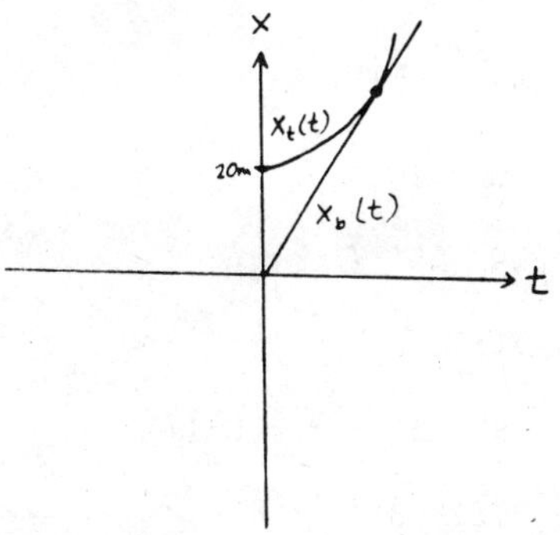

$$x_b(t) = v_b t$$

$$x_t(t) = \tfrac{1}{2} at^2 + x_{ot}$$

When the boy reaches the truck,

$$x_b = x_t$$

$$v_b t = \tfrac{1}{2} at^2 + x_{ot}$$

$$\tfrac{1}{2} at^2 - v_b t + x_{ot} = 0$$

Since the boy just overtakes the truck, rather than passing it, there can only be one solution. Thus,

$$t = \frac{v_b \pm \sqrt{v_b^2 - 2ax_{ot}}}{a}$$

Because there is only one solution,

$$\sqrt{v_b^2 - 2ax_{ot}} = 0$$

$$v_b = \sqrt{2ax_{ot}}$$

$$v_b = 5.66 \text{ m/s}$$

2-43 We have

$$x_t(t) = v_t t = (80 \text{ km/h})t = (22.2 \text{ m/s})t$$

$$x_c(t) = \tfrac{1}{2} at^2 + v_o t = (0.6 \text{ m/s}^2)\ t^2 + v_o t$$

The car passes the truck at 500 m; thus,

$$x_t = 500\text{ m} = (22.2\text{ m/s})t$$

$$t = 22.5\text{ s}$$

Now,

$$x_c = 500\text{ m} = (0.6\text{ m/s}^2)(22.5\text{ s})^2 + v_o(22.5\text{ s})$$

$$v_o = 8.72\text{ m/s}$$

2-44 The time required for each complete bounce is equal to twice the time required for a ball dropped from 10 cm to hit the floor.

Thus,

$$x = 10\text{ cm} = \tfrac{1}{2}gt^2$$

$$t = \sqrt{\frac{2x}{g}} = 0.14\text{ s}$$

$$T = \text{period of each full bounce}$$

$$= 2t = 0.28\text{ s}$$

$$F = \text{bounces per minute}$$

$$= \frac{60}{T} = 214\ \text{min}^{-1}$$

2-45 Using Equation 2-16,

At $\lambda = 30°\ 15'$ (Austin),

$$g_A = 9.79344$$

At $\lambda = 59°\ 15'$ (Stockholm),

$$g_S = 9.81858$$

We have

$$s - s_o = \frac{v^2 - v_o^2}{2a}$$

In Austin:

$$15\text{ m} = \frac{0 - v_o^2}{-2g_A}$$

$$v_o = \sqrt{(30\text{ m})g_A} = 17.1407\text{ m/s}$$

In Stockholm:

$$s = \frac{0 - v_o^2}{-2g_S} = 14.9616\text{ m}$$

2-46 a) $y_1(t) = -\tfrac{1}{2}gt^2 + v_{01}t$

$$y_2(t) = -\tfrac{1}{2}g(t - 1)^2 + v_{02}(t - 1)$$

When the balls collide,

$$y_1 = y_2$$

$$-\tfrac{1}{2} gt^2 + v_{01}t = -\tfrac{1}{2}g(t - 1)^2 + v_{02}(t - 1)$$

$$-\tfrac{1}{2}gt^2 + v_{01}t = -\tfrac{1}{2}gt^2 + gt - \tfrac{1}{2}g + v_{02}t - v_{02}$$

$$(g + v_{02} - v_{01})t = \tfrac{1}{2} g + v_{02}$$

$$t = \frac{\tfrac{1}{2} g + v_{02}}{g + v_{02} - v_{01}} = 1.51 \text{ s}$$

b) $y = y_1(1.51 \text{ s}) = -\tfrac{1}{2} g(1.51 \text{ s})^2 + (12 \text{ m/s})(1.51 \text{ s})$

$y = 6.95 \text{ m}$

c) The velocity of the first ball indicates the direction of motion.

$v_1(1.51 \text{ s}) = -g(1.51 \text{ s}) + (12 \text{ m/s})$

$v_1(1.51 \text{ s}) = -2.8 \text{ m/s}$

Thus, the first ball is falling.

2-47 y_a = height reached during acceleration

$y_a = \tfrac{1}{2} at^2 = \tfrac{1}{2}(40 \text{ m/s}^2)(3 \text{ s})^2$

$y_a = 180 \text{ m}$

v_a = velocity reached during acceleration

$v_a = at = (40 \text{ m/s}^2)(3 \text{ s})$

$v_a = 120 \text{ m/s}$

y = maximum height reached

$$= \frac{v^2 - v_a^2}{-2g} + y_a = \frac{v_a^2}{2g} + y_a = 914.7 \text{ m}$$

2-48 $y = -\tfrac{1}{2} gt^2 + v_o t$

$\frac{h}{2} = -\tfrac{1}{2} g(3.5 \text{ s})^2 + v_o(3.5 \text{ s})$

$h = (7.0 \text{ s})v_o - (120.0 \text{ m})$

Also,

$$h = \frac{0 - v_o^2}{-2g}$$

$$h = \frac{v_o^2}{(19.6 \text{ m/s}^2)}$$

Substituting,

$$\frac{v_o^2}{(19.6 \text{ m/s}^2)} - (7.0 \text{ s})v_o + (120.0 \text{ m}) = 0$$

Solving for v_o,

$$v_o = \frac{7.0 \pm \sqrt{49.0 - 24.5}}{0.102}$$

$v_o = 20.1$ m/s , 117 m/s

The actual v_o produces a veloctiy which is negative at t = 3.5 s. Thus, using $v_o = 20.1$ m/s,

$v(3.5) = -g(3.5 \text{ s}) + (20.1 \text{ m/s}) = -14.2 \text{ m/s}$

Thus,

$v_o = 20.1$ m/s

$h = (7.0\text{ s})v_o - (120.0 \text{ m}) = 20.7 \text{ m}$

2-49 a) $x_1(t) = -\frac{1}{2} gt^2 - (8 \text{ m/s})t + (40 \text{ m})$

$x_2(t) = -\frac{1}{2} gt^2 + (12 \text{ m/s})t$

When they collide,

$x_1 = x_2$

$-\frac{1}{2} gt^2 - (8 \text{ m/s})t + (40 \text{ m}) = -\frac{1}{2} gt^2 + (12 \text{ m/s})t$

$(20 \text{ m/s})t = 40 \text{ m}$

$t = 2 \text{ s}$

b) $x = x_2(2 \text{ s}) = -\frac{1}{2} g(2 \text{ s})^2 + (12 \text{ m/s})(2 \text{ s}) = 4.4 \text{ m}$

c) $v_2(2 \text{ s}) = -gt + (12 \text{ m/s}) = -7.6 \text{ m/s}$

The second ball is moving downward.

2-50 a) $a(t) = \alpha t^3 + \beta t$

$v(t) = \int a(t)dt = \int (\alpha t^3 + \beta t)dt$

$v(t) = \frac{\alpha}{4} t^4 + \frac{\beta}{2} t^2 + C$

$v_o = 0$, so $C = 0$

$s(t) = \int v(t)dt = \int (\frac{\alpha}{4} t^4 + \frac{\beta}{2} t^2)dt$

$s(t) = \frac{\alpha}{20} t^5 + \frac{\beta}{6} t^3 + C'$

$s_o = 0$, so $C' = 0$

$s(t) = \frac{\alpha}{20} t^5 + \frac{\beta}{6} t^3$

b) $a(t) = A \sin(\omega t + \delta)$

$v(t) = \int a(t)dt = \int A \sin(\omega t + \delta)dt$

$= \frac{-A}{\omega} \cos(\omega t + \delta) + C$

Since $v_o = 0$, $C = \frac{A}{\omega} \cos(\delta)$

$v(t) = \frac{-A}{\omega} \cos(\omega t + \delta) + \frac{A}{\omega} \cos(\delta)$

$s(t) = \int v(t)dt = \int [\frac{-A}{\omega} \cos(\omega t + \delta) + \frac{A}{\omega} \cos(\delta)]dt$

$= \frac{-A}{\omega^2} \sin(\omega t + \delta) + [\frac{A}{\omega} \cos(\delta)]t + C$

Since $s_o = 0$, $C = \frac{A}{\omega^2}\sin(\delta)$

$$s(t) = \frac{-A}{\omega^2}\sin(\omega t + \delta) + [\frac{A}{\omega}\cos(\delta)]t + \frac{A}{\omega^2}\sin(\delta)$$

c) $a(t) = Ate^{-\mu t}$

$$v(t) = \int a(t)dt = \int (Ate^{-\mu t})dt$$

$$= \int At\, d(\frac{-1}{\mu} e^{-\mu t})$$

$$= \frac{-A}{\mu} te^{-\mu t} - \int \frac{-A}{\mu} e^{-\mu t} dt$$

$$= \frac{-A}{\mu} te^{-\mu t} - \frac{A}{\mu^2} e^{-\mu t} + C$$

Since $v(0) = 0$, $C = \frac{A}{\mu^2}$

$$v(t) = \frac{-A}{\mu} te^{-\mu t} - \frac{A}{\mu^2} e^{-\mu t} + \frac{A}{\mu^2}$$

$$s(t) = \int v(t)dt = \int (\frac{-A}{\mu} t\, e^{-\mu t} - \frac{A}{\mu^2} e^{-\mu t} + \frac{A}{\mu^2})dt$$

$$= \frac{-A}{\mu}\int (t\, e^{-\mu t})dt + \frac{A}{\mu^3} e^{-\mu t} + \frac{A}{\mu^2} t$$

$$= \frac{-A}{\mu} (-\frac{1}{\mu} t\, e^{-\mu t} + \frac{1}{\mu}\int e^{-\mu t} dt) + \frac{A}{\mu^3} e^{-\mu t} + \frac{A}{\mu^2} t$$

$$= \frac{A}{\mu^2} t\, e^{-\mu t} + \frac{A}{\mu^3} e^{-\mu t} + \frac{A}{\mu^3} e^{-\mu t} + \frac{A}{\mu^2} t + C$$

Since $s(0) = 0$, $C = \frac{-2A}{\mu^3}$

$$s(t) = \frac{A}{\mu^2} t\, e^{-\mu t} + \frac{2A}{\mu^3} e^{-\mu t} + \frac{A}{\mu^2} t - \frac{2A}{\mu^3}$$

2-51 Let D_i be the distance the trucks are apart at the beginning of each trip by the roadrunner. Let d_i be the distance traveled by the roadrunner on each trip. For an infinite number of trips,

$$d_T = d_o + d_1 + \ldots = \sum_{i=0}^{\infty} d_i$$

We have

$$d_i = D_i - x_i$$

where

$$\frac{D_i - x_i}{70 \text{ km/h}} = \frac{x_i}{50 \text{ km/h}}$$

$$x_i = \frac{50}{120} D_i$$

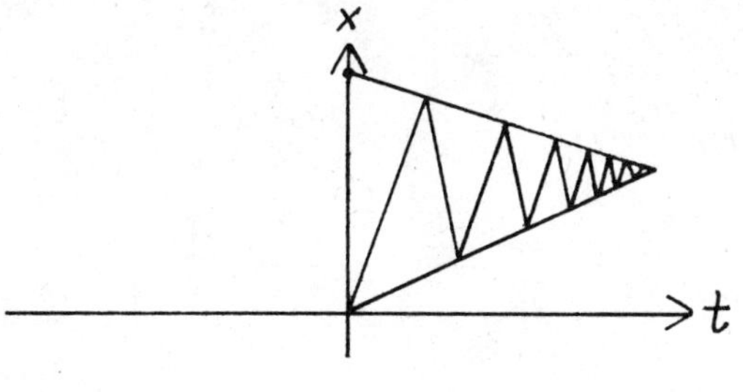

Thus,

$$d_i = \frac{70}{120} D_i$$

Now, using the distance traveled by the trucks during the trip,

$$D_i = D_{i-1} - 2x_{i-1} = (d_{i-1} + x_{i-1}) - 2x_{i-1} = d_{i-1} - x_{i-1} = d_{i-1} - (50\ \text{km/h})t_{i-1}$$

$$d_i = \frac{70}{120} D_i = \frac{70}{120} [d_{i-1} - (50\ \text{km/h})t_{i-1}]$$

$$= \frac{70}{120}\left[d_{i-1} - \frac{50\ \text{km}}{\text{h}} \left(\frac{d_{i-1}}{70\ \text{km/h}}\right)\right] = \frac{1}{6} d_{i-1}$$

This is in the form of a geometric series which can be written as

$$d_i = d_o\left(\frac{1}{6}\right)^i$$

Now,

$$d_o = \frac{70}{120} D_o = \frac{70}{120}\ 20\ \text{km} = \frac{35}{3}\ \text{km}$$

Thus,

$$d_i = \frac{35}{3} \left(\frac{1}{6}\right)^i\ \text{km}$$

Using Eq. A-4,

$$d_T = \frac{35}{3} \frac{1}{1 - \left(\frac{1}{6}\right)}\ \text{km} = 14\ \text{km}$$

2-52 $x(t) = \dfrac{2t^2 + (t + 1)\, e^{\alpha(t-1)}}{1 + e^{\alpha(t - 1)}}$

a)

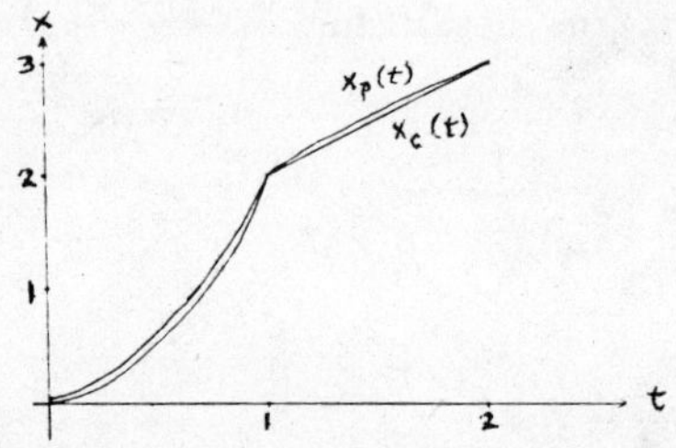

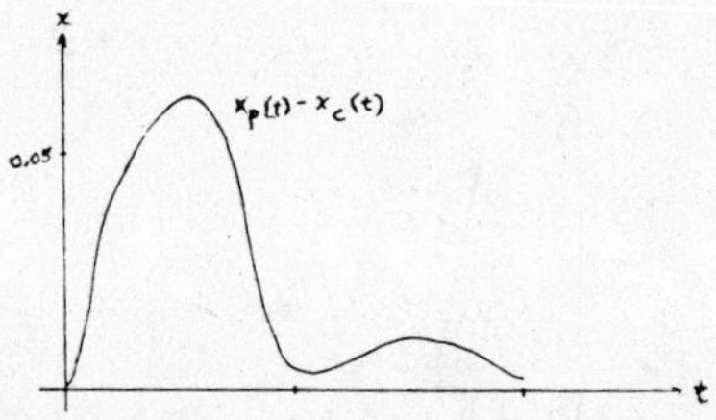

b) $$\frac{dx_p}{dt} = \frac{\left(1 + e^{\alpha(t-1)}\right)\left(4t + e^{\alpha(t-1)} + \alpha(t+1)e^{\alpha(t-1)}\right) - \left(2t^2 + (t+1)e^{\alpha(t-1)}\right)\alpha e^{\alpha(t-1)}}{\left(1 + e^{\alpha(t-1)}\right)^2}$$

$$v_p(1) = \frac{(1 + 1)(4 + 1 + 2\alpha) - (2 + 2)\alpha}{4} = 2.5 \text{ m/s}$$

From the left,

$$\frac{dx_c}{dt} = 4t$$

$$v_c(1) = 4 \text{ m/s}$$

From the right,

$$\frac{dx_c}{dt} = 1$$

$$v_c(1) = 1 \text{ m/s}$$

c) Let $\alpha(t-1) = x$, $2 - 8t = 4$, $2t^2 - t - 1 = z$

$$\frac{dv_p}{dt} = \frac{4 + e^x(\alpha y - \alpha^2 z + 8) + e^{2x}(\alpha y + \alpha^2 z + 4)}{(1 + e^x)^3}$$

At $t = 1$, $x = 0$, $y = -6$, and $z = -1$

$$a = \frac{4 + (-6\alpha + \alpha^2 + 8) + (-6\alpha - \alpha^2 + 4)}{2^3}$$

$$= \frac{4 - 12\alpha + 12}{8} = 2 - 1.5\,\alpha$$

2-53 $x(t) = +\sqrt{(t + 1)^2 + E^2}$ $\quad E < < 1$

a) For $E = 0.1$, the greatest difference between the two functions is 0.2 which occurs at $t = -1$.

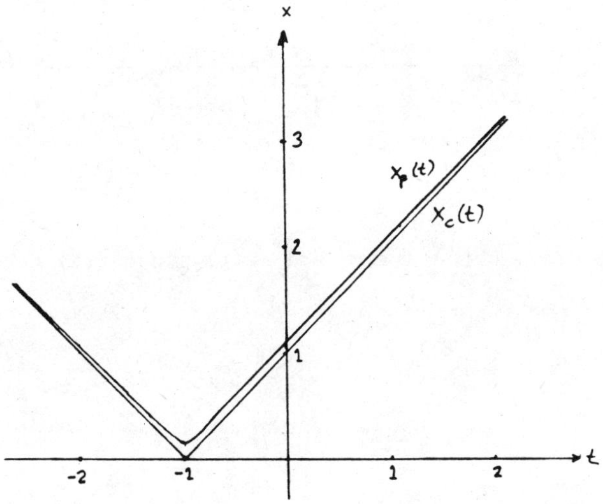

b) $$\frac{dx}{dt} = \frac{1}{2}\,\frac{1}{\sqrt{(t + 1)^2 + E^2}}\;2(t + 1)$$

$$v(t) = \frac{t + 1}{\sqrt{(t + 1)^2 + E^2}}$$

$$v(-1\ s) = \frac{0}{\sqrt{E^2}} = 0$$

c) $$\frac{dv}{dt} = \frac{1}{\sqrt{(t+1)^2 + E^2}} + (t+1)\left(-\tfrac{1}{2}\right)\frac{1}{((t+1)^2 + E^2)^{3/2}}\,2(t+1)$$

$$a(t) = \frac{(t+1)^2 + E^2}{((t+1)^2 + E^2)^{3/2}} - \frac{(t+1)^2}{((t+1)^2 + E^2)^{3/2}}$$

$$= \frac{E^2}{[(t+1)^2 + E^2]^{3/2}}$$

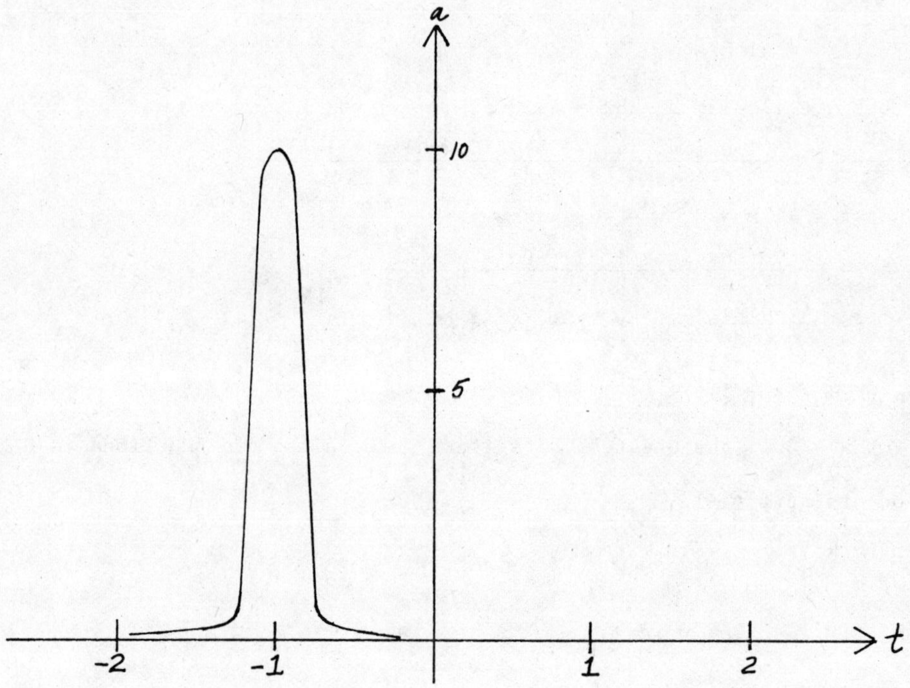

d) A real particle could not cross between x(t) and -x(t) since x(t) never reaches zero, and thus the two functions never cross.

3-1 Using graphical analysis, it can be seen that R is the resultant displacement.

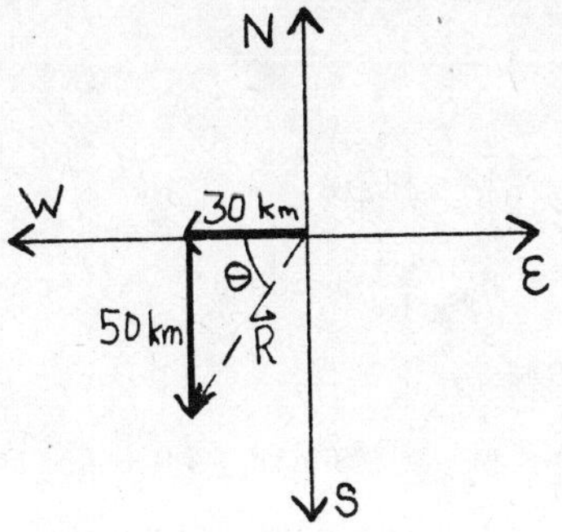

Analytically,

$$|\vec{R}| = \sqrt{(30\text{ km})^2 + (50\text{ km})^2} = 58.3\text{ km} = \text{net displacement}$$

$$\theta = \tan^{-1}\frac{50\text{ km}}{30\text{ km}} = 59.0°$$

That is, a displacement of 58.3 km, 59.0° south of west.

3-2 a) $|\vec{R}| = \sqrt{(1.6)^2 + (2.4)^2}$

$= 2.88$ units

$\theta = \tan^{-1}\frac{1.6}{2.4}$

$= 33.69°$ west of north

b) $|\vec{R}| = \sqrt{[3(1.6)]^2 + (2.4)^2}$

$= 5.37$ units

$\theta = \tan^{-1}\frac{3(1.6)}{2.4}$

$= 63.43°$ east of north

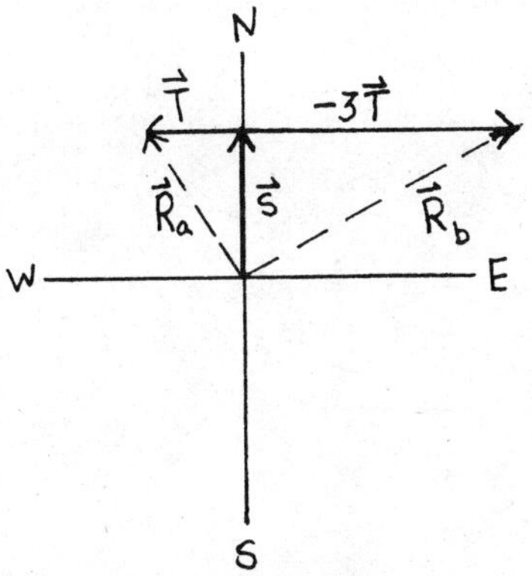

3-3 Using Eq. 3-4, we have

$$C_u = C\cos\theta_u = 3\cos 227° = -2.046$$

3-4 We have,

$$\vec{a} = 3\hat{i} + 2\hat{j}$$

$$\vec{b} = -\hat{i} + 7\hat{j}$$

and $\vec{a} + \vec{b} + \vec{c} = 0$

Then,

$$\vec{c} = -\vec{a} - \vec{b} = (-3\hat{i} - 2\hat{j}) - (-\hat{i} + 7\hat{j})$$

$$= -2\hat{i} - 9\hat{j}$$

3-5 $|\vec{A}| = 3, \ |\vec{B}| = 4$

$$R_x = A_x + B_x = 3 \cos 15° + 4 \cos 75°$$

$$= 3.93$$

$$R_y = A_y + B_y = 3 \sin 15° + 4 \sin 75°$$

$$= 4.64$$

$$|R| = \sqrt{(3.93)^2 + (4.64)^2} = 6.08 \text{ units}$$

$$\theta = \tan^{-1} \frac{R_y}{R_x} = 49.7°$$

The angle between R and A is

$$\phi = \theta - 15° = 34.7°$$

This agrees exactly with Example 3-5, showing that the choice of coordinate frames does not affect the results.

3-6 $C_x = A_x + B_x = 5 \cos 20° + 4 \cos 50°$

$$= 7.27$$

$$C_y = A_y + B_y = 5 \sin 20° + 4 \sin 50°$$

$$= 4.77$$

$$|\vec{C}| = \sqrt{C_x^2 + C_y^2} = 8.70 \text{ units}$$

$$\theta = \tan^{-1} \frac{C_y}{C_x} = 33.3°$$

3-7 We have $\vec{a} + \vec{b} = \vec{s}$ and $a + b = s$

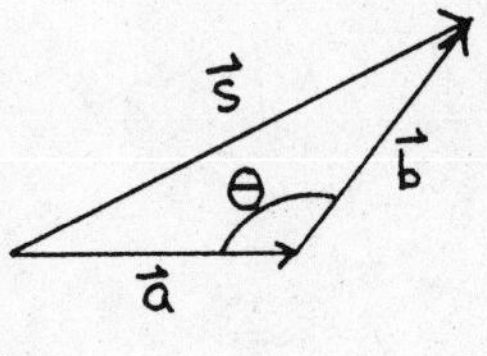

The law of cosines is

$$s^2 = a^2 + b^2 - 2ab \cos \theta$$

Now,

$$s = a + b$$

$$s^2 = (a + b)^2 = a^2 + 2ab + b^2$$

For both equations to hold, $\cos \theta = -1$.

Thus, $\theta = 180°$

$\vec{a}$ and $\vec{b}$ must be parallel, and in the same direction.

3-8 Given $\vec{A} = 2\hat{i} + 3\hat{j}$ and $\vec{B} = 3\hat{i} - 4\hat{j}$

a) $\vec{C} = \vec{A} + \vec{B}$

$$= 5\hat{i} - \hat{j}$$

$$|\vec{C}| = \sqrt{(5)^2 + (1)^2} = 5.10 \text{ units}$$

$$\theta = \tan^{-1}\left(\frac{-1}{5}\right) = -11.31° = 348.69°$$

b) $\vec{D} = 3\vec{A} - 2\vec{B} = 0\hat{i} + 17\hat{j}$

$$|\vec{D}| = 17 \text{ units}$$

$$\theta = 90°$$

3-9 Since $\vec{A}$ is perpendicular to $\vec{B}$,

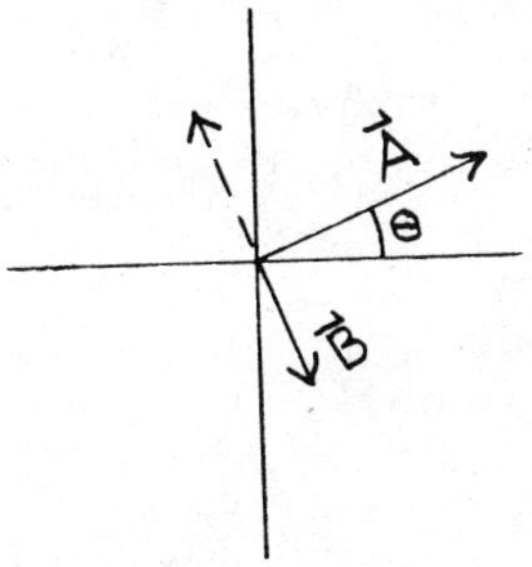

$$A_x = |\vec{A}|\cos\theta$$
$$B_x = |\vec{B}|\cos(\theta \pm \tfrac{\pi}{2}) = \mp |\vec{B}| \sin\theta$$
$$A_y = |\vec{A}|\sin\theta$$
$$B_y = |\vec{B}|\sin(\theta \pm \tfrac{\pi}{2}) = \pm |\vec{B}| \cos\theta$$

$$A_xB_x = \mp|\vec{A}|\,|\vec{B}|\cos\theta\,\sin\theta$$
$$A_yB_y = \pm|\vec{A}|\,|\vec{B}|\cos\theta\,\sin\theta$$
$$A_xB_x = -A_yB_y$$

3-10 We have

$$\vec{A} = 5\hat{i} + \hat{j} \quad , \quad \vec{B} = -2\hat{i} + 4\hat{j}$$

$$\theta_A = \tan^{-1}\frac{A_y}{A_x} = 11.3°$$

$$\theta_B = \tan^{-1}\frac{B_y}{B_x} = -63.4°$$

$$\theta_{AB} = \theta_A - \theta_B = 74.7°$$

3-11 $\vec{A} = 2\hat{i} + 3\hat{j}$

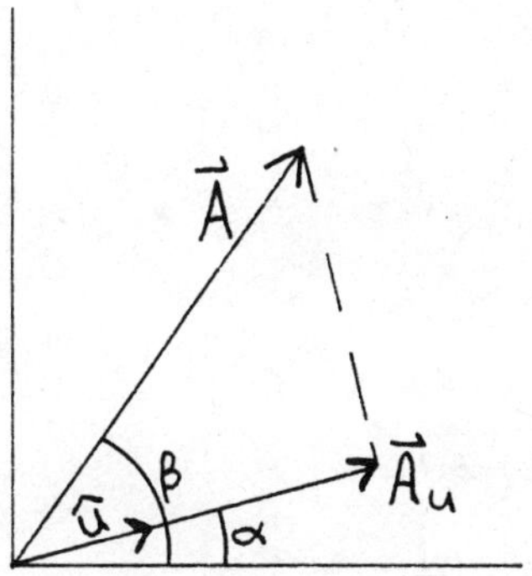

$$|\vec{A}| = \sqrt{(2)^2 + (3)^2} = \sqrt{13} = 3.606$$

$$\hat{u} = \frac{1}{\sqrt{5}}(4\hat{i} + \hat{j})$$

$$\alpha = \tan^{-1}\frac{1}{4} = 14.04°$$

$$\beta = \tan^{-}\frac{3}{2} = 56.31°$$

$$A_u = A\cos(\beta - \alpha) = \sqrt{13}\cos 42.27°$$

$$= 2.668 \text{ units}$$

3-12 We can write

$$\vec{a} = -3\hat{i} - 2\hat{j} + 7\hat{k}$$

$$\vec{b} = \hat{i} + 3\hat{j} + 3\hat{k}$$

$$\vec{c} = -5\hat{k}$$

$$\vec{a}+\vec{b}+\vec{c} = -2\hat{i} + \hat{j} + 5\hat{k}$$

3-13 $\vec{r} = 2000\ \hat{i} + 1000\ \hat{j} - 300\ \hat{k}$

$$|\vec{r}| = \sqrt{(2000)^2 + (1000)^2 + (300)^2}\ \text{m}$$

$$= 2256\ \text{m}$$

$$\tan\theta = \frac{2000\ \text{m}}{1000\ \text{m}} = 2$$

$$\theta = 63.4^\circ\ \text{east of north}$$

$$\tan\phi = \frac{300\ \text{m}}{\sqrt{(2000)^2 + (1000)^2}}$$

$$= 0.134$$

$$\phi = 7.6^\circ\ \text{below the surface}$$

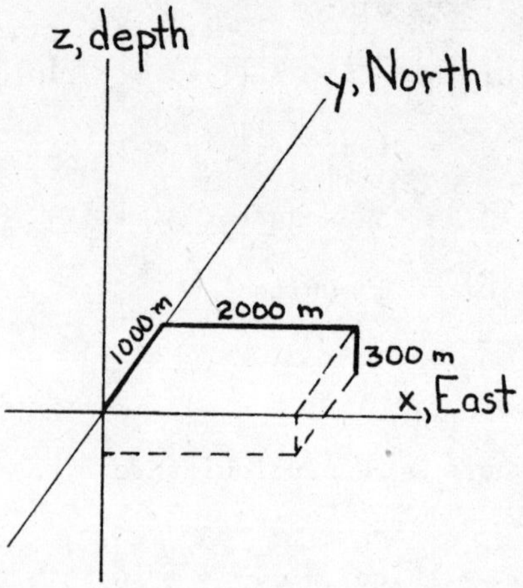

3-14 Given

$$\vec{A} = 3\hat{i} - 2\hat{j} + 2\hat{k}$$

$$\vec{B} = 4\hat{i} + 3\hat{j} - 2\hat{k}$$

$$\vec{C} = 3\hat{i} + 2\hat{j} + 3\hat{k}$$

$$\vec{D} = \vec{A} - \vec{B} + \vec{C}$$

$$= 2\hat{i} - 3\hat{j} + 7\hat{k}$$

In projecting $\vec{D}$ onto the xy-plane, the z-component is set equal to zero:

$$D_{xy} = 2\hat{i} - 3\hat{j}$$

$$D_{xy} = \sqrt{(2)^2 + (3)^2} = 3.6\ \text{units}$$

$$\theta = \tan^{-1}\left(\frac{-3}{2}\right) = -56.31^\circ = 303.69^\circ$$

3-15 We have

$$\vec{A} + \vec{B} + \vec{C} + \vec{E} = 0$$

This can be expressed as

$$\vec{E} = -\vec{A} - \vec{B} - \vec{C}$$

$$-\vec{A} = -3\hat{i} + 2\hat{j} - 2\hat{k}$$

$$-\vec{B} = -4\hat{i} - 3\hat{j} + 2\hat{k}$$

$$-\vec{C} = -3\hat{i} - 2\hat{j} - 3\hat{k}$$

$$\vec{E} = -10\hat{i} - 3\hat{j} - 3\hat{k}$$

3-16 $\vec{B} \cdot \vec{D} = (6)\ (-1) + (1)\ (3) + (-2)\ (0)$

$$= -6 + 3 + 0 = -3$$

$$|\vec{B}| = \sqrt{(6)^2 + (1)^2 + (2)^2} = \sqrt{41} = 6.40$$

$$|\vec{D}| = \sqrt{(1)^2 + (3)^2} = \sqrt{10} = 3.16$$

Using the definition of the dot product, we have

$$\cos\theta_{BD} = \frac{\vec{B}\cdot\vec{D}}{|\vec{B}|\ |\vec{D}|} = -0.148$$

$$\theta_{BD} = 98.5°$$

3-17 For $\vec{a}$ and $\vec{b}$ to be perpendicular, we must have $\theta_{ab} = 90°$. So,

$$\vec{a}\cdot\vec{b} = |\vec{a}|\ |\vec{b}|\ \cos 90° = 0$$

In this problem,

$$\vec{a}\cdot\vec{b} = (9)(3) + (1)(-7) + (-4)(5)$$

$$= 27 - 7 - 20 = 0$$

Thus, $\vec{a}$ must be perpendicular to $\vec{b}$.

3-18 We have

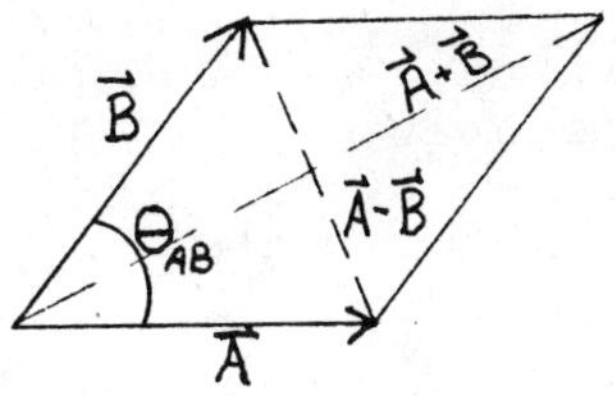

$$|\vec{A}+\vec{B}| = 2|\vec{A}-\vec{B}|$$

$$|\vec{A}+\vec{B}|^2 = 2^2\ |\vec{A}-\vec{B}|^2$$

But $|\vec{x}|^2 = \vec{x}\cdot\vec{x}$; thus,

$$(\vec{A}+\vec{B})\cdot(\vec{A}+\vec{B}) = 4(\vec{A}-\vec{B})\cdot(\vec{A}-\vec{B})$$

$$|\vec{A}|^2 + 2\vec{A}\cdot\vec{B} + |\vec{B}|^2 = 4|\vec{A}|^2 + 4|\vec{B}|^2 - 8\vec{A}\cdot\vec{B}$$

$$|\vec{A}|^2 + 2|\vec{A}|\ |\vec{B}|\ \cos\theta_{AB} + |\vec{B}|^2 = 4|\vec{A}|^2 + 4|\vec{B}|^2 - 8|\vec{A}|\ |\vec{B}|\ \cos\theta_{AB}$$

$$10|\vec{A}|\,|\vec{B}|\ \cos\theta_{AB} = 3|\vec{A}|^2 + 3|\vec{B}|^2$$

Now, $|\vec{A}| = |\vec{B}|$ since the parallelogram is equilateral;

then,

$$10|A|^2\cos\theta_{AB} = 6|A|^2$$

$$\theta_{AB} = \cos^{-1}\left(\frac{6}{10}\right) = 53.1°$$

3-19 $\vec{C} = 2\hat{i} - \hat{j}$

$\vec{D} = \hat{i} + 2\hat{j} - 3\hat{k}$

$$\vec{C}\times\vec{D} = \begin{vmatrix} \hat{i} & \hat{j} & \hat{k} \\ 2 & -1 & 0 \\ 1 & 2 & -3 \end{vmatrix}$$

$$= \hat{i}[(-1)(-3) - (2)(0)] - \hat{j}[(2)(-3) - (0)(1)]$$

$$+ \hat{k}[(2)(2) - (1)(-1)]$$

$$= 3\hat{i} + 6\hat{j} + 5\hat{k}$$

By definition,

$$|\vec{C} \times \vec{D}| = |\vec{C}|\ |\vec{D}| \sin\theta_{CD}$$

$$= \sqrt{(3)^2 + (6)^2 + (5)^2} = \sqrt{70} = 8.37$$

$$\vec{C} = \sqrt{(2)^2 + (1)^2} = \sqrt{5} = 2.24$$

$$\vec{D} = \sqrt{(1)^2 + (2)^2 + (3)^2} = \sqrt{14} = 3.74$$

$$\sin\theta_{CD} = \frac{|\vec{C} \times \vec{D}|}{|\vec{C}|\,|\vec{D}|} = 1$$

$$\theta_{CD} = 90°$$

3-20 $\vec{C} = 2\hat{i} - 4\hat{j} + 2\hat{k}$

$$\vec{D} = -3\hat{i} + 6\hat{j} - 3\hat{k}$$

$$\vec{C} \times \vec{D} = \begin{vmatrix} \hat{i} & \hat{j} & \hat{k} \\ 2 & -4 & 2 \\ -3 & 6 & -3 \end{vmatrix}$$

$$= \hat{i}(12 - 12) + \hat{j}(-6 + 6) + \hat{k}(12 - 12)$$

$$= 0$$

$\vec{C}$ and $\vec{D}$ are antiparallel.

3-21 To compute the angle between $\vec{A} \times \vec{B}$ and the z-axis, we find

$$\cos\theta = \hat{k} \cdot \frac{\vec{A} \times \vec{B}}{|\vec{A} \times \vec{B}|}$$

$$\vec{A} \times \vec{B} = \begin{vmatrix} \hat{i} & \hat{j} & \hat{k} \\ 2 & 3 & 1 \\ 1 & 3 & 2 \end{vmatrix} = 3\hat{i} - 3\hat{j} + 3\hat{k}$$

$$|\vec{A} \times \vec{B}| = \sqrt{(3)^2 + (3)^2 + (3)^2} = 3\sqrt{3} = 5.20$$

So,

$$\cos\theta = \frac{\hat{k} \cdot (3\hat{i} - 3\hat{j} + 3\hat{k})}{5.20} = \frac{3}{5.20} = 0.577$$

Thus,

$$\theta = 54.8°$$

3-22 We have

$$\vec{C} = \vec{A} \times \vec{B}$$

$$= \begin{vmatrix} \hat{i} & \hat{j} & \hat{k} \\ A_x & A_y & A_z \\ B_x & B_y & B_z \end{vmatrix}$$

Let

$$\frac{A_x}{B_x} = \frac{A_y}{B_y} = \frac{A_z}{B_z} = \psi$$

Multiplying the last row of the determinant by ψ we obtain

$$\vec{C} = \frac{1}{\psi}\begin{vmatrix} \hat{i} & \hat{j} & \hat{k} \\ A_x & A_y & A_z \\ A_x & A_y & A_z \end{vmatrix} = 0$$

This is true because a determinant with two identical rows must equal zero.

From this, the direction cosines are related by

$$\cos\theta_{A_x} = \cos\theta_{B_x}$$

$$\cos\theta_{A_y} = \cos\theta_{B_y}$$

$$\cos\theta_{A_z} = \cos\theta_{B_z}$$

3-23 Area $= \frac{1}{2}(\text{base}) \times (\text{height})$

$$= \tfrac{1}{2}\,|\vec{A}|\,|\vec{B}|\sin\theta$$

$$= \tfrac{1}{2}\,|\vec{A} \times \vec{B}|$$

$$\vec{A} \times \vec{B} = \begin{vmatrix} \hat{i} & \hat{j} & \hat{k} \\ 2 & 3 & 0 \\ 1 & -2 & 1 \end{vmatrix} = 3\hat{i} - 2\hat{j} - 7\hat{k}$$

$$|\vec{A} \times \vec{B}| = \sqrt{(3)^2 + (2)^2 + (7)^2} = \sqrt{62}$$

$$\text{Area} = \frac{\sqrt{62}}{2} = 3.937$$

Unit vectors that are normal to the triangle are:

$$\hat{u}_N = \pm\frac{1}{\sqrt{62}}(3\hat{i} - 2\hat{j} - 7\hat{k})$$

3-24 $\alpha = 78°\ 30' + 32°\ 30' = 111°$

$R = 6.38 \times 10^6$ m

a) $S = R\alpha$

$= (6.38 \times 10^6\ \text{m})(111°)(0.01745\ \text{rad/deg})$

$= 1.236 \times 10^7$ m

$= 12{,}360$ km

b) $d = 2\,R\sin\frac{\alpha}{2}$

$= 2(6.38 \times 10^6\ \text{m})\sin 55.5°$

$= 1.0516 \times 10^7$ m

$= 10{,}516$ km

3-25 We have

$$\vec{r}_1 = 3\hat{i} + 7\hat{j} - 2\hat{k}$$

$$\vec{r}_2 = -2\hat{i} + 6\hat{k}$$

$$\vec{r} = \vec{r}_1 - \vec{r}_2$$

$$= (3 + 2)\hat{i} + 7\hat{j} + (-2-6)\hat{k}$$

$$= 5\hat{i} + 7\hat{j} - 8\hat{k}$$

$$|\vec{r}| = \sqrt{(5)^2 + (7)^2 + (8)^2} = 11.75$$

3-26 Using the coordinate frame shown,

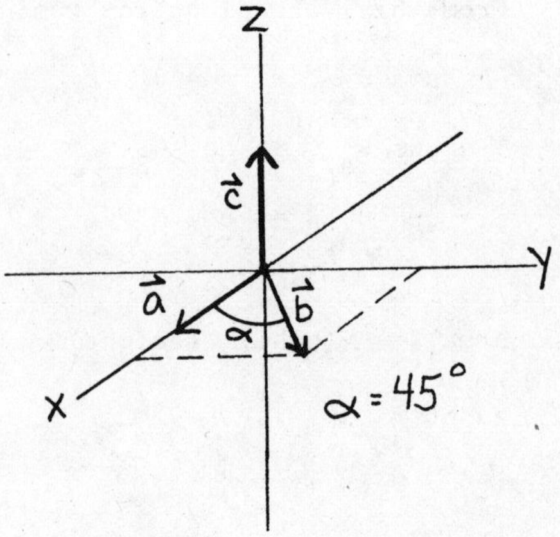

$$\vec{a} = 5\hat{i}$$

$$\vec{b} = 6\hat{i}\cos 45° + 6\hat{j}\sin 45°$$

$$= 4.24\hat{i} + 4.24\hat{j}$$

$$\vec{c} = 7\hat{k}$$

Let $\vec{d} = \vec{a} + \vec{b} + \vec{c}$

$$= (5 + 4.24)\hat{i} + 4.24\hat{j} + 7\hat{k}$$

$$= 9.24\hat{i} + 4.24\hat{j} + 7\hat{k}$$

$$|\vec{d}| = \sqrt{(9.24)^2 + (4.24)^2 + (7)^2} = 12.34$$

To find direction of $\vec{d}$ relative to $\vec{a}$ and $\vec{b}$,

$$\cos\theta_{ad} = \frac{\vec{a}\cdot\vec{d}}{|\vec{a}||\vec{d}|} = \frac{46.2}{61.7} = 0.75$$

$$\theta_{ad} = 41.5°$$

$$\cos\theta_{bd} = \frac{\vec{b}\cdot\vec{d}}{|\vec{b}||\vec{d}|} = \frac{57.2}{74.0} = 0.77$$

$$\theta_{bd} = 39.4°$$

3-27 The three vectors form a triangle as shown.

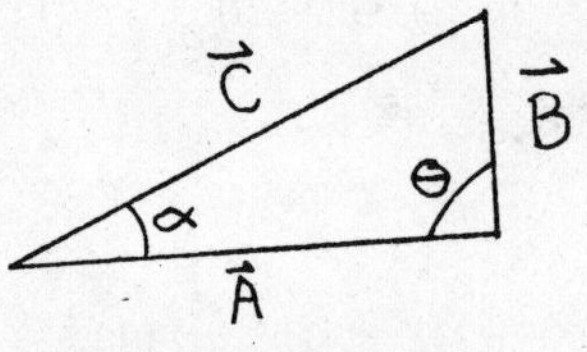

$$A^2 + B^2 - 2AB\cos\theta = C^2$$

$$2AB\cos\theta = A^2 + B^2 - C^2 = 0$$

$$\theta = 90°$$

To find the angle α,

$$\tan\alpha = \frac{7}{24}; \ \alpha = 16.3°$$

$|\vec{A}| = 24$; $|\vec{B}| = 7$; $|\vec{C}| = 25$

3-28

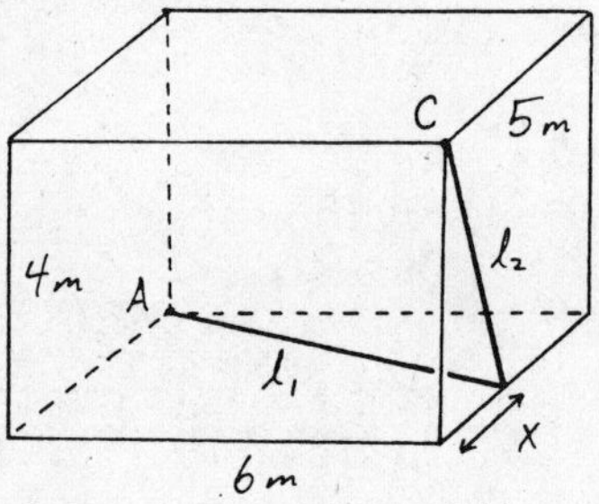

$$\ell_1^2 = 6^2 + (5 - x)^2$$

$$\ell_2^2 = 4^2 + x^2$$

$$L = \ell_1 + \ell_2$$

Require that $dL/dx = 0$; thus,

$$L = \sqrt{6^2 + (5 - x)^2} + \sqrt{4^2 + x^2}$$

$$\frac{dL}{dx} = \frac{x - 5}{\sqrt{6^2 + (5 - x)^2}} + \frac{x}{\sqrt{4^2 + x^2}} = 0$$

Simplifying,

$$x^2 + 8x - 20 = 0$$

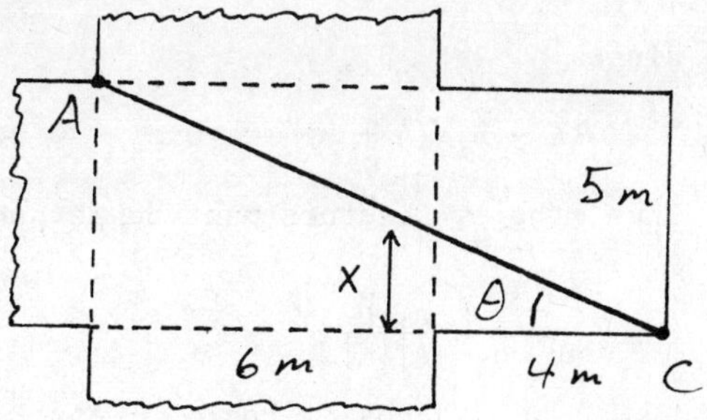

This gives x = 2 m

Alternate method: Fold down sides so that

$$\tan\theta = \frac{5}{10} = \frac{x}{4}$$

Thus,

$$x = 2 \text{ m}$$

b) $L = \sqrt{5^2 + 10^2} = 11.18$ m

Flying distance $= \sqrt{4^2 + 5^2 + 6^2} = 8.77$ m

Path difference = 2.41 m

3-29 We have

$$\vec{r}_1 = 2\hat{i} + \hat{j} + 3\hat{k}$$

$$\vec{r}_2 = 4\hat{i} - 2\hat{j} + 2\hat{k}$$

$$\cos\theta = \frac{\vec{r}_1 \cdot \vec{r}_2}{|\vec{r}_1||\vec{r}_2|}$$

$$= \frac{(2)(4) + (1)(-2) + (3)(2)}{\sqrt{[(2)^2 + (1)^2 + (3)^2]\ [(4)^2 + (2)^2 + (2)^2]}}$$

$$= \frac{12}{\sqrt{336}} = 0.655$$

$$\theta = 49.1°$$

3-30 $\vec{a} = \hat{i}\cos\alpha + \hat{j}\sin\alpha$

$\vec{b} = \hat{i}\cos\beta + \hat{j}\sin\beta$

$\vec{a}\cdot\vec{b} = \cos(\alpha - \beta) = \cos\alpha\cos\beta + \sin\alpha\sin\beta$

Next,

$$\vec{a}\times\vec{b} = \begin{vmatrix} \hat{i} & \hat{j} & \hat{k} \\ \cos\alpha & \sin\alpha & 0 \\ \cos\beta & \sin\beta & 0 \end{vmatrix} = (\cos\alpha\sin\beta - \sin\alpha\cos\beta)k$$

$|\vec{a}\times\vec{b}| = \sin(\alpha - \beta) = \cos\alpha\sin\beta - \sin\alpha\cos\beta$

3-31 We have $|\vec{B}| = |\vec{A}|$

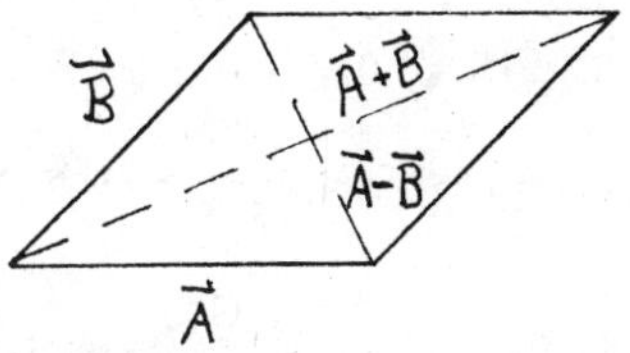

For $\vec{A} - \vec{B}$ to be perpendicular to $\vec{A} + \vec{B}$, we must have

$$(\vec{A} - \vec{B})\cdot(\vec{A} + \vec{B}) = 0$$

$$(\vec{A} - \vec{B})\cdot(\vec{A} + \vec{B}) = |\vec{A}|^2 - |\vec{B}|^2$$

Since $|\vec{A}| = |\vec{B}|$,

$$(\vec{A} - \vec{B})\cdot(\vec{A} + \vec{B}) = 0$$

Thus, the two vectors must be perpendicular.

3-32 $\cos\theta_{ab} = \dfrac{\vec{a}\ \vec{b}}{|\vec{a}||\vec{b}|}$

$$\cos\theta_{ab} = \frac{\cos\alpha_a\cos\alpha_b + \cos\beta_a\cos\beta_b + \cos\gamma_a\cos\gamma_b}{\sqrt{(\cos^2\alpha_a + \cos^2\beta_a + \cos^2\gamma_a)(\cos^2\alpha_b + \cos^2\beta_b + \cos^2\gamma_b)}}$$

$$= \cos\alpha_a\cos\alpha_b + \cos\beta_a\cos\beta_b + \cos\gamma_a\cos\gamma_b$$

3-33 $\vec{a} = \hat{i} - 2\hat{j} + 3\hat{k}$

$\vec{b} = 2\hat{i} + \hat{j} - \hat{k}$

Area $= |\vec{a} \times \vec{b}|$

$$= \begin{vmatrix} \hat{i} & \hat{j} & \hat{k} \\ 1 & -2 & 3 \\ 2 & 1 & -1 \end{vmatrix} = |-\hat{i} + 7\hat{j} + 5\hat{k}|$$

$$= \sqrt{(1)^2 + (7)^2 + (5)^2} = \sqrt{75} = 8.66$$

The unit vector in the direction of $\vec{a} \times \vec{b}$ is

$$\hat{u} = \frac{\vec{a} \times \vec{b}}{|\vec{a} \times \vec{b}|} = \frac{1}{8.66}(-\hat{i} + 7\hat{j} + 5\hat{k})$$

3-34 $\vec{B} \times \vec{C} = \begin{vmatrix} \hat{i} & \hat{j} & \hat{k} \\ B_x & B_y & B_z \\ C_x & C_y & C_z \end{vmatrix}$

$$= (B_yC_z - B_zC_y)\hat{i} + (B_zC_x - B_xC_z)\hat{j} + (B_xC_y - B_yC_x)\hat{k}$$

$$\vec{A}\cdot(\vec{B} \times \vec{C}) = (B_yC_z - B_zC_y)A_x + (B_zC_x - B_xC_z)A_y + (B_xC_y - B_yC_x)A_z$$

$$= \begin{vmatrix} A_x & A_y & A_z \\ B_x & B_y & B_z \\ C_x & C_y & C_z \end{vmatrix}$$

3-35 $\vec{C} = \vec{A} \times \vec{B}$

$$= (A_yB_z - A_zB_y)\hat{i} + (A_zB_x - A_xB_z)\hat{j} + (A_xB_y - A_yB_x)\hat{k}$$

For $\vec{C}$ to be perpendicular to $\vec{A}$ and $\vec{B}$, then

$\vec{C}\cdot\vec{A} = 0$ and $\vec{C}\cdot\vec{B} = 0$

$$\vec{C}\cdot\vec{A} = (A_yB_zA_x - A_zB_yA_x) + (A_zB_xA_y - A_xB_zA_y) + (A_xB_yA_z - A_yB_xA_z)$$

$$= 0$$

$$\vec{C}\cdot\vec{B} = (A_yB_zB_x - A_zB_yB_x) + (A_zB_xB_y - A_xB_zB_y) + (A_xB_yB_z - A_yB_xB_z)$$

$$= 0$$

Thus, C is perpendicular to both A and B.

3-36 To find the total volume we use

$$V = h\,A_{base}$$

Now,

$$A_{base} = |\vec{B} \times \vec{C}|$$

$$h = \hat{u}_{\vec{B}\times\vec{C}} \cdot \vec{A}$$

where $\hat{u}_{\vec{B}\times\vec{C}}$ is the unit vector in the direction of $\vec{B} \times \vec{C}$.

$$h = \frac{\vec{B} \times \vec{C}}{|\vec{B} \times \vec{C}|} \cdot \vec{A}$$

Thus,

$$V = \left(\vec{A} \cdot \frac{\vec{B} \times \vec{C}}{|\vec{B} \times \vec{C}|}\right) |\vec{B} \times \vec{C}|$$

$$= \vec{A} \cdot (\vec{B} \times \vec{C})$$

3-37 Using the method derived in Problem 3-36, we have

$$V = \vec{a} \cdot (\vec{b} \times \vec{c})$$

$$\vec{b} \times \vec{c} = \begin{vmatrix} \hat{i} & \hat{j} & \hat{k} \\ 1 & 2 & -1 \\ 0 & 0 & 3 \end{vmatrix} = 6\hat{i} - 3\hat{j}$$

$$V = (3\hat{i} - 2\hat{j}) \cdot (6\hat{i} - 3\hat{j}) = 18 + 6$$

$$= 24$$

3-38 $\vec{B} \times \vec{C}$ is, by definition, perpendicular to the plane formed by vectors $\vec{B}$ and $\vec{C}$. If $\vec{A}$ is also in this plane, then $\vec{B} \times \vec{C}$ must also be perpendicular to $\vec{A}$. Thus,

$$\vec{A} \cdot (\vec{B} \times \vec{C}) = 0$$

By the relation shown in Problem 3-34,

$$\begin{vmatrix} A_x & A_y & A_z \\ B_x & B_y & B_z \\ C_x & C_y & C_z \end{vmatrix} = 0$$

4-1 We have

$$\vec{r}(t) = 2t\hat{i} + (8t^2 + 1)\hat{j}$$

and

$$y = ax^2 + bx + c$$

$$= a(2t)^2 + b(2t) + c$$

$$= 4at^2 + 2bt + c = 8t^2 + 1$$

Equating coefficients of powers of t, we have

$$4a = 8 \text{ or } a = 2$$

$$2b = 0 \text{ or } b = 0$$

$$c = 1$$

Hence,

$$y = 2x^2 + 1$$

4-2 We have

$$\vec{r}(t) = 6\left(\cos 2t\right)\hat{i} + 6\left(\sin 2t\right)\hat{j}$$

Then,

$$\vec{r}_P = \vec{r}(0) = 6\hat{i}$$

$$\vec{r}_Q = \vec{r}(\tfrac{\pi}{4}) = 6\hat{j}$$

Thus,

$$\Delta\vec{r} = \vec{r}_Q - \vec{r}_P$$

$$= -6\hat{i} + 6\hat{j}$$

4-3 We have

$$\vec{r}_1 = (6 \text{ m})\hat{i} + (7 \text{ m})\hat{j}$$

and

$$\vec{r}_2 = (4 \text{ m})\hat{i} - (2 \text{ m})\hat{j} \text{ after 1 s.}$$

Then,

$$\bar{\vec{v}} = \frac{\vec{r}_2 - \vec{r}_1}{\Delta t}$$

$$= (-2 \text{ m/s})\hat{i} - (9 \text{ m/s})\hat{j}$$

$$\text{average speed} = |\bar{\vec{v}}|$$

$$= \sqrt{(2\ \text{m/s})^2 + (9\ \text{m/s})^2}$$

$$= 9.22\ \text{m/s}$$

4-4 a) $\vec{r}(t) = \alpha t^3\hat{i} + \beta t^2\hat{j}$

$$\vec{v}(t) = \frac{d}{dt}\vec{r}(t) = 3\alpha t^2\hat{i} + 2\beta t\hat{j}$$

$$\vec{v}(1) = 3\alpha\hat{i} + 2\beta\hat{j}$$

b) $|\vec{r}(1)| = \sqrt{\alpha^2 + \beta^2} = \sqrt{5}\ \text{m}$

$$|\vec{v}(1)| = \sqrt{9\alpha^2 + 4\beta^2} = 5\ \text{m/s}$$

Thus, we have two equations containing α^2 and β^2, namely,

$$9\alpha^2 + 4\beta^2 = 25$$

$$\alpha^2 + \beta^2 = 5$$

Substituting,

$$9\alpha^2 + 4(-\alpha^2 + 5) = 25$$

$$5\alpha^2 = 5$$

$$\alpha = \pm 1$$

Then,

$$1 + \beta^2 = 5$$

$$\beta = \pm 2$$

Therefore,

$$|\vec{r}(2)| = (\pm 8\ \hat{i} \pm 8\ \hat{j})\text{m}$$

$$|\vec{r}(2)| = \sqrt{(\pm 8)^2 + (\pm 8)^2}\ \text{m} = 8\sqrt{2}\ \text{m} = 11.3\ \text{m}$$

4-5 $\vec{r}(t) = ct\ \hat{i} + (bt - at^2)\hat{j}$

Since $\theta(0) = 45°$, $v_x = v_y$ at $t = 0$

$$v_x(0) = c$$

$$v_y(0) = b$$

Thus, $c = b$

and we can write

$$\vec{r}(t) = bt\ \hat{i} + (bt - at^2)\hat{j}$$

To find t when $y(t) = h$, we have

$$\frac{dy(t)}{dt} = 0 = b - 2at$$

$$t = \frac{b}{2a}$$

Then,

$$y(\frac{b}{2a}) = \frac{b^2}{4a} = h$$

Therefore,

$$x(\frac{b}{2a}) = \frac{b^2}{2a} = 2h$$

4-6 $\vec{r}(t) = x(t)\hat{i} + y(t)\hat{j} + z(t)\hat{k}$

with $x(t) = a \sin \omega t$

$y(t) = a \cos \omega t$

$z(t) = bt$

$\vec{r}(0) = a\hat{j}$

$\vec{r}(\frac{2\pi}{\omega}) = a\hat{j} + \frac{2\pi b}{\omega}\hat{k}$

$\bar{\vec{v}} = \frac{\Delta\vec{r}}{\Delta t} = \frac{2\pi b/\omega}{2\pi/\omega}\hat{k} = b\hat{k}$

$\vec{v}(t) = \frac{d}{dt}\vec{r}(t) = a\omega(\cos \omega t)\hat{i} - a\omega(\sin \omega t)\hat{j} + b\hat{k}$

$|\vec{v}(t)| = \sqrt{a^2\omega^2(\cos^2 \omega t + \sin^2 \omega t) + b^2} = \sqrt{a^2\omega^2 + b^2}$

$\vec{r}(t)$ describes a helical path.

4-7 We have

$x(t) = 3t^2 + 2t$

$y(t) = 2t^2 + 4t + 3$

Then,

$$\vec{v}(t) = \frac{dx}{dt}\hat{i} + \frac{dy}{dt}\hat{j}$$

$$= (6t + 2)\hat{i} + (4t + 4)\hat{j}$$

and

$$\vec{a}(t) = \frac{dv_x}{dt}\hat{i} + \frac{dv_y}{dt}\hat{j}$$

$$= 6\hat{i} + 4\hat{j}$$

Since $\vec{a}(t)$ does not depend on time, we have

$\vec{a}(2) = 6\hat{i} + 4\hat{j}$

$|\vec{a}| = \sqrt{6^2 + 4^2} = \sqrt{52} = 7.21 \text{ m/s}^2$

$\theta = \tan^{-1}\frac{4}{6} = 33.69°$

4-8 We have

$\vec{r}(t) = (t^3 + 1)\hat{i} + (t^3 - 6t^2 + 12)\hat{j} + (3t - 6)\hat{k}$

a) $\vec{v}(t) = \frac{d}{dt}\vec{r}(t) = 3t^2\hat{i} + (3t^2 - 12t + 12)\hat{j} + 3\hat{k}$

$\vec{a}(t) = \frac{d}{dt}\vec{v}(t) = 6t\hat{i} + (6t - 12)\hat{j}$

(b) $v_y(t) = 3t^2 - 12t + 12$

For v_y to be minimum, $\frac{dv_y}{dt} = 0$ and $\frac{d^2v_y}{dt^2} > 0$

$$\frac{dv_y}{dt} = 6t - 12 = 0$$

Thus, $t = 2$ s

$$\frac{d^2v_y}{dt} = 6 > 0$$

Therefore, at $t = 2$ s, v_y is minimum.

(c) $\vec{r}(2) = 9\hat{i} + 8\hat{j}$

$$|\vec{r}| = \sqrt{9^2 + 8^2} = \sqrt{145} = 12.0 \text{ m}$$

$\vec{r}$ is in the x – y plane.

$$\vec{v}(2) = 12\hat{i} + 3\hat{k}$$

$$|\vec{v}| = \sqrt{12^2 + 3^2} = \sqrt{153} = 12.4 \text{ m/s}$$

$\vec{v}$ is in the x – z plane.

$$\vec{a}(2) = 12\hat{i}$$

$$|\vec{a}| = 12 \text{ m/s}^2$$

$\vec{a}$ is along the x-axis.

4-9 $\vec{r}(t) = R(\omega t - \sin \omega t)\hat{i} + R(1 - \cos \omega t)\hat{j}$

$$\vec{v}(t) = \frac{d}{dt}\vec{r}(t) = R(\omega - \omega \cos \omega t)\hat{i} + R\omega(\sin \omega t)\hat{j}$$

$$\vec{a}(t) = \frac{d}{dt}\vec{v}(t) = R\omega^2(\sin \omega t)\hat{i} + R\omega^2(\cos \omega t)\hat{j}$$

The particle will be at rest when v_x and v_y are 0. Thus, $t = 2n\pi, n = 0, 1, 2...,$ and $t = \frac{2n\pi}{\omega}$

At these times,

$$\vec{r}\left(\frac{2n\pi}{\omega}\right) = R(2n\pi - \sin 2n\pi)\hat{i} + R(1 - \cos 2n\pi)$$

$$= 2\pi Rn\hat{i}$$

$$\vec{a}\left(\frac{2n\pi}{\omega}\right) = (R\omega^2 \sin 2n\pi)\hat{i} + (R\omega^2 \cos 2n\pi)\hat{j} = R\omega^2\hat{j}$$

$$|\vec{a}| = \sqrt{(R\omega^2 \sin \omega t)^2 + (R\omega^2 \cos \omega t)^2}$$

$$= \sqrt{R^2\omega^4(\sin^2 \omega t + \cos^2 \omega t)}$$

$$= R\omega^2$$

The magnitude of $\vec{a}(t)$ does not depend on time.

$$\vec{v} \cdot \vec{a} = v_x a_x + v_y a_y = R^2\omega^3 \sin \omega t$$

$$|\vec{v}| = \sqrt{R^2\omega^2 - 2\omega^2R^2 \cos \omega t + R^2\omega^2} = \omega R\sqrt{2(1 - \cos \omega t)}$$

Using the relation; $\frac{1}{2}(1 - \cos \omega t) = \sin^2\left(\frac{\omega t}{2}\right)$, we have

$$|\vec{v}| = 2\omega R \sin(\frac{\omega t}{2})$$

$$\cos\theta_{va} = \frac{\vec{v}\cdot\vec{a}}{|\vec{v}||\vec{a}|} = \frac{\omega^3 R^2 \sin\omega t}{2\omega^3 R^2 \sin\omega t/2}$$

Using the relation; $\sin\omega t = 2(\sin\frac{\omega t}{2})(\cos\frac{\omega t}{2})$, we have

$$\cos\theta_{va} = \frac{2(\sin\omega t/2)(\cos\omega t/2)}{2\sin\omega t/2} = \cos\frac{\omega t}{2}$$

from which

$$\theta_{va} = \frac{\omega t}{2}$$

4-10 To catch up to the ball, the velocity of the boy must be

$$v = v_o \cos\alpha$$

$$\cos\alpha = \frac{v}{v_o} = (\frac{20\text{ m}}{3\text{ s}})(\frac{1}{20\text{ m/s}}) = 0.33$$

$$\alpha = 70.5°$$

$$v_{oy} = v_o \sin\alpha = 18.9\text{ m/s}$$

Using Eq. 4-21, we find

$$y_m = \frac{(v_o \sin\alpha)^2}{2g} = \frac{(18.9)^2}{2(9.8)} = 18.2\text{ m}$$

4-11 (a) Using Eq. 4-21, we have

$$Z_m = \frac{v_o^2}{2g}\sin^2\alpha$$

but since the starting point is 2 m from the ground

$$Z_m = 2\text{ m} + \frac{v_o^2}{2g}\sin^2\alpha = 7.10\text{ m}$$

(b) We have

$$Z = Z_o + v_{oz} t + \frac{1}{2} at^2$$

$$= Z_m + (v_o \sin\alpha)t - \frac{1}{2} gt^2$$

and

$$y = (v_o \cos\alpha)t$$

$$t = \frac{y}{v_o \cos\alpha}$$

Substituting,

$$Z = 2\text{ m} + v_o(\sin\alpha)\frac{y}{v_o\cos\alpha} - \frac{1}{2} g \frac{y^2}{v_o^2\cos^2\alpha}$$

When the ball strikes the ground,

$$0 = 2\text{ m} + y\tan\alpha - \frac{gy^2}{2v_o^2\cos^2\alpha}$$

$$= (-0.0163)y^2 + (0.577)y + 2\text{ m}$$

Solving for y, we obtain

$$y = \frac{-0.577 \pm \sqrt{(0.577)^2 + (0.130)}}{-0.0326} = 38.57\text{ m}$$

4-11 (c) We know that

$$v_z^2 - v_{oz}^2 = 2az$$

$$v_z^2 - v_o^2 \sin^2 \alpha = -2g(-2\text{ m}) \text{ at impact}$$

$$v_z^2 = 139.2(\text{m/s})^2$$

$$v_z = -11.8\text{ m/s}$$

$$v_{impact} = \sqrt{v_z^2 + v_y^2} = \sqrt{v_z^2 + v_o^2 \cos^2\alpha} = 20.96\text{ m/s}$$

To find the time of impact, set

$$v_z - v_{oz} = at$$

$$(-11.8\text{ m/s}) - v_o \sin \alpha = -gt_{impact}$$

$$t_{impact} = 2.22\text{ s}$$

4-12 From Eq. 4-18, we have

$$R = \frac{v_o^2}{g} \sin 2\alpha = 180\text{ m}$$

From Eq. 4-17, when the ball clears the tree, we have

$$15\text{ m} = -\left(\frac{g}{2\, v_o^2 \cos^2\alpha}\right)(30\text{ m})^2 + (\tan \alpha)(30\text{ m})$$

Substuting,

$$15\text{ m} = -\left(\frac{\sin 2\alpha}{2\, R \cos^2\alpha}\right)(30\text{ m})^2 + (\tan \alpha)(30\text{ m})$$

$$15\text{ m} = -(\tan \alpha)\frac{(30\text{ m})^2}{R} + (\tan \alpha)(30\text{ m})$$

$$\tan \alpha = 0.6$$

$$\alpha = 30.9°$$

Again from Eq. 4-18,

$$v_o = \sqrt{\frac{Rg}{\sin 2\alpha}} = 44.7\text{ m/s}$$

4-13 We have from Eq. 4-18,

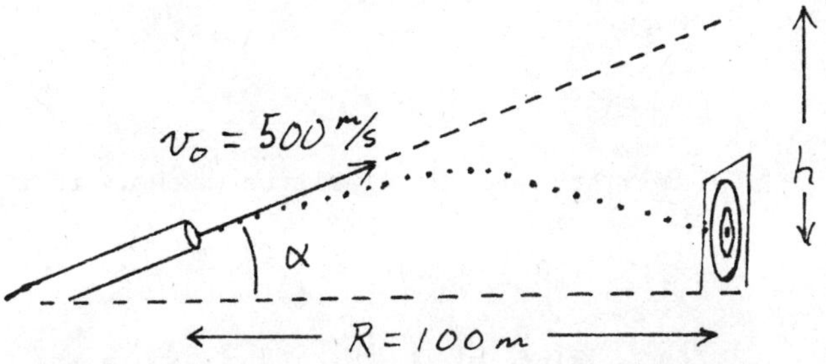

$$R = \frac{v_o^2}{g} \sin 2\alpha$$

$$100\text{ m} = \frac{(500\text{ m/s})^2}{(9.8\text{ m/s}^2)} \sin 2\alpha$$

$$\sin 2\alpha = 3.92 \times 10^{-3}$$

$$\alpha = 0.112°$$

Now, we have

$$\tan \alpha = \frac{h}{R}$$

$$h = (100 \text{ m})\tan(0.112°) = 0.195 \text{ m}$$

4-14 For the two paths to be equal, the ball must strike the wall when it is at the very top of its path. Therefore, $R = 2(6 \text{ m}) = 12 \text{ m}$.

Now,

$$R = \frac{v_o^2}{g} \sin 2\alpha$$

$$v_o = \sqrt{\frac{gR}{\sin 2\alpha}}$$

We can see that the minimum v_o occurs when $\sin 2\alpha = 1$ or $\alpha = 45°$.

In this case,

$$v_o = \sqrt{gR} = 10.8 \text{ m/s}$$

4-15 Using Eq. 4-18, we have

$$R\pm = \frac{v_o^2}{g} \sin(90° \pm 2\beta)$$

But $\sin(90° + 2\beta) = \cos 2\beta = \sin(90° - 2\beta)$

So $$R_+ = R_- = \frac{v_o^2}{g} \cos 2\beta.$$

Now,

$$v_y = v_o \sin(45° \pm \beta) - gt$$

gives twice the time to the top as

$$t_\pm = \frac{2\, v_o}{g}\sin(45° \pm \beta)$$

or

$$\Delta t = t_+ - t_- = \frac{2\, v_o}{g}[\sin(45° + \beta) - \sin(45° - \beta)]$$

Now use $\sin A - \sin B = 2 \sin \frac{1}{2}(A - B)\cos \frac{1}{2}(A + B)$ to give $\Delta t = \frac{2\sqrt{2}\, v_o}{g} \sin \beta$

When $\beta = 45°$, $\vec{v}_o$ is either vertical or horizontal and $\Delta t = 2v_o/g$, which is the time required for a vertical projectile to rise and fall.

4-16 From Eq. 4-21,

$$z_m = \frac{v_o^2}{2g} \sin^2 \alpha$$

Clearly, the distance will be maximum if z_m occurs over the wall. Thus,

$$(20 \text{ m} - 4 \text{ m}) = \frac{(22 \text{ m/s})^2}{2(9.8 \text{ m/s}^2)} \sin^2 \alpha$$

$$\sin^2\alpha = 0.648$$

$$\alpha = 53.6°$$

To find R, we again use Eq. 4-21,

$$z_m = \frac{1}{4} R \tan \alpha$$

$$R = \frac{4 z_m}{\tan \alpha} = 47.18 \text{ m}$$

The wall is at a distance R/2, so the distance to the wall = 23.59 m.

4-17 From Eq. 4-17,

$$z = -\left(\frac{g}{2 v_o^2 \cos^2\alpha}\right)y^2 + (\tan \alpha)y$$

Using the relationship,

$$1 + \tan^2\alpha = \frac{1}{\cos^2 \alpha},$$

we find

$$z = \frac{-g}{2 v_o^2}(1 + \tan^2 \alpha)y^2 + (\tan \alpha)y$$

$$\frac{-gy^2}{2 v_o^2}(\tan^2 \alpha) + y(\tan \alpha) - \left(\frac{gy^2}{2 v_o^2} + z\right) = 0$$

Solving for $\tan \alpha$,

$$\tan \alpha = \frac{-y \pm \sqrt{y^2 - \left(\frac{g^2y^4}{v_o^4} + \frac{2 gy^2z}{v_o^2}\right)}}{-gy^2/v_o^2}$$

$$= \frac{-10 \pm 4.85}{-9.8} = 1.52,\ 0.525$$

$$\alpha = 56.6°,\ 27.7°$$

4-18 From Eq. 4-17,

$$z = -\left(\frac{g}{2v_o^2 \cos^2\alpha}\right)y^2 + (\tan \alpha)y$$

$$-0.6 \text{ m} = -\left(\frac{9.786 \text{ m/s}^2}{2 v_o^2 \cos^2 25°}\right)(8.90 \text{ m})^2 + (\tan 25°)(8.90 \text{ m})$$

$$v_o^2 = 99.5 \text{ (m/s)}^2$$

$$v = 9.97 \text{ m/s}$$

$$v_{ox} = \text{running speed} = v_o \cos 25° = 9.04 \text{ m/s}$$

4-19 We have $R = 0.2$ m. Also,

$$\nu = 1000 \text{ rpm} = 16.67 \text{ s}^{-1}$$

$$\omega = 2\pi\nu = 104.7 \text{ s}^{-1}$$

$$v = \omega R = 20.94 \text{ m/s}$$

$$a = R\omega^2 = 2194 \text{ m/s}^2$$

4-20 We have

$$a = 7\,g = R\omega^2$$

$$7(9.8\text{ m/s}^2) = (5\text{ m})\omega^2$$

$$\omega = 3.70\text{ s}^{-1}$$

$$\nu = \frac{\omega}{2\pi} = 0.589\text{ s}^{-1} = 35.4\text{ rpm}$$

Now,

$$\omega = \alpha t = \frac{a_\phi}{R}t$$

For $a_\phi = \frac{g}{30} = 0.327\text{ m/s}^2$, we find

$$t = \frac{\omega R}{a_\phi} = 56.6\text{ s}$$

4-21 We have $R = 0.1$ m, $\alpha_0 = \pi\text{ s}^{-2}$, $\phi_0 = \pi/2$, and $\omega_0 = 0$. At $t = 1$ s, we find

(a) $$\phi = \frac{1}{2}\alpha_0 t^2 + \omega_0 t + \phi_0$$

$$= \frac{1}{2}(\pi\text{ s}^{-2})(1\text{ s})^2 + \frac{\pi}{2} = \pi = 3.14\text{ rad}$$

(b) $$\omega = \alpha_0 t + \omega_0 = (\pi\text{ s}^{-2})(1\text{ s}) = \pi\text{ s}^{-1} = 3.14\text{ s}^{-1}$$

(c) $$v = \omega R = (0.1\text{ m})(\pi\text{ s}^{-1}) = (\frac{\pi}{10})\text{ m/s} = 0.314\text{ m/s}$$

(d) $$a_\phi = R\alpha_0 = (0.1\text{ m})(\pi\text{ s}^{-2}) = 0.314\text{ m/s}^2$$

$$a_\rho = R\omega^2 = (0.1\text{ m})(\pi\text{ s}^{-1})^2 = 0.987\text{ m/s}^2$$

(e) $$|\vec{a}| = \sqrt{a_\rho^2 + a_\phi^2} = \sqrt{(0.987)^2 + (0.314)^2}\text{ m/s}^2$$

$$= 1.036\text{ m/s}^2$$

4-22 From Eqs. 4-34 and 4-35,

$$\phi = \frac{1}{2}\alpha_o t^2 + \omega_o t + \phi_o$$

$$\frac{1}{2}\alpha_o t^2 + \omega_o t - (\phi - \phi_o) = 0$$

$$t = \frac{-\omega_o \pm \sqrt{\omega_o^2 + 2\alpha_0(\phi - \phi_0)}}{\alpha_0} = \frac{\omega - \omega_0}{\alpha}$$

$$\omega = \pm\sqrt{\omega_0^2 + 2\alpha_0(\phi - \phi_0)}$$

$$\omega^2 = \omega_0^2 + 2\alpha_0(\phi - \phi_0)$$

$$\omega^2 - \omega_0^2 = 2\alpha_0(\phi - \phi_0)$$

4-23 (a) $$\phi(t) = 2t^3 - 3t^2 + \frac{\pi}{2}$$

$$\omega(t) = \frac{d\phi}{dt} = 6t^2 - 6t$$

$$a_\rho(t) = R\omega^2 = 0.2(6t^2 - 6t)^2$$

$$a_\phi(t) = R\frac{d^2\phi}{dt^2} = 1.2(2t - 1)$$

(b) For $\vec{a}$ to be entirely radial, $a_\phi = 0$; thus,

$$1.2(2t - 1) = 0$$

This gives

$$t = 0.5 \text{ s}$$

$$a_\rho(t = 0.5) = 0.2(1.5 - 3)^2 = 0.45 \text{ m/s}^2$$

(c) For $\vec{a}$ to be entirely tangential, $a_\rho = 0$; thus,

$$0 = R\omega^2$$

$$\omega = 0 = 6t^2 - 6t$$

Solving for t, we find $t = 0, 1$ s. At $t = 0$,

$$a = a_\phi = 1.2(2t - 1) = 1.2 \text{ m/s}^2$$

At $t = 1$ s,

$$a = 1.2(2 - 1) = 1.2 \text{ m/s}^2$$

4-24 $\rho = R_E \cos\lambda$

$v = \rho\omega = (2\pi R_E/T)\cos\lambda$

To find the centripetal acceleration,

$$\frac{a_c}{g} = \frac{v^2}{\rho g} = \frac{4\pi^2 R_E}{T^2 g}\cos\lambda$$

$$= 3.43 \times 10^{-3} \cos\lambda$$

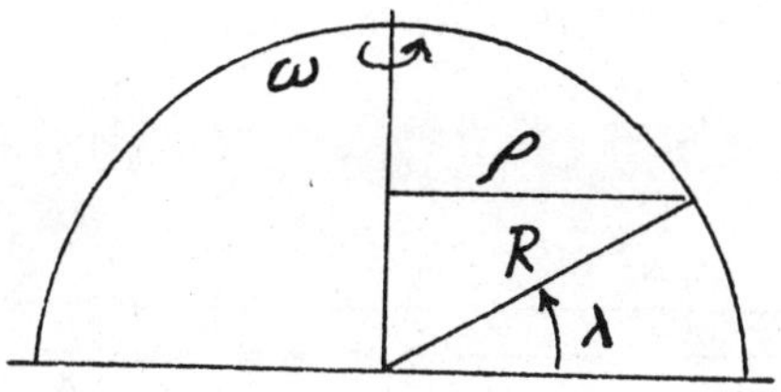

4-25 Let L be the length of the escalator.

$$v_s = \frac{L}{30 \text{ s}}$$

$$v_m = \frac{L}{20 \text{ s}}$$

$$v_f = v_s + v_m = \frac{L}{30 \text{ s}} + \frac{L}{20 \text{ s}} = \frac{(20 \text{ s})L + (30 \text{ s})L}{600 \text{ s}^2}$$

$$t = \frac{L}{v_f} = \frac{600 \text{ s}^2}{20 \text{ s} + 30 \text{ s}} = 12 \text{ s}$$

4-26 v = river speed; v_o = canoe speed in still water

$v_o + v = 2.9$ m/s; $v - v_o = -1.2$ m/s

Adding and dividing by 2 gives

$$v = 0.85 \text{ m/s}$$

4-27 v = river speed. The distance traveled is the same, so

$d = (5\ \text{km/h} + v)T = (10\ \text{km/h} - v)(1.15\ T)$

from which

$$v = 3.02\ \text{km/h}$$

4-28 $v_t = 12\ \text{m/s}$

Let v_r be the velocity of the raindrop with respect to the ground.

$$\frac{v_t}{v_r} = \tan\alpha$$

$$v_r = \frac{v_t}{\tan\alpha} = \frac{12\ \text{m/s}}{\tan 27°} = 23.6\ \text{m/s}$$

The speed of the raindrops with respect to the train is

$$v = \sqrt{v_r^2 + v_t^2} = 26.5\ \text{m/s}$$

4-29 Let v_g be the velocity with respect to the ground.

$$\vec{v}_g = \vec{v}_p + \vec{v}_\omega$$

$$v_p \sin\alpha = v_\omega$$

$$\alpha = 16.6° \text{ west of north}$$

$$v_g = v_p \cos\alpha = 335\ \text{km/h}$$

4-30

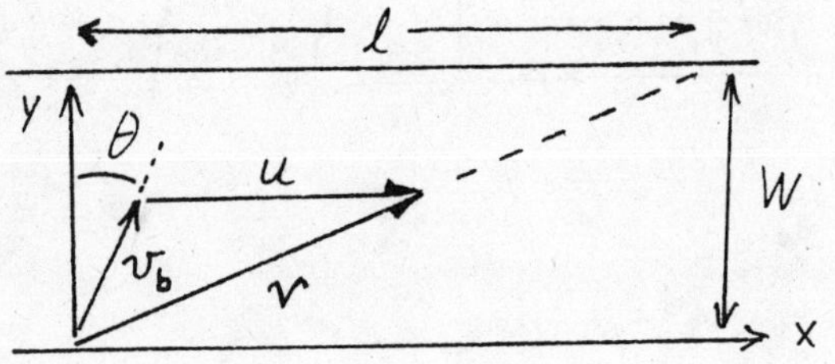

$w = 150$ m

u = water's speed

$u = 2$ m/s

v_b = boat's speed relative to water

$v_b = 3$ m/s

$$\vec{v} = \vec{v}_b + \vec{u}$$

$v_x = v_b \sin\theta + u$ $\qquad$ $v_y = v_b \cos\theta$

$w = v_y t = v_b t \cos\theta$

$\cos\theta = \dfrac{150\ \text{m}}{(120\ \text{s})(3\ \text{m/s})}$ $\qquad$ $\theta = \pm 65.4°$

$\ell = v_x t = (u \pm v_b \sin 65.4°)(120\ \text{s}) = 567, -87.3\ \text{m}$

4-31 $\vec{r}(t) = x(t)\hat{i} + y(t)\hat{j} + z(t)\hat{k}$

$x(t) = a \sin\omega t$

$y(t) = b \cos\omega t$

$z(t) = c \cos\omega t$

and $a^2 = b^2 + c^2$

$$\vec{v}(t) = \frac{d\vec{r}}{dt} = (a\omega \cos \omega t)\hat{i} - (b\omega \sin \omega t)\hat{j} - (c\omega \sin \omega t)\hat{k}$$

$$\text{speed} = |\vec{v}(t)| = \sqrt{a^2\omega^2 \cos^2 \omega t + b^2\omega^2 \sin^2 \omega t + c^2\omega^2 \sin^2 \omega t}$$

$$= \sqrt{\omega^2(a^2 \cos^2 \omega t + (b^2 + c^2)\sin^2 \omega t)}$$

$$= \sqrt{\omega^2 a^2(\cos^2\omega t + \sin^2 \omega t)} = \omega a$$

$$\vec{a}(t) = \frac{d\vec{v}}{dt} = -(a\omega^2 \sin \omega t)\hat{i} - (b\omega^2 \cos \omega t)\hat{j} - (c\omega^2 \cos \omega t)\hat{k}$$

$$\vec{a} \cdot \vec{v} = -(a^2\omega^3 \sin \omega t \cos \omega t) + (b^2\omega^3 \cos \omega t \sin \omega t) + (c^2\omega^3 \sin \omega t \cos \omega t)$$

$$= \omega^3 (\sin \omega t)(\cos \omega t) (-a^2 + b^2 + c^2) = 0$$

This is in agreement with Ex. 4-5. The path is a circle of radius a.

4-32

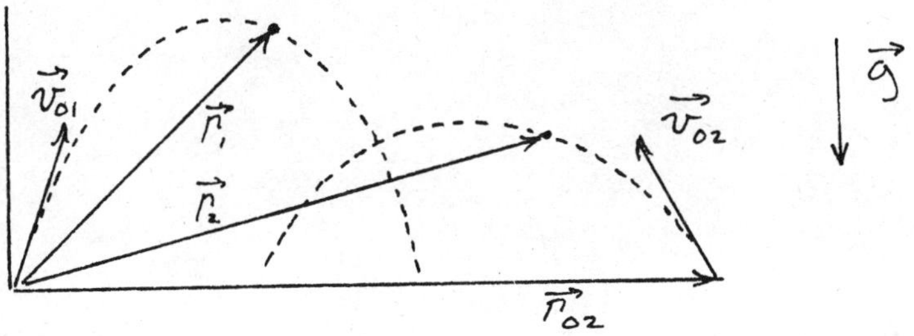

$$\vec{r}_1(t) = v_{01}\, t - \frac{1}{2}\vec{g}t^2$$

$$\vec{r}\ (t) = \vec{r}_{02} + \vec{v}_{02}\, t - \frac{1}{2}\vec{g}t^2$$

A collision requires

$$\vec{r}_1(t') = \vec{r}_2(t')$$

or

$$\vec{v}_{01}\, t' = \vec{v}_{02}\, t' + \vec{r}_{02}$$

But this is also the impact condition for $g = 0$.

$$t' = \frac{r_{02}}{|\vec{v}_{01} - \vec{v}_{02}|}$$

4-33 $\vec{a}(t) = [(2\ \text{m/s}^3)t + (1\ \text{m/s}^2)]\hat{i}$

$$\vec{v}(1) = (2\ \text{m/s})\hat{i} + (7\ \text{m/s})\hat{k}$$

$$\vec{r}(0) = (3\ \text{m})\hat{j}$$

$$v(t) = \int \vec{a}(t)dt = (t^2 + t + C_1)\hat{i} + C_2\hat{j} + C_3\hat{k}$$

Using $\vec{v}(1)$ to find C_1, C_2, and C_3,

$$\vec{v}(t) = (t^2 + t)\hat{i} + 7\hat{k}$$

$$\vec{r}(t) = \int \vec{v}(t)dt = (\frac{t^3}{3} + \frac{t^2}{2} + C_4)\hat{i} + C_5\hat{j} + (7t + C_6)\hat{k}$$

Using $\vec{r}(0)$ to find C_4, C_5, and C_6,

$$\vec{r}(t) = (\frac{t^3}{3} + \frac{t^2}{2})\hat{i} + 3\hat{j} + (7t)\hat{k}$$

4-34 From Eq. 4-18

$$R = \frac{v_o^2}{g} \sin 2\alpha = 40.8 \text{ m}$$

The time for the ball to hit the ground is

$$t = \frac{R}{v_o \cos \alpha} = 2.88 \text{ s}$$

This time must equal the time for the receiver to run to the same spot.

$$t = \frac{60 \text{ m} - R}{v_R}$$

$$v_R = \frac{60 \text{ m} - 40.8 \text{ m}}{2.88 \text{ s}}$$

$$= 6.67 \text{ m/s}$$

4-35 $R = 60$ m

From Eq. 4-18,

$$R = \frac{v_o^2}{g} \sin 2\alpha$$

$$\sin 2\alpha = \frac{Rg}{v_o^2}$$

$$\alpha = 10.8°$$

Using Eq. 4-17,

$$z = -\left(\frac{g}{2 v_o^2 \cos^2\alpha}\right)y^2 + (\tan \alpha)y + 1 \text{ m}$$

At $y = 45$ m,

$$z = 3.16 \text{ m}$$

4-36 (a) Using Eq. 4-17,

$$z = -\left(\frac{g}{2 v_o^2 \cos^2\alpha}\right)y^2 + (\tan \alpha)y$$

At $y = 40$ m,

$$z = 2.68 \text{ m}$$

The ball will not clear the goal post.

(b) Using the relationship $1/\cos^2 \alpha = 1 + \tan^2 \alpha$, we have

$$z = \frac{-gy^2}{2 v_o^2} \tan^2\alpha + (\tan \alpha)y - \frac{gy^2}{2 v_o^2}$$

$$\frac{-gy^2}{2 v_o^2} \tan^2\alpha + (\tan \alpha)y - \left(\frac{gy^2}{2 v_o^2} + z\right) = 0$$

Solving for $\tan \alpha$,

$$\tan \alpha = \frac{-y \pm \sqrt{y - \left(\frac{g^2 y^4}{v_o^4} + \frac{2gzy^2}{v_o^2}\right)}}{-gy^2/v_o^2}$$

$$\tan \alpha = 1.33, \ .880$$

So $41.4° \le \alpha \le 53.0°$

4-37 $z(t) = z_o + v_o t \sin\theta - \frac{1}{2} gt^2 = 0$

The time of impact is

$$t = \frac{v_o \sin\theta}{g} + \sqrt{\left(\frac{v_o \sin\theta}{g}\right)^2 + \frac{z_o}{g}}$$

Note that the negative root is not chosen as it yields a negative value for time.

The difference in time of impact for the two wall heights is

$$t_{20} - t_{10} = \sqrt{\frac{(v_o \sin\theta)^2}{g^2} + \frac{2(20\text{ m})}{g}} - \sqrt{\frac{(v_o \sin\theta)^2}{g^2} + \frac{2(10\text{ m})}{g}}$$

$$= 0.47\text{ s}$$

$$R_{20} - R_{10} = (v_o \cos\theta)(t_{20} - t_{10}) = 7.2\text{ m}$$

4-38 (a) The centripetal acceleration at the top of the path must be equal to g.

Thus,

$$R = \frac{v^2}{a} = \frac{v_o^2 \cos^2\alpha}{g}$$

(b) Using Eq. 4-17,

$$z = -\left(\frac{g}{2 v_o^2 \cos^2\alpha}\right) y^2 + (\tan\alpha) y$$

$$\frac{dz}{dy} = -\left(\frac{g}{v_o^2 \cos^2\alpha}\right) y + (\tan\alpha)$$

$$\frac{d^2z}{dy^2} = -\frac{g}{v_o^2 \cos^2\alpha}$$

$$R = \left|\frac{d^2z}{dy^2}\right|^{-1} = \frac{v_o^2 \cos^2\alpha}{g}$$

The two values of R are identical.

4-39 For the particle not to slide on the hemisphere, the radius of curvature of the path must be larger than the radius of the hemisphere.

$$R \le R_c = \frac{v_o^2}{g}$$

$$v \ge \sqrt{Rg}$$

The minimum value for v is therefore

$$v_o = \sqrt{Rg}$$

4-40 $v = R\omega = 2\pi R\nu = 0.349\text{ m/s}$

$$s = vt = (0.349\text{ m/s})\left(\frac{1}{20{,}000}\text{ s}\right) = 1.75 \times 10^{-5}\text{ m}$$

4-41 From Eq. 4-19,

$$R_m = \frac{v_o^2}{g}$$

$$v_o^2 = gR_m$$

Now, using Eq. 6-21, we have,

$$z_m = \frac{v_o^2}{2g} \sin \alpha$$

Clearly, this will be a maximum when

$$\sin^2\alpha = 1$$

Thus,

$$z_m = \frac{v_o^2}{2g} = \frac{gR_m}{2g} = \frac{1}{2} R_m$$

4-42 $a_\rho = 500{,}000 \text{ g}$

$R = 0.01 \text{ m}$

$$\omega = \sqrt{\frac{a}{R}} = 2.21 \times 10^4 \text{ s}^{-1}$$

$$\nu = \frac{\omega}{2\pi} = 3.52 \times 10^3 \text{ Hz} = 2.11 \times 10^5 \text{ rpm}$$

Using Eq. 4-34,

$$\omega = \omega_o + \alpha t$$

$$t = \frac{\omega}{\alpha} = \frac{\omega}{a_\phi / R}$$

Using Eq. 4-35,

$$\phi = \phi_o + \omega_o t + \frac{1}{2}\alpha t^2$$

$$= \frac{1}{2}\left(\frac{a_\phi}{R}\right)\left(\frac{\omega}{a_\phi/R}\right)$$

$$= \frac{1}{2}\frac{R\omega^2}{a_\phi} = \frac{1}{2}\frac{a_\rho}{a_\phi} = \frac{5 \times 10^5 \text{ g}}{20 \text{ g}}$$

$$= 2.5 \times 10^4 \text{ rad} = 3980 \text{ rev}$$

4-43 (a) Let v_b = velocity of the boat in water and v_w = velocity of the water

On the trip upstream,

$$v_b - v_w = 4 \text{ km/h}$$

To find the total travel time,

$$\frac{10 \text{ km}}{v_b + v_w} + 1 \text{ h} = \frac{6 \text{ km}}{v_w}$$

Combining the two equations above, we obtain

$$v_w^2 + v_w - 12 = 0$$

$$v_w = \frac{-1 \pm \sqrt{1 + 48}}{2}$$

$$= 3, -4 \text{ km/h}$$

The value of v_w must be positive since we already know its direction. Thus,

$$v_w = 3 \text{ km/h}$$

(b) $$v_b = (4 \text{ km/h}) + v_w = 7 \text{ km/h}$$

4-44 $$y_e = v_e t$$

$$y_b = y_o + v_e t - \frac{1}{2} gt^2$$

(a) When the ball hits the elevator floor,

$$y_e = y_b$$

$$\frac{1}{2} gt^2 = y_o$$

$$t = \sqrt{\frac{2\ y_o}{t}} = 0.78 \text{ s}$$

(b) The velocity of the bolt with respect to the elevator is, at time of collision,

$$v_{be} = -gt = -7.64 \text{ m/s}$$

(c) $$v_b = v_o - gt = -5.64 \text{ m/s}$$

(d) In the elevator, the distance is 3 m. To the outside observer,

$$d = y_o - y_e(0.78 \text{ s}) = y_o - v_e t = 1.44 \text{ m}$$

4-45 $$d_1^2 + d_2^2 = (300 \text{ m})^2$$

$$\frac{d_1}{v_1} = \frac{d_2}{v_2}$$

$$\frac{d_1}{v_1} = \frac{\sqrt{(300 \text{ m})^2 - d_1^2}}{v_2}$$

$$\frac{d_1^2}{v_1^2} = \frac{(300 \text{ m})^2 - d_1^2}{v_2^2}$$

$$\frac{d_1^2(v_2^2 + v_1^2)}{v_1^2\ v_2^2} = \frac{(300 \text{ m})^2}{v_2^2}$$

$$d_1 = \sqrt{\frac{v_1^2(300 \text{ m})^2}{v_2^2 + v_1^2}}$$

$$= 180 \text{ m}$$

$$t = \frac{d_1}{v_1} = \frac{180 \text{ m}}{3 \text{ m/s}} = 60 \text{ s}$$

$$d_2 = \frac{v_2}{v_1} d_1 = 240 \text{ m}$$

4-46 $$v_b = 20 \text{ km/h}$$

$$\tan 15° = \frac{v_w}{v_b}$$

$$v_w = v_b \tan 15° = 5.36 \text{ km/h}$$

4-47 Let $\vec{v}$ be the velocity of the plane in the ground frame, $\vec{v}_p$ be its velocity with respect to the air, and $\vec{v}_w$ be the wind speed.

For $\vec{v}_w = 0$, the round-trip time is

$$t = \frac{2\,d}{v_p}$$

where d is the distance between the airports. On the first half of the trip,

$$v_1 = v_p \cos\phi_1 + v_w \cos\theta \qquad (1)$$

$$v_p \sin\phi_1 = v_w \sin\theta \qquad (2)$$

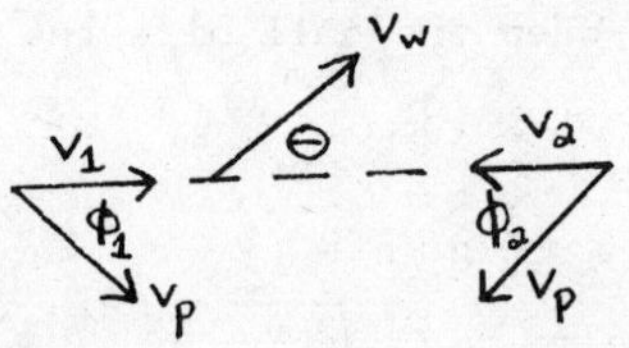

On the return trip,

$$v_2 = v_p \cos\phi_2 - v_w \cos\theta \qquad (3)$$

$$v_p \sin\phi_2 = v_w \sin\theta \qquad (4)$$

From Eqs. 3 and 4, $\phi = \phi_1 = \phi_2$. The time for the round-trip is, using Eqs. 1 and 3,

$$t' = \frac{d}{v_p \cos\phi + v_w \cos\theta} + \frac{d}{v_p \cos\phi - v_w \cos\theta}$$

$$= \frac{2dv_p \cos\phi}{v_p^2 \cos^2\phi - v_w^2 \cos^2\theta}$$

$$\frac{t'}{t} = \frac{v_p^2 \cos\phi}{v_p^2 \cos^2\phi - v_w^2 \cos^2\theta} = \frac{\cos\phi}{\cos^2\phi - (v_w^2 \cos^2\theta)/v_p}$$

$$= \frac{\cos\phi}{\cos^2\phi - v_w^2(1 - \sin^2\theta)/v_p^2}$$

Using Eq. 2,

$$\frac{t'}{t} = \frac{\cos\phi}{\cos^2\phi + \sin^2\phi - (v_w/v_p)^2} = \frac{\cos\phi}{1 - (v_w/v_p)^2}$$

$$= \frac{(1 - \sin^2\phi)^{1/2}}{1 - (v_w/v_p)^2} = \frac{[1 - (v_w \sin\theta/v_p)^2]^{1/2}}{1 - (v_w/v_p)^2}$$

Since

$$1 - (v_w/v_p)^2 < 1-(v_w \sin\theta/v_p)^2$$

and

$$1 - (v_w \sin\theta/v_p)^2 < 1,$$

$$\frac{t'}{t} > 1$$

Chapter 5

5-1 To find the acceleration during the first 10 s use Newton's second law.

$$a = \frac{F}{m} = \frac{3\ N}{2\ kg} = \frac{3}{2}\ m/s^2 \text{ (in direction of applied force)}$$

Since the force is constant, the acceleration is constant. Eqs. 2-9 and 2-10 can be used to determine velocity and position of object.

$$v(t = 10\ s) = at + v_o = 3/2\ (10) + 0 = 15\ m/s$$

$$s(t = 10\ s) = \frac{1}{2}at^2 + v_o t + s_o = \frac{1}{2}\left(\frac{3}{2}\right)10^2 + 0 + 0 = 75\ m$$

After 10 s, no force is applied, so there is no acceleration; the object moves with a constant velocity

$$s(t = 20\ s) = s(t = 10\ s) + v(t = 10\ s)t$$

$$= 75\ m + 15\ m/s(10\ s) = 225\ m$$

5-2 Assuming constant acceleration,

$$v = v_o + at.$$

The first two points give

$$5 = v_o + a \text{ and } 8 = v_o + 2\ a$$

Solving for v_o and a,

$$v_o = 2\ m/s \text{ and } a = 3\ m/s^2.$$

So $v(t) = 2 + 3t$ m/s (with t in seconds)

Using Newton's second law,

$$F = ma = (4\ kg)(3\ m/s^2) = 12\ N$$

5-3 From Newton's second law,

$$\vec{F} = m\vec{a} = 0.5(4\hat{i} - 2\hat{j})N = 2\hat{i} - \hat{j}\ N$$

The magnitude of F is

$$F = \sqrt{F_x^2 + F_y^2} = \sqrt{4 + 1}\ N = \sqrt{5}\ N$$

5-4 To find force acting on bullet, first determine bullet's deceleration in block.

Assuming constant deceleration, Eq. 4-11 can be used.

$$a = \frac{v^2 - v_o^2}{2(s-s_o)} = \frac{0 - 400^2}{2(0.15)}\ \frac{m}{s^2} = -5.33 \times 10^5\ m/s^2$$

From Newton's second law,

$$F = ma = (0.012\ kg)(-5.33 \times 10^5\ m/s^2) = -6.4 \times 10^3\ N.$$

In this problem, the negative sign indicates that the force is in a direction opposite to the motion of the particle.

5-5 The tension of the string is the only force holding the object in circular motion. By Newton's second law, the acceleration corresponding to a force of 100 N is

$$a = \frac{F}{m} = \frac{100\ N}{0.5\ kg} = 200\ m/s^2$$

Since the motion is curcular, the acceleration is centripetal. Using Eq. 4-26,

$$v = \sqrt{a\ R} = \sqrt{200(0.5)}\ m/s = 10\ m/s$$

5-6 First, find the acceleration. Assuming constant acceleration,

$$a = \frac{v - v_o}{t} = \left(\frac{90\ km/hr}{12.2\ s}\right)\left(\frac{10^3\ m}{km}\right)\left(\frac{hr}{3600\ s}\right)$$

$$= 2.05\ m/s$$

Using Newton's second law,

$$F = ma = (2000)(2.05)\ N = 4.10 \times 10^3\ N$$

5-7 As above,

$$a = \frac{v - v_o}{t} = \frac{10{,}000\ km/hr}{yr}\left(\frac{yr}{365\ days}\right)\left(\frac{day}{24\ hr}\right)\left(\frac{hr}{3600\ s}\right)^2\left(\frac{10^3\ m}{km}\right)$$

$$= 8.81 \times 10^{-5}\ m/s^2$$

$$F = ma = (5000)(8.81 \times 10^{-5})N = 0.44\ N$$

5-8 Use Newton's second law to find the deceleration due to the braking force:

$$a = \frac{F}{m} = \frac{6000\ N}{2200\ kg} = 2.73\ m/s^2$$

To find the time required to come to a stop, use Eq. 4-11.

$$s = \frac{v^2 - v_o^2}{2a} = \frac{-(32\ m/s)^2}{2(-2.73\ m/s^2)} = 188\ m$$

5-9 Since velocity is constant, $\vec{a} = 0$. This means $\Sigma F = 0$. Assuming the x-direction to be the velocity direction,

$$0 = \Sigma F_x = F_{1x} + F_{2x} + F_{3x} = (8\ N)(\cos 35°)+(12\ N)(\cos 75°)+ F_{3x}$$

$$0 = \Sigma F_y = F_{1y} + F_{2y} + F_{3y} = (8\ N)(\sin 35°)-(12\ N)(\sin 75°) + F_{3y}$$

This gives $F_{3x} = -9.66\ N$

$F_{3y} = 7.00\ N$

$$F_3 = \sqrt{F_{3x}^2 + F_{3y}^2} = 11.93 \text{ N}$$

$$\theta_3 = \tan^{-1}\left(\frac{F_{3y}}{F_{3x}}\right) = 144°$$

5-10 All cars accelerate at the same rate, $a = \frac{F}{100\text{ m}}$, where m is the mass of a single car. The tension in the front coupling of the leading car obviously equals the force being delivered to the coupling; $T_1 = 10^6$ N. For the last car, $T_{100} = ma = m\left(\frac{10^6\text{N}}{100\text{ m}}\right) = 10^4$ N. The tension in the front coupling of the nth car is the force necessary to pull the nth car plus all cars behind it at the given acceleration.

$$T_n = ma = [(100-(n-1)m]\frac{F}{100\text{ m}} = (101-n)10^4 \text{ N}$$

5-11 The rope-block combination accelerates at

$$a = \frac{F}{m + M} = \frac{12 \text{ N}}{4.8 \text{ kg}} = 2.5 \text{ m/s}^2$$

The force of the rope on the block is the only force acting on the block.

$$F = Ma = (4 \text{ kg})(2.5 \text{ m/s}^2) = 10 \text{ N}$$

5-12 For circular motion, acceleration is centripetal.

$$F = ma = mR\omega^2 = mR\left(\frac{4\pi^2}{T^2}\right) = \frac{4\pi^2(0.80)(2)}{4^2} \text{ N} = 3.95 \text{ N}$$

$$v = \frac{2\pi R}{T} = \pi \text{ m/s}$$

So $\alpha = \pi$ m/s

For circular motion, $\vec{a}$ is perpendicular to $\vec{v}$.

$\vec{F} = \pm 3.95\hat{j}$ N, the + or − is chosen according to whether the motion is clockwise or counterclockwise.

5-13 The spring scale measures the normal force being exerted on the object.

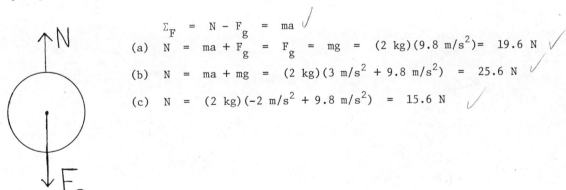

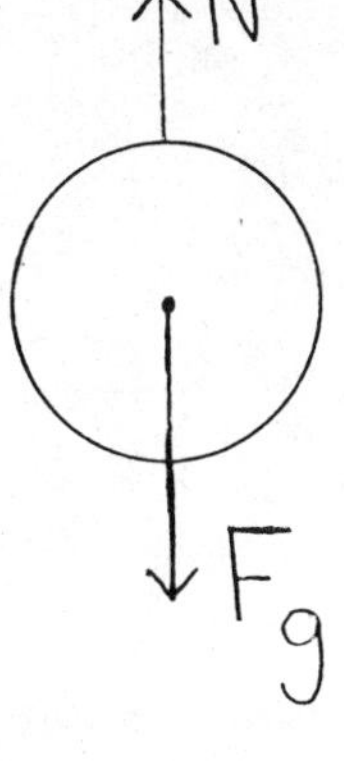

$$\Sigma_F = N - F_g = ma$$

(a) $N = ma + F_g = F_g = mg = (2 \text{ kg})(9.8 \text{ m/s}^2) = 19.6 \text{ N}$

(b) $N = ma + mg = (2 \text{ kg})(3 \text{ m/s}^2 + 9.8 \text{ m/s}^2) = 25.6 \text{ N}$

(c) $N = (2 \text{ kg})(-2 \text{ m/s}^2 + 9.8 \text{ m/s}^2) = 15.6 \text{ N}$

5-14 The inertial observer will see the ball move with a constant velocity and will see the spaceship accelerate.

$$\text{Transit time} = \frac{5\text{ m}}{15\text{ m/s}} = \frac{1}{3}\text{ s}$$

During this time the ship moves

$$s = \frac{1}{2}at^2 = \frac{1}{2}(9.8)(\frac{1}{3})^2 = 0.544\text{ m}$$

An observer cannot distinguish his accelerating frame from an inertial frame in a gravitational field.

5-15 $a_1 = 3.5\text{ m/s}^2\text{(down)}$ $\quad a_2 = 2.5\text{ m/s}^2\text{(up)}$

$$\Sigma F = W - mg = ma$$

$$W_1 = m(g - a_1)$$

$$W_2 = m(g + a_2)$$

$$\text{Percent Profit} = \frac{W_2 - W_1}{W_1}(100) = \frac{a_2 + a_1}{g - a_1}(100) = 95.2\%$$

5-16 Eq. 7-16 describes the force acting on a body in pure radial motion with respect to a plane that rotates with constant angular speed about a fixed axis.

$$\vec{F} = m(\frac{dv_\rho}{dt} - \rho\omega^2)\hat{u}_\rho + 2\,mv_\rho\omega\hat{u}_\phi$$

v_ρ constant means $\frac{dv_\rho}{dt} = 0$

$$\vec{F} = (0.5\text{ g})(-15\text{ cm})(33.3)^2(2\pi)^2(\frac{1}{60\text{ s}})^2\hat{u}_\rho + 2(0.5\text{ g})(5\text{ cm/s})(33.3)(2\pi)(\frac{1}{60\text{ s}})\hat{u}_\phi$$

$$= (-9.12\,\hat{u}_\rho + 1.74\,\hat{u}_\phi)\times 10^{-4}\text{ N}$$

The force is frictional; it is due to contact between the bug's legs and the turntable.

5-17 In 30 minutes, the earth rotates through $\frac{1}{48}$ of its 24-hour-long rotation.

$$\text{Angular deviation} = \frac{1}{48}(360°) = 7.5°$$

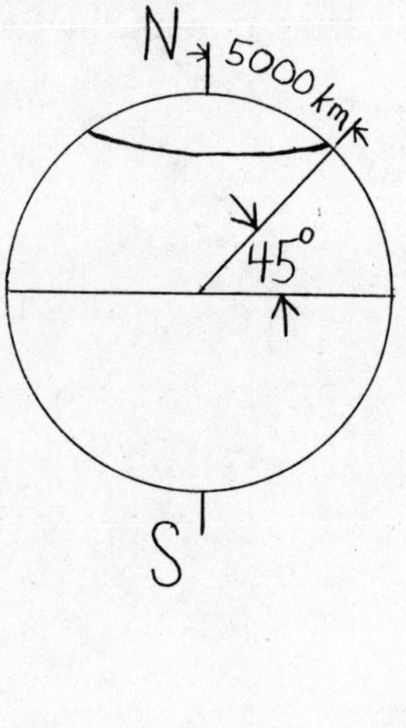

The circumference of the earth is

$$C_e \cong (2\pi)(6.37\times 10^3\text{ km}) = 4.00\times 10^4\text{ km}$$

Thus, 5000 km corresponds to a latitude of 45° north. At 45° north latitude, the distance travelled during one rotation is:

$$D = 2\pi R_e\cos 45° = 2.83\times 10^4\text{ km}$$

Distance between point aimed for and impact point =

$$\frac{1}{48}(2.83\times 10^4)\text{km} = 589\text{ km}$$

5-18 $T - mg = ma$ ✓ $a = \frac{T}{m} - g = 6.87 \text{ m/s}$ ✓

5-19 $v_f^2 = 2ax = 2x\left(\frac{F}{m}\right)$ ✓

$F = \frac{mv_f^2}{2x} = 1.66 \times 10^6 \text{ N}$ ✓

5-20 As above, $F = \frac{mv_f^2}{2x} = 147 \text{ N}$ ✓

5-21 $v_f^2 - v_o^2 = 2ax = 2x\left(\frac{F}{m}\right)$

$0 - (10)^2 = 2x\left(\frac{0.5}{0.5}\right)$ ✓

$x = 50 \text{ m}$ ✓

5-22 $v = v_o - at = v_o - \left(\frac{F}{m}\right)t$ ✓

$= 5 - 2t \text{ m/s}$ ✓

Tension $= T(t) = \frac{v^2 m}{R}$ ✓

MINUS

$T(t) = 8t^2 \boxed{=} 40t + 50 \text{ N}$ ✓

$\bigotimes$

5-23 $\frac{dv}{dt} = a = \frac{F}{m} = \frac{12x}{3} s^{-2} = 4x \, s^{-2}$

$a(x = 5 \text{ m}) = 20 \text{ m/s}^2$ ✓

$\frac{dv}{dt} = \frac{dv}{dx}\frac{dx}{dt} = v\frac{dv}{dx} = 4x$

$\int_0^v v \, dv = \int_3^x 4x \, dx$

$\frac{v^2}{2} = 2x^2 - 18$

$v = \frac{dx}{dt} = (4x^2 - 36)^{\frac{1}{2}}$

$\int_3^x \frac{dx}{(4x^2 - 36)^{\frac{1}{2}}} = \int_0^t dt$

$\frac{1}{6}\int_3^x \frac{dx}{\left(\frac{x^2}{9} - 1\right)^{\frac{1}{2}}} = t$

Use the substitution

$x = 3 \sec u$

$\frac{x^2}{9} - 1 = \sec^2 u - 1 = \tan^2 u$

$\frac{2x}{9} dx = 2 \sec^2 u \tan u \, du$

$\frac{3 \sec u}{9} dx = \sec^2 u \tan u \, du$

$dx = 3 \sec u \tan u \, du$

The integral becomes

Quicker: $\frac{d^2x}{dt^2} = \frac{dv}{dt} = 4x$

$\ddot{x} - 4x = 0$

$\therefore x = \alpha e^{2t} + \beta e^{-2t} \Rightarrow @ \, t = 0, \; 3 = x_0 = \alpha + \beta \left\{ \begin{array}{l} \alpha = \beta \\ = \frac{3}{2} \end{array} \right.$

$\dot{x} = 2(\alpha e^{2t} - \beta e^{-2t}) \Rightarrow @ \, t = 0, \; \dot{x} = 0 = 2(\alpha - \beta)$

$V := \dot{x} = 2\left(\frac{3}{2}e^{2t} - \frac{3}{2}e^{-2t}\right) = 3(e^{2t} - e^{-2t})$ ANS

$$\frac{1}{2}\int_0^{\cos^{-1}\frac{3}{x}} \frac{\sec u \tan u \, du}{\tan u} = t$$

$$\log\left|\sec u + \tan u\right|\Big]_0^{\cos^{-1}\frac{3}{x}} = 2t$$

$$\log\left|\frac{x}{3} + \frac{x}{3}\sin(\cos^{-1}\frac{3}{x})\right| - \log\left|1 + 0\right| = 2t$$

$$\log\left|\frac{x}{3} + \frac{x}{3}(1 - \frac{9}{x^2})^{\frac{1}{2}}\right| = 2t$$

$$\frac{x}{3} + (\frac{x^2}{9} - 1)^{\frac{1}{2}} = e^{2t}$$

$$\frac{x^2}{9} - 1 = (e^{2t} - \frac{x}{3})^3$$

$$\frac{x^2}{9} - 1 = e^{4t} - \frac{2x}{3}e^{2t} + \frac{x^2}{9}$$

$$\frac{2x}{3}e^{2t} = e^{4t} + 1$$

$$x = \frac{3}{2}(e^{2t} + e^{-2t}) \text{ m}$$

$$v = \frac{dx}{dt} = 3(e^{2t} - e^{-2t}) \text{ m/s}$$

5-24 (a) $F_1 = \frac{Gm_1m_2}{R_o^2} = \frac{(6.67 \times 10^{-11})(15)^2}{5^2}$ N

$= 6.00 \times 10^{-10}$ N

(b) $a_o = \frac{F_1}{m_1} = 4.00 \times 10^{-11}$ m/s^2

(c) The acceleration is not constant

$a_f = \frac{Gm_1}{R_f^2} = \frac{(6.67 \times 10^{-11})(15)}{(4.99)^2} = 4.02 \times 10^{-11}$ m/s^2

So $\bar{a} = \frac{a_f + a_o}{2} = 4.01 \times 10^{-11}$ m/s^2

$x = \frac{1}{2}\bar{a}t^2$

$t = \sqrt{\frac{2x}{\bar{a}}} = \sqrt{\frac{2(0.005 \text{ m})}{4.01 \times 10^{-11} \text{ m/s}^2}} = 1.6 \times 10^4 \text{ s} = 4.44 \text{ h}$

5-25 (a) $t_5 = \frac{5 \text{ m}}{1.6 \text{ m/s}} = 3.125$ s

The particle escapes at $t_{10} = \frac{10 \text{ m}}{1.6 \text{ m/s}} = 6.25$ s

(b) $\Delta t = t_{10} - t_5 = 3.125$ s

$y(t_{10}) = \frac{1}{2}a(\Delta t)^2 = \frac{1}{2}(\frac{F_y}{m})(\Delta t)^2$

$= 15.6$ m

(c) $v_x = 1.6$ m/s

$$v_y = \left(\frac{F_y}{m}\right)\Delta t = 10 \text{ m/s} \checkmark$$

$$v = \sqrt{v_x^2 + v_y^2} = 10.13 \text{ m/s} \checkmark$$

$$\tan \theta = \frac{v_y}{v_x} \qquad \theta = 80.9° \checkmark$$

5-26 $a = \underset{\text{(2 R)}^2}{\dfrac{GM_e}{}} = \dfrac{GM_e}{4 R^2} = \dfrac{g}{4}$

$\to (3R)^2 \text{ i.e. } (1R+2R)^2 \qquad |a| = \left|\dfrac{-g}{9}\right| = 1.09 \text{ m/s}^2 \quad \bigotimes$

$$= 2.45 \text{ m/s}^2$$

5-27 (a) $g' = 7.32 \text{ m/s}^2$

$$F_g = Mg' = (2 \times 10^3)(7.32)\text{N} = 1.46 \times 10^4 \text{ N} \checkmark$$

$$f_g = mg' = (90)(7.32)\text{N} = 659 \text{ N} \checkmark$$

$$g' = \frac{v^2}{r} \qquad v = \sqrt{rg'} = \sqrt{(7.38 \times 10^6)(7.32)} \checkmark$$

$$v = 7350 \text{ m/s} \checkmark$$

(b) $v_2 = (0.80)v = 5880 \text{ m/s} \checkmark$

$$mg' - W = m\frac{v_2^2}{R} \qquad \qquad \text{How, why?}$$

$$W = m(g' - \frac{v_2^2}{R}) = 237 \text{ N}$$

5-28 $\vec{T} + m\vec{g} = m\vec{a}$

horizontally $T \sin \alpha = ma$

vertically $T \cos \alpha - mg = 0$

Squaring both equations and adding gives

$$T^2(\cos^2\alpha + \sin^2\alpha) = m^2(g^2 + a^2)$$

$$T = m\sqrt{g^2 + a^2} = (0.25)\sqrt{(9.8)^2 + (2)^2} = 2.50 \text{ N} \checkmark$$

Multiplying the horizontal equation by $\cos \alpha$ and the vertical equation by $\sin \alpha$, then subtracting gives

$$ma \cos \alpha = mg \sin \alpha$$

So $\alpha = \tan^{-1}\left(\dfrac{a}{g}\right) = \tan^{-1}\left(\dfrac{2.0}{9.8}\right)$

$$= 11.5° \checkmark$$

5-29 $\vec{T} + m\vec{g} = m\vec{a} \checkmark$

horizontally $T \sin 10° = m\dfrac{v^2}{R}$

vertically $T \cos 10° - mg = 0$

Multiply first equation by $\cos 10°$ and the second by $\sin 10°$, then subtract.

$$mg \sin 10° = m\frac{v^2}{R}\cos 10°$$

$$R = \frac{v^2}{g}\cot 10° = \frac{(30 \times 10^3/3600)^2}{9.8}\cot 10°$$

$$= 40.2 \text{ m}$$ ✓

5-30 Let t_m be the time it takes to reach maximum height.

$$v_o \sin\theta = gt_m$$

$$(v_o\cos\theta)2\,t_m = 2$$

$$(v_o\sin\theta)t_m - \frac{1}{2}gt_m^2 = 1$$

$$gt_m^2 - \frac{1}{2}gt_m^2 = 1$$

$$t_m = \sqrt{2/g}$$

$$\tan\theta = \frac{v_o\sin\theta}{v_o\cos\theta} = \frac{gt_m}{1/t_m} = gt_m^2 = 2$$

$$\theta = 63.4°$$ ✓

$$v_o = \frac{gt_m}{\sin 63.4°} = \frac{\sqrt{2g}}{\sin 63.4°} = 4.95 \text{ m/s}$$ ✓

To get the acceleration,

$$v_o^2 = 2ax + 0^2$$ ✓

$$a = \frac{v_o^2}{2x} = \frac{(4.95)^2}{2(0.015)} = 817 \text{ m/s}^2$$ ✓

Therefore, $\frac{F}{\text{weight}} = \frac{ma}{mg} = 83.4$ ✓

For a 150-pound person this would be like lifting about $6\frac{1}{4}$ tons. $= \frac{150 \times 83\frac{1}{3}}{2000}$ Tons

5-31 $W = \sqrt{(mg)^2 + F_c)^2} = \sqrt{(mg)^2 + (m\omega^2R)^2}$

$$= m\sqrt{g^2 + \omega^4R^2}$$

(X)

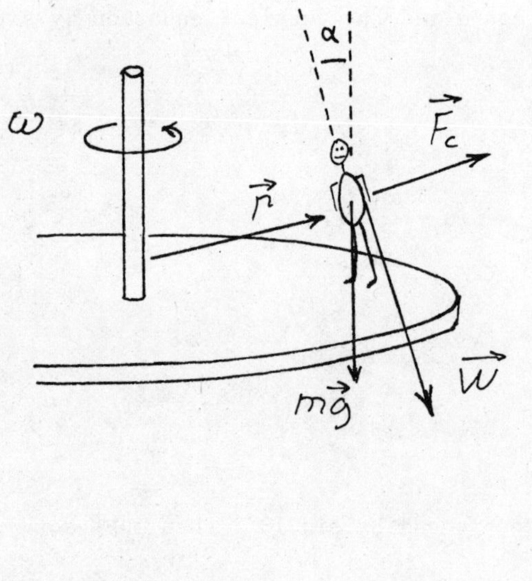

$$\tan\theta = \frac{F_c}{mg}$$ (θ circled; α written below)

$$\tan\theta = \frac{\omega^2R}{g}$$

r	W	α
1 m	846 N	22.2°
2 m	1012 N	39.2°
4 m	1501 N	58.5°
8 m	2680 N	73.0°
12 m	3920 N	78.5°

Each plant will grow along the local "up" direction giving them an inward slope which is slightly concave.

5-32 Weight $= mg - m\omega^2 R = m(g - (\frac{2\pi}{T})^2 R)$

$$= m(g - \frac{4\pi^2 R}{T^2}) = (1.0\ [9.8066 - \frac{4\pi^2(6.38 \times 10^6)}{(24 \times 3600)^2}]$$

$$= 9.77\ \text{N}$$

5-33 Weight $= m\omega^2 R = mR(\frac{2\pi}{T})^2$

$$= \frac{4\pi^2 mR}{T^2} = \frac{4\pi^2(80)(200)}{(60)^2}$$

$$= 175.5\ \text{N}$$

5-34 (a) $a = \frac{v^2}{R}$

$$R = \frac{v^2}{a} = \left[(\frac{600\ \text{km/h}}{3600\ \text{s/h}})(\frac{1000\ \text{m}}{\text{km}})\right] \frac{1}{7(9.8\ \text{m/s}^2)} = 405\ \text{m}$$

(b) s = length of arc $= \pi R = 1272$ m

$$t = \frac{s}{v} = \frac{1272\ \text{m}}{(600\ \text{km/h})(1000\ \text{m/km})}(\frac{3600\ \text{s}}{\text{h}}) = 7.6\ \text{s}$$

(c) The only force aside from gravity is the force due to air pressure on wings, F_{air}.

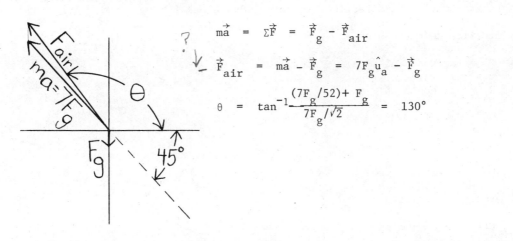

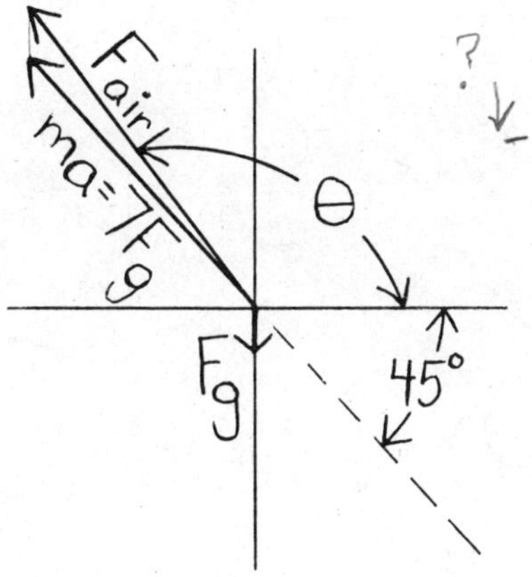

$$m\vec{a} = \Sigma\vec{F} = \vec{F}_g - \vec{F}_{air}$$

$$\vec{F}_{air} = m\vec{a} - \vec{F}_g = 7F_g\hat{u}_a - \vec{F}_g$$

$$\theta = \tan^{-1}\frac{(7F_g/52) + F_g}{7F_g/\sqrt{2}} = 130°$$

6-1 (a) Apply $\vec{F} = m\vec{a}$ to each diagram.

$$T' - 2T = 0$$

$$T - m_1 g = m_1 a$$

$$T - m_2 g = -m_2 a$$

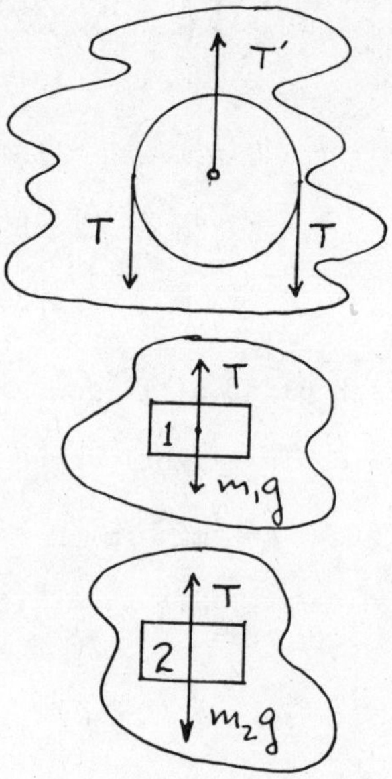

Subtract the third equation from the second to get

$$(m_2 - m_1)g = (m_1 + m_2)a$$

or

$$a = \frac{-m_1 + m_2}{m_1 + m_2} g = 1.960 \text{ m/s}^2$$

(b) $T = m_1(g+a) = 23.52$ N

(c) $T' = 2T = 47.04$ N

6-2 The force equations are:

$$T - m_1 g \sin 45° = m_1 a$$

$$-T + m_2 g \sin 60° = m_2 a$$

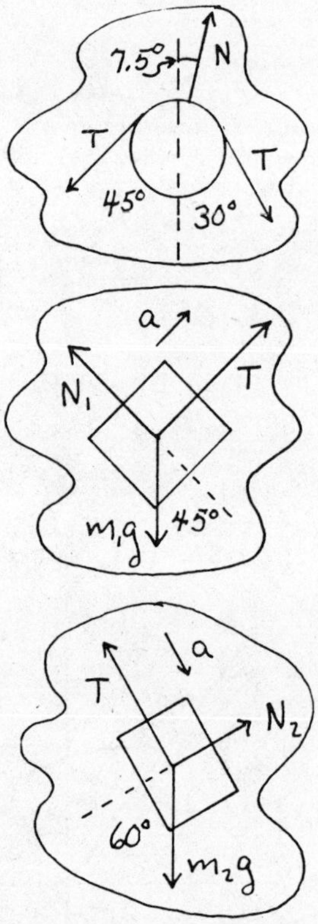

(a) Add to get

$$\left(\frac{\sqrt{3}}{2} m_2 - \frac{\sqrt{2}}{2} m_1\right)g$$

$$= (m_2 + m_1)a$$

or $a = -1.148$ N

The fact that this answer came out negative indicates that the a vectors in the free-body diagram are pointing in the wrong direction; in actuality, block 1 will slide down the slope.

(b) $T = m_1(g \sin 45° + a) = 28.91$ N

(c) $N = 2T \cos 37.5° = 45.87$ N

6-3 (a) The force equations are

(1) $F - T_1 - m_1 g = m_1 a$

(2) $T_1 - T_2 - m_s g = m_s a$

(3) $T_2 - m_2 g = m_2 a$

Add (1) + (2) + (3) to get

$$F - (m_1 + m_2 + m_s)g$$

$$= (m_1 + m_2 + m_s)a$$

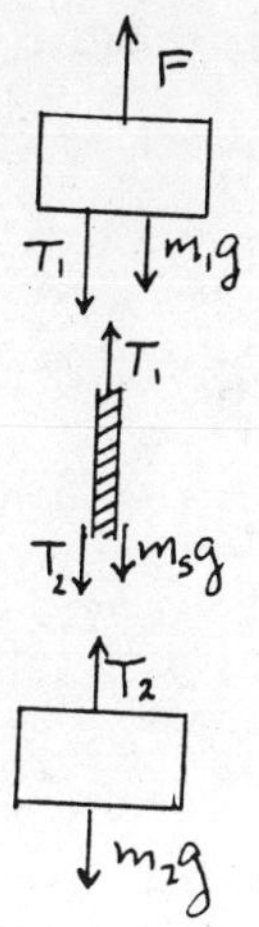

or

$$a = \frac{F}{m_1 + m_2 + m_s} - g = 3.836 \text{ m/s}^2$$

(b) From (1),

$$T_1 = F - m_1(g+a) = 68.18 \text{ N}$$

(c) From (3),

$$T_2 = m_2(g+a) = 54.55 \text{ N}$$

(d) The force equation for the lower portion of the system is

$$T_m - (m_2 + \frac{1}{2} m_s)g = (m_2 + \frac{1}{2} m_s)a$$

or

$$T_m = (m_2 + \frac{1}{2} m_s(g+a) = 61.36 \text{ N}$$

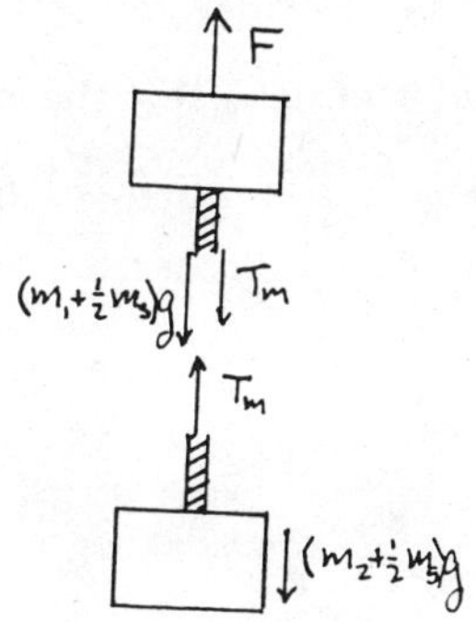

6-4 The mechanical advantage is contained in the fact that

$$4F - T' = 0$$

($ma = 0$ because $m = 0$ for the ropes and pulleys.) Thus,

$$T' = 4F$$

In words, the force available to do work is four times the applied force.

The second diagram shows that

$$T' - Mg = Ma$$

Using our other equation,

$$F = \tfrac{1}{4}M(g+a) = 750 \text{ N}$$

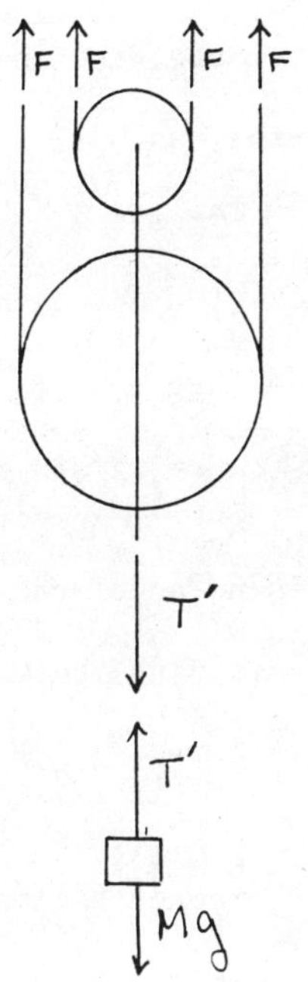

6-5 Taking components,

$$F_x = T \sin 40°$$

$$F_y = 2T + T \cos 40°$$

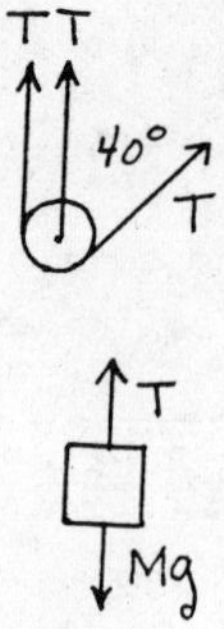

with

$$T = Mg$$

Thus

$$F_x = 11.34 \text{ N}$$

$$F_y = 48.79 \text{ N}$$

So

$$F = \sqrt{F_x^2 + F_y^2} = 50.09 \text{ N}$$

The angle from the positive x-axis is

$$\theta = \tan^{-1}(F_y/F_x) = 76.9°$$

6-6 In the static case, the component equations are

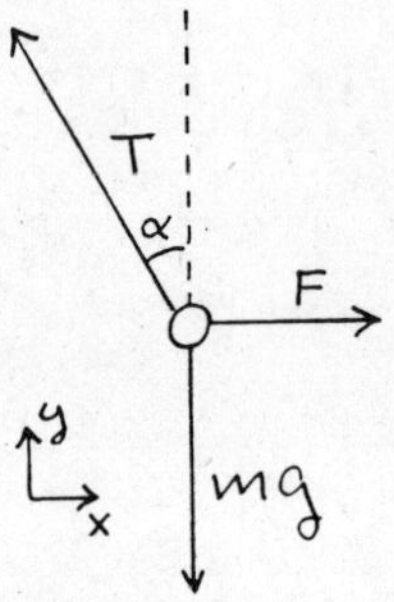

$$F - T \sin\alpha = 0$$

$$T \cos\alpha - mg = 0$$

or

$$T \sin\alpha = F$$

$$T \cos\alpha = mg$$

Divide and get

$$F = mg \tan\alpha = 5.350 \text{ N}$$

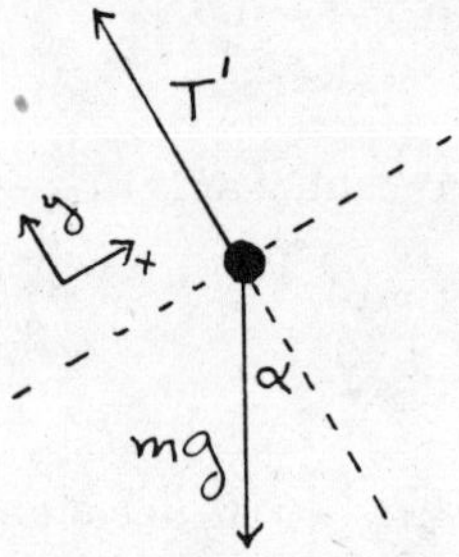

After release, the initial motion must be perpendicular to the string, so

$$-mg \sin\alpha = -ma$$

$$T' - mg \cos\alpha = 0$$

(Observe that $T' \neq T$.) The first of these yields the desired result:

$$a = g \sin\alpha = 3.352 \text{ m/s}^2$$

6-7 There is no acceleration, so the tension at this point is simply the weight of the man plus the weight of the rope below the point:

$$T = W_{MAN} + \frac{6}{12} W_{ROPE} = 862.4 \text{ N}$$

6-8 Taking components along the directions of motion,

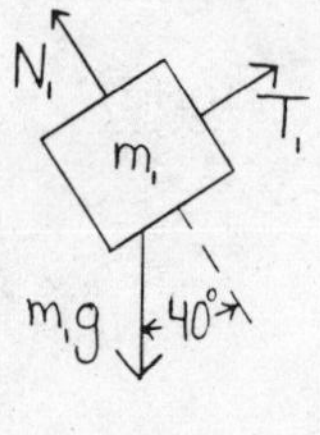

(1) $T_1 - m_1 g \sin 40° = m_1 a$

(2) $T_2 - T_1 = m_2 a$

(3) $m_3 g - T_2 = m_3 a$

Add (1) + (2) + (3) to get

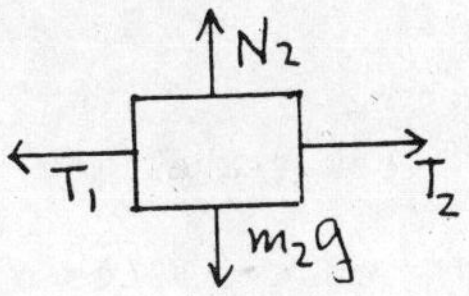

$$(m_3 - m_1 \sin 40°)g = (m_1 + m_2 + m_3)a$$

or $$a = \frac{m_3 - m_1 \sin 40°}{m_1 + m_2 + m_3} g$$

$$= 2.256 \text{ m/s}^2$$

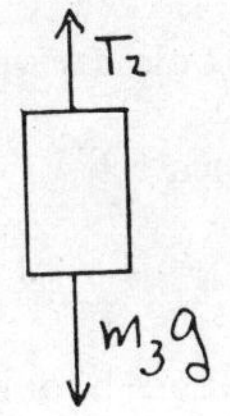

Use (1) and (3), respectively, to get

$$T_1 = m_1(g \sin 40° + a) = 51.33 \text{ N}$$

$$T_2 = m_3(g-a) = 60.35 \text{ N}$$

6-9 At maximum acceleration, the frictional force is $u_s N$, as shown. $\vec{F} = m\vec{a}$ gives

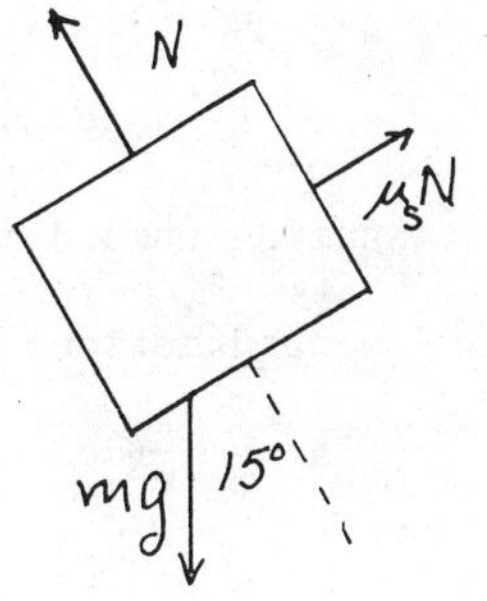

$$u_s N - mg \sin 15° = ma$$

$$N - mg \cos 15° = 0$$

Solving the second for N and inserting yields

$$a = (u_s \cos 15° - \sin 15°)g$$

$$= 1.250 \text{ m/s}^2$$

6-10 The acceleration is

$$a = \frac{F}{m} = u_k g$$ = a CONSTANT (OVER TIME DOMAN)

Recall from kinematics that

$$v^2 - v_o^2 = 2as$$

so $$s = \frac{-v_o^2}{2a} = \frac{v_o^2}{2u_k g} = 27.21 \text{ m}$$

6-11 (a) $\vec{F} = m\vec{a}$ gives

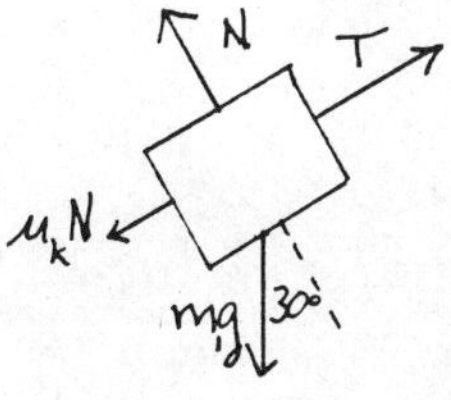

(1) $T - u_k N - m_1 g \sin 30° = m_1 a$

(2) $N - m_1 g \cos 30° = 0$

(3) $m_2 g - T = m_2 a$

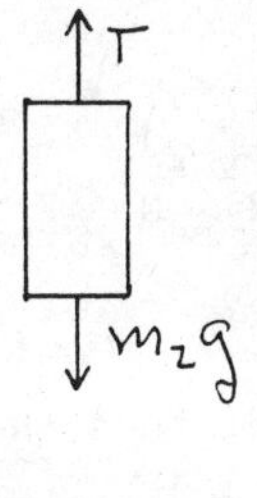

Solve (2) for N and (3) for T, insert into (1) to get

$$m_2(g-a) - u_k m_1 g \cos 30° - m_1 g \sin 30° = m_1 a$$

or

$$a = \frac{m_2 - m_1(u_k \cos 30° + \sin 30°)}{m_1 + m_2} g$$

$$= 4.876 \text{ m/s}^2$$ 2 OR 3 TSPO

Insert this result into (3) to get

(b) $T = m_2(g-a) = 24.64\ \text{N}$

6-12 For a block travelling up the slope,

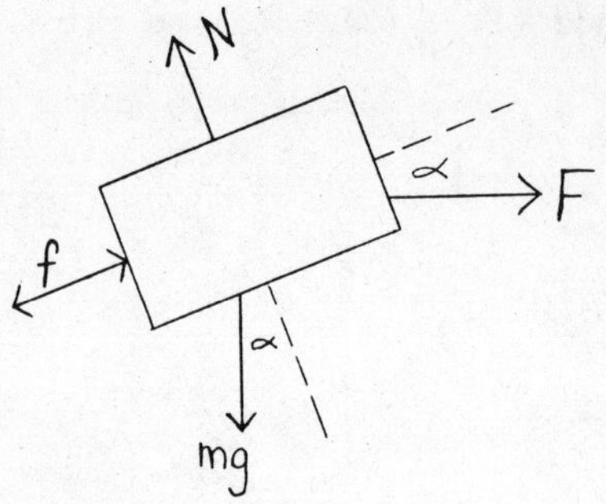

$$ma = F\cos\alpha - mg\sin\alpha - \mu_k N$$

If the block is moving down,

$$ma = F\cos\alpha - mg\sin\alpha + \mu_k N$$

$$N = mg\cos\alpha + F\sin\alpha$$

Assuming the block starts from rest and moves up the plane,

$$a = \frac{F}{m}(\cos\alpha - \mu_k\sin\alpha) - g(\mu_k\cos\alpha + \sin\alpha)$$

(a) $$a = \frac{20\ \text{N}}{2\ \text{kg}}(\cos 10° - 0.40\sin 10°) - \frac{9.8\ \text{m}}{\text{s}^2}(0.40\cos 10° + \sin 10°)$$

$$= 3.59\ \text{m/s}^2 \text{ \underline{up} the plane}$$

(b) Assuming upward motion, the above expression gives $a < 0$, which indicates <u>downward</u> motion. This reverses the direction of f.

merely change signs of all the μ_k-terms

$$a = \frac{F}{m}(\cos\alpha + \mu_k\sin\alpha) - g(-\mu_k\cos\alpha + \sin\alpha)$$

$$= \frac{20\ \text{N}}{2\ \text{kg}}(\cos 70° + 0.40\sin 70°) - \frac{9.8\ \text{m}}{\text{s}^2}(-0.40\cos 70° + \sin 70°)$$

$$= -0.689\ \text{m/s}^2 \text{ \underline{down} the plane}$$

6-13 (1) $F\cos\alpha - f = ma$

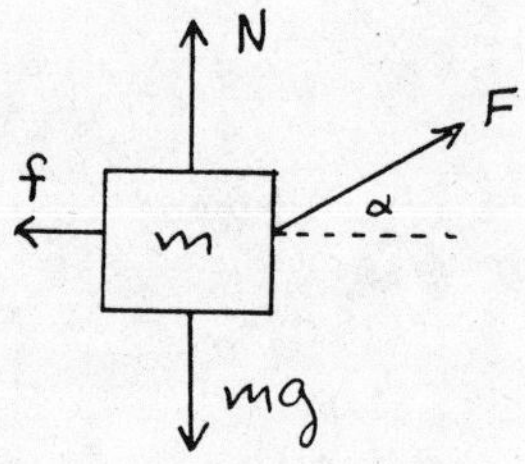

(2) $N + F\sin\alpha - mg = 0$

(3) $f = \mu_k N$

Eliminating f and N gives

$$a = \frac{F}{m}(\cos\alpha + \mu_k\sin\alpha) - \mu_k g$$

Now we seek the value of α which maximizes a; so set

$$\frac{da}{d\alpha} = 0 = \frac{F}{m}(-\sin\alpha + \mu_k\cos\alpha)$$

so

$$\sin\alpha = \mu_k\cos\alpha$$

or

$$\tan\alpha = \mu_k$$

6-14 (1) $F_{max}\cos\alpha - mg\sin\alpha - f = 0$

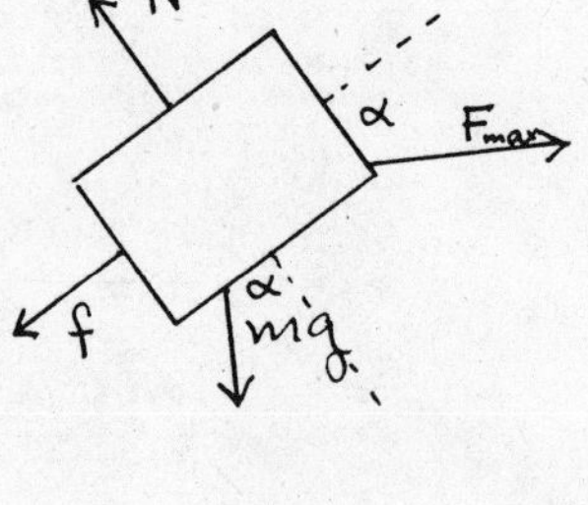

(2) $N - F_{max}\sin\alpha - mg\cos\alpha = 0$

(3) $f = \mu_s N$

This has the solution

$$F_{max} = \frac{\sin\alpha + \mu_s\cos\alpha}{\cos\alpha - \mu_s\sin\alpha}mg = 29.69\ \text{N}$$

For the other case,

(1) $F_{min} \cos\alpha - mg \sin\alpha + f = 0$

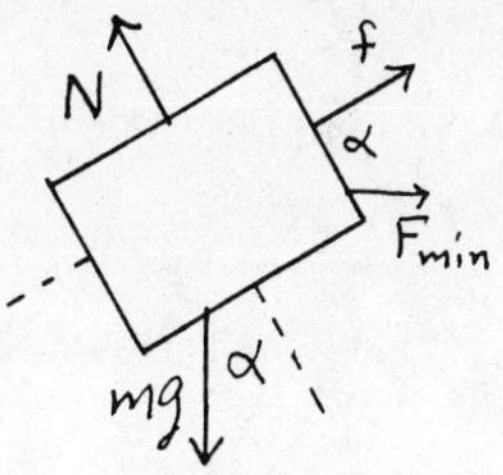

(2) and (3) are unchanged. The solution here is

$$F_{min} = \frac{\sin\alpha - \mu_s \cos\alpha}{\cos\alpha + \mu_s \sin\alpha} mg = 1.176 \text{ N}$$

6-15 For rotational motion, $a = \omega^2 r$.

(a) $N = m\omega_o^2 r$

$f - mg = 0$

$f = \mu_s N$

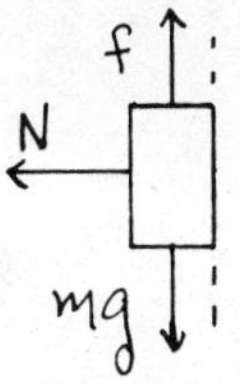

The solution is

$$\omega_o^2 = \frac{g}{\mu_s r}$$

or

(b) $\omega_o = \sqrt{\frac{g}{\mu_s r}} = 3.130 \frac{\text{radian}}{\text{s}} = 29.89 \frac{\text{rev}}{\text{min}}$

6-16 (a) The maximum force per wheel which can be supplied by static friction is

$$f = \frac{1}{4} \mu_s mg = 490 \text{ N}$$

Thus, 80% of the maximum for two wheels is

$$(\frac{4}{5})(2)f = 784 \text{ N}$$

The rate of deceleration is

$$a = 0.784 \text{ m/s}^2$$

From kinematics, with $v_o = \frac{50}{9}$ m/s,

$$s = \frac{v_o^2}{2a} = 19.68 \text{ m}$$

$$t = \frac{v_o}{a} = 7.086 \text{ s}$$

(b) Now, the force per wheel is

$$f' = \frac{1}{4} \mu_k mg = 245 \text{ N}$$

The total force for two wheels is

$$2f' = 490 \text{ N}$$

So $a = 0.490 \text{ m/s}^2$

This gives

$$s = 31.49 \text{ m}$$

$$t = 11.34 \text{ s}$$

(c) This gives

$$a = (\frac{2}{100})(0.10)g = 0.0196 \text{ m/s}^2$$

whence

$$s = 787.4 \text{ m}$$

$$t = 283.4 \text{ s}$$

(d) If we are interested in an accuracy of ~10%, then we can neglect rolling friction.

6-17 The radial and tangential components of $\vec{F} = m\vec{a}$ give

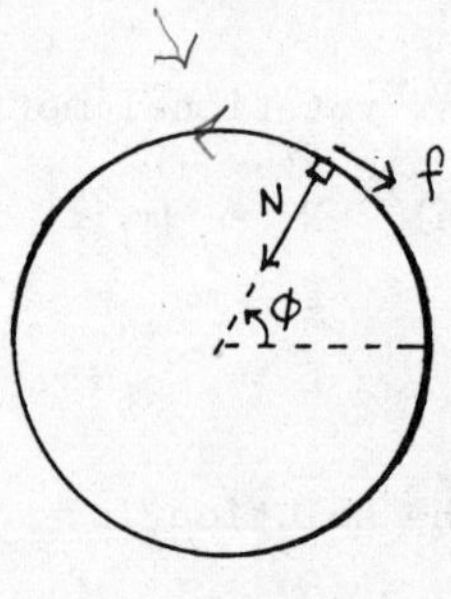

$$N = m\omega^2 R$$

$$f = -mR\frac{d\omega}{dt}$$

Use $f = \mu_k N$

to reduce these to

$$\frac{d\omega}{dt} = -\mu_k \omega^2$$

Integrate to get

$$\frac{1}{\omega_o} - \frac{1}{\omega} = -\mu_k t$$

or

$$\omega = \frac{\omega_o}{1 + \mu_k \omega_o t}$$

or, using $v = \omega R$,

$$v = \frac{v_o}{1 + \frac{\mu_k v_o}{R} t} = \frac{2}{1 + \frac{2}{9}t} \text{ m/s}$$

Observe that v never becomes zero, but approaches zero as t increases.

6-18 Integrating Stokes' Law gives

$$v(t) = v_\infty(1 - e^{-(b/m)t})$$

where $v_\infty = \frac{mg}{b}$

Thus, $\frac{b}{m} = \frac{g}{v_\infty}$

so

$$\frac{v}{v_\infty} = 0.99 = 1 - e^{-(g/v_\infty)t}$$

or

$$t = -\frac{v_\infty}{g}\ln(1 - 0.99) = 0.3759 \text{ s}$$

6-19 (a) $m\frac{dv}{dt} = mg - bv$

$$\frac{dv}{v - \frac{mg}{b}} = -\frac{b}{m}dt$$

Now integrate, observing that

$$v(0) = v_o$$

to get

$$(1)\quad \ell n(v - \frac{mg}{b}) - \ell n(v_0 - \frac{mg}{b}) = -\frac{b}{m}t$$

where

$$v_\infty = \frac{mg}{b}$$

So

$$\frac{v - v_\infty}{v_0 - v_\infty} = e^{-(b/m)t}$$

or

$$v = v_\infty + (v_0 - v_\infty)e^{-(b/m)t}$$

(b) Noting that $\frac{b}{m} = \frac{g}{v_\infty}$, use (1) above to get

$$t = \frac{v_\infty}{g}\,\ell n\,\frac{v_0 - v_\infty}{v - v_\infty} = 3.552 \text{ s}$$

6-20 At terminal velocity the net force is zero, so

$$mg - \alpha v_\infty - \beta v_\infty^2 = 0$$

which has the solution

$$v_\infty = \sqrt{(\frac{\alpha}{2\beta})^2 + \frac{mg}{\beta}} - \frac{\alpha}{2\beta}$$

6-21 The force equations give

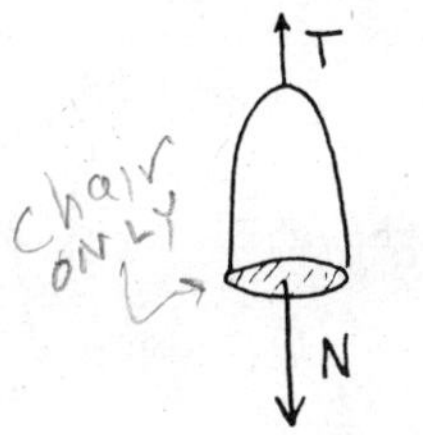

$$T - N = 0,\ a = 0$$

$$T + N - mg = ma$$

thus

$$T = \frac{1}{2}m(g+a)$$

$$= 400 \text{ N}$$

Since T = N there is no limit to the acceleration; that is, he will never pull himself out of the seat.

6-22 (a) The force equations are

(1) $2T_0 - T_1 = 0$ (Note that the mass is zero)

(2) $m_2 g - T_1 = m_2 a_2$ ($\vec{a}_2$ is taken to be downward)

(3) $T_2 - 2T_0 = 0$

(4) $T_0 - m_1 g = m_1 a_1$ ($\vec{a}_1$ is taken to be up)

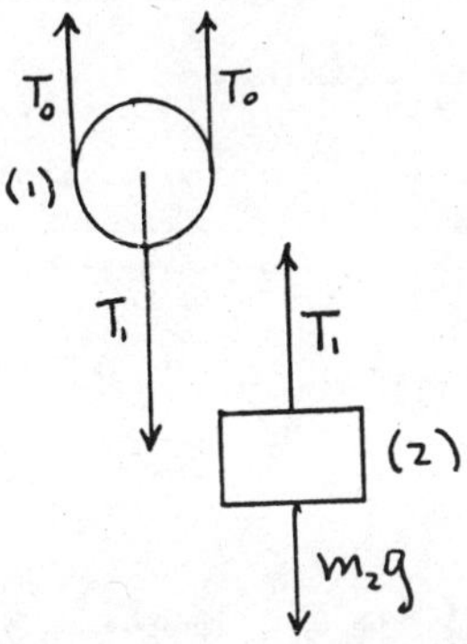

The fixed length of the cord implies that

(5) $a_2 = \frac{1}{2}a_1$

Add (2) to twice (4) to get

$$(m_2 - 2m_1)g + (2T_0 - T_1) = m_2 a_2 + 2m_1 a_1$$

Now use (1) on the left and (5) on the right and rearrange:

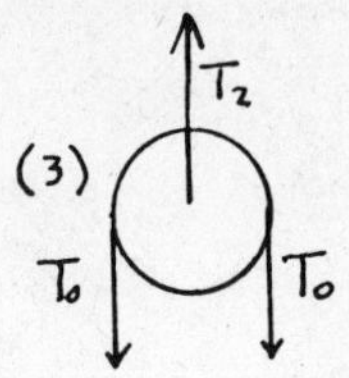

$$a_1 = \frac{m_2 - 2m_1}{\frac{1}{2}m_2 + 2m_1} g = 1.508 \text{ m/s}^2$$

$$a_2 = \tfrac{1}{2}a_1 = 0.7538 \text{ m/s}^2$$

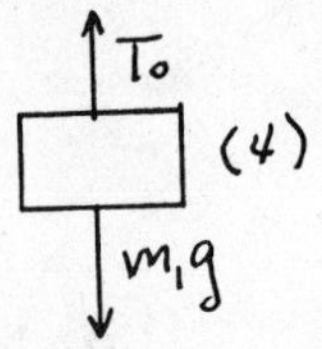

(b) From (4),

$$T_0 = m_1(g+a_1) = 22.62 \text{ N}$$

From (1) and (3),

$$T_1 = T_2 = 2T_0 = 45.23 \text{ N}$$

6-23 (a) Let a_3 be the (downward) acceleration of block 3, let a_B be the (downward) acceleration of block 2 relative to pulley B. Then the (downward) absolute accelerations are:

$$a_3$$

$$a_2 = -a_3 + a_B$$

$$a_1 = -a_3 - a_B$$

The force equations now give

(1) $m_1 g - T_B = m_1(-a_3 - a_B)$

(2) $m_2 g - T_B = m_2(-a_3 + a_B)$

(3) $m_3 g - T_A = m_3 a_3$

(4) $T_A - 2\,T_B = 0$

Add $\frac{(1)}{m_1} + \frac{(2)}{m_2}$, apply (4) and then apply (3) to get:

$$a_3 = \frac{\frac{1}{m_1} + \frac{1}{m_2} - \frac{4}{m_3}}{\frac{1}{m_1} + \frac{1}{m_2} + \frac{4}{m_3}} g = \frac{1}{49} g = 0.2 \text{ m/s}^2$$

Now (2)-(1) gives

$$a_B = \frac{m_2 - m_1}{m_2 + m_1}(g + a_3) = 2 \text{ m/s}^2$$

Summarizing,

$a_1 = 2.2 \text{ m/s}^2$ up

$a_2 = 1.8$ " down

$a_3 = 0.2$ " down

(b) $T_B = 24$ N $\quad T_A = 48$ N

6-24 (a) $T_a = \frac{1}{2} F = 17.5$ N

$m_1 g = 19.6\ \text{N}, \quad m_2 g = 49\ \text{N}$

$T_a < m_1 g < m_2 g$

So there is no motion.

(b) $T_b = 35\ \text{N}$

$m_1 g < T_b < m_2 g$

Here block 2 stays put, so

$$m_1 a_1 = T_b - m_1 g$$

$$a_1 = 7.7\ \text{m/s}^2$$

$$a_2 = 0$$

$$a_p = \frac{1}{2} a_1 = 3.85\ \text{m/s}^2$$

(c) $m_1 g < m_2 g < T_c = 70\ \text{N}$

$$m_1 a_1 = T_c - m_1 g \qquad m_2 a_2 = T_c - m_2 g$$

$$a_1 = 25.2\ \text{m/s}^2$$

$$a_2 = 4.2\ \text{m/s}^2$$

$$a_p = \frac{1}{2}(a_1 + a_2) = 14.7\ \text{m/s}^2$$

6-25 $f_1 = \mu_s N_1$

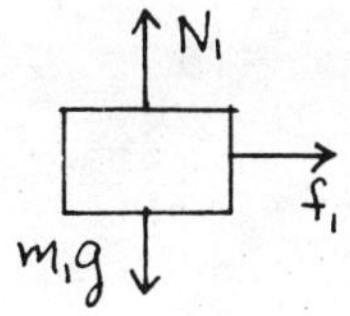

$f_z = \mu_k N_2$

$N_1 = m_1 g$

$f_1 = m_1 a$

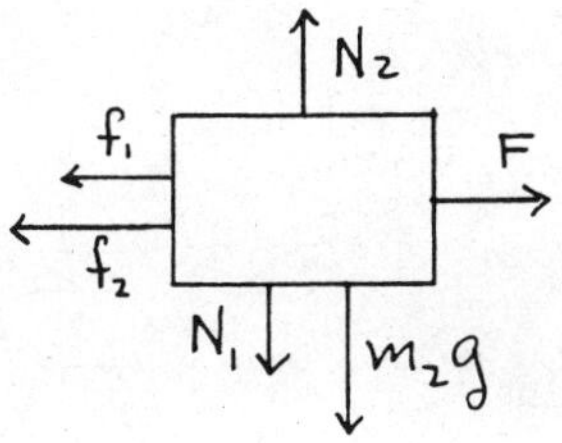

$N_2 = N_1 + m_2 g$

$F - f_1 - f_2 = m_2 a$

These can be solved and an answer obtained, but a more intuitive approach gives

$$\mu_s m_1 g = m_1 a$$

$$F - \mu_k (m_1 + m_2) g = (m_1 + m_2) a$$

or

$$\mu_s = \frac{F}{(m_1 + m_2) g} - \mu_k = 0.6003$$

6-26 $f = \mu_s N$

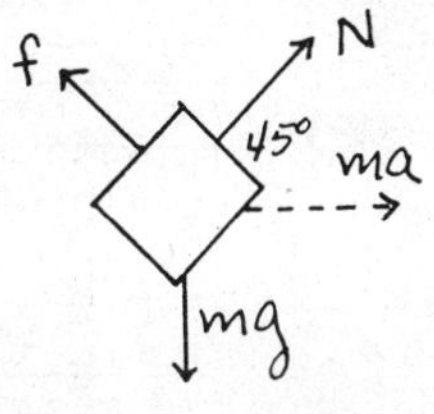

$$\frac{1}{\sqrt{2}}(f + N) = mg = \frac{1}{\sqrt{2}} N(\mu_s + 1)$$

$$\frac{1}{\sqrt{2}}(N - f) = ma = \frac{1}{\sqrt{2}} N(1 - \mu_s)$$

So

$$a = \frac{1 - \mu_s}{1 + \mu_s} g = \frac{3}{7} g$$

6-27 $f = \mu_k N$

$N = mg \sin 45° = \frac{1}{\sqrt{2}} mg$

$\frac{1}{\sqrt{2}} mg - f = ma$

So

$a = \frac{1}{\sqrt{2}}(1 - \mu_k)g = 6.375 \text{ m/s}^2$

$v = \sqrt{2as} = 35.71 \text{ m/s}$

$t = \sqrt{2s/a} = 5.601 \text{ s}$

6-28 First, obtain a.

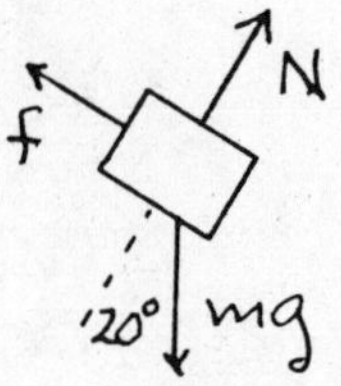

$N = mg \cos 20°$

$-mg \sin 20° + f = ma$

$f = \mu_k N$

So

$a = (\mu_k \cos 20° - \sin 20°)g = 2.788 \text{ m/s}^2$

Now, use

$v^2 - v_0^2 = 2as$

to get

$s = 1.614 \text{ m}$

6-29 First, find the velocity reached by the twig at the edge of the roof:

$v_0 \cos 30° \, t = 2$

$v_0 \sin 30° \, t + \frac{1}{2} gt^2 = 4$

Eliminate t to get

$2 \tan 30° + \frac{2g}{v_0^2 \cos^2 30°} = 4$

or

$v_0 = \frac{1}{\cos 30°}\sqrt{\frac{g}{2 - \tan 30°}} = 3.031 \text{ m/s}$

Now we can find the acceleration

$a = \frac{v_0^2}{2s} = 3.062 \text{ m/s}^2$

Next use $\vec{F} = m\vec{a}$ to write a in terms of μ_k:

$N - mg \cos 30° = 0$

$mg \sin 30° - f = ma$

$f = \mu_k N$

So

$a = (\sin 30° - \mu_k \cos 30°)g$

Finally, solve for μ_k.

$\mu_k = \frac{g \sin 30° - a}{g \cos 30°} = 0.2166$

6-30 (a)

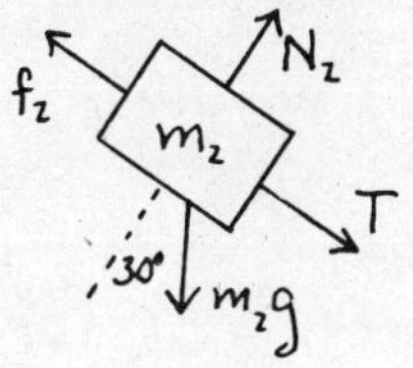

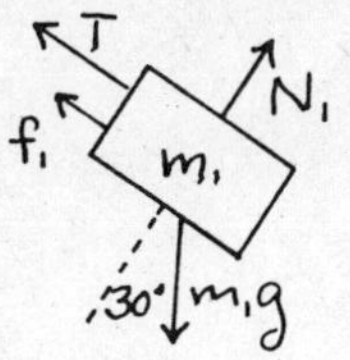

$$f_2 = \mu_2 N_2$$

$$N_2 - m_2 g \cos 30° = 0$$

$$T + m_2 g \sin 30° - f_2 = m_2 a$$

$$f_1 = \mu_1 N_1$$

$$N_1 - m_1 g \cos 30° = 0$$

$$-T + m_1 g \sin 30° - f_1 = m_1 a$$

Solving and inserting numbers,

$$T + 7.062 = 3a$$

$$-T + 16.01 = 5a$$

giving

$$a = 2.884 \text{ m/s}^2 \qquad T = 1.591$$

(b) Block 1 would slide into block 2, after which the acceleration would be the same as determined above.

6-31 $F \sin \alpha + mg - N = 0$

$F \cos \alpha - f = ma$

$f = \mu_k N$

Solving,

$$ma = F(\cos \alpha - \mu_k \sin \alpha) - \mu_k mg$$

or

$$F = \frac{m(a + \mu_k g)}{\cos \alpha - \mu_k \sin \alpha}$$

Note that when the denominator vanishes, $F \to \infty$; so the critical angle is

$$\alpha_c = \tan^{-1}\left(\frac{1}{\mu_k}\right)$$

6-32 (1) $T - m_1 g \sin 15° - \mu_s m_1 g \cos 15° = 0$

(2) $T - m_1 g \sin 63° + \mu_s m_1 g \cos 63° = 0$

(3) $T = m_2 g$

Substitute (3) into (1) and (2). Divide (1) and (2) by m_1; this gives two unknowns, μ_s and $\frac{m_2}{m_1}$.

(a) $\mu_s = \dfrac{\sin 63° - \sin 15°}{\cos 63° + \cos 15°} = 0.445$

(b) From (1),

$$m_2 = m_1(\sin 15° + \mu_s \cos 15°)$$
$$= 1.38 \text{ kg}$$

6-33 $g \sin\theta = ma_T = m\dfrac{dv_T}{dt} = R\omega\dfrac{d\omega}{d\theta}$

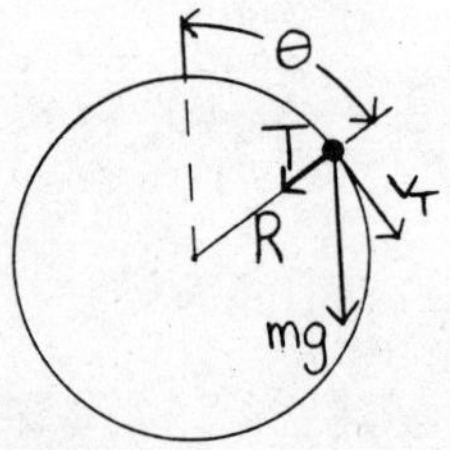

$$\frac{g}{R}\int_0^\theta \sin\theta\, d\theta = \int_{\omega_0}^{\omega} \omega\, d\omega$$

$$\frac{-g}{R}(\cos\theta - 1) = \frac{1}{2}(\omega^2 - \omega_0^2)$$

$$\omega^2 = \omega_0^2 + \frac{2g}{R}(1 - \cos\theta)$$

6-34 $\omega = 3.49 \text{ s}^{-1}$

$$a = \omega^2 R$$

$$f_{s,max} = \mu_s N = (0.25)mg$$

$$ma = m\omega^2 R = 0.25 \text{ mg}$$

$$R = \frac{0.25 \text{ g}}{\omega^2} = 0.20 \text{ m}$$

6-35 If the force exerted by the floor is entirely perpendicular to the floor, f = 0.

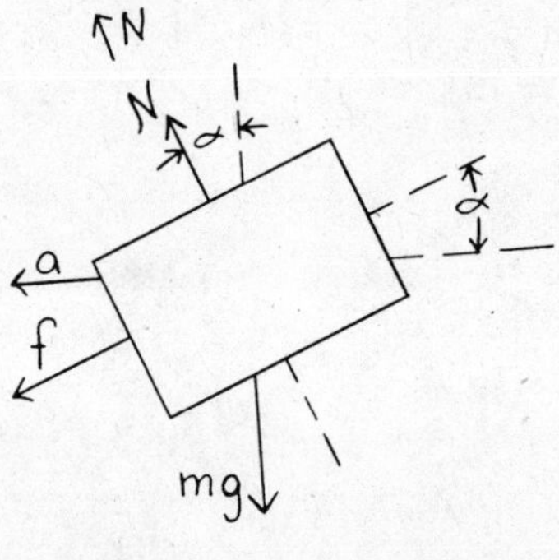

$$ma = \frac{mv^2}{R} = N\sin\alpha$$

$$N\cos\alpha = mg$$

$$\tan\alpha = \frac{v^2}{gR}$$

$$\alpha = 25.7°$$

6-36 $y(t) = \dfrac{m^2}{b^2}g(e^{-(b/m)t} - 1) + \dfrac{mg}{b}t$

$$\cong \frac{m^2}{b^2}g\left(1 - \frac{b}{m}t + \frac{1}{2}\frac{b^2}{m^2}t^2 - \frac{1}{6}\frac{b^3}{m^3}t^3 - 1\right) + \frac{mg}{b}t$$

$$= \frac{1}{2}gt^2 - \frac{1}{6}\frac{b}{m}gt^3$$

$$\Delta y(\text{lead-paper}) \cong \frac{1}{6}gt^3\left(\left(\frac{b}{m}\right)_{paper} - \left(\frac{b}{m}\right)_{lead}\right)$$

$$= \frac{1}{6}g^2t^3\left(\frac{1}{v_{\infty\ paper}} - \frac{1}{v_{\infty\ lead}}\right)$$

$$= \frac{h}{3}gt\left(\frac{1}{v_{\infty\ paper}} - \frac{1}{v_{\infty\ lead}}\right)$$

For h $= \frac{1}{2} gt^2$,

$$t = \sqrt{\frac{2h}{g}} = \sqrt{\frac{40}{9.8}} = 2.02 \text{ s}$$

So,

$$\Delta y \cong \left(\frac{20 \text{ m}}{3}\right)\left(\frac{9.8 \text{ m}}{s^2}\right)(2.02 \text{ s})\left(\frac{1}{60 \text{ m/s}} - \frac{1}{600 \text{ m/s}}\right)$$

$$= 1.98 \text{ m}$$

6-37 Using the numerical method,

$$v_{i+1} = v_i + \Delta v_i$$

where

$$\Delta v_i = g\left(1 - \frac{v_i}{v_\infty}\right)\Delta t_i$$

Exact values are

$$v_i = v_\infty(1 - e^{-(b/m)t_i}).$$

t_i	v_i, $\frac{m}{s}$	$\frac{v_i}{v_\infty}$	Δv_i, $\frac{m}{s}$	Numerical Method v_{i+1}, $\frac{m}{s}$	Exact v_{i+1}, $\frac{m}{s}$
0	0	0	19.60	19.60	17.80
2	19.60	0.1960	15.76	35.36	32.43
4	35.36	0.3536	12.67	48.03	44.46
6	48.03	0.4803	10.19	58.22	54.34
8	58.22	0.5822	8.19	66.41	62.47
10	66.41	0.6641	6.58	72.99	69.15
12	72.99	0.7299	5.29	78.28	74.64
14	78.28	0.7828	4.26	82.54	79.15
16	82.54	0.8254	3.42	85.96	82.86
18	85.96	0.8596	2.75	88.71	85.91

The approximation could be improved by using a smaller Δt value.

6-38 (a) $v_f = v_i - g\mu_k(v_i)\Delta t = v_i - a_i\Delta t$

$$\mu_k = 0.27\left(\frac{1 + 0.0044\, v_i}{1 + 0.064\, v_i}\right)$$

t_i, s	v_i, m/s	μ_k	$a_i\Delta t$, m/s	v_f, m/s	Δt, s
0	300.00	0.0310	30.39	269.61	100
100	269.61	0.0323	31.69	237.92	100
200	237.92	0.0341	33.38	204.54	100
300	204.54	0.0364	35.69	168.86	100
400	168.86	0.0399	39.06	129.80	100
500	129.80	0.0456	44.67	85.13	100
600	85.13	0.0576	28.20	56.93	50
650	56.93	0.0727	17.81	39.12	25
675	39.12	0.0903	11.06	28.06	12.5
687.5	28.06	0.1085	6.65	21.41	6.25
693.75	21.41	0.1246	7.63	13.76	6.25
700.00	13.76	0.1523	4.66	9.10	3.125
703.125	9.10	0.1775	5.43	3.67	3.125
706.250	3.67	0.2222	3.40	0.27	1.5625

(b) $\bar{\mu}_k = 0.27\left(\frac{1 + 0.0044(150)}{1 + 0.064\ (150)}\right) = 0.0423$

$$v = v_o + \bar{a}t = v_o - \bar{\mu}_k g t$$

$$t = \frac{v_o}{\bar{\mu}_k g} = \frac{300\ \text{m/s}}{0.0423(9.8\ \text{m/s}^2)} = 724\ \text{s}$$

Using $\mu_k = 0.27$, the low-speed limit of μ_k,

$$t = \frac{300}{0.27(9.8)} = 113\ \text{s}$$

Thus, the low-speed limit of μ_k is not a good substitute for $\bar{\mu}_k$

III-1 (a) $\int_1^2 (x^2 + 3)dx = [\frac{x^3}{3} + 3x]_1^2 = \frac{8}{3} + 6 - \frac{1}{3} - 3 = \frac{16}{3}$

(b) $\int_0^{\pi} \sin\theta\, d\theta = [-\cos\theta]_0^{\pi} = 1 + 1 = 2$

(c) $\int_{-1}^{+1} y^3\, dy = [\frac{1}{4} y^4]_{-1}^{1} = \frac{1}{4} - \frac{1}{4} = 0$

Note that this is an odd integrand evaluated over a symmetric interval.

(d) $\int_0^1 xe^x dx = \int_0^1 d(xe^x) - \int_0^1 e^x dx$

$= xe^x\Big|_0^1 - e^x\Big|_0^1$

$= (e - 0) - (e - 1)$

$= 1$

III-2 If $u = 5 - t$, $du = -dt$

$$\int_1^t \frac{dt}{\sqrt{5-t}} = \int_4^{5-t} \frac{-du}{\sqrt{u}} = -2\sqrt{u}\Big|_4^{5-t} = -2\sqrt{5-t} + 2\sqrt{4}$$

So $y(t) = 4 - 2\sqrt{5-t}$

III-3 (a) $\int_0^{\theta} \sin^2\theta\, d\theta = \int_0^{\theta} \frac{1}{2}(1 - \cos 2\theta)d\theta$

$= \frac{\theta}{2} - \frac{1}{2}\int_0^{\theta} \frac{2}{2}\cos 2\theta d\theta$

$= \frac{\theta}{2} - \frac{1}{4}\sin 2\theta$

(b) $\int_0^{\theta} \cos^2\theta d\theta = \int_0^{\theta} (1 - \sin^2\theta)d\theta$

$= \theta - \int_0^{\theta} \sin^2\theta d\theta = \frac{\theta}{2} + \frac{1}{4}\sin 2\theta$

III-4 $\int_0^{\pi} \sin^3\theta\, d\theta = \int_0^{\pi} (1 - \cos^2\theta)\sin\theta\, d\theta$

$x = \cos\theta$, $dx = -\sin\theta\, d\theta$

To change the limits of integration,

$\cos\pi = -1$, $\cos 0 = 1$

$$\int_0^{\pi} \sin^3\theta d\theta = \int_1^{-1} -(1 - x^2)dx = \int_{-1}^{1} (1 - x^2)dx$$

$$= [x - \frac{x^3}{3}]_{-1}^{1} = 1 - \frac{1}{3} + 1 - \frac{1}{3} = \frac{4}{3}$$

III-5 Let $\cos\beta = a(x^2 + a^2)^{-\frac{1}{2}}$

$-\sin\beta\, d\beta = -ax(x^2 + a^2)^{-3/2}dx$

$-x(x^2 + a^2)^{-\frac{1}{2}}d\beta = -ax(x^2 + a^2)^{-3/2}dx$

$$\frac{1}{a}(x^2 + a^2)d\beta = dx$$

$$\frac{1}{a}(a^2/\cos^2 \beta)d\beta = dx$$

$$a \cos^{-2}\beta \; d\beta = dx$$

When $x = 0$, $\beta = 0$

When $x = a$, $\beta = \frac{\pi}{4}$

So

$$\int_0^a \frac{a^2 dx}{(x^2 + a^2)^{3/2}} = \int_0^{\pi/4} \frac{a^3/\cos^2\beta \; d\beta}{a^3/\cos^3\beta}$$

$$= \int_0^{\pi/4} \cos \beta d\beta = \sin \frac{\pi}{4} = \frac{1}{\sqrt{2}}$$

III-6 $kx\,dx = mv\,dv$

$$\int_0^x kx\,dx = \int_{v_0}^{v} mv\,dv$$

$$\frac{1}{2} kx^2 = \frac{1}{2} m(v^2 - v_0^2)$$

III-7 $y^2\,dy = e^{-at}dt$

$$\int_0^y y^2\,dy = \int_0^t e^{-at}dt$$

$$\frac{1}{3} y^3 = -\frac{1}{a}(e^{-at} - 1)$$

$$y = [\frac{3}{a}(1 - e^{-at})]^{1/3}$$

III-8 (a) $\displaystyle\int_1^{+\infty} \frac{dx}{x^2} = \lim_{t\to+\infty} \int_1^t \frac{dx}{x^2}$

$$= \lim_{t\to\infty} \left[-\frac{1}{x}\right]_1^t = \lim_{t\to\infty}(1 - \frac{1}{t}) = 1$$

(b) $\displaystyle\int_0^{\infty} ve^{-v^2}dv = -\frac{1}{2}\lim_{t\to\infty} \int_0^t e^{-v^2}(-2vdv)$

$$= -\frac{1}{2}\lim_{t\to\infty} [e^{-v^2}]_0^t = \frac{1}{2}$$

(c) $\displaystyle\int_{-\infty}^{+\infty} \frac{2x\,dx}{(1 + x^2)^2} = \int_{-\infty}^{0} \frac{2x\,dx}{(1 + x^2)^2} + \int_0^{\infty} \frac{2x\,dx}{(1 + x^2)^2}$

$$= \lim_{t\to\infty}\int_{-t}^{0} \frac{2x\,dx}{(1 + x^2)^2} + \lim_{s\to\infty}\int_0^{s} \frac{2x\,dx}{(1 + x^2)^2}$$

Let $u = 1 + x^2$; $du = 2x\,dx$

$$\int_{-\infty}^{+\infty} \frac{2x\,dx}{(1 + x^2)^2} = \lim_{t\to\infty}\int_{1+t^2}^{1} u^{-2}du + \lim_{s\to\infty}\int_1^{1+s^2} u^{-2}du$$

$$= \lim_{t\to\infty}\left[\frac{-1}{u}\right]_{1+t}^{1} + \lim_{s\to\infty}\left[\frac{-1}{u}\right]_1^{1+s}$$

$$= -1 + 0 + 0 + 1 = 0$$

III-9 (a) $\displaystyle A_1 = \int_0^1 x^2dx = [\frac{1}{3}x^3]_0^1 = \frac{1}{3}$

(b) $A_2 = \int_0^1 e^x dx = [e^x]_0^1 = e - 1$

(c) $A_3 = \int_1^e \frac{4dx}{x} = 4[\ln x]_1^e = 4$

(d) $A_4 = \int_0^{16} (x + 9)^{-\frac{1}{2}} dx = 2[(x + 9)^{\frac{1}{2}}]_0^{16} = 4$

(e) $A_5 = \int_0^{\frac{3\pi}{2}} \sin x dx = [-\cos x]_0^{\frac{3\pi}{2}} = 1$

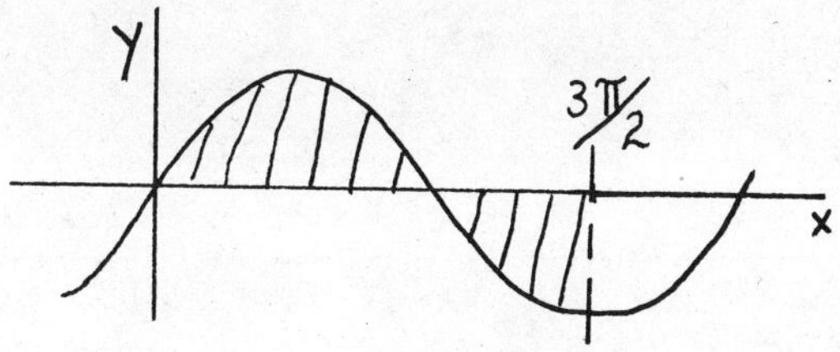

III-10 When $f(-x) = f(x)$, the function is even.

$$\int_{-a}^{a} f(x)dx = \int_{-a}^{0} f(x)dx + \int_{0}^{a} f(x)dx$$
$$= \int_{a}^{0} f(-x)(-dx) + \int_{0}^{a} f(x)dx$$
$$= -\int_{a}^{0} f(x)dx + \int_{0}^{a} f(x)dx$$
$$= \int_{0}^{a} f(x)dx + \int_{0}^{a} f(x)dx = 2\int_{0}^{a} f(x)dx$$

When $f(-x) = -f(x)$ the function is odd.

$$\int_{-a}^{a} f(x)dx = \int_{-a}^{0} f(x)dx + \int_{0}^{a} f(x)dx$$
$$= \int_{a}^{0} f(-x)(-dx) + \int_{0}^{a} f(x)dx$$
$$= -\int_{0}^{a} f(x)dx + \int_{0}^{a} f(x)dx = 0$$

III-11 $y = \frac{x}{1 + x^2}$ and $y = \frac{x}{5}$

The two curves intersect at $x = 0$ and $x = 2$.

$$\text{Area} = \int_0^2 \left(\frac{x}{1 + x^2} - \frac{x}{5}\right) dx = \frac{1}{2}\int_1^5 \frac{du}{u} - \frac{1}{10}[x^2]_0^2$$
$$= \frac{1}{2}\ln(5) - \frac{2}{5} = 0.40$$

III-12 $y = \sqrt{32x}$ and $y = x^3$

The curves intersect at $x = 0$ and $x = 2$

$$\text{Area} = \int_0^2 (\sqrt{32x} - x^3)dx = \left[\frac{2\sqrt{32}}{3} x^{3/2} - \frac{1}{4} x^4\right]_0^2 = \frac{20}{3}$$

III-13 $f(x) = \dfrac{8a^3}{(x^2 + 4a^2)}$

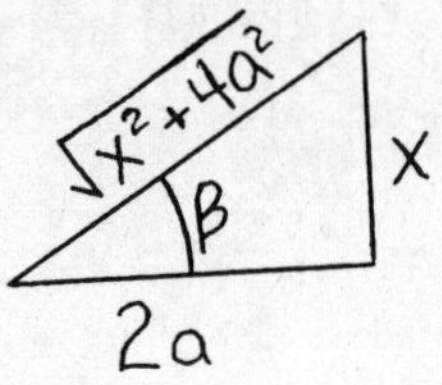

Let $\cos\beta = \dfrac{2a}{\sqrt{x^2 + 4a^2}}$

$$-\sin\beta\, d\beta = \frac{-2ax\, dx}{(x^2 + 4a^2)^{3/2}}$$

$$-\sin\beta\, d\beta = \frac{-2a\, dx}{(x^2 + 4a^2)} \sin\beta$$

$$d\beta = \frac{2a\, dx}{(x^2 + 4a^2)}$$

When $x \to +\infty$, $\beta \to \frac{\pi}{2}$

When $x \to -\infty$, $\beta \to -\frac{\pi}{2}$

So, $\displaystyle\int_{-\infty}^{\infty} \frac{8a^3 dx}{(x^2 + 4a^2)} = 4a^2 \int_{-\frac{\pi}{2}}^{\frac{\pi}{2}} d\beta$

$$\text{Area} = 4\pi a^2$$

III-14 $\displaystyle\int_{\text{circle}} y\, dx = 4 \int_{\substack{\text{first} \\ \text{quadrant}}} y\left(\frac{dx}{d\varphi}\right) d\varphi$

$$= 4 \int_{\frac{\pi}{2}}^{0} R \sin\varphi(-R\sin\varphi) d\varphi$$

$$= -4R^2 \int_{\frac{\pi}{2}}^{0} \sin^2\varphi\, d\varphi = 2R^2 \int_{\frac{\pi}{2}}^{0} (\cos 2\varphi - 1) d\varphi$$

$$= 2R^2 \left[\frac{1}{2}\sin 2\varphi - \varphi\right]_{\frac{\pi}{2}}^{0} = \pi R^2 = \text{Area}$$

III-15 $\displaystyle\text{Area} = 4 \int_0^R y\, dx = 4 \int_0^R (R^2 - x^2)^{\frac{1}{2}} dx$

Let $x = R\cos\varphi$, $dx = -R\sin\varphi\, d\varphi$

When $x = 0$, $\varphi = \frac{\pi}{2}$

When $x = R$, $\varphi = 0$

So, $\displaystyle\text{Area} = 4 \int_{\frac{\pi}{2}}^{0} R(1 - \cos^2\varphi)^{\frac{1}{2}}(-R\sin\varphi\, d\varphi)$

$$= -4 \int_{\frac{\pi}{2}}^{0} R^2 \sin^2\varphi\, d\varphi = \pi R^2$$

This method and the method in problem III-14 lead to the same integration.

Chapter 7

7-1 a) Work done on the block by the rope

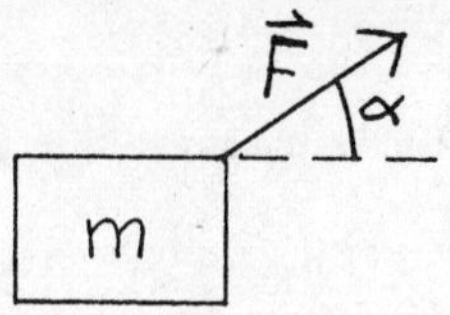

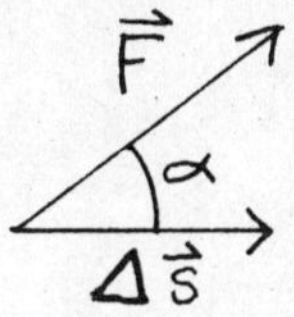

$$W = \vec{F}\cdot\Delta\vec{S} = F\Delta S \cos 30°$$

$$= 3(10) \cos 30° = 26.0 \text{ J}$$

b) Work done on the rope by the block

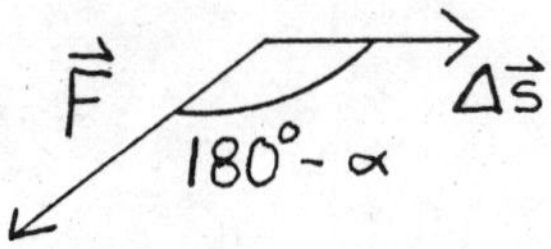

$$W = \vec{F}\cdot\Delta\vec{S} = F\Delta S \cos 150° = -26.0 \text{ J}$$

c) Work done on the student by the rope

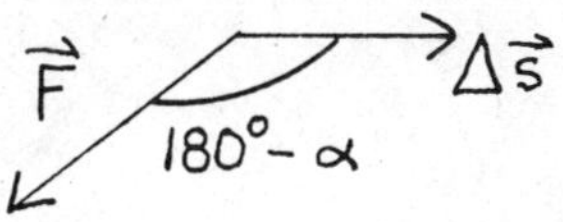

$$W = F\Delta S \cos 150° = -26.0 \text{ J}$$

d) Work done on the rope by the student

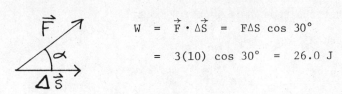

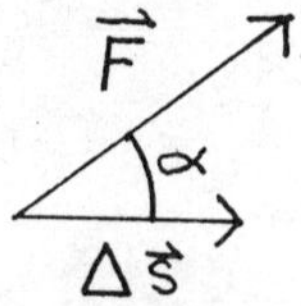

$$W = F\Delta S \cos 30° = 26.0 \text{ J}$$

The net work on the rope is zero. This is reasonable because the rope is massless (no work is needed to move it anywhere) and thus all of the student's work goes into moving the block. If the rope had nonzero mass, the student's work would be divided between moving the rope and moving the block.

7-2 $W_{g1} = -m_1 g\Delta s = (10\text{ kg})(-9.8\text{ m/s}^2)(-0.5\text{ m}) = 49.0\text{ J}$

$W_{g2} = -m_2 g\Delta s = (8\text{ kg})(-9.8\text{ m/s}^2)(0.5\text{ m}) = -39.2\text{ J}$

(1) $m_1 a = m_1 g - T$

(2) $m_2 a = T - m_2 g$

Solve (2) for T and put into (1).

$$m_1 a = m_1 g - m_2(a + g)$$

$$a = \frac{m_1 - m_2}{m_1 + m_2} g = 1.09 \text{ m/s}^2$$

The equations for total work are

$W_1 = F_1\Delta s = m_1 a\Delta s = (10\ kg)(-1.09\ m/s^2)(-0.5\ m) = 5.45\ J$

$W_2 = F_2\Delta s = m_2 a\Delta s = (8\ kg)(1.09\ m/s^2)(0.5\ m) = 4.36\ J$

$-(W_1 - W_{g1}) = W_2 - W_{g2}$

The above relationship means that the rope does no total work.

$$W_1 - W_{g1} = W_{rope\ 1} \text{ and } W_2 - W_{g2} = W_{rope\ 2}$$

$$W_{rope\ 1} + W_{rope\ 2} = 0$$

7-3 $W_s = \vec{F}\cdot\vec{s} = (-mg)(-h) = mgh$

There is also a normal force $\vec{N}$ acting on the block. It acts perpendicular to the motion and thus does no work.

7-4 $W_g = \vec{F}\cdot\vec{s} = (-mg)(h-h') = mg(h'-h)$

7-5 $W_{g1} = (-m_1 g)h = -m_1 gh$

$W_{g2} = (-m_2 g)(-h \sin 30°) = m_2 gh \sin 30° = \frac{m_2 gh}{2}$

$W_g = \frac{m_2 gh}{2} - m_1 gh$

7-6 A slab of dirt of thickness dx has a mass of

$$dm = (40\ m)(15\ m)(4 g/cm^3)dx$$

$$= 2.4 \times 10^6\ dx\ kg/m$$

The force applied to lift the slab g dm and the distance lifted is x + 3 m.

$$W = \int_0^{10} (2.40 \times 10^6\ dx)(9.8)(x + 3)$$

$$= 2.35 \times 10^7 \left(\frac{x^2}{2} + 3x\right)\Big]_0^{10}\ N$$

$$= 1.88 \times 10^9\ N$$

7-7

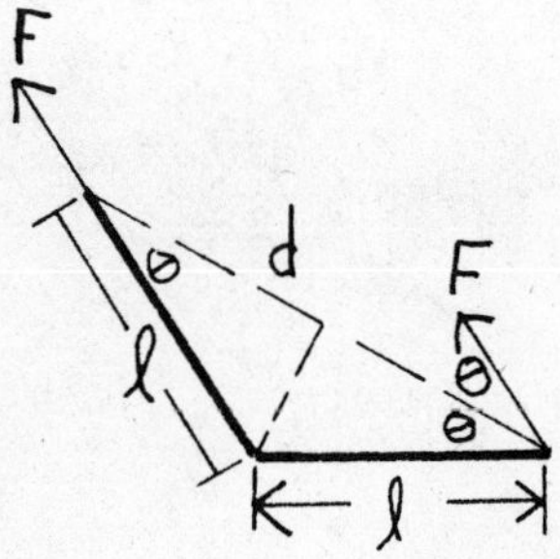

Constant force means constant direction as well as magnitude.

To strip the tape off,

$$\theta < 90°$$

The work done to strip the tape off is:

$$W = (F \cos \theta)d$$

Since $\frac{d}{2} = \ell \cos \theta$,

$$W = (F \cos \theta)(2\ \ell \cos \theta)$$

$$W = 2\ell F \cos^2 \theta$$

For F to be a minimum,

$$\cos^2 \theta = 1$$

$\theta = 0°.$

$$F_{min} = \frac{W}{2\ell \cos^2 \theta} = \frac{4\ J}{2(.15\ m)} = 13.3\ N$$

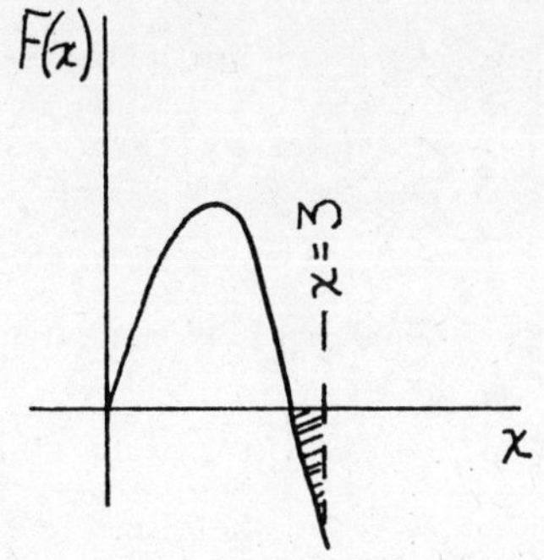

7-8 $W = \int_1^3 \vec{F}(x) \cdot d\vec{x} = \int_1^3 (5x - 2x^2)dx$

$$= (\frac{5x^2}{2} - \frac{2x^3}{3})\Big|_1^3 = (\frac{45}{2} - \frac{54}{3}) - (\frac{5}{2} - \frac{2}{3}) = \frac{40}{2} - \frac{52}{3}$$

$$= 20 - 17\frac{1}{3} = \boxed{2\frac{2}{3}\ J}$$

This answer takes into account the fact that F(x) changes sign in the interval by subtracting the area above the curve in the negative F(x) region (shaded in diagram) automatically in the integration.

7-9 $W = \int_{1\ m}^{3\ m} \left[(\frac{3N}{m^2})x^2 - \frac{2N}{m}\ x\right] dx$

$$= x^3\Big]_1^3 - x^2\Big]_1^3\ N \cdot m$$

$$= 26 - 8\ J = 18\ J$$

7-10 $0 = \int_0^x (\vec{F}(x) + \vec{f}) \cdot dx\ \hat{i}$

$$= \int_0^x \left[(10\ \frac{N}{m})(2m - x) - (0.20)(2\ kg)(9.8\ m/s^2)\right]dx$$

$$= [(20 - 3.92)N]\,x - \left(5\ \frac{N}{m}\right)x^2$$

Solving for x,

$$x = 0\ m \text{ and } x = 3.22\ m$$

$$0 = F_{total} = F(x) - f = 20\ N - 10\ x\ \frac{N}{m} - 3.92\ N$$

7-11 $W_F = \int_{0\ m}^{1\ m} 10\frac{N}{m}\ (2\ m - x)dx$

$$= [20\,N\,x - \frac{5\ N}{m}\ x^2]_{0\ m}^{1\ m}$$

$$= 15\ J$$

$$W_f = \int_{0\ m}^{1\ m} f\ dx = (0.20)(2\ kg)(9.8\ m/s^2)(1\ m)$$

$$= 3.92\ J$$

7-12 $W = (-mg)(-h) = (0.25\ kg)(9.8\ m/s^2)(10\ m) = 24.5\ J$

$$\Delta K = W = 24.5\ J$$

$$\Delta K = \frac{1}{2}\ m\ (v^2 - v_o^2)$$

$$v = \sqrt{\frac{2(24.5)J}{0.25\ kg}} = 14\ m/s$$

If the mass of the object changes, W will change. Since $W = \Delta K$, ΔK changes.

v_f will not change.

7-13 $W = \Delta K = \frac{1}{2} m(v^2 - v_o^2) = \frac{1}{2} m(0 - v_o^2) = -\frac{1}{2} mv_o^2$

$$W = \int_0^{x_B} \vec{F} \cdot d\vec{s} = \int_0^{x_B} -\kappa x \, dx = -\frac{1}{2} \kappa x_B^2$$

$$x_B = \sqrt{\frac{m}{\kappa}}\, v_o$$

7-14 $\Delta K = \int_0^x -\mu_k \, mg \, dx$

$$-\frac{1}{2} mv_o^2 = -\mu_k \, mgx$$

$$x = \frac{v_o^2}{2\mu_k g}$$

7-15 Choose the coordinate system so that y increases as the block is lowered. Then, the net force in the vertical direction can be written

$$F = mg - T - mg - ky^2$$

$$W = \int_0^{10} mg - ky^2 \, dy = mgy - \frac{ky^3}{3}\Big|_0^{10} = 10 \text{ mg} - \frac{1000 \text{ k}}{3}$$

$$W = (10 \text{ m})(10 \text{ kg})(9.8 \text{ m/s}^2) - (1000 \text{ m}^3)(2 \frac{\text{N}}{\text{m}^2})(\frac{1}{3})$$

$$= 313 \text{ J}$$

$$\frac{1}{2} mv^2 = \Delta K = W$$

$$v = \sqrt{\frac{2(313)\text{J}}{10 \text{ kg}}} = 7.91 \text{ m/s}$$

7-16 The farthest descent occurs when v reaches zero.

$$\frac{1}{2} mv^2 = 0 = \Delta K = W = mgy - \frac{ky^3}{3}$$

Solving for y,

$$y = 0 \text{ or } y = \pm\sqrt{\frac{3 \text{ mg}}{k}}$$

The farthest descent is

$$y = \sqrt{\frac{3 \text{ mg}}{k}} = \sqrt{\frac{3(10 \text{ kg})(9.8 \text{ m/s}^2)}{2 \text{ N/m}^2}} = 12.1 \text{ m}$$

The speed is a maximum when W is maximum. Set the derivative $\frac{dW}{dy}$ equal to zero and solve to find position of maximum speed.

$$0 = \frac{dW}{dy} = mg - ky^2$$

$$y = \pm\sqrt{\frac{mg}{k}} = \pm 7.0 \text{ m}$$

7-17 $\frac{1}{2} m(v^2 - v_o^2) = \Delta K = 0$

$$\Delta K = W = \int_A^B \vec{F} \cdot d\vec{s}$$

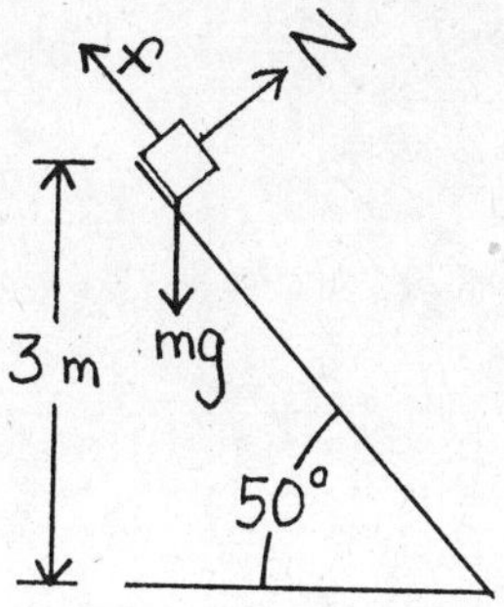

Region 1

$$\frac{3 \text{ m}}{\sin 50°} = 3.9 \text{ m}$$

$$W_1 = \int_0^{3.9 \text{ m}} [(mg \sin 50°) - (\mu_k mg \cos 50°)]ds$$

$$= mgs(\sin 50° - \mu_k \cos 50°)\Big]_0^{3.9}$$

$$= (3.9 \text{ m})(2 \text{ kg})(9.8 \text{ m/s}^2)(0.77 - 0.10)$$

$$= 51.2 \text{ J}$$

For the level region,

$$W_2 = \int_0^{6 \text{ m}} - \mu_k mg\, dx = -(0.15)(2 \text{ kg})(9.8 \text{ m/s}^2)(6 \text{ m}) = -17.6 \text{ J}$$

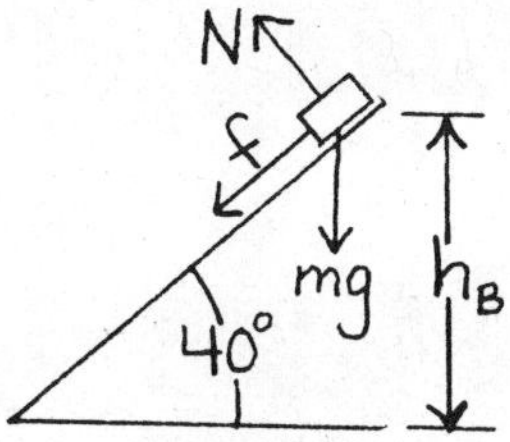

Region 3

$$W_3 = \int_0^{h_B} [- mg \sin 40° - \mu_k mg \cos 40°]ds$$

$$= -mg(\sin 40° - \mu_k \cos 40°)h_B$$

$$= -10.3\ h_B \text{ N}$$

The work should add to zero.

$$0 = W_1 + W_2 + W_3 = (51.2 - 17.6)\text{J} - 10.3\ h_B\text{N}$$

$$h_B = \frac{51.2 - 17.6}{10.3} \text{ m} = 3.26 \text{ m}$$

7-18

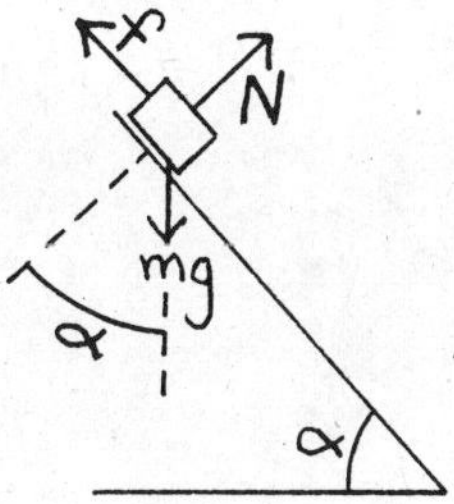

$$\frac{1}{2} m(v^2 - v_o^2) = \Delta K = W = \int_0^{\ell} (mg \sin\alpha - \mu_k mg \cos \alpha)ds$$

$$- \frac{1}{2} mv_o^2 = \ell mg (\sin\alpha - \mu_k \cos \alpha)$$

$$\sin \alpha - \mu_k \cos \alpha = - \frac{1}{2} \frac{v_o^2}{\ell g}$$

$$\mu_k = \tan\alpha + \frac{v_o^2}{2\ell g \cos \alpha}$$

7-19 In the vertical direction,

$$ma = 0 = -F \sin 20° - mg + N$$

$$N = F \sin 20° + mg$$

In the horizontal direction, from x = 0 to x = 6 m,

$$F_1 = F\cos 20° - \mu_k(mg + F\sin 20°)$$

As the block passes 6 m, the direction of the horizontal force is reversed. Until the object reverses its direction of motion,

$$F_2 = -F\cos 20° - \mu_k(mg + F\sin 20°)$$

Once the object reverses direction the frictional force is reversed.

$$F_3 = -F\cos 20° + \mu_k(mg + F\sin 20°)$$

In the first region

$$W_1 = \int_0^6 F_1\,dx = [F(\cos 20° - \mu_k \sin 20°) - \mu_k mg](6\text{ m})$$

$$= 9.80\text{ J}$$

To find the stopping point of the block,

$$\Delta K = 0 = W_1 + W_2$$

$$W_2 = -W_1$$

$$W_2 = \int_6^x [-F(\cos 20° + \mu_k \sin 20°) - \mu_k mg]dx$$

$$W = -28.43(x-6)\text{ J}$$

$$x - 6 = \frac{-W_1}{-28.43} = \frac{9.80}{28.43}$$

$$= 6.34\text{ m}$$

To bring the block back to x = 0,

$$W_3 = \int_{6.34}^0 [F(\mu_k \sin 20° - \cos 20°) + \mu_k mg]dx$$

$$= 10.36\text{ J}$$

$$\Delta K = \frac{1}{2}mv_f^2 = W_3$$

$$v_f = -\sqrt{\frac{2(10.36\text{ J})}{4\text{ kg}}} = -2.28\text{ m/s}$$

7-20 $\frac{1}{2}m(v^2 - v_o^2) = \int_0^{6\text{ m}} F\,dx = \int_0^{6\text{ m}}\left(2x\,\frac{\text{N}}{\text{m}} + 4\text{ N}\right)dx$

$$\frac{1}{2}(0.5\text{ kg})\left(v^2 - \frac{1.5^2\text{ m}^2}{\text{s}^2}\right) = (36 + 24)\text{ J}$$

$$\frac{0.5\text{ kg}}{2}v^2 = 60.56\text{ J}$$

$$v = \sqrt{\frac{(60.56)(2)}{0.5}}\text{ m/s} = 15.56\text{ m/s}$$

$$\frac{1}{2}mv^2 = \frac{1}{2}mv_o^2 + \int_0^x F dx = \frac{1}{2}mv_o^2 + \int_0^x (2x + 4)dx = \frac{1}{2}mv_o^2 + x^2 + 4x$$

$$v = \sqrt{v_o^2 + \frac{2}{m}(x^2 + 4x)}$$

7-21 Notice that the force at x = 0 is acting in the $-\hat{i}$ direction. The force does not become positive until x > 2. To see whether the particle reaches x = 2, solve for v as a function of x.

$$\frac{1}{2}mv^2 = \frac{1}{2}mv_o^2 + \int_0^x F\,dx$$

$$= \frac{1}{2}mv_o^2 + x^2 - 4x$$

$$v = \sqrt{v_o^2 + \frac{2}{m}(x^2 - 4x)}$$

The positive root is chosen since the initial conditions specify a positive velocity. Set v = 0 to find out whether the particle stops before it reaches x = 2.

$$0 = v_o^2 + \frac{2}{m}(x^2 - 4x)$$

$$x = 2 - \sqrt{4 - \frac{mv_o^2}{2}} = (2 - 1.323)\text{m} = 0.677\text{ m}$$

The velocity of the particle reaches zero at x = 0.677 m, at which point the particle reverses direction since the force is acting in the negative x-direction.

7-22 For the block to stay in contact at the top of the loop, which is the place it loses contact if it is just barely short of the necessary energy,

$$\frac{v^2}{R} \geq g$$

$$\Delta K = \frac{1}{2}m(v^2 - v_o^2) = mg(h - 2R)$$

$$v^2 = 2g(h - 2R)$$

$$\frac{2g(h - 2R)}{R} \geq g$$

$$h \geq \frac{5}{2}R$$

7-23 Placement of the block effects the position of the center of gravity of the cart system. The force of gravity acting on the cart is considered to be acting at the center of gravity of the cart. Thus, a center of gravity positioned further back in the car allows the force of gravity to act over a longer distance.

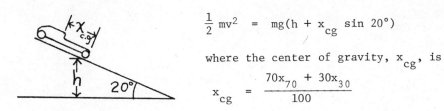

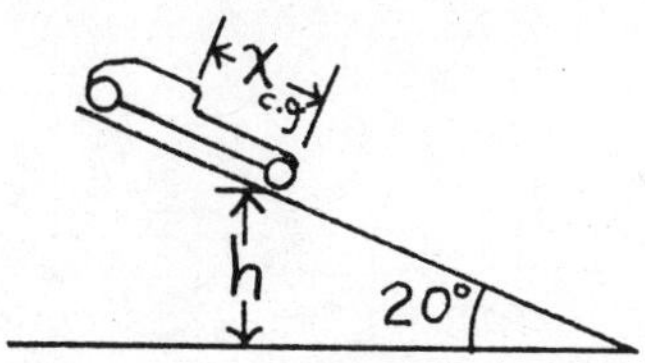

$$\frac{1}{2}mv^2 = mg(h + x_{cg}\sin 20°)$$

where the center of gravity, x_{cg}, is

$$x_{cg} = \frac{70x_{70} + 30x_{30}}{100}$$

x_{70} is the distance of the center of gravity of the 70-lb cart and driver from the front of the cart. x_{30} is the distance from the front of the cart that the 30-lb block is placed. To maximize velocity, x_{30} and x_{70} should be maximized; both block and driver should be as far back on the cart as possible.

When the block is placed in the middle of the cart, $x_{30} = 1$ m.

$$\frac{1}{2}mv^2 = mg\left(h + \frac{70x_{70} + 30\text{ m}}{100}\right) = \frac{1}{2}m\left(\frac{10\text{ m}}{\text{s}}\right)^2$$

$$h + \frac{70x_{70}}{100} = \frac{100}{2g}\frac{\text{m}^2}{\text{s}^2} - \frac{30\text{ m}}{100}$$

$$= 4.8\text{ m}$$

So the expression for velocity can be written

$$v = \sqrt{2g\left(4.8\text{ m} + \frac{30x_{30}}{100}\right)}$$

The maximum value of x_{30} is 2 m, the length of the car. So

$$v_{max} = \sqrt{2g\left(4.8 + \frac{60}{100}\right)\text{m}} = 10.3\text{ m/s}$$

$$v_{min} = v(x_{30} = 0) = 9.7\text{ m/s}$$

7-24

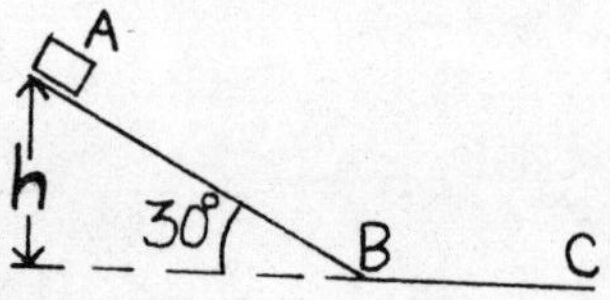

Using Eq. 10-10, in the region between A and B,

$$U(x_B) + \frac{1}{2}mv_B^2 = \frac{1}{2}mv_B^2 = U(x_A) + \frac{1}{2}mv_A^2 + \int_0^{h/\sin 30°} -f\,dx$$

$$\frac{1}{2}mv_B^2 = mgh - \mu_k mg\cos 30°\left(\frac{h}{\sin 30°}\right)$$

In the region between B and C,

$$U(x_C) + \frac{1}{2}mv_C^2 = 0 = U(x_B) + \frac{1}{2}mv_B^2 + \int_0^x -f\,dx$$

$$0 = mgh - \mu_k mgh\cot 30° - \mu_k mgx$$

$$x = \frac{h - \mu_k h\cot 30°}{\mu_k} = 1.96\text{ m}$$

7-25 (a) $\Delta K = \frac{1}{2}m(v^2 - v_o^2) = -\frac{1}{2}mv_o^2 = -160$ J

(b) $\Delta U = mg(3\text{ m})\sin 30° = 73.5$ J

(c) The energy lost to friction is 86.5 J.

$$f = \frac{86.5\text{ J}}{3\text{ m}} = 28.8\text{ N}$$

(d) $f = \mu_k N = \mu_k mg \cos 30° = 28.8 \text{ N}$

$$\mu_k = \frac{28.8 \text{ N}}{(5 \text{ kg})(9.8 \text{ m/s}^2) \cos 30°} = 0.68$$

7-26

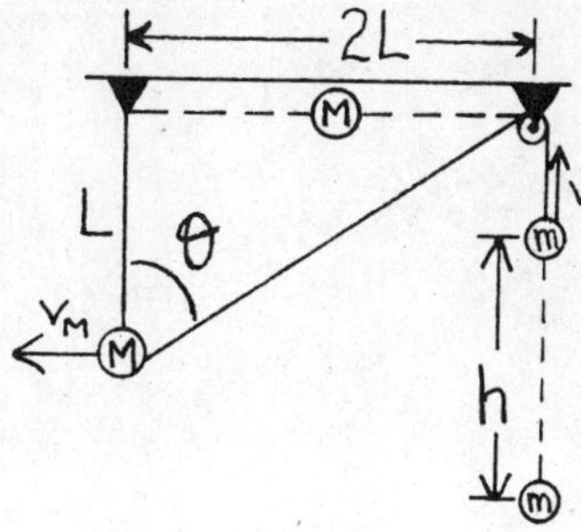

$$\theta = \tan^{-1} \frac{2L}{L} = 63.4°$$

The relationship between the final velocities is

$$v_m = v_M \sin \theta = 0.89\ v_M$$

Assuming the initial configuration has no total energy, the final configuration will likewise have $E = 0$.

$$E = 0 = mgh - MgL + \frac{1}{2} mv_m^2 + \frac{1}{2} Mv_M^2$$

$$= mg(\sqrt{5} - 1)L - MgL + \frac{1}{2} m(0.89\ v_M)^2 + \frac{1}{2} Mv_M^2$$

$$v_M = \sqrt{\frac{2gL[(1 - \sqrt{5})m + M]}{0.89^2\ m + M}} = 2.60 \text{ m/s}$$

7-27

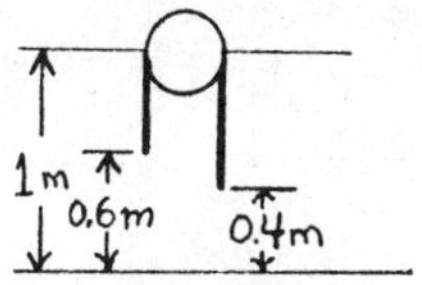

$$dm = \frac{0.2 \text{ kg}}{1 \text{ m}} dx$$

$$U_o = \frac{0.2 \text{ kg}}{1 \text{ m}} \left[g \int_0^{0.40 \text{ m}} 0.60\ dx + g \int_0^{0.60 \text{ m}} 0.40 \text{ m} + x\ dx \right]$$

$$= 0.2g(0.24 + \frac{0.16}{2} + 0.24 + \frac{0.36}{2}) \text{kg} \cdot \text{m}$$

$$= 1.45 \text{ J}$$

$$U_f = \frac{0.2 \text{ kg}}{1 \text{ m}} g \int_0^{1.0 \text{ m}} x\ dx = 0.98 \text{ J}$$

Since there is no friction, the total mechanical energy is constant.

$$E = U_o = U_f + \frac{1}{2} mv^2$$

$$v = \sqrt{\frac{2}{m}(U_o - U_f)} = 2.17 \text{ m/s}$$

7-28 (a)

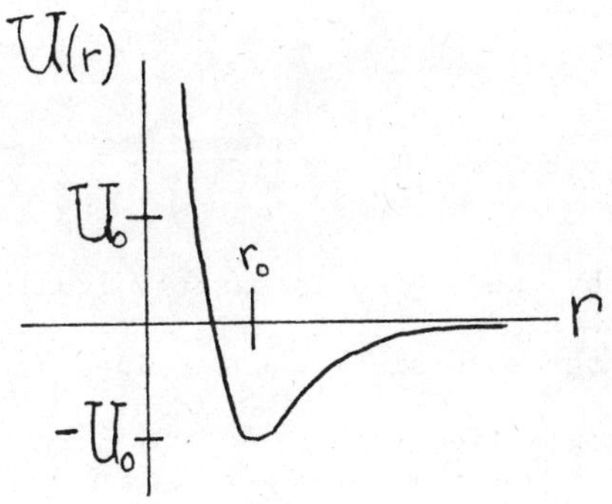

(b) $0 = \frac{dU}{dr} = \frac{12U_o}{r}\left[-\left(\frac{r_o}{r}\right)^{12} + \left(\frac{r_o}{r}\right)^6\right] = \frac{12U_o}{r}\left(\frac{r_o}{r}\right)^6\left[1 - \left(\frac{r_o}{r}\right)^6\right]$

$0 = 1 - \left(\frac{r_o}{r}\right)^6$

$r = r_o$

$U(r = r_o) = U_o\left[\left(\frac{r_o}{r_o}\right)^{12} - 2\left(\frac{r_o}{r_o}\right)^6\right] = -U_o$

(c) $-\frac{1}{4}U_o = U(r) = U_o\left[\left(\frac{r_o}{r}\right)^{12} - 2\left(\frac{r_o}{r}\right)^6\right]$

Let $r' = \left(\frac{r}{r_o}\right)^6$

$-\frac{1}{4} = \left(\frac{1}{r'}\right)^2 - \frac{2}{r'}$

$\frac{r'^2}{4} - 2r' + 1 = 0$

$r' = 4 \pm 2\sqrt{3} = 0.54,\ 7.46$

$\frac{r}{r_o} = (r')^{1/6} = 0.90,\ 1.40$

7-29 $P_M = \frac{\Delta W}{\Delta t} = \frac{mgh + \frac{1}{2}mv^2}{\Delta t}$

$v = \frac{150\ m}{60\ s} = 2.5\ m/s$

$P_M = \frac{(2000\ kg)(9.8\ m/s\ )(150\ m) + \frac{1}{2}(2000\ kg)(2.5\ \frac{m}{s})^2}{60\ s}$

$= 4.91\ kW$

$= 0.35\ P_E$

$P_E = \frac{4.91\ kW}{.35} = 14.0\ kW$

Ignoring the kinetic energy term is reasonable. It is small compared to the potential energy gained by the material.

$P'_M = \frac{(2000\ kg)(9.8\ m/s\ )(150\ m)}{60\ s} = 4.90\ kW$

If the load were hoisted very quickly, v would be much greater and could not be ignored.

7-30 $P = 45\ hp = 33.6\ kW = Fv$

The effective force being delivered by the engine is exactly equal to the component of the force of gravity parallel to the road since the car does not accelerate.

$P = 33.6\ kW = (400\ kg)(9.8\ m/s^2)(\sin\theta)v$

$v = 33\ m/s = 74\ mi/h$

Note that this solution assumes no power being used to overcome internal friction or air resistance.

7-31 $P = F_x v_x = (ma_x)v_x = mv_x \frac{dv_x}{dt} = \frac{d}{dt}(\frac{1}{2} mv_x^2)$

$$\int_0^v d(\frac{1}{2} mv_x^2) = \int_0^t Pdt$$

$$\frac{1}{2} mv^2 = Pt$$

$$\frac{ds}{dt} = v = \sqrt{\frac{2P}{m}}\, t^{1/2}$$

$$\int_0^s ds = \int_0^t \sqrt{\frac{2P}{m}}\, t^{1/2}$$

$$s = \sqrt{\frac{2P}{m}}\, \frac{2}{3}\, t^{3/2} = \sqrt{\frac{8Pt^3}{9}}$$

7-32

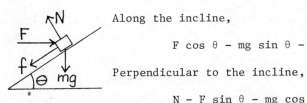

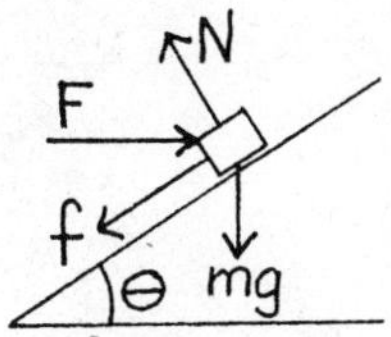

Along the incline,

$$F \cos\theta - mg \sin\theta - f = ma$$

Perpendicular to the incline,

$$N - F \sin\theta - mg \cos\theta = 0$$

$$f = \mu_k N = (0.20)(F \sin\theta + mg \cos\theta) = 12.6 \text{ N}$$

$$(5 \text{ kg})\, a = (50 \text{ N}) \cos 20° - (5 \text{ kg})(9.8 \text{ m/s}^2)\sin 20° - 12.6 \text{ N}$$

$$a = 3.53 \text{ m/s}^2$$

$$v_f^2 - v_0^2 = v_f^2 = 2as = 2(3.53 \text{ m/s}^2)(0.5 \text{ m})$$

$$v(s = 0.5 \text{ m}) = 1.88 \text{ m/s}$$

$$P_f(s = 0.5 \text{ m}) = fv = (12.6 \text{ N})(1.88 \text{ m/s}) = 23.7 \text{ W}$$

From Problem 7-31,

$$P = \frac{d}{dt}(\frac{1}{2} mv_x^2) = Fv$$

$$\frac{dK}{dt} = Fv = (F \cos\theta - mg \sin\theta - f)v$$

$$= 33.1 \text{ W}$$

7-33 $W_s = \int_0^{-0.25 \text{ m}} (2 \text{ kg})\, \frac{4}{5}g\, dx = -3.92 \text{ J}$

$$W_g = -\int_0^{-0.25 \text{ m}} (2 \text{ kg})g\, dx = 4.90 \text{ J}$$

$$\frac{1}{2} mv^2 = W_{total} = 0.98 \text{ J}$$

$$v = \sqrt{\frac{2(0.98 \text{ J})}{2 \text{ kg}}} = 0.99 \text{ m/s}$$

7-34 $W = \Delta U = mg\bar{h} = (50 \text{ kg})(9.8 \text{ m/s}^2)(1.5 \text{ m}) = 735 \text{ J}$

7-35 The block will be raised by

$$h = \frac{\ell}{\cos\alpha_f} - \frac{\ell}{\cos\alpha_o} = 2.6\ \text{m}$$

$$W = \int_0^{2.6\ \text{m}} (3\ \text{kg})(9.8\ \text{m/s}^2)dx = 76.3\ \text{J}$$

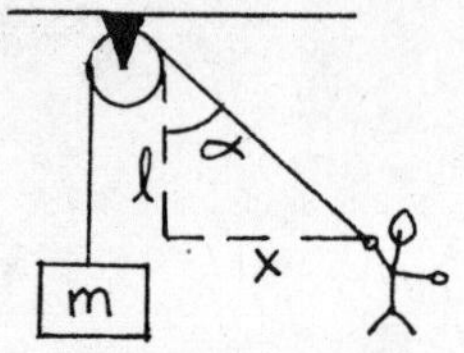

$$W = \int_{x_{30°}}^{x_{45°}} F\,dx = \int_{x_{30°}}^{x_{45°}} -mg\ \sin\alpha\ dx$$

$$x = \frac{\ell}{\cos\alpha}\sin\alpha = \ell\tan\alpha$$

$$dx = \ell\sec^2\alpha\ d\alpha$$

$$W = \int_{30°}^{45°} mg\ell\frac{\sin\alpha}{\cos^2\alpha}d\alpha = \frac{-mg\ell}{\cos\alpha}\Big]_{30°}^{45°} = 76.3\ \text{J}$$

7-36

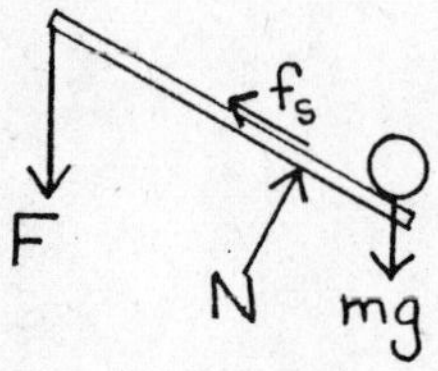

$$W_{student} = F\Delta s = F(3h)$$

$$W_{total} = \Delta U = Mgh$$

As the pivot point does not move, the forces at the pivot point do no work.

The only work done is that of the student.

$$3Fh = Mgh$$

$$F = \frac{Mg}{3}$$

7-37 $v_{ox} = v_{oy} = v_o(\frac{\sqrt{2}}{2})$

The time to reach maximum height is

$$t = \frac{v_{oy}}{g} = \frac{\sqrt{2}}{2}\frac{v_o}{g}$$

The time to reach ground is twice that to reach maximum height.

$$t = \sqrt{2}\frac{v_o}{g}$$

$$x = v_o(\frac{\sqrt{2}}{2})(\sqrt{2}\frac{v_o}{g}) = \frac{v_o^2}{g}$$

$$v_o = \sqrt{gx}$$

$$v_{0,22} = 14.7\ \text{m/s} \qquad K_{22} = \frac{1}{2}(7.3\ \text{kg})(14.7\ \text{m/s})^2 = 789\ \text{J}$$

$$v_{0,71} = 26.4\ \text{m/s} \qquad K_{71} = \frac{1}{2}(2.0\ \text{kg})(26.4\ \text{m/s})^2 = 697\ \text{J}$$

$$v_{0,95} = 30.5\ \text{m/s} \qquad K_{95} = \frac{1}{2}(0.80\ \text{kg})(30.5\ \text{m/s})^2 = 372\ \text{J}$$

The greater the mass of the object, the greater the kinetic energy imparted to the object. The efficiency of the muscles seems to depend on the mass of the object thrown.

7-38 $U_o = 0$

$$U_f = (80\text{ kg})(0.01\text{ m})g + (40\text{ kg})\left(\frac{0.02\text{ m} - 0\text{ m}}{2}\right)g$$

$$= 11.76\text{ J}$$

$$W = Fd = T(0.01\text{ m}) + T(0.02\text{ m}) = T(0.03\text{ m})$$

$$T = \frac{11.76\text{ J}}{0.03\text{ m}} = 392\text{ N}$$

If the tension necessary to raise the plank off the ground is greater than the student's weight, the student will not be able to raise it.

$$T_{max} = (80\text{ kg})g$$

Let M be the mass of the plank. Let x be the distance the center of the plank is raised.

$$U_f = (80 + M)xg$$

$$W = Fd = T(3x) = (80\text{ kg})(g)(3x)$$

$$(80\text{ kg} + M)xg = (80\text{ kg})g(3x)$$

$$M = 160\text{ kg}$$

7-39 $dm = \frac{0.5\text{ kg}}{1\text{ m}}dx$

$$W = \int_0^{1/3\text{ m}} \frac{0.5\text{ kg}}{\text{m}}gx\,dx = \frac{(0.5\text{ kg})}{1\text{ m}}(9.8\text{ m/s}^2)\frac{1}{2}\left(\frac{1}{3}\text{ m}\right)^2 = 0.27\text{ J}$$

7-40 Since potential energy is undefined to within an additive constant, let U(table level) = 0.

$$U_o = -\int_0^{-1/3\text{ m}} \frac{0.5\text{ kg}}{\text{m}}gx\,dx = -0.27\text{ J}$$

$$U_f = -\int_0^{-1\text{ m}} \frac{0.5\text{ kg}}{\text{m}}gx\,dx = -2.45\text{ J}$$

$$U_f + \frac{1}{2}mv^2 = U_o$$

$$v = \sqrt{\frac{2}{m}(U_o - U_f)} = 2.95\text{ m/s}$$

7-41 The block loses contact with the sphere when the normal force between the block and sphere goes to zero.

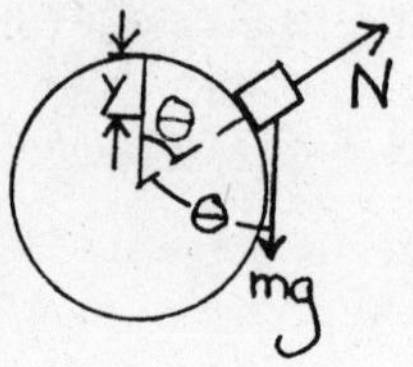

$$N - mg\cos\theta = -\frac{mv^2}{R}$$

So, when the block loses contact, $g\cos\theta = \frac{v^2}{R}$

But $y = R - R\cos\theta = R(1 - \cos\theta)$

$$U_o = mgy = \frac{1}{2}mv^2$$

$$g\cos\theta = \frac{2gy}{y}(1 - \cos\theta)$$

$$\theta = 48°$$

$$y = R(1 - \cos 48°) = \frac{1}{3}R$$

7-42 The minimum velocity the ball may have at the point directly above the peg satisfies the equation

$$m\frac{v_A^2}{\ell-d} = mg$$

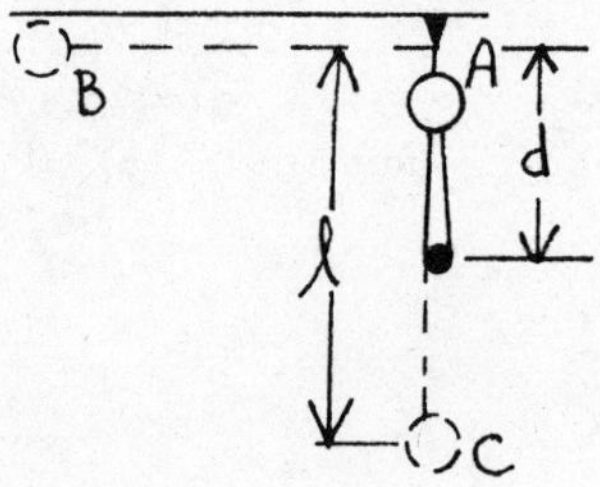

So $v_A^2 = g(\ell-d)$

In order to have the velocity v_A, the ball must satisfy the following conservation of energy requirement:

$$\frac{1}{2}mv_B^2 + U_B = 0 = \frac{1}{2}mv_A^2 - mg[\ell-2(\ell-d)]$$

$$v_A^2 = 2g(2d-\ell)$$

$$2g(2d-\ell) = g(\ell-d)$$

$$d = \frac{3\ell}{5}$$

7-43

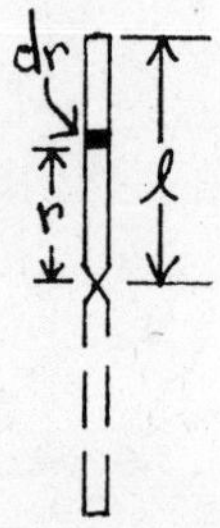

$$K_f = U_o - U_f = mg\ell$$

$$dm = \frac{m}{\ell}dr$$

$$v_{dm} = \omega r$$

$$K_{dm} = \frac{1}{2}dm\, v_{dm}^2 = \frac{m\omega^2}{2\ell}r^2dr$$

$$K_f = \int_o^\ell \frac{m\omega^2}{2\ell}r^2dr = \frac{m\omega^2}{2\ell}\frac{\ell^3}{3} = \frac{m\omega^2\ell^2}{6}$$

So,

$$mg\ell = \frac{m\omega^2\ell^2}{6}$$

$$\omega = \sqrt{\frac{6g}{\ell}}$$

The velocity at the bottom of the swing is

$$v = \omega\ell = \sqrt{6g\ell}$$

For any segment dr of the rod, Newton's second law can be written

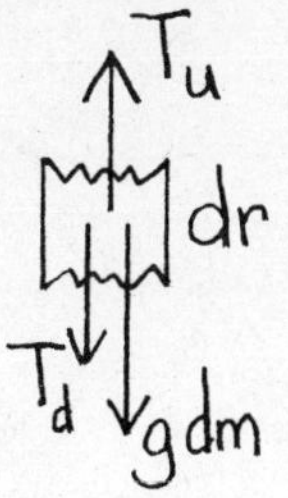

$$a\,dm = \frac{\omega^2 r^2}{r}dm = T_u - T_d - g\,dm$$

Integrating over the length of the rod,

$$\int_0^{\ell} T_u - T_d = T = \int_0^{\ell} \omega^2 r(\frac{m}{\ell}\,dr) + \int_0^{\ell} g(\frac{m}{\ell}\,dr)$$

$$T = \frac{m\omega^2}{\ell}\,(\frac{\ell^2}{2}) + \frac{mg}{\ell}(\ell)$$

$$T = m(\frac{6g}{\ell})\,\frac{\ell}{2} + mg = 4\ mg$$

7-44 Let $E_{top} = \frac{1}{2}\,mv^2_{top}$

Then $E_{bot} = \frac{1}{2}\,mv^2_{bot} - 2mgR$

Since energy is a constant,

$$v^2_{top} = v^2_{bot} - 4gR$$

From Newton's second law,

$$T_{top} + mg = \frac{mv^2_{top}}{R} = \frac{m}{R}(v^2_{bot} - 4gR)$$

$$T_{bot} - mg = \frac{mv^2_{bot}}{R}$$

$$T_{bot} - T_{top} = 2\ mg + 4\ mg = 6\ mg$$

7-45

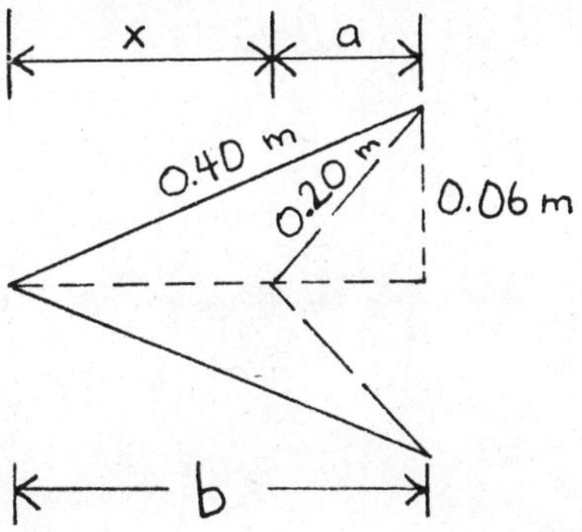

$$U_o = Fx = (50\ N)x$$

$$x = a - b$$

$$a^2 + (0.06\ m)^2 = (0.4\ m)^2 \qquad a = 0.40\ m$$

$$b^2 + (0.06\ m)^2 = (0.2\ m)^2 \qquad b = 0.19\ m$$

$$U_o = (50\ N)(0.21\ m) = 10.5\ J$$

$$E = U_o = \frac{1}{2}\,mv^2$$

$$v = \sqrt{\frac{2U_o}{m}} = \sqrt{\frac{2(10.5\ J)}{0.1\ kg}} = 14.5\ m/s$$

7-46

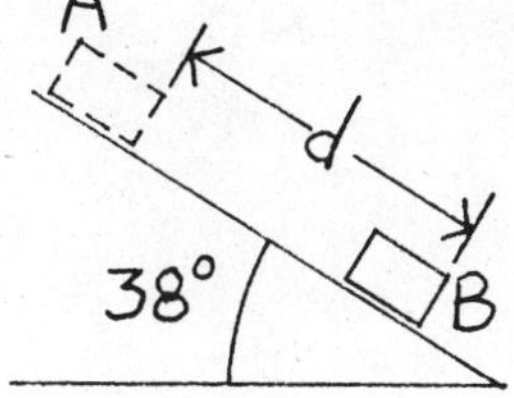

$$\frac{1}{2}\,mv^2 = mgd\,\sin 38° + \int_0^d -\mu_k mg\,\cos 38°\,dx$$

$$\mu_k = \frac{\frac{1}{2}\,mv^2 - mgd\,\sin 38°}{-mgd\,\cos 38°}$$

$$= \frac{-\frac{1}{2}v^2 + gd\,\sin 38°}{gd\,\cos 38°} = 0.38$$

7-47

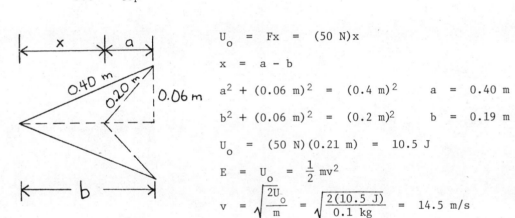

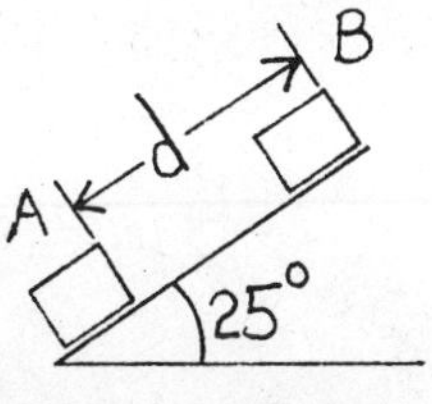

$$U_B = mgd\,\sin 25° = \frac{1}{2}\,mv_o^2 - \int_0^d \mu_k mg\,\cos 25°\,dx$$

$$mgd\,\sin 25° = \frac{1}{2}\,mv_o^2 - \mu_k mgd\,\cos 25°$$

$$d = \frac{v_0^2}{2(g \sin 25° + \mu_k g \cos 25°)} = 0.49 \text{ m}$$

The block will not slide down the plane if

$$mg \sin 25° < \mu_s mg \cos 25°$$

$$\tan 25° < \mu_s$$

$$0.47 < 0.50$$

The block will not slide back down the plane.

7-48

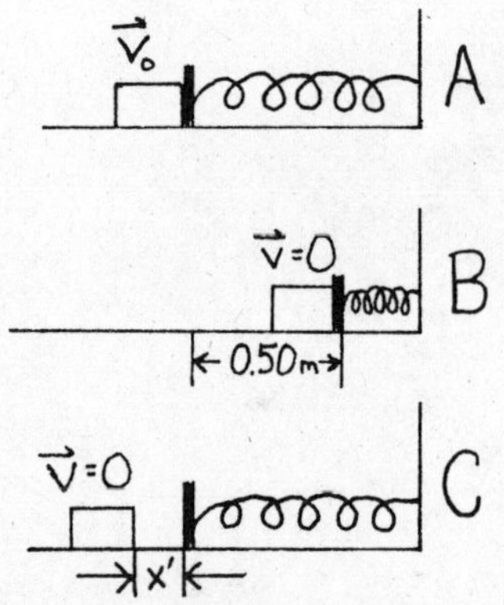

$$U_B = \frac{1}{2}\kappa x^2 = \frac{1}{2}mv_0^2 - \int_0^x f\,dx$$

$$v_0^2 = \frac{\kappa x^2 + 2\mu_k mgx}{m}$$

$$v_0 = 2.68 \text{ m/s}$$

$$U_c = 0 = U_B + \int_0^{-(x'+0.5)m} f\,dx$$

$$0 = \frac{1}{2}\kappa(0.5 \text{ m})^2 - \mu_k mg(x' + 0.5 \text{ m})$$

$$x' = \frac{\frac{1}{2}\kappa(0.5 \text{ m})^2}{\mu_k mg} - 0.5 \text{ m}$$

$$= 0.05 \text{ m}$$

7-49 $F = \frac{dU}{dx} = 2\alpha x + 3\beta x^2$

$$F = 0 = 2x - \frac{3x^2}{2}$$

$$2x = \frac{3}{2}x^2$$

$$x = 0, \frac{4}{3} \text{ m}$$

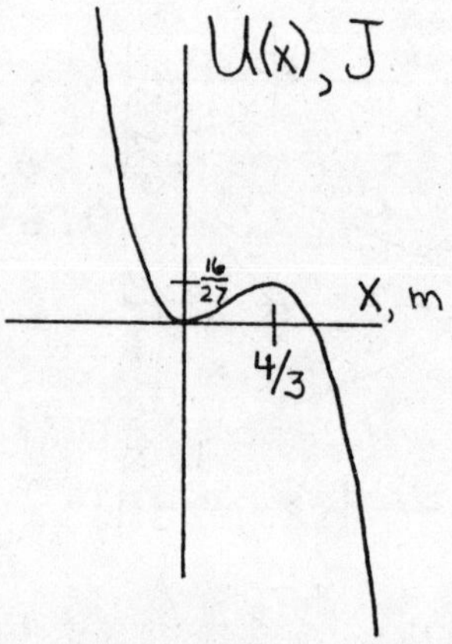

At $x = 0$ the equilibrium is stable.

At $x = 4/3$ m it is not stable.

$$U(x = \tfrac{4}{3}) = (\tfrac{4}{3}\text{ m})^2(\tfrac{1J}{m^2} - \tfrac{x}{2}\tfrac{J}{m}) = \frac{16}{27} \text{ J}$$

The minimum total energy a particle must have to escape from the potential well is $\frac{16}{27}$ J.

7-50 $U(r) = U_0\left[\left(\frac{r_0}{r}\right)^5 - a\left(\frac{r_0}{r}\right)^3\right]$

$$\frac{dU}{dr} = \frac{U_0}{r}\left[3a\left(\frac{r_0}{r}\right)^3 - 5\left(\frac{r_0}{r}\right)^5\right]$$

Set $r = r_0$. Set $\frac{dU}{dr} = 0$.

$$0 = \frac{U_0}{r}(3a - 5)$$

$$a = \frac{5}{3}$$

$$U(r_o) = U_o(1 - (\frac{5}{3})) = -\frac{2}{3} U_o$$

$$U(r) = -\frac{1}{4} U_o = U_o\left[\left(\frac{r_o}{r}\right)^5 - \frac{5}{3}\left(\frac{r_o}{r}\right)^3\right]$$

Let x_1 be the larger $\frac{r_o}{r}$ value. Let x_2 be the smaller $\frac{r_o}{r}$ value.

To determine x_1,

$$-\frac{1}{4} = x_1^5 - \frac{5}{3} x_1^3$$

$$x_1^5 = \frac{5}{3} x_1^3 - \frac{1}{4}$$

To determine x_2,

$$-\frac{1}{4} = x_2^5 - \frac{5}{3} x_2^3$$

$$\frac{5}{3} x_2^3 = x_2^5 + \frac{1}{4}$$

To find x_1, guess that $x_1 = 10$

old x_1	$(\frac{5}{3} x_1^3 - \frac{1}{4})$	new x_1	$\left\lvert\frac{\text{old } x_1 - \text{new } x_1}{\text{new } x_1}\right\rvert$
10.000	1666.4	4.409	1.27
4.409	142.6	2.697	0.63
2.697	32.45	2.006	0.34
2.006	13.19	1.675	0.20
1.675	7.586	1.500	0.12
1.500	5.371	1.400	0.07
1.400	4.320	1.340	0.04

To find x_2, guess that $x_2 = 0.1$

old x_2	$\frac{3}{5}(x_2^5 + \frac{1}{4})$	new x_2	$\left\lvert\frac{\text{old } x_2 - \text{new } x_2}{\text{new } x_2}\right\rvert$
0.1000	0.1500	0.5313	0.81
0.5313	0.1754	0.5598	0.05
0.5598	0.1830	0.5677	0.01

IV-1 (a) $\frac{\partial w}{\partial x} = \frac{\partial}{\partial x}(6x^2y) + \frac{\partial}{\partial x}(2x) + \frac{\partial}{\partial x}(-7y^2z) = 12xy + 2 + 0 = 12xy + 2$

$\frac{\partial w}{\partial y} = \frac{\partial}{\partial y}(6x^2y) + \frac{\partial}{\partial y}(2x) + \frac{\partial}{\partial y}(-7y^2z) = 6x^2 + 0 - 14yz = 6x^2 - 14yz$

$\frac{\partial w}{\partial z} = \frac{\partial}{\partial z}(6x^2y) + \frac{\partial}{\partial z}(2x) + \frac{\partial}{\partial^2}(-7y^2z) = 0 + 0 - 7y^2 = -7y^2$

(b) $\frac{\partial f}{\partial y} = \frac{(x-y)-y(-1)}{(x-y)^2} = \frac{x}{(x-y)^2}$

$\frac{\partial f}{\partial x} = y\frac{\partial}{\partial x}(\frac{1}{x-y}) = y\frac{-1}{(x-y)^2} = \frac{-y}{(x-y)^2}$

(c) $\frac{\partial \rho}{\partial \theta} = \cos\phi\frac{\partial}{\partial \theta}(\sin\theta) = \cos\theta\cos\phi$

$\frac{\partial \rho}{\partial \phi} = \sin\theta\frac{\partial}{\partial \phi}(\cos\phi) = -\sin\theta\sin\phi$

IV-2 $\frac{\partial f}{\partial x} = 3x^2y - y^3$ $\qquad \frac{\partial f}{\partial y} = x^3 - 3xy^2$

$\frac{\partial^2 f}{\partial x^2} = \frac{\partial}{\partial x}(\frac{\partial f}{\partial x}) = 6xy$ $\qquad \frac{\partial^2 f}{\partial y^2} = \frac{\partial}{\partial y}(\frac{\partial f}{\partial y}) = -6xy$

$\frac{\partial^2 f}{\partial x^2} + \frac{\partial^2 f}{\partial y^2} = 6xy - 6xy = 0$

IV-3 $\frac{\partial^2 g}{\partial x\partial z} = \frac{\partial}{\partial x}(\frac{\partial g}{\partial z}) = \frac{\partial}{\partial x}(\frac{1}{x} - 2y^4) = -\frac{1}{x^2}$

$\frac{\partial^2 g}{\partial z\partial x} = \frac{\partial}{\partial z}(\frac{\partial g}{\partial x}) = \frac{\partial}{\partial z}(18x^2y^2 - \frac{z}{x^2}) = -\frac{1}{x^2}$

IV-4 $\frac{\partial r}{\partial x} = \frac{\partial}{\partial x}(x^2 + y^2 + z^2)^{\frac{1}{2}} = \frac{1}{2}(x^2 + y^2 + z^2)^{-\frac{1}{2}}(2x) = \frac{x}{r}$

$\frac{\partial r}{\partial y} = \frac{\partial}{\partial y}(x^2 + y^2 + z^2)^{\frac{1}{2}} = \frac{1}{2}(x^2 + y^2 + z^2)^{-\frac{1}{2}}(2y) = \frac{y}{r}$

$\frac{\partial r}{\partial z} = \frac{\partial}{\partial z}(x^2 + y^2 + z^2)^{\frac{1}{2}} = \frac{1}{2}(x^2 + y^2 + z^2)^{-\frac{1}{2}}(2z) = \frac{z}{r}$

IV-5 $\frac{\partial u}{\partial x} = \frac{\partial u}{\partial r}\frac{\partial r}{\partial x} = (ar)(\frac{x}{r}) = ax$

$\frac{\partial u}{\partial y} = \frac{\partial u}{\partial r}\frac{\partial r}{\partial y} = (ar)(\frac{y}{r}) = ay$

$\frac{\partial u}{\partial z} = \frac{\partial u}{\partial r}\frac{\partial r}{\partial z} = (ar)(\frac{z}{r}) = az$

IV-6 (a) $du = \frac{\partial u}{\partial x}dx + \frac{\partial u}{\partial y}dy + \frac{\partial u}{\partial z}dz$

$= 6y\,dx + (6x + 2y + 2z)dy + 2y\,dz$

(b) $du = \frac{\partial u}{\partial \rho}d\rho + \frac{\partial u}{\partial \phi}d\phi$

$= -\alpha e^{-\alpha\rho}(\cos\phi)d\rho - e^{-\alpha\rho}\sin\phi$

(c) $dw = \frac{\partial w}{\partial \alpha}d\alpha + \frac{\partial w}{\partial \beta}d\beta$

$= \cos(\alpha-\beta)d\alpha - \cos(\alpha-\beta)d\beta$

IV-7 $\Delta V = (x \pm \Delta x)(y \pm \Delta y)(z \pm \Delta z) - V$

$= \pm xy\Delta z \pm x\Delta yz \pm \Delta xyz \pm \Delta x\Delta yz \pm \Delta xy\Delta z \pm x\Delta y\Delta z + \Delta x\Delta y\Delta z$

$\Delta V_{max} = 0.07 + 0.10 + 0.23 + 0.002 + 0.001 + 0.0005 + 0.000008$

$= 0.404\ m^3$

$$\frac{\Delta V_{max}}{V} = \frac{0.404}{(1.21)(2.87)(4.02)} = 2.9\%$$

Using the differential approximation,

$$dV = \frac{\partial V}{\partial x} dx + \frac{\partial V}{\partial y} dy + \frac{\partial V}{\partial z} dz$$

$= yz\ dx + xz\ dy + xy\ dz$

$= 0.23 + 0.10 + 0.07\ m^3$

$= 0.40\ m^3$

The differential approximation is accurate to 1%.

IV-8 $V = \pi r^2 h$

$$dV = \frac{\partial v}{\partial r} dr + \frac{\partial v}{\partial h} dh$$

$= (2\pi rh)dr + (\pi r^2)dh$

$= 2\pi(3.5)(12)(0.05) + \pi(3.5)^2(0.1)\ cm^3$

$= 17.0\ cm^3$

8-1

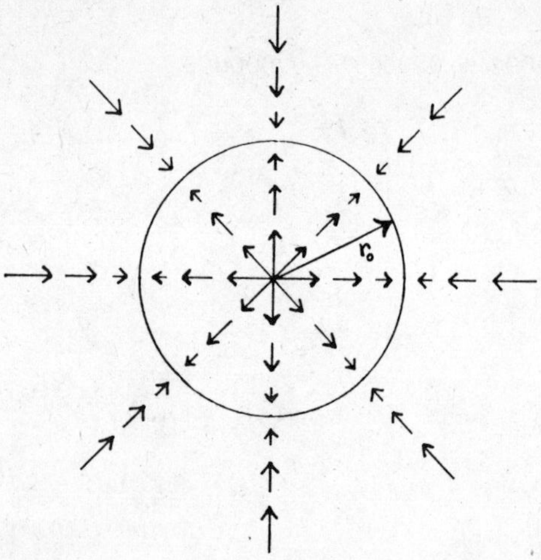

8-2 $\vec{F} = 3Axy\hat{i} + 3Ax^2\hat{j} + \frac{1}{8}Axz\hat{k}$ with $A = \frac{10}{3}\frac{N}{m^2}$

$$\oint_P^Q \vec{F}\cdot d\vec{s} = 3A\oint_P^Q xy\,dx + 3A\oint_P^Q x^2 dy + \frac{A}{8}\oint_P^Q xz\,dz$$

$$= 3Aa^3\int_{t=0}^{t=2}\frac{1}{2}t(t+1)\frac{1}{2}\,dt + 3Aa^3\int_{t=0}^{t=2}(\frac{1}{2}t)^2 dt + \frac{Aa^3}{8}\int_{t=0}^{t=2}\frac{1}{2}t(2t)2\,dt$$

$$= \frac{3Aa^3}{4}(\frac{t^3}{3} + \frac{t^2}{2})\Big]_0^2 + \frac{3Aa^3}{4}\frac{t^3}{3}\Big]_0^2 + \frac{Aa^3}{4}\frac{t^3}{3}\Big]_0^2$$

$$= \frac{37}{6}Aa^3 = 20\tfrac{5}{9}\text{ J}$$

The closed-path integral of $\vec{F}\cdot d\vec{s}$ does not generally equal zero since the value of the integral is path dependent. This path dependence can be seen by comparing the result of this problem with that of Example 8-1.

8-3 From Example 8-2,

$$\oint \vec{F}\cdot d\vec{s} = \int_{(0,1,0)}^{(0,2,0)} F_y dy + \int_{(0,2,0)}^{(0,2,3)} F_z dz + \int_{(0,2,3)}^{(2,2,3)} F_x dx$$

$$= \int_{(0,1,0)}^{(0,2,0)} a^2Ax\,dy + \int_{(0,2,0)}^{(0,2,3)} 0\,dz + \int_{(0,2,3)}^{(2,2,3)} -a^2Ay\,dx$$

$$= a^2A(0)y\Big]_1^2 + 0 - a^2A(2)x\Big]_0^2 = 0 + 0 - 4Aa^2$$

$$= (-4\text{ m}^2)A$$

It has already been shown that this line integral is path-dependent. This result supports that fact since it does not equal the value of any of the integrals evaluated in the example. From Example 8-3,

$$\oint \vec{F}\cdot d\vec{s} = \int_{(0,1,0)}^{(0,2,0)} a^2Ax\ dy + \int_{(0,2,0)}^{(0,2,3)} 0\ dz + \int_{(0,2,3)}^{(2,2,3)} a^2Ay\ dx$$

$$= a^2A(0)\Big]_1^2 + 0 + a^2A(2)x\Big]_0^2 = 0 + 0 + 4Aa^2$$

$$= (4\ m^2)A$$

This result is the same as that of the three other paths used in the example. This suggests the work done by the force is <u>not</u> path-dependent; if so, this force is a conservative force.

8-4 $\oint (F_x dx + F_y dy) = A\oint y\ dx + B\oint x^2 dy$

$$= A\int_0^1 a(t^2 - 1)2a\ dt + B\int_0^1 (2\ at)^2 2at\ dt$$

$$= 2Aa^2(\frac{t^3}{3} - t)\Big|_0^1 + 8\ Ba^3\ \frac{t^4}{4}\Big|_0^1$$

$$= 2(\frac{1}{3} - 1) + 8(\frac{1}{4}) = \frac{2}{3}\ J$$

8-5 $\oint_P^Q (F_x dx + F_y dy) = A\oint_P^Q (x + y)dx + B\oint_P^Q x^2y\ dy$

The path is given by

$$y = ax^2 + b$$

Knowing it passes through P = (1,0) and Q = (2,2), <u>a</u> and <u>b</u> can be evaluated.

Using P,

$$0 = a + b$$

Using Q,

$$2 = 4a + b$$

Solving for <u>a</u> and <u>b</u>,

$$a = -b = \frac{2}{3}$$

$$\oint_P^Q (F_x dx + F_y dy) = A\int_1^2 (x + \frac{2}{3}x^2 - \frac{2}{3})dx + B\int_0^2 (\frac{3y}{2} + 1)y\ dy$$

$$= A(\frac{x^2}{2} + \frac{2}{9}x^3 - \frac{2}{3}x)\Big]_1^2 + B(\frac{y^3}{2} + \frac{y^2}{2})\Big]_0^2$$

$$= 7\ \frac{7}{36}\ J$$

8-6 $\oint_P^Q (F_x dx + F_y dy) = \int_{(1,0)}^{(2,0)} Ax\ dx + \int_{(2,0)}^{(2,2)} B(2)^2y\ dy$

$$= A\ \frac{x^2}{2}\Big|_1^2 + 4B\ \frac{y^2}{2}\Big|_0^2 = A(\frac{3}{2}) + 4B(2)$$

$$= 8\frac{3}{4}\ J$$

8-7 $A_1\oint_P^Q xyz\ dx + A_2\oint_P^Q (9x^2 - y^2)dy + A_3\oint_P^Q (3x - y - 3z)dz$

$$= A_1\oint_0^2 (at)a(3t - 2)(\frac{1}{2}at^2)a\ dt + A_2\int_0^2 [9(at)^2 - a^2(3t - 2)^2]3a\ dt$$

$$+A_3\int_0^2 [3at - a(3t - 2) - \frac{3}{2}at^2]at\,dt$$

$$= \frac{1}{2}A_1 a^4\left(\frac{3t^5}{5} - \frac{t^4}{2}\right)\Big]_0^2 + 3A_2 a^3\left(3t^3 - \frac{1}{9}(3t - 2)^3\right)\Big]_0^2 + A_3 a^2\left(t^2 - \frac{3t^4}{8}\right)\Big|_0^2$$

$$= 60\text{ J}$$

The Cartesian coordinates of the endpoints are

$P(t = 0) = (x(0), y(0), z(0)) = (0, -2a, 0)$

$Q(t = 2) = (2a, 4a, 2a)$

To find the projection onto the x-y plane, use the following equations.

$$x = at \qquad\qquad y = a(3t - 2)$$

Solve for $y(x)$.

$$y = [3(\frac{x}{a}) - 2] = 3x - 2a$$

This is an equation for a straight line.

In the x-z plane,

$$x = at \qquad\qquad z = \frac{1}{2}at^2$$

So, $\quad z = \frac{1}{2}a(\frac{x}{a})^2 = \frac{1}{2a}x^2$

This projection forms a parabola.

In the y-z plane,

$$y = a(3t - 2) \qquad z = \frac{1}{2}at^2$$

So, $\quad z = \frac{1}{2}a(\frac{y+2a}{3a})^2 = \frac{1}{18a}(y + 2a)^2$

This is also a parabola.

8-8 $\int_{(0,0)}^{(1,0)}(F_x dx + F_y dy) + \int_{(1,0)}^{(1,1)}(F_x dx + F_y dy) + \int_{(1,1)}^{(0,1)}(F_x dx + F_y dy) + \int_{(0,1)}^{(0,0)}(F_x dx + F_y dy)$

$$= A\int_0^1 x(0)dx + A\int_0^1 (1^2 - y^2)dy + A\int_1^0 x(1)dx + A\int_1^0 (0^2 - y^2)dy$$

$$= 0 + A(y - \frac{y^3}{3})\Big]_0^1 + A\frac{x^2}{2}\Big]_1^0 - A\frac{y^3}{3}\Big]_1^0$$

$$= A(\frac{2}{3} - \frac{1}{2} + \frac{1}{3}) = \frac{1}{2}A$$

8-9 Assuming there is not tangential acceleration,

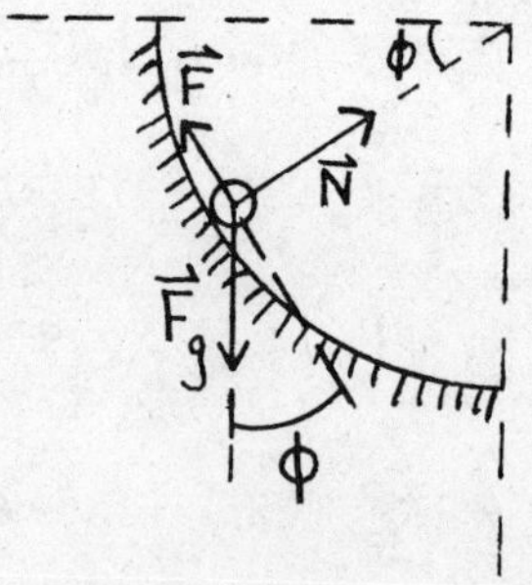

$$mg\sin\phi - N = -mR\omega^2$$

$$mg\cos\phi - F = 0$$

$$W_F(P\to Q) = \oint \vec{F}\cdot d\vec{s} = -\int_0^{\frac{\pi}{2}} mg\cos\phi\, R\, d\phi$$

$$= -mgR\sin\phi\Big|_0^{\pi/2} = -mgR$$

8-10 $F_x = 3Axy$ $\qquad F_y = 3Ax^2$ $\qquad F_z = \frac{A}{8} xz$

$\frac{\partial F_x}{\partial y} = 3Ax$ $\qquad \frac{\partial F_y}{\partial x} = 6A_x$ $\qquad \frac{\partial F_z}{\partial x} = \frac{A}{8} z$

$\frac{\partial F_x}{z} = 0$ $\qquad \frac{\partial F_y}{z} = 0$ $\qquad \frac{\partial F_z}{y} = 0$

Since $\frac{\partial F_x}{y} \neq \frac{\partial F_y}{x}$, the field is not conservative.

8-11 $\vec{F}_{total} = \vec{F}_1 + \vec{F}_2 + \vec{F}_3 + \vec{F}_4$

$$= -\kappa x\hat{i} - \kappa x\hat{i} - \kappa y\hat{j} - \kappa y\hat{j}$$

$$= -2\kappa x\hat{i} - 2\kappa y\hat{j}$$

$$\frac{\partial F_x}{\partial y} = 0 \qquad \frac{\partial F_y}{\partial x} = 0$$

The force is conservative.

$$U(x,y) = -\int (F_x dx + Fydy)$$

$$= 2\int \kappa x\, dx + 2\int \kappa y\, dy$$

$$= \kappa(x^2 + y^2) + \text{constant}$$

To transform $\vec{F}$ and U into polar coordinates, use

$$x = \rho \cos \phi$$

$$y = \rho \sin \phi$$

$$\hat{i} = \hat{u}_\rho \cos \phi - \rho\hat{u}_\phi \sin \phi$$

$$\hat{j} = \hat{u}_\rho \sin \phi - \rho\hat{u}_\phi \cos \phi$$

$$F = -2\kappa\rho \cos \phi(\hat{u}_\rho \cos \phi - \hat{u}_\phi \sin \phi) - 2\kappa\rho \sin \phi(\hat{u}_\rho \sin \phi + \hat{u}_\phi \cos \phi)$$

$$= -2\kappa\rho\hat{u}_\rho$$

$$U = \kappa(\rho^2 \cos^2 \phi + \rho^2 \sin^2 \phi) + \text{constant}$$

$$= \kappa\rho^2 + \text{constant}$$

8-12 $U(x,y) = -\int (F_x dx + F\, dy)$

$$= \int \kappa x\, dx + \int \kappa y\, dy$$

$$= \frac{\kappa}{2}(x^2 + y^2) + \text{constant}$$

Taking the constant to be equal to zero,

$$U(x,y) = \frac{\kappa}{2}(x^2 + y^2)$$

$$N(x,y) = +\gamma(x^2 + y^2)^{\frac{1}{2}}$$

$$U_o + \frac{1}{2} mv_o^2 = U_f + \frac{1}{2} mv_f^2$$

At the farthest excursion, $v_f = 0.$

$$\frac{1}{2} mv_o^2 = U_f = \frac{\kappa}{2}(x^2 + y^2) = \frac{\kappa}{2} r^2$$

So,

$$r = v_o\sqrt{\frac{m}{\kappa}}$$

8-13 $U_f = \frac{1}{2} mv_0^2 - u_k\gamma\int_0^r r\,dr = \frac{1}{2} mv_0^2 - \frac{1}{2} u_k\gamma r^2$

$$(\kappa + u_k\gamma)r^2 = mv_0^2$$

$$r = v_0\sqrt{\frac{m}{\kappa + u_k\gamma}}$$

8-14 (a) $W = \oint_P^Q F_x dx + \oint_P^Q F_y dy$

$$= B\int_2^3 a(t - 2)a^2(t^2 - 2t)^2 a\,dt + B\int_2^3 a^2(t - 2)^2 a(t^2 - 2t)a(2t - 2)dt$$

$$= Ba^4\int_2^3 (3t^5 - 20t^4 + 48t^3 - 48t^2 + 16t)dt$$

$$= Ba^4(\frac{t^6}{2} - 4t^5 + 12t^4 - 16t^3 + 8t^2)\Big|_2^3$$

$$= Ba^4(\frac{9}{2} - 0) = -\frac{9}{2}\text{ J}$$

(b) First require that the endpoints satisfy the given equation for y.

$$y_P = 0 = m(0) + b$$

$$y_Q = 3 = m(1) + b$$

This gives the following path equation.

$$y = 3x$$

$$W = \oint_P^Q (F_x dx + F_y dy)$$

$$= \int_0^3 [(B\frac{y^3}{3})\frac{1}{3}dy + (B\frac{y^3}{9})dy]$$

$$= B\frac{y^4}{18}\Big|_0^3 = -\frac{9}{2}\text{ J}$$

(c) $W = \int_0^1 F_x dx + \int_0^3 F_y dy$

$$= \int_0^1 Bx(0)^2 dx + \int_0^3 By(1)^2 dy = \int_0^3 By\,dy$$

$$= B\frac{y^2}{2}\Big|_0^3 = -\frac{9}{2}\text{ J}$$

8-15 Since the bead is constrained to move only in the x-direction, the net force in the y-direction must be zero.

$$N - F_y = N + Axy = 0$$

The energy equation is

$$\frac{1}{2} mv_0^2 + U_0 - \int_0^x \mu_k N\,dx = \frac{1}{2} mv_f^2 + U_f$$

$$\frac{1}{2} mv_0^2 + A\mu_k\int_0^x xy\,dx = \frac{1}{2} Ax^2y$$

$$x = v_0\sqrt{\frac{m}{Ay(1 - \mu_k)}} = 12.9\text{ m}$$

The forces acting at the point of rest are

$$F_x = -Axy = -\frac{1}{5}(2)(12.9) = -5.16 \text{ N}$$

$$F_y = N = -Axy = -5.16 \text{ N}$$

$$F_z = 0$$

To prevent the bead from beginning to move again the force of static friction must equal F_x.

$$\mu_s N = \mu_s(-5.16) = F_x = -5.16 \text{ N}$$

The coefficient of static friction must be at least unity in order to prevent the bead from beginning to move.

8-16 Assuming no change in the kinetic energy of the barge, the only work done by the winches is that done against F_b.

$$W = -\oint \vec{F}_b \cdot d\vec{r} = -\oint F_b dy$$

$$= \int_{(0,0)}^{(0,L)} v_o\left[1 - \frac{x^2}{a^2}\right]dy + \int_{(0,L)}^{(a,L)} 0\, dx + \int_{(a,L)}^{(a,0)} 0\, dy + \int_{(a,0)}^{(0,0)} 0\, dx$$

$$= \int_0^L v_o dy = v_o L$$

The force on the barge is not conservative since the value of the closed-path line integral does not equal zero.

8-17 (a) $\vec{F} = F_x\hat{i} + F_y\hat{j}$

The slope of the tangent to this force is

$$\tan\theta = \frac{F_y}{F_x}$$

However, knowing the equation for the line of force, another expression for the slope of the tangent line can be obtained.

$$\tan\theta = \frac{dy}{dx}$$

(b) $\frac{dy}{dx} = \frac{F_y}{F_x} = \frac{x}{y}$

$$y\, dy = x\, dx$$

Integrating to find the equation for the lines of force,

$$y^2 = x^2 + \text{constant}$$

This is the equation for a family of hyperbolas with $y = \pm x$ as asymptotes.

(c) $F_x = -\frac{\partial U}{\partial x} = Ay \qquad F_y = -\frac{\partial U}{\partial y} = Ax$

$\vec{F}_+ = Ay\hat{i} + Ax\hat{j}$

$U = -Axy = \text{constant}$

Transforming U to x' - y' coordinates, which are a 45° rotation of the x - y coordinates,

$$U = -A(x'\cos\theta - y'\sin\theta)(x'\sin\theta + y'\cos\theta) + \text{constant}$$

$$= -A(\sin\theta)(\cos\theta)(x'^2 - y'^2) + \text{constant}$$

$$= -\frac{1}{2}A(x'^2 - y'^2) + \text{constant}$$

Since the potential is only defined to within an additive constant,

$$U = -\frac{1}{2}A(x'^2 - y'^2)$$

This is an equation for a family of hyberbolas with $y' = \pm x'$ as asymptotes.

The equation for the slope of the tangent line to any of these hyperbolas can be found as follows.

$$x'^2 - y'^2 = -\frac{2U}{A}$$

$$2x'dx' - 2y'dy' = 0$$

$$\frac{dy'}{dx'} = \frac{x'}{y'}$$

This can be transformed back to the original coordinate system by using the following equations.

$$x' = x\cos\theta + y\sin\theta$$

$$dx' = dx\cos\theta + dy\sin\theta$$

$$y' = -x\sin\theta + y\cos\theta$$

$$dy' = -dx\sin\theta + dy\cos\theta$$

$$\frac{dy'}{dx'} = \frac{-dx\sin\theta + dy\cos\theta}{dx\cos\theta + dy\sin\theta} = \frac{x\cos\theta + y\sin\theta}{-x\sin\theta + y\cos\theta}$$

In this instance, $\theta = 45°$. So,

$$\frac{dy - dx}{dy + dx} = \frac{y + x}{y - x}$$

$$(dy - dx)(y - x) = (dy + dx)(y + x)$$

$$dy[(y - x) - (y + x)] = dx[(y + x) + (y - x)]$$

$$\frac{dy}{dx} = -\frac{y}{x}$$

Since this is the negative reciprocal of $\frac{F_y}{F_x}$, the U(x,y) hyperbolas are everywhere perpendicular to the force hyperbolas.

(d) $\frac{F_y}{F_x} = \frac{dy}{dx} = \frac{Ax}{-Ay} = \frac{-x}{y}$

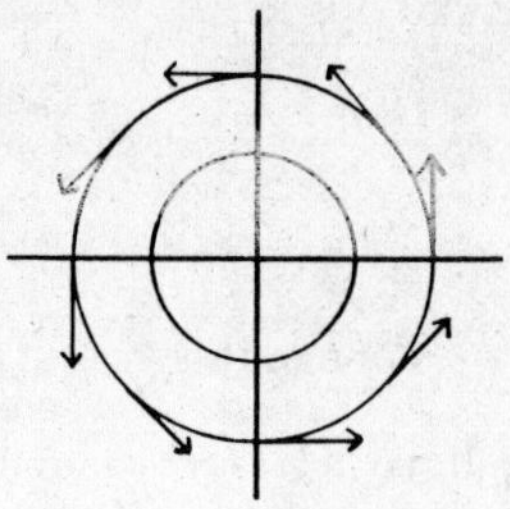

$y\,dy = -x\,dx$

$y^2 = -x^2 + \text{constant}$

This is the equation for a family of concentric circles with centers at the origin.

(e) From the diagram above it can be seen that the line integral around a circle centered on the origin is not equal to zero.

$$\oint \vec{F} \cdot d\vec{s} \neq 0$$

Since this force is not conservative, it cannot be derived from a potential U(x,y). The derivation of Eqs. 12-11 imply the existence of such a potential. Since this potential doesn't exist, Eqs. 2-11 will not be true.

9-1 $\vec{F} = m\frac{d\vec{v}}{dt} + \vec{v}\frac{dm}{dt} = 0 + (25\ \frac{m}{s})(-0.60\ \frac{kg}{s}) = -15\ N$

In each increment dt of time, an increment dm of mass leaves the nozzle.

To balance this force of the nozzle backward, a force must be exerted by the gardener equal in magnitude but opposite in direction (and thus in the same direction as v).

$$F_{gardener} = 15\ N$$

9-2 $F_{in} = m_i\frac{dv_i}{dt} + v_i\frac{dm_i}{dt}$

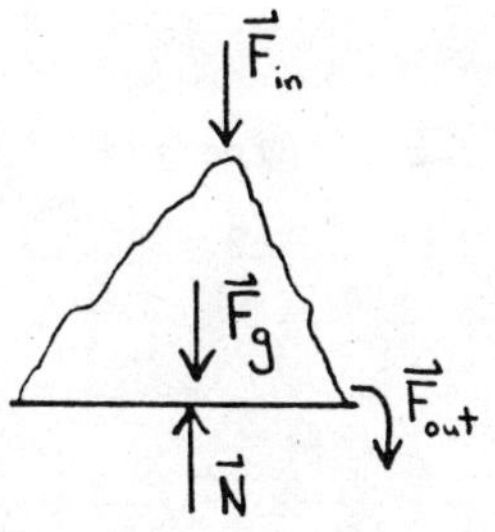

$F_{out} = v_\ell\frac{dm_\ell}{dt} + m_\ell\frac{dv_\ell}{dt} = m_\ell\frac{dv_\ell}{dt}$

Total Force $= 0 = N-(F_{in} + mg - F_{out})$

$N = v_i\frac{dm_i}{dt} + m_i\frac{dv_i}{dt} + mg - m_\ell\frac{dv_\ell}{dt}$

At equilibrium the mass m_i that is poured on is equal to the mass m_ℓ that is spilled off, and $\frac{dv_i}{dt} = g = \frac{dv_\ell}{dt}$

$$N = v_i\frac{dm_i}{dt} + mg$$

From $y = \frac{1}{2}a_yt^2$ and $v_y = a_yt$, we get

$$v_i = \sqrt{2a_yy} = 4.4\ m/s$$

So,

$$N = (4.4\ \frac{m}{s})(\frac{0.5\ kg}{s}) + (9.8\ N)$$

$$= 12.0\ N$$

9-3 Initial momentum $\vec{p}_i = +mv\hat{i}$

Final momentum $\vec{p}_f = -mv\hat{i}$

$\Delta\vec{p} = -2mv\hat{i}$

$\bar{\vec{F}} = \frac{\Delta\vec{p}}{\Delta t} = \frac{-2m\vec{v}}{\Delta t} = \frac{-2(0.15\ kg)(18\ m/s)}{1.5\times 10^{-3}\ s}\ \hat{i}$

$= -3600\hat{i}\ N$

9-4 (a) $\vec{J} = \Delta\vec{p} = m\vec{v} = (0.057\ kg)(40\ m/s)$

$= 2.3\ \frac{kg\cdot m}{s}$ in the direction of the stroke

(b) $\bar{\vec{F}} = \frac{\Delta\vec{p}}{\Delta t} = \frac{2.3}{5\times 10^{-3}}\ N = 460\ N$ in the direction of the stroke

9-5 The travel time going down is

$$t_1 = \sqrt{\frac{2h}{g}}$$

The velocity at impact is

$$v_1 = -gt_1 = -\sqrt{2gh}$$

The time to go back up is

$$t_2 = \sqrt{\frac{2\eta h}{g}}$$

The velocity right after the impact is

$$v_2 = \sqrt{2\eta gh}$$

To find the impulse time interval,

$$\Delta t = \frac{\Delta p}{\overline{F}} = \frac{mv_2 - mv_1}{\overline{F}} = \frac{m\sqrt{2gh}(1 + \sqrt{\eta})}{\overline{F}}$$

$$\frac{t_1 + t_2}{\Delta t} = \frac{\sqrt{\frac{2h}{g}} + \sqrt{\frac{2\eta h}{g}}}{\frac{m\sqrt{2gh}(1 + \sqrt{\eta})}{\overline{F}}} = \frac{\overline{F}}{mg}$$

9-6 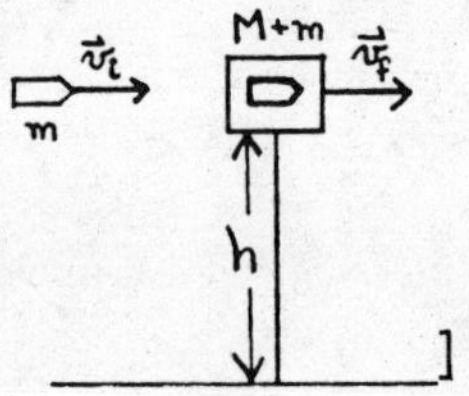

At the instant of the collision, the motion is perpendicular to the gravitational force. Using conservation of momentum,

$$mv_i = (m + M)v_f$$

Applying the equation of motion in the y-direction,

$$h = \frac{1}{2}gt^2$$

$$t = \sqrt{\frac{2h}{g}}$$

In the x-direction,

$$x = v_f t = v_f\sqrt{\frac{2h}{g}} = \frac{m}{m + M}v_i\sqrt{\frac{2h}{g}}$$

So the initial velocity is

$$v_i = \frac{m + M}{m}x\sqrt{\frac{g}{2h}} = \frac{0.61\text{ kg}}{0.01\text{ kg}}(4\text{ m})\sqrt{\frac{9.8}{2(2)\text{ s}^2}}$$

$$= 382\text{ m/s}$$

Note that conservation of energy cannot be used since the collision is inelastic.

9-7 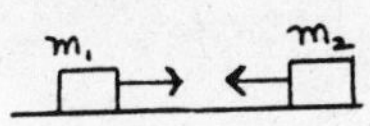

(a) By conservation of momentum,

$$m_1 v_1 + m_2 v_2 = (m_1 + m_2) v_f$$

$$v_f = \frac{m_1 v_1 + m_2 v_2}{m_1 + m_2} = \frac{(2\ \text{kg})(5\ \text{m/s}) + (3\ \text{kg})(-4\ \text{m/s})}{5\ \text{kg}}$$

$$= \frac{-2}{5}\ \text{m/s}$$

(b) $$\Delta K = K_{final} - K_{initial}$$

$$= \frac{1}{2}(m_1 + m_2)v_f^2 - \frac{1}{2} m_1 v_1^2 - \frac{1}{2} m_2 v_2^2$$

$$= \frac{1}{2}(5\ \text{kg})(-\frac{2}{5}\ \text{m/s})^2 - \frac{1}{2}(2\ \text{kg})(5\ \text{m/s})^2 - \frac{1}{2}(3\ \text{kg})(4\ \text{m/s})^2$$

$$= -48\frac{3}{5}\ \text{J}$$

This energy is lost as heat.

9-8 Use conservation of momentum to determine the velocity of the putty-and-block system at the moment of initial compression.

$$v_i = \frac{m_p v}{m_p + m_b}$$

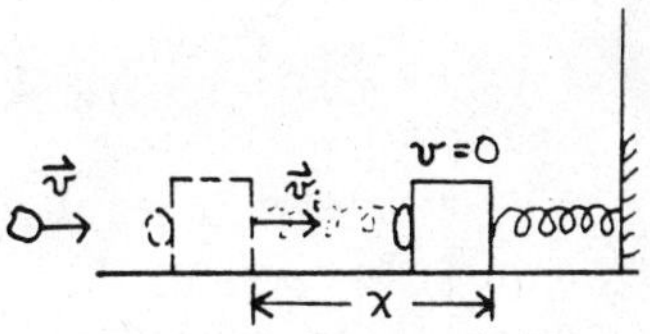

By conservation of energy,

$$\frac{1}{2}(m_p + m_b)v_i^2 = \frac{1}{2}\kappa x^2$$

So, the initial velocity can be written

$$v = \frac{m_p + m_b}{m_p} v_i = \frac{m_p + m_b}{m_p}\sqrt{\frac{\kappa}{m_p + m_b}}\ x$$

$$= \frac{2.5\ \text{kg}}{0.5\ \text{kg}}\left(\sqrt{\frac{250}{2.5}}\ \text{s}^{-1}\right)(0.10\ \text{m})$$

$$= 5\ \text{m/s}$$

9-9 $$m_1\vec{v}_1 + m_2\vec{v}_2 = (m_1 + m_2)\vec{v}_f$$

$$v_f = \frac{m_1}{m_1 + m_2}\vec{v}_1 + \frac{m_2}{m_1 + m_2}\vec{v}_2 = \frac{1500}{3500}(90\ \text{km/h})\hat{j} + \frac{2000}{3500}(60\ \text{km/h})\hat{i}$$

$$= 34.3\ \hat{i}\ \text{km/h} + 38.6\ \hat{j}\ \text{km/h}$$

$$v_f = \sqrt{38.6^2 + 34.3^2} = 51.6\ \text{km/h}$$

$$\theta = \tan^{-1}\frac{38.6}{34.3} = 48.4°$$

The final velocity is directed 48.4° north of east.

9-10

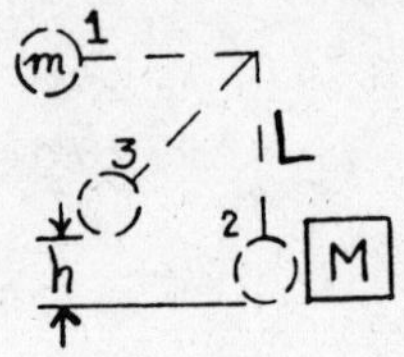

(a) Applying conservation of energy of the ball between situations 1 and 2,

$$mgL = \frac{1}{2} mv^2 \qquad (1)$$

At 2, both conservation of momentum and conservation of energy can be used.

$$mv = Mv_M - mv_m = 2\, mv_M - mv_m \qquad (2)$$

$$\frac{1}{2} mv^2 = \frac{1}{2} Mv_M^2 + \frac{1}{2} mv_m^2 = mv_M^2 + \frac{1}{2} v_m^2 \qquad (3)$$

Applying conservation of energy to the ball after its collision,

$$\frac{1}{2} mv_m^2 = mgh \qquad (4)$$

Substituting (1) and (4) into (2) and (3), and dividing by m,

$$\sqrt{2\, gL} = 2\, v_m - \sqrt{2\, gh}$$

$$gL = v_M^2 + gh$$

Solving the above equations for h,

$$0 = 9\, h^2 - 10\, Lh + L^2$$

$$h = \frac{5 \pm 4}{9} L = L, \frac{1}{9} L$$

The h = L solution corresponds to the initial configuration. The height to which the ball rises after collision is $\frac{1}{9}$ L.

(b) $v_M = \frac{1}{2}(\sqrt{2\, gL} + \sqrt{2\, gh}) = \frac{1}{2}[\sqrt{2(9.8)(0.50)} + \sqrt{2(9.8)(0.056)}]$ m/s

$= 2.09$ m/s

9-11 $\frac{1}{2} mv_o^2 = 8$ J (1)

$mv_1 \cos\alpha = mv_o = mv_2 \cos\beta$ (2)

$mv_1 \sin\alpha = mv_2 \sin\beta$ (3)

$\frac{1}{2} mv_1^2 + \frac{1}{2} mv_2^2 = 6$ J (4)

Dividing (2) and (3) by m, squaring each, and adding,

$$v_1^2 = v_o^2 - 2\, v_o\, v_2 \cos\beta + v_2^2$$

From (1) and (4),

$$v_o^2 = v_1^2 + v_2^2 + \frac{4\ J}{m}$$

So, $\quad v_o^2 - v_2^2 - \frac{4\ J}{m} = v_o^2 - 2\, v_o v_2 \cos\beta + v_2^2$

Solving for v_2,

$$v_2 = v_0 \cos\beta \pm \sqrt{v_0^2 \cos^2\beta - \frac{8\,J}{m}}$$

Clearly,

$$v_0^2 \cos^2\beta \geq \frac{8\,J}{m}$$

So,

$$\beta_{max} = \cos^{-1}\sqrt{\frac{4\,J}{\frac{1}{2}mv_0^2}} = \cos^{-1}\sqrt{\frac{1}{2}} = 45°$$

It is clear from symmetry that this is also the maximum angle for α.

9-12 Let v_g and v_p be the velocities of the girl and the plank, respectively, relative to the ice. Since there is no net external force acting on the girl-plank system, conservation of momentum can be used.

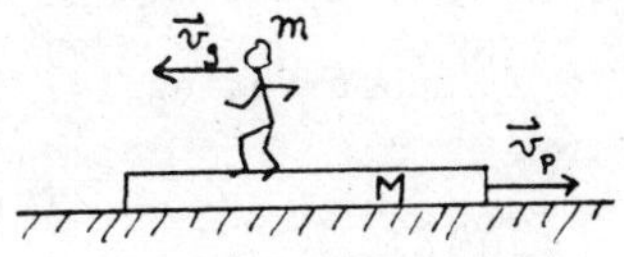

$$p_{initial} = 0 = p_{final} = Mv_p - mv_g$$

The velocity of the girl relative to the plank is

$$v_{g\ rel} = 1.5\ \text{m/s} = v_g + v_p$$

(a) To find v_g,

$$v_g = \frac{M}{m}v_p = \frac{M}{m}(v_{g\ rel} - vg)$$

$$v_g = \frac{v_{g\ rel}}{1 + \frac{m}{M}} = \frac{(1.5\ \text{m/s})}{(1 + \frac{3}{10})} = 1.15\ \text{m/s}$$

(b) $v_p = \frac{m}{M}v_g = \frac{3}{10}(1.15\ \text{m/s}) = 0.345\ \text{m/s}$

9-13 $\varepsilon = \dfrac{v_2}{v_1}$

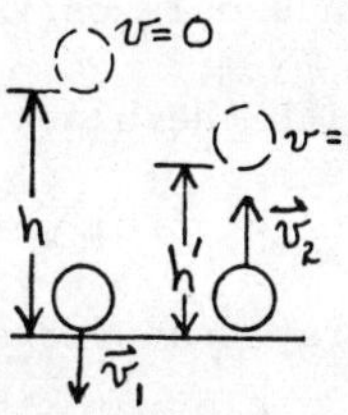

Using $h = \frac{1}{2}gt_1^2$ and $v_1 = gt_1$,

$$v_1 = \sqrt{2gh}$$

Using $h' = v_2t_2 - \frac{1}{2}gt_2^2$ and $v_2 = gt_2$,

$$v_2 = \sqrt{2gh'}$$

$$\varepsilon = \frac{\sqrt{2gh'}}{\sqrt{2gh}}$$

This gives

$$h' = \varepsilon^2 h$$

9-14 Because the collision is elastic, both conservation of energy and conservation of momentum can be applied.

$$\frac{1}{2} mv_i^2 = \frac{1}{2} mv_m^2 + \frac{1}{2} Mv_M^2 \qquad (1)$$

$$mv_i = -mv_m + Mv_M \qquad (2)$$

To find the ratio $\frac{v_M}{v_i}$, divide (1) by $\frac{1}{2} mv_i^2$ and (2) by mv_i.

$$1 = \frac{v_m^2}{v_i^2} + \frac{M}{m}\frac{v_M^2}{v_i^2} \qquad (3)$$

$$1 = -\frac{v_m}{v_i} + \frac{M}{m}\frac{v_M}{v_i} \qquad (4)$$

Substituting (4) into (3),

$$1 = \left(\frac{M}{m}\frac{v_M}{v_i} - 1\right)^2 + \frac{M}{m}\frac{v_M^2}{v_i^2}$$

Solving to find the ratio and discarding the case where it equals zero,

$$\frac{v_M}{v_i} = \frac{2}{\frac{M}{m} - 1}$$

Thus, the fraction of the particle's original kinetic energy that is transferred to the object is

$$\frac{\frac{1}{2} Mv_M^2}{\frac{1}{2} mv_i^2} = \frac{M}{m}\left(\frac{2}{\frac{M}{m} - 1}\right)^2$$

(a) For $\frac{M}{m} = 500$, the fraction is

$$500 \left(\frac{2}{499}\right)^2 = 0.008$$

(b) Notice that the fraction of the block's original energy that is transferred to the earth depends only on the relative masses of the two objects. This means the height from which the block is dropped does not effect the value of the ratio. However, the height from which the block is dropped will effect the absolute amount of kinetic energy transferred to the earth.

$$\frac{Mv_M^2}{mv_i^2} = \frac{6.0 \times 10^{24}\text{kg}}{100\text{ kg}}\left[\frac{2}{(6.0 \times 10^{22}) - 1}\right]^2 = 6.7 \times 10^{-23}$$

9-15 The center of mass of the four particles at the base of the pyramid is at the center of the base. The center of mass of the five-particle system lies on the line perpendicular to the base and passing through the peak of the pyramid; obviously, this line also passes through the C.M. of the four particles at the base. The height of the five-particle system's C.M. from the base is

$$x = \frac{mh + 4m(0)}{5m} = \frac{1}{5}h$$

9-16 $$\text{C.M.} = \frac{(329{,}390M_E)\quad(0) + M_E(1.496 \times 10^8 \text{ km})}{329{,}390M_E + M_E}$$

$$= 455 \text{ km}$$

The center of mass of the sun-earth system is well inside the sun.

9-17

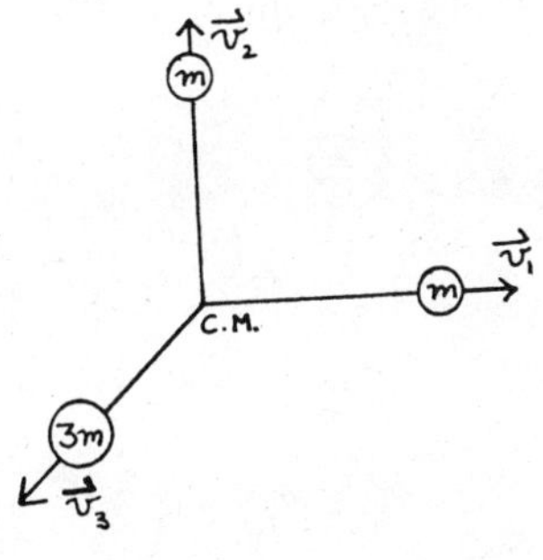

Using conservation of momentum in the center of mass system,

$$0 = mv\hat{i} + mv\hat{j} + 3m\,\vec{v}_3$$

$$\vec{v}_3 = \frac{-v}{3}\hat{i} - \frac{v}{3}\hat{j}$$

To find the velocity of the third particle relative to the first particle,

$$\vec{v}_3' = \vec{v}_3 - \vec{v}_1 = -\frac{v}{3}\hat{i} - \frac{v}{3}\hat{j} - v\hat{i} = -\frac{4}{3}v\hat{i} - \frac{1}{3}v\hat{j}$$

Relative to the second particle,

$$\vec{v}_3'' = \vec{v}_3 - \vec{v}_2 = -\frac{v}{3}\hat{i} - \frac{v}{3}\hat{j} - v\hat{j} = -\frac{1}{3}v\hat{i} - \frac{4}{3}v\hat{j}$$

9-18 $$\vec{r}_1 = 3t^2\hat{i} + 2t\hat{j} \qquad m_1 = 1 \text{ kg}$$

$$\vec{r}_2 = 3\hat{i} + (2t^3 + 5t)\hat{j} + 9\hat{k} \qquad m_2 = 2 \text{ kg}$$

(a) $$\vec{R} = \frac{(1 \text{ kg})(3t^2\hat{i} + 2t\hat{j}) + (2 \text{ kg})[3\hat{i} + (2t^3 + 5t)\hat{j} + 9\hat{k}]}{3 \text{ kg}}$$

$$= (t^2 + 2)\hat{i} + (\frac{4}{3}t^3 + 4t)\hat{j} + 6\hat{k}$$

(b) $$\frac{d\vec{R}}{dt} = 2t\hat{i} + (4t^2 + 4)\hat{j} = 2t\hat{i} + 4(t^2 + 1)\hat{j}$$

(c) $\frac{d^2\vec{R}}{dt^2} = 2\hat{i} + 8t\hat{j}$

9-19 The velocity of the center of mass remains

$$V = 10 \text{ m/s}$$

(a) At the time the first fragment hits the target, the center of mass is a distance d from the target.

$$d = 100 \text{ m} - (10 \text{ m/s})(8 \text{ s}) = 20 \text{ m}$$

Since the two fragments are of equal mass, the second fragment must be 40 m from the target.

(b) The second fragment is travelling at

$$v_2 = \frac{60 \text{ m}}{8 \text{ s}} = 7.5 \text{ m/s}$$

The time between fragments arriving at the target is

$$t = \frac{2d}{v_2} = \frac{40 \text{ m}}{7.5 \text{ m/s}} = 5.3 \text{ s}$$

9-20 (a) The velocity of the boulder relative to the ground is

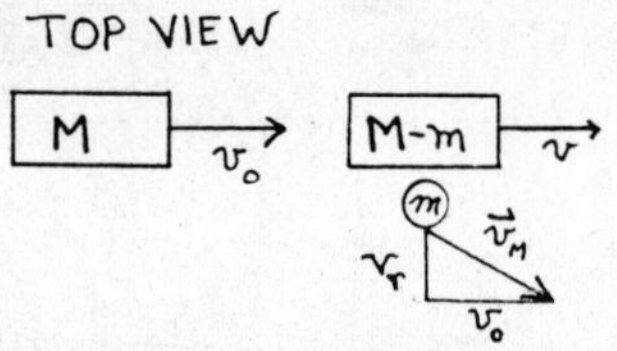

$$\vec{v}_m = v\hat{i} - v_r\hat{j}$$

Applying conservation of momentum in the x-direction,

$$Mv = mv_0 + (M-m)v$$

So $\quad v = v_0 = 2 \text{ m/s}$

(b) Using 9-17 in the x-direction,

$$M\Delta\vec{v} = \vec{F}^e_x \Delta t + \vec{v}_{r,x}\Delta M$$

Since $\vec{v}_r$ and the external force of the tracks are perpendicular to the x-direction, they have no x-components.

$$M\Delta\vec{v} = 0 + 0$$

$$\Delta\vec{v} = 0$$

So $\quad v = v_0 = 2 \text{ m/s}$

(c) In the x-direction,

$$Mv_0 = (M-m)v$$

$$v = \frac{M}{M - m} v_0 = \frac{400}{350} (2 \text{ m/s}) = 2.3 \text{ m/s}$$

Using 9-17,

$$M\Delta v = v_r\Delta M = -v\Delta M = vm$$

$$v = v_0 + \Delta v = v_0 + \frac{m}{M} v$$

$$v = \frac{M}{M - m} v_o = 2.3 \text{ m/s}$$

(d) These calculations are based on conservation of momentum in the x-direction only. A change in the z-component of the boulder's velocity is not important since, for the two cases considered, it will not affect the x-component of velocity.

(e) The magnitude of the boulder's velocity is irrelevant because it is ejected perpendicular to the track in either the center of mass frame or the handcar frame. This allows for determination of velocity in the x-direction without any knowledge of the magnitude of the velocity.

9-21 Using Equation 9-17 with $\frac{d\vec{v}}{dt} = 0$,

$$0 = \vec{F}^e + \vec{v}_r \frac{dM}{dt}$$

$$\frac{dM}{dt} = \frac{-F^e}{v_r} = \frac{-(160{,}000 \text{ kg})(9.8 \text{ m/s}^2)}{3000 \text{ m/s}} = -523 \text{ kg/s}$$

For $\frac{d\vec{v}}{dt} = 2g$,

$$\frac{dM}{dt} = \frac{-(2\ Mg + Mg)}{v_r} = -1570 \text{ kg/s}$$

In the first case, the thrust is

$$\Omega = v_r \frac{dM}{dt} = F^e = Mg = 1.57 \times 10^6 \text{ N}$$

For an acceleration of 2g, the thrust is

$$\Omega = v_r \frac{dM}{dt} = 2\ Mg + F_e = 3\ Mg = 4.70 \times 10^6 \text{ N}$$

9-22 Using 9-17,

$$M \frac{dv}{dt} = -Mg + v_r \frac{dM}{dt}$$

$$dv = -g\ dt + v_r \frac{dM}{M}$$

Integrating,

$$v = -gt_b + v_r \int_{M_i}^{M_f} \frac{dM}{M}$$

$$= -gt_b + v_r[\ln M_f - \ln M_i]$$

$$= -v_r \ln \frac{M_i}{M_f} - gt_b$$

Since the rate of fuel consumption is constant, the time to burn the fuel is

$$t_b = \frac{M_i - M_f}{\frac{dm}{dt}}$$

9-23 (a) $\frac{dv}{dt} = -g + \frac{v_r}{M}\frac{dM}{dt} = (-9.80 \text{ m/s}^2) + \frac{1}{145 \times 10^3 \text{ kg}}(1.92 \times 10^6 \text{ N})$

$$= 3.44 \text{ m/s}^2$$

(b) $$\frac{dM}{dt} = \frac{\Omega}{v_r} = \frac{1.92 \times 10^6 \text{ N}}{-2400 \text{ m/s}} = -800 \text{ kg/s}$$

The gas is ejected at 800 kg/s.

(c) $$v = -v_r \ln \frac{M_i}{M_f} - gt$$

$$= -(-2400 \text{ m/s}) \ln \frac{145 \times 10^3 \text{ kg}}{(145 \times 10^3 \text{ kg}) - (800 \text{ kg/s})(150 \text{ s})} - (9.8 \text{ m/s}^2)(150 \text{ s})$$

$$= 2750 \text{ m/s}$$

9-24 From 9-22,

$$v = -v_r \ln \frac{M_i}{M_f} - gt_b$$

In deep space the effect of gravity is not felt.

$$v = -v_r \ln \frac{M_i}{M_f}$$

(a) $$M_i = e^{-v/vr} M_f = e^5 (3.0 \times 10^3 \text{ kg}) = 4.45 \times 10^5 \text{ kg}$$

The mass of fuel and oxidizer is

$$\Delta M = M_i - M_f = (445 - 3) \times 10^3 \text{ kg} = 442 \text{ metric tons}$$

(b) $$\Delta M = e^2 (3 \text{ metric tons}) - 3 \text{ metric tons} = 19.2 \text{ metric tons}$$

Because of the exponential, a relatively small increase in fuel and/or engine efficiency causes a large change in the amount of fuel and oxidizer required.

9-25 (a) $$\Omega = -v_r \frac{dm}{dt} = (1200 \text{ m/s})(5 \text{ kg/s}) = 6000 \text{ N}$$

(b) The acceleration due to the propeller alone is

$$a_p = \frac{v^2}{2s} = \frac{(200 \text{ km/h})^2}{0.800 \text{ km}} = 3.86 \text{ m/s}^2$$

The total acceleration is

$$a = a_p + \frac{\Omega}{m} = 3.86 + 2.00 = 5.86 \text{ m/s}^2$$

The distance required for lift-off is

$$x = \frac{3.86}{5.86} x_p = 263 \text{ m}$$

9-26 $$F = \frac{dp}{dt} = v \frac{dm}{dt}$$

To find $\frac{dm}{dt}$,

$$\frac{dm}{dt} = \left(\frac{10^3 \text{ liters}}{\text{min}}\right)\left(\frac{10^{-3} \text{ m}^3}{\text{liter}}\right)\left(\frac{\text{min}}{60 \text{ s}}\right)\left(\frac{10^3 \text{ kg}}{\text{m}^3}\right)$$

$$= 16.7 \text{ kg/s}$$

To find v,

$$v = \left(\frac{10^3 \text{ liter}}{\text{min}}\right)\left(\frac{10^{-3} \text{ m}^3}{\text{liter}}\right)\left(\frac{1}{\pi(0.50)^2\text{m}^2}\right)\left(\frac{\text{min}}{60 \text{ s}}\right)$$

$$= 2.12 \text{ m/s}$$

$$F = (2.12 \text{ m/s})(16.7 \text{ kg/s}) = 35.4 \text{ N}$$

9-27 Since the angle is 45°, Equation 4-19 can be used to determine v_o.

$$v_o = \sqrt{Rg} = 44.3 \text{ m/s}$$

$$\overline{F} = \frac{mv_o}{\Delta t} = \frac{(0.046 \text{ kg})(44.3 \text{ m/s})}{7 \times 10^{-3} \text{ s}} = 291 \text{ N}$$

9-28 (a) $\overline{F} = \dfrac{(4 \text{ kg})(2.9 \text{ m/s})}{0.07 \text{ s}} = 166 \text{ N}$

(b) Using conservation of momentum,

$$m_{gun} v_{gun} = m_{shot} v_{shot}$$

$$v_{shot} = \frac{(4 \text{ kg})(2.9 \text{ m/s})}{0.035 \text{ kg}} = 331 \text{ m/s}$$

(c) $s = \overline{v}t = \dfrac{v}{2} t$

$$t = \frac{2s}{v} = \frac{2(0.75 \text{ m})}{331 \text{ m/s}} = 4.5 \times 10^{-3} \text{ s} = 4.5 \text{ ms}$$

This is much less than 0.07 s.

9-29 Applying conservation of momentum to the collision,

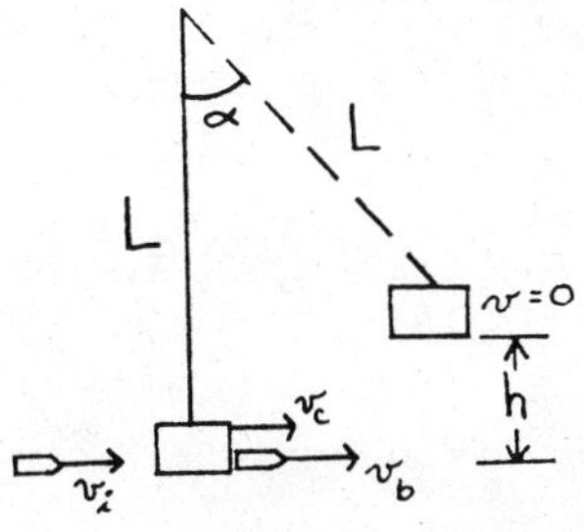

$$m_b v_i = m_b v_b + m_c v_c$$

The kinetic energy of the can becomes gravitational potential energy.

$$\frac{1}{2} m_c v_c^2 = m_c gh = m_c gL(1 - \cos \alpha)$$

So,

$$m_b v_i = m_b v_b + m_c \sqrt{2gL(1 - \cos \alpha)}$$

$$v_b = v_i - \frac{m_c}{m_b} \sqrt{2gL(1 - \cos \alpha)}$$

$$= 331 \text{ m/s}$$

9-30 $\dfrac{\Delta m}{\Delta t} = \left(\dfrac{160}{4 \text{ s}}\right)(0.130 \text{ kg}) = 5.2 \text{ kg/s}$

(a) $\overline{F} = \dfrac{v\Delta m}{\Delta t} = (800 \text{ m/s})(5.2 \text{ kg/s}) = 4160 \text{ N}$

Note that, since the velocity of the shells is much greater than the velocity of the helicopter, it is not necessary to use a relative velocity for this calculation.

(b) $M\Delta V = mv$

$$\Delta V = \frac{m}{M} v = \frac{(160)(0.130 \text{ kg})}{4000 \text{ kg}}(800 \text{ m/s}) = 4.16 \text{ m/s}$$

9-31 (a) Applying conservation of momentum in the x-direction,

$$MV = m(v_r \cos 60° - V)-(M-m)V$$

$$V = \frac{mv_r \cos 60°}{2M} = \frac{(50 \text{ kg})(300 \text{ m/s})(\frac{1}{2})}{40{,}000 \text{ kg}} = 0.19 \text{ m/s}$$

(b) $(v_r \sin 60°)^2 = 2gh$

$$h = \frac{(300 \text{ m/s})^2(\frac{3}{4})}{2(9.8 \text{ m/s}^2)} = 34444 \text{ m}$$

9-32

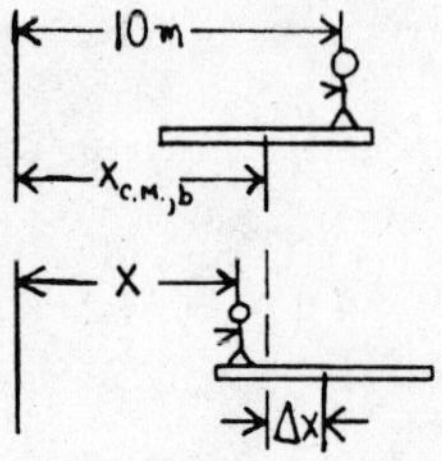

Since there are no external forces acting on the system and since $v_{C.M.} = 0$, the position of the center of mass does not change.

$$X = \frac{x_m m_m + x_{CM,b} m_b}{m_m + m_b}$$

Initially,

$$X = \frac{(10 \text{ m})(75 \text{ kg}) + x_{CM,b}(225 \text{ kg})}{300 \text{ kg}}$$

Finally,

$$X = \frac{x(75 \text{ kg})+(x_{CM,b} + \Delta x)(225 \text{ kg})}{300 \text{ kg}}$$

The distance the man has moved with respect to the ground is

$$(10 \text{ m})- x = (2 \text{ m})- \Delta x$$

Substituting for Δx and equating initial and final x_{CM},

$$\frac{750 \text{ m} + x_{CM,b}(225)}{300} = \frac{x(75) + x_{CM,b}(225)+[x - (8\text{m})](225)}{300}$$

So $x = 8.5$ m

9.33 By conservation of momentum,

$$mv = (m + M)v'$$

The kinetic energy of the bullet-block system is lost to friction,

$$\frac{1}{2}(M+m)\, v'^2 = \mu_k(M+m)gx$$

To determine μ_k,

$$0.80 \text{ N} = \mu_k Mg$$

$$\mu_k = \frac{0.80 \text{ N}}{(5 \text{ kg})(9.8 \text{ m/s}^2)} = 0.016$$

Thus,

$$v = \frac{m + M}{m} v' = \frac{m + M}{m} \sqrt{2\mu_k gx}$$

$$= \frac{5.02 \text{ kg}}{0.02 \text{ kg}} \sqrt{2(0.016)(9.8 \text{ m/s}^2)(1.5 \text{ m})}$$

$$= 172 \text{ m/s}$$

9-34 (a) Since there is no external force acting on the spring-mass system, conservation of momentum gives

$$m_A v_A = m_B v_B$$

$$v_B = \frac{m_A}{m_B} v_A = \frac{1}{2} (2 \text{ m/s}) = 1 \text{ m/s}$$

(b) The potential energy stored in the spring, U_s, becomes the kinetic energy of the blocks.

$$U_s = \frac{1}{2} m_A v_A^2 + \frac{1}{2} m_B v_B^2$$

$$= \frac{1}{2}(1 \text{ kg})(2 \text{ m/s})^2 + \frac{1}{2}(2 \text{ kg})(1 \text{ m/s})^2$$

$$= 3 \text{ J}$$

(c) $\frac{K_A}{U_s} = \frac{2 \text{ J}}{3 \text{ J}} = \frac{2}{3}$

$\frac{K_B}{U_s} = \frac{1 \text{ J}}{3 \text{ J}} = \frac{1}{3}$

The force applied to each block by the spring is equal. Since $a = \frac{F}{m}$, the more massive block accelerates less and hence gains less energy.

(d) $U_s = \frac{1}{2} \kappa x^2$

$$\kappa = \frac{2 U_s}{x^2} = \frac{2(3 \text{ J})}{(0.50 \text{ m})^2} = 24 \text{ N/m}$$

9-35 (a) $J = m\Delta v = m(v - 0) = mv$

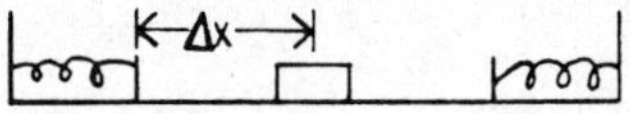

$$v = \frac{J}{m} = 5 \text{ m/s}$$

$$K_o = \frac{1}{2} mv^2 = 2.5 \text{ J}$$

$$f = \mu_k mg = 0.39 \text{ N}$$

(b) $K_1 = K_o - f \Delta x = 2.5 - 0.2 = 2.3 \text{ J}$

Assuming no loss of energy due to friction while the puck is in contact with the spring, the potential energy of the spring when the puck is momentarily at rest is

$$U_s = \frac{1}{2}\kappa x_1^2 = K_1$$

$$x = \sqrt{\frac{2\ K_1}{\kappa}} = 0.15\ \text{m}$$

(c) Again assuming no energy loss to friction during contact with the spring,

$$K_2 = K_1 - 2f\Delta x = 2.3 - 2(0.39)(0.5) = 1.9\ \text{J}$$

$$x_2 = \sqrt{\frac{2\ K_2}{\kappa}} = 0.14\ \text{m}$$

(d) At the nth collision with a spring,

$$K_n = K_o - (2n - 1)(\Delta x)f$$

To determine the number of collisions that occur before the puck comes to rest,

$$K_n = 0 = K_o - (2n - 1)(\Delta x)f$$

$$n = \frac{1}{2}\left(\frac{K_o}{(\Delta x)f} + 1\right) = \frac{1}{2}\left(\frac{2.5\ \text{J}}{(0.5\ \text{m})(0.39\ \text{N})}\right) + 1 = 6.9$$

The distance travelled after the sixth collision is

$$x' = v_o t + \frac{1}{2}at^2 = \sqrt{\frac{2\ K_6}{m}}\ t - \frac{1}{2}\frac{f}{m}\ t^2$$

To find t,

$$v = v_o + at = v_o - \frac{f}{m}t = 0$$

$$t = \frac{mv_o}{f} = \frac{m}{f}\sqrt{\frac{2\ K_6}{m}}$$

So, $$x' = \frac{2\ K_6}{f} - \frac{1}{2}\left(\frac{f}{m}\right)\left(\frac{m^2}{f^2}\right)\frac{2\ K_6}{m}$$

$$= \frac{K_6}{f} = \frac{1}{f}(K_o - 11(\Delta x)f)$$

$$= 0.91\ \text{m}$$

The total distance travelled by the puck is

$$D = (2n - 1)\Delta x + \sum_{n=1}^{6} x_n + x'$$

$$= 11(0.50\ \text{m}) + \sum_{n=1}^{6}\sqrt{\frac{2\ K_n}{\kappa}} + (0.91\ \text{m})$$

$$= (6.4\ \text{m}) + \frac{2}{\kappa}\sum_{n=1}^{6}\sqrt{K_o - (2n - 1)(\Delta x)f}$$

$$= (6.4\ \text{m}) + \sqrt{\frac{2}{\kappa}}\ (6.66\ \text{J}^{\frac{1}{2}})$$

$$= (6.4\ \text{m}) + (6.66\sqrt{\frac{2}{200}}\ \text{m})$$

$$= 7.1\ \text{m}$$

9-36 Since there are no forces acting in the x-direction, conservation of momentum can be used. Let v_r be the velocity of the block relative to the wedge.

$$MV = m(v_r \cos\alpha - V)$$

Using conservation of energy,

$$mgh = \frac{1}{2}MV^2 + \frac{1}{2}mv^2$$

To find v^2 in terms of v_r and V,

$$v^2 = v_x^2 + v_y^2$$

$$= (v_r \cos\alpha - V)^2 + (v_r \sin\alpha)^2$$

$$= v_r^2 - 2Vv_r \cos\alpha + V^2$$

So,

$$mgh = \frac{1}{2}MV^2 + \frac{1}{2}m(v_r^2 - 2Vv_r \cos\alpha + V^2)$$

$$= \frac{1}{2}MV^2 + \frac{1}{2}m[(\frac{M+m}{m\cos\alpha}V)^2 - 2\frac{M+m}{m}V^2 + V^2]$$

$$= V^2[\frac{1}{2}M + \frac{(M+m)^2}{2m\cos^2\alpha} - (M+m) + \frac{1}{2}m]$$

$$= V^2[\frac{(M+m)^2 - m\cos^2\alpha(M+m)}{2m\cos^2\alpha}]$$

Therefore,

$$V = [\frac{2m^2gh\cos^2\alpha}{(M+m)^2 - (m+M)m\cos^2\alpha}]^{1/2}$$

9-37

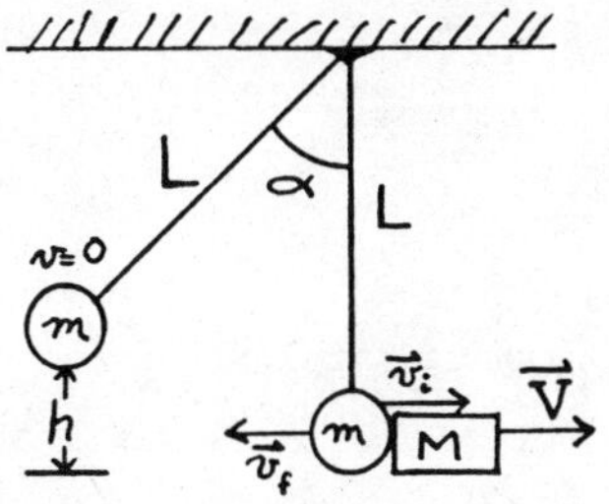

By conservation of energy,

$$\frac{1}{2}mv_i^2 = mgh = mgL(1 - \cos\alpha_1)$$

$$v_i = \sqrt{2gL(1 - \cos\alpha_1)} = 2\sqrt{gL}\sin\frac{\alpha_1}{2}$$

Applying conservation of energy after the collision,

$$v_f = 2\sqrt{gL}\sin\frac{\alpha_2}{2}$$

(a) By conservation of momentum,

$$mv_i = MV - mv_f$$

$$V = \frac{m}{M}(v_i + v_f) = \frac{2m}{M}\sqrt{gL}(\sin\frac{\alpha_1}{2} + \sin\frac{\alpha_2}{2})$$

$$= \frac{4m}{M}\sqrt{gL}(\sin\frac{\alpha_1+\alpha_2}{4}\cos\frac{\alpha_1-\alpha_2}{4})$$

(b) $$\varepsilon = \frac{v_f + V}{v_i} = \frac{\sin\frac{\alpha_2}{2} + \frac{2m}{M}(\sin\frac{\alpha_1+\alpha_2}{4}\cos\frac{\alpha_1-\alpha_2}{4})}{\sin\frac{\alpha_1}{2}}$$

9-38 By conservation of momentum,

$$P = m_1v_1 + m_2v_2 = m_1v_1' + m_2v_2'$$

$$v_2' = \frac{P - m_1 v_1'}{m_2}$$

The respective kinetic energies are

$$K = \frac{1}{2} m_1 v_1^2 + \frac{1}{2} m_2 v_2^2$$

$$K' = \frac{1}{2} m_1 v_1'^2 + \frac{1}{2} m_2 v_2'^2$$

$$= \frac{1}{2} m_1 v_1'^2 + \frac{1}{2} m_2 \left(\frac{P^2 - 2P\, m_1 v_1' + m_1^2 v_1'^2}{m_2^2}\right)$$

$$\frac{d\Delta K}{dv_1'} = \frac{dK'}{dv_1'} - \frac{dK}{dv_1'} = \frac{dK'}{dv_1'} = m_1 v_1' + \frac{1}{2\, m_2}(-2\, Pm_1 + 2\, m_1^2 v_1')$$

To find the maximum change in kinetic energy, set

$$\frac{d\Delta K}{dv_1'} = 0$$

$$0 = m_1 v_1' + \frac{1}{m_2}(-Pm_1 + m_1^2\, v_1')$$

$$= m_1 v_1' + \frac{1}{m_2}[-m_1 (m_1 v_1' + m_2 v_2') + m_1^2 v_1']$$

$$= m_1 v_1' = m_1 v_2'$$

In order for the equality to hold,

$$v_1' = v_2'$$

9-39 By conservation of momentum,

$$mv_o = Mv_M - mv_m \qquad (1)$$

The energy equation is

$$\frac{1}{2} mv_o^2 - E = \frac{1}{2} Mv_M^2 + \frac{1}{2} mv_m^2 \qquad (2)$$

Solving (1) for v_m^2 and substituting into (2),

$$\frac{1}{2} mv_o^2 - E = \frac{1}{2} Mv_M^2 + \frac{1}{2} m\left(\frac{Mv_M - mv_o}{m}\right)^2$$

The following quadratic equation is obtained from the equation above.

$$(1 + \frac{M}{m}) v_M^2 - 2v_o v_M + \frac{2E}{M} = 0$$

Solving for v_M,

$$v_M = \frac{2v_o \pm \sqrt{4v_o^2 - 4(1 + \frac{M}{m})(\frac{2E}{M})}}{2(1 + \frac{M}{m})}$$

Requiring v_M to be real gives

$$v_o^2 \geq (1 + \frac{M}{m})(\frac{2E}{M})$$

$$v_o \geq \sqrt{(1 + \frac{M}{m})(\frac{2E}{M})}$$

9-40 $m_1v_1 = m_1v_1' \cos\theta_1 + m_2v_2' \cos\theta_2$ (1)

$m_1v_1' \sin\theta_1 = m_2v_2' \sin\theta_2$ (2)

$$\frac{v_1'}{v_2'} = \frac{m_2 \sin\theta_2}{m_1 \sin\theta_1} = \frac{4\, m_1(0.82)}{m_1(0.91)} = 3.6$$

From (1),

$$m_1v_1 = m_1v_1' \cos\theta_1 + m_2\left(\frac{v_1'}{3.6}\right)\cos\theta_2$$

$$\frac{v_1'}{v_1} = \frac{m_1 \cos\theta_1 + \frac{4m_1}{3.6}\cos\theta_2}{m_1} = 0.42 + 0.64 = 1.1$$

$$K_i = \frac{1}{2} m_1v_1^2 = \frac{1}{2} m_1\left(\frac{v_1'}{1.1}\right)^2 = 0.41\, m_1v_1'^2$$

$$K_f = \frac{1}{2} m_1v_1'^2 + \frac{1}{2} m_2v_2'^2 = \frac{1}{2} m_1v_1'^2 + 2\, m\left(\frac{v_1'}{3.6}\right)^2$$

$$= 0.65\, m_1v_1'^2$$

Since $K_i \neq K_f$, kinetic energy is not conserved. The collision is not elastic.

9-41 $m_1v_1 - m_1v_1' \cos\theta_1 = m_2v_2' \cos\theta_2$ (1)

$m_1v_1' \sin\theta_1 = m_2v_2' \sin\theta_2$ (2)

$\frac{1}{2} m_1v_1^2 = \frac{1}{2} m_1v_1'^2 + \frac{1}{2} m_2v_2'^2$ (3)

Squaring (1) and (2) and adding,

$$m_1^2v_1^2 - 2m_1^2v_1v_1' \cos\theta_1 + m_1^2v_1'^2 = m_2^2v_2'^2$$

Use (3) to derive an expression for v_2' and substitute it into the above equation.

$$m_1^2v_1^2 = 2m_1^2v_1v_1' \cos\theta_1 + m_1^2v_1'^2 = m_2^2\left(\frac{m_1(v_1^2 - v_1'^2)}{m_2}\right)$$

This is a quadratic equation for v_1.

$$(m_1 - m_2)v_1^2 - (2m_1v_1' \cos\theta_1)v_1 + (m_1 + m_2)v_1'^2 = 0$$

$$v_1 = \frac{2m_1v_1' \cos\theta_1 \pm \sqrt{4m_1^2v_1'^2 \cos^2\theta_1 - 4(m_1 - m_2)(m_1 + m_2)v_1'^2}}{2(m_1 - m_2)}$$

Requiring v_1 to be real,

$$m_1^2v_1'^2 \cos^2\theta_1 \geq (m_1 - m_2)(m_1 + m_2)v_1'^2$$

$$\cos^2\theta_1 \geq \frac{m_1^2 + m_2^2}{m_1^2}$$

For $0 < \theta < \frac{\pi}{2}$, when θ_1 is a maximum, $\cos\theta_1$ is a minimum. Thus,

$$\cos^2\theta_{1\,max} = \frac{m_1^2 + m_2^2}{m_1^2} = 1 + \frac{m_2^2}{m_1^2}$$

9-42 (a) To find the time at which the first collision occurs,

$$s_b = s_{0,b} - \frac{1}{2} gt^2$$

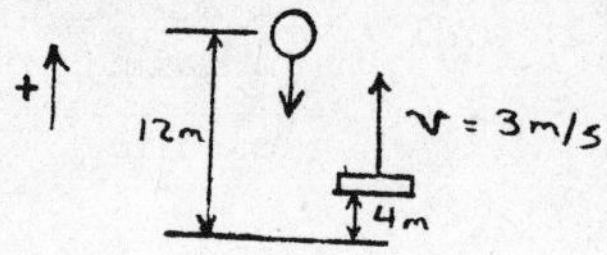

$$s_p = s_{0,p} + vt$$

At the time of collision,

$$s_b = s_p$$

So, $$12 - \frac{1}{2}(9.8)t^2 = 4 + 3t$$

$$4.9\ t^2 + 3t - 8 = 0$$

Discarding $t < 0$

$$t = 1.01\ s$$

The height of the ball and platform is then

$$s_b = s_{0,p} + vt = 4 + 3(1.01) = 7.03\ m$$

The velocity of the ball is then

$$v_b = at = (9.8)(1.01)\ m/s = 9.90\ m/s$$

Since the collision is elastic, kinetic energy is conserved. Assuming no change in the velocity of the platform (i.e. an infinitely massive platform), energy conservation is calculated in the platform's rest frame. The velocity of the ball at impact relative to the platform is

$$v_i = 9.90 + 3 = 12.90\ m/s$$

Using conservation of energy,

$$\frac{1}{2} mv_i^2 = \frac{1}{2} mv_f^2$$

So the velocity of the ball relative to the platfrom after impact is the same as the initial velocity. Returning to the reference frame of the fixed surface, the ball is moving up immediately after impact at the following velocity

$$v_{0,b} = v_f + 3 = v_i + 3 = 15.90\ m/s$$

The equations of motion for the ball after the first collision are

$$v_b = v_{0,b} - g(t - 1.01) = 15.90 - g(t - 1.01)$$

$$s_b = 7.03 + 15.90(t - 1.01) - \frac{1}{2} g(t - 1.01)^2$$

Solving the velocity equation for $v = 0$ gives the time at which maximum height is reached.

$$t = \frac{15.90}{9.8} + 1.01 = 2.63\ s$$

Solving the position equation for t = 2.63 s gives the maximum height reached by the ball.

$$s_{max} = 7.03 + 15.90(1.62) - \frac{9.8}{2}(1.62)^2 = 19.9 \text{ m}$$

(b) $s_b = s_p$

$$7.03 + 15.90\,(t - 1.01) - \frac{1}{2}\,g(t - 1.01)^2 = 4 + 3t$$

$$t = 3.67 \text{ s}$$

9-43

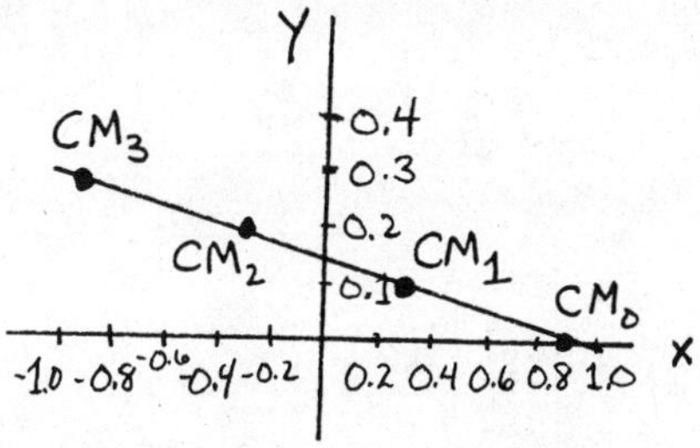

Time, s	Position m_1	m_2	m_3
0	(4, 2)	(-3, 2)	(2, -2)
1	(4, 5)	(-5, 2)	(2, -3)
2	(4, 8)	(-7, 2)	(2, -4)
3	(4, 11)	(-9, 2)	(2, -5)

$$C.M._0 = \frac{[4(2\text{ kg}) - 3(3\text{ kg}) + 2(5\text{ kg})]\hat{i} + [2(2\text{ kg}) + 2(3\text{ kg}) - 2(5\text{ kg})]\hat{j}}{10\text{ kg}}$$

$$= (0.9,\ 0)$$

$$C.M._1 = \frac{[4(2\text{ kg}) - 5(3\text{ kg}) + 2(5\text{ kg})]\hat{i} + [5(2\text{ kg}) + 2(3\text{ kg}) - 3(5\text{ kg})]\hat{j}}{10\text{ kg}}$$

$$= (0.3,\ 0.1)$$

$$C.M._2 = \frac{[4(2\text{ kg}) - 7(3\text{ kg}) + 2(5\text{ kg})]\hat{i} + [8(2\text{ kg}) + 2(3\text{ kg}) - 4(5\text{ kg})]\hat{j}}{10\text{ kg}}$$

$$= (-0.3,\ 0.2)$$

$$C.M._3 = \frac{[4(2\text{ kg}) - 9(3\text{ kg}) + 2(5\text{ kg})]\hat{i} + [11(2\text{ kg}) + 2(3\text{ kg}) - 5(5\text{ kg})]\hat{j}}{10\text{ kg}}$$

$$= (-0.9,\ 0.3)$$

$$\vec{v}_{C.M.} = -0.6\,\hat{i} + 0.1\,\hat{j}$$

Since the velocity of the C.M. is constant, no net force acts on the system.

9-44 (a) Let the x-direction be parallel to the ground. Applying 9-17 in the x-direction,

$$M\frac{dv_x}{dt} = F_x^e + v_{r,x}\frac{dM}{dt}$$

There are no external forces in the x-direction. The relative velocity of the oil is in the y-direction only.

$$M\frac{dv_x}{dt} = 0$$

So, v_x = constant = 2 m/s

The distance travelled is

$$x = v_x t = (2 \text{ m/s})(30 \text{ min})(60 \text{ s/min})$$

$$= 3600 \text{ m}$$

$$= 3.6 \text{ km}$$

(b) Again using 9-17 in the x-direction,

$$M \frac{dv_x}{dt} = 0 + v_{r,x} \frac{dM}{dt}$$

$$= 0 - v_x \frac{dM}{dt}$$

This gives

$$\frac{dv_x}{v_x} = \frac{-dM}{M}$$

Integrating,

$$\ell n\, v_x = -\ell n\, CM$$

where C is a constant.

$$\ell n\, CMv_x = 0$$

$$CMv_x = 1$$

At $t = 0$,

$$M = M_o = 4.5 \times 10^4 \text{ kg}$$

$$v_x = v_o = 2 \text{ m/s}$$

So, $$v_x = \frac{M_o v_o}{M} = \frac{M_o v_o}{M_o - \frac{dm}{dt} t} = \frac{M_o v_o}{M_o - \alpha t}$$

The distance travelled is

$$x = \int_0^t v_x dt = M_o v_o \int_0^t \frac{dt}{M_o - \alpha t}$$

$$= \frac{-M_o v_o}{\alpha} \ell n\, (M_o - \alpha t)\Big|_0^t$$

$$= \frac{-M_o v_o}{\alpha} \ell n \frac{M_o - \alpha t}{M_o}$$

$$= \frac{M_o v_o}{\alpha} \ell n \frac{M_o}{M_o - \alpha t}$$

$$= \frac{M_o v_o}{\alpha} \ell n \frac{M_o}{M_o - m}$$

where m is the mass of the oil lost.

$$x = \frac{M_o v_o}{\alpha} \ell n \frac{4.5}{1.5}$$

$$= \frac{(4.5 \times 10^4 \text{ kg})(2 \text{ m/s})}{(3 \times 10^4 \text{ kg/1800 s})} \ell n\, 3$$

$$= 5.9 \text{ km}$$

9-45 (a) Eq. 9-16 is

$$\frac{d}{dt}(M\vec{v}) = \vec{F}^e + \vec{u}\frac{dM}{dt}$$

Let M be the mass of chain that is already off the table and y be the length of chain that has been lifted off the table.

$$M = \frac{y}{\ell} m$$

The relative velocity $\vec{u}$ is zero since the velocity of $\frac{dM}{dt}$ is equal to the velocity of M. M is constant. In the y-direction,

$$\frac{d}{dt}(Mv) = v\frac{dM}{dt} + M\frac{dv}{dt} = v\frac{dM}{dt} = F^e$$

The net external forces include gravity acting on the length of chain raised above the table and the force F(y) necessary to raise the chain.

$$v\frac{dM}{dt} = -Mg + F(y)$$

$$F(y) = v\frac{dM}{dt} + Mg$$

$$= v\frac{d}{dt}\left(\frac{y}{\ell} m\right) + \left(\frac{y}{\ell} m\right)g$$

$$= v\frac{m}{\ell}\frac{dy}{dt} + \frac{mg}{\ell} y$$

$$= \frac{m}{\ell}(v^2 + gy)$$

(b) Using conservation of energy, all energy gained must be due to the work done by F(y).

$$\int F(y)dy = \frac{1}{2} Mv^2 + Mg\left(\frac{y}{2}\right)$$

$$= \frac{1}{2}\frac{mv^2}{\ell} y + \frac{mg}{2\ell} y^2$$

$$F(y) = \frac{d}{dy}\left(\frac{1}{2}\frac{mv^2}{\ell} y + \frac{mg}{2\ell} y^2\right)$$

$$= \frac{1}{2}\frac{mv^2}{\ell} + \frac{mg}{\ell} y$$

$$= \frac{m}{\ell}\left(\frac{1}{2} v^2 + gy\right)$$

This expression does <u>not</u> correspond to that obtained in part (a) of this problem. The difference in the two forces occurs becuase conservation of energy is not applicable in this problem; some energy is lost to internal friction. That energy is

$$E = \Delta F \times \ell = \frac{mv^2}{2\ell} \times \ell = \frac{mv^2}{2}$$

9-46 (a) $M\frac{dv_y}{dt} = F^e_y - v_{r,y}\frac{dM}{dt} = F^e_y$

The mass of chain hanging off of the table is m; its length is y.

$$m\frac{y}{\ell}\frac{d^2y}{dt^2} = m\frac{y}{\ell}g - T$$

where T is the tension in the chain. The tension in the chain is the force associated with the acceleration of the chain remaining on the table.

$$T = \frac{m}{\ell}(\ell - y)\frac{d^2y}{dt^2}$$

So, $$m\frac{y}{\ell}\frac{d^2y}{dt^2} = m\frac{y}{\ell}g - \frac{m}{\ell}(\ell-y)\frac{d^2y}{dt^2}$$

Solving for the acceleration,

$$\ell\frac{d^2y}{dt^2} = gy$$

(b) Assume the energy remains at the constant value E.

$$E = \frac{1}{2}mv^2 - Mg\frac{y}{2}$$

$$= \frac{1}{2}mv^2 - \frac{mg}{2\ell}y^2$$

$$= \frac{1}{2}m\left(\frac{dy}{dt}\right)^2 - \frac{mg}{2\ell}y^2$$

Differentiating with respect to time,

$$0 = m\left(\frac{dy}{dt}\right)\frac{d^2y}{dt^2} - \frac{mg}{\ell}y\frac{dy}{dt}$$

So, $$\ell\frac{d^2y}{dt^2} = gy$$

(c) Conservation of energy applies in this problem because the chain is straight and, thus, the entire chain is in motion at the same time. If the chained were coiled, energy would not be conserved. As an example, (a) and (b) are reworked assuming only half of the length of chain on the table is in motion.

(a) $$\frac{my}{\ell}\frac{d^2y}{dt^2} = \frac{my}{\ell}g - \frac{1}{2}\frac{m}{\ell}(\ell-y)\frac{d^2y}{dt^2}$$

$$\frac{\ell + y}{2}\frac{d^2y}{dt^2} = gy$$

(b) $$E = \frac{1}{2}\frac{m}{\ell}\left(y + \frac{1}{2}(\ell-y)\right)v^2 - \frac{mg}{2}y^2$$

Differentiating,

$$0 = \frac{\ell + y}{2}\frac{d^2y}{dt^2} + \frac{1}{4}\frac{d^2y}{dt^2} - yg$$

$$\frac{\ell + y}{2}\frac{d^2y}{dt^2} + \frac{1}{4}\frac{d^2y}{dt^2} = gy$$

This is not the same as in (a).

9-47 (a) At the end of the first burn the velocity is

$$v_1 = -v_r \ln\frac{m_1 + m_2 + M_1 + M_2}{M_1 + M_2 + m_2}$$

Since there is no separation velocity, the velocity of the second stage remains at

v_1 after the first stage is jettisoned. After the second burn,

$$v = v_1 - v_r \ \ell n \frac{M_2 + m_2}{M_2}$$

So,

$$v = -v_r \left(\ell n \frac{m_1 + m_2 + M_1 + M_2}{M_1 + M_2 + m_2} + \ell n \frac{M_2 + m_2}{M_2} \right)$$

$$= -v_r \ \ell n \frac{(m_1 + m_2 + M_1 + M_2)(M_2 + m_2)}{(M_1 + M_2 + m_2)M_2}$$

(b) $$v = (2500 \text{ m/s}) \ell n \frac{(30{,}000)(5{,}000)}{(10{,}000)(2{,}000)}$$

$$= 5040 \text{ m/s}$$

(c) $$v' = -v_r \ \ell n \frac{M_1 + M_2 + m_1 + m_2}{M_1 + M_2}$$

$$= 3640 \text{ m/s}$$

(d) $$v'' = -v_r \ \ell n \frac{M_1 + M_2 + m_1 + m_2 - 1000 \text{ kg}}{M_1 + M_2 - 1000 \text{ kg}}$$

$$= 3940 \text{ m/s}$$

The two-stage burn is much more efficient. As would be expected, the one-stage burn without excess mass produces a greater velocity than the same burn including redundant engine and accesories.

10-1 (a) $\ell = rp \sin\theta = mvr \sin\theta = (4\text{ kg})(30\text{ m/s})(6\text{ m}) = 720\ \frac{\text{kg}\cdot\text{m}^2}{\text{s}}$

(b) $a = \frac{dv}{dt} = 15\text{ m/s}^2$

$\tau = rF \sin\theta = mar \sin\theta = (4\text{ kg})(15\text{ m/s}^2)(6\text{ m}) = 360\text{ N}\cdot\text{m}$

10-2 (a) Using the identity,

$$\vec{\ell} = m\vec{r} \times (\vec{\omega} \times \vec{r}) = m[(\vec{r}\cdot\vec{r})\vec{\omega} - (\vec{r}\cdot\vec{\omega})\vec{r}]$$

But in this frame $\vec{r} \perp \vec{\omega}$, giving $\vec{r}\cdot\vec{\omega} = 0$.

Thus,

$$\vec{\ell} = m(\vec{r}\cdot\vec{r})\vec{\omega} = mr^2\omega$$

(b) In this case $\vec{r}\cdot\vec{\omega} = -r\omega\cos\alpha$.

$$\vec{\ell} = m[r^2\vec{\omega} - (r\omega\cos\alpha)\vec{r}] = mr[r\vec{\omega} - (\omega\cos\alpha)\vec{r}]$$

10-3 $\ell = mr^2\omega = (\text{Earth mass})(1\text{ A.U.})^2(\frac{2\pi}{1\text{ year}})$

$= 2.65 \times 10^{40}\ \frac{\text{kg}\cdot\text{m}^2}{\text{s}}$

10-4 From origin O,

$$\vec{\ell} = \vec{r}_1 \times \vec{p}_1 + \vec{r}_2 \times \vec{p}_2$$

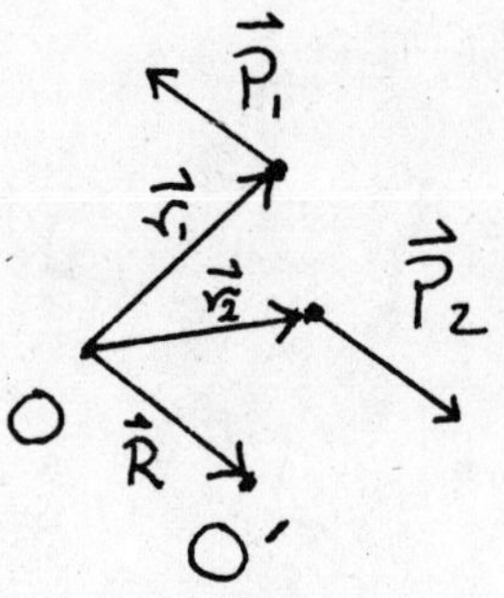

But since $\vec{p}_2 = -\vec{p}_1$,

$$\vec{\ell} = (\vec{r}_1 - \vec{r}_2) \times \vec{p}_1$$

Similarly from O',

$$\vec{\ell}' = (\vec{r}_1' - \vec{r}_2') \times \vec{p}_1$$

But $\vec{r}_1' - \vec{r}_2' = (\vec{r}_1 - \vec{R}) - (\vec{r}_2 - \vec{R}) = \vec{r}_1 - \vec{r}_2$,

so $\vec{\ell}' = \vec{\ell}$ for any $\vec{R}$.

10-5 At the origin O',

$$\tau_{O'} = (N - mg\cos\alpha)r_{O'} = 0$$

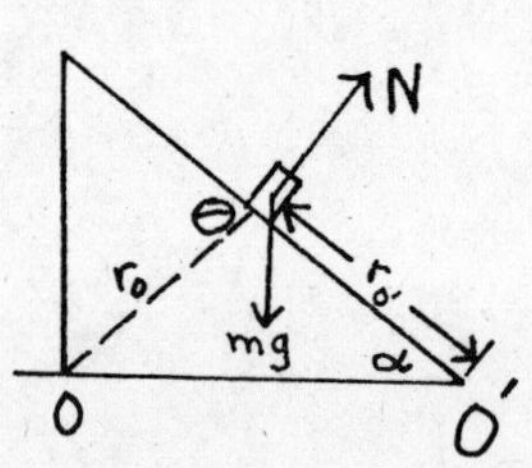

This gives

$$N = mg\cos\alpha$$

At O,

$$\tau_O = [(N - mg\cos\alpha)r_0 \cos\theta] + (mg\sin\alpha)r_0 \sin\theta$$

$$= (mg\sin\alpha)r_0 \sin\theta$$

But $\tau_O = \frac{d\ell_O}{dt} = \frac{d}{dt}(mvr_0 \sin\theta) = mar_O \sin\theta$

Equating the two expressions for τ_0,

$$a = g \sin \alpha$$

10-6 (a) Conservation of angular momentum about the pivot gives

$$m_1 v_1 \ell = m_1 v_1' \ell \cos \alpha + m_2 \ell^2 \omega$$

or $$\omega = \frac{m_1}{m_2} \frac{v_1 - v_1' \cos \alpha}{\ell} = 7.628 \text{ s}^{-1}$$

(b) Conservation of energy gives

$$\frac{1}{2} m_2 \omega^2 \ell^2 = m_2 g \ell (1 - \cos \beta)$$

$$\beta = \cos^{-1} [1 - \frac{\omega^2 \ell}{2g}] = 100.8°$$

(c) $$E_{initial} = \frac{1}{2} m_1 v_1^2 = 18 \text{ J}$$

$$E_{final} = \frac{1}{2} m_1 v_1'^2 + \frac{1}{2} m_2 \omega^2 \ell = 15.79 \text{ J}$$

$$E_i - E_f = 2.210 \text{ J}$$

Thus the collision is not elastic. Linear momentum is not conserved because the rod exerts an external force. Note that the rod contributes no torque since we are computing moments about the pivot; therefore angular momentum is conserved.

10-7 (a) $$\vec{F}_1 = m_1 \frac{d^2\vec{r}_1}{dt^2} = 6\hat{i} + 12t\,\hat{j}$$

$$\vec{F}_2 = m_2 \frac{d^2\vec{r}_2}{dt^2} = 12t\,\hat{i} + 4\,\hat{j}$$

(b) Using Equation 10-13,

$$\vec{R} = \frac{m_1\vec{r}_1 + m_2\vec{r}_2}{m_1 + m_2} = (t^3 + \frac{3}{2}t^2 - \frac{5}{2}t)\hat{i} + (t^3 + t^2)\hat{j}$$

$$\frac{d^2\vec{R}}{dt^2} = (6t + 3)\hat{i} + (6t + 2)\hat{j} = \vec{a}_{CM}$$

Using Equation 9-13,

$$\vec{F}^e = \vec{F}_1^e + \vec{F}_2^e = \vec{F}_{12} + \vec{F}_1 + \vec{F}_{21} + \vec{F}_2 = \vec{F}_1 + \vec{F}_2$$

$$= (12t + 6)\hat{i} + (12t + 4)\hat{j}$$

$$= (m_1 + m_2)\vec{a}_{CM}$$

$$\vec{a}_{CM} = (6t + 3)\hat{i} + (6t + 2)\hat{j}$$

(c) $$\vec{v}_{12} = (6t^2 - 6t + 5)\hat{i} + (-6t^2 + 4t)\hat{j}$$

$$\vec{a}_{12} = (12t - 6)\hat{i} + (-12t + 4)\hat{j}$$

(d) No. We know that

$$\vec{F}_1 = \vec{F}_1^e + \vec{F}_{12}$$

$$\vec{F}_2 = \vec{F}_2^e + \vec{F}_{21}$$

$$\vec{F}_{12} = -\vec{F}_{21}$$

We have three equations and four unknowns. Thus, $\vec{F}_{12}$ cannot be determined.

10-8 (a)

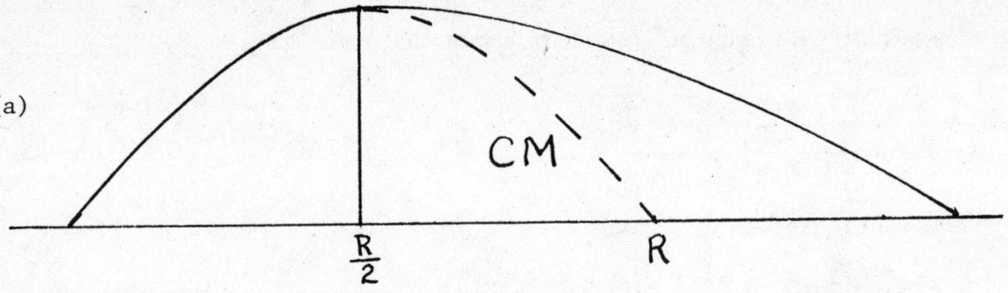

(b) Since one framgent fell from rest, we know that the explosion imparted no vertical acceleration to either particle; thus they will hit ground at the same time. If R is the range of the CM, obviously the range of particle 1 is $\frac{1}{2}$ R, so that of particle 2 is $\frac{3}{2}$ R. Using Equation 4-18 we get, for $\alpha = 45°$,

$$\frac{3}{2} R = \frac{3}{2} \frac{v_o^2}{g} = 38.3 \text{ km}$$

(c) After the explosion,

$$t = \frac{v_o}{g} \sin \alpha = 36.1 \text{ s}$$

10-9 Define a coordinate system whose origin moves with the balloon, $\hat{i}$ = east, $\hat{j}$ = north and $\hat{k}$ = up. The position of the CM of the system which was dropped is, in SI units,

$$\vec{R}(t) = \frac{-1}{2} gt^2 \hat{k} \qquad \vec{R}(20) = -1960 \hat{k}$$

We are given that, for t = 20 s,

$$\vec{r}_1 = -200 \hat{j} - 2500 \hat{k}$$

Solving equation 10-13 for $\vec{r}_2$ with $m_2 = 2 m_1$ gives

$$\vec{r}_2 = 100 \hat{j} - 1690 \hat{k}$$

10-10 $F(r) = -U(r) = -\kappa(n + 1)r^n$

Also, for $r = r_o$,

$$F(r_o) = \frac{-\mu v^2}{r_o}$$

Equating the two expressions for force,

$$\frac{-\mu v^2}{r_o} = -\kappa(n + 1)r_o^n$$

$$K = \frac{\mu v^2}{2} = \frac{\kappa(n+1)r_o^{n+1}}{2} = \frac{n+1}{2} U(r_o)$$

10-11 Using Equation 10-9,

$$\tau = |\vec{r} \times \vec{F}| = RF \sin \phi = mgR \sin \phi$$

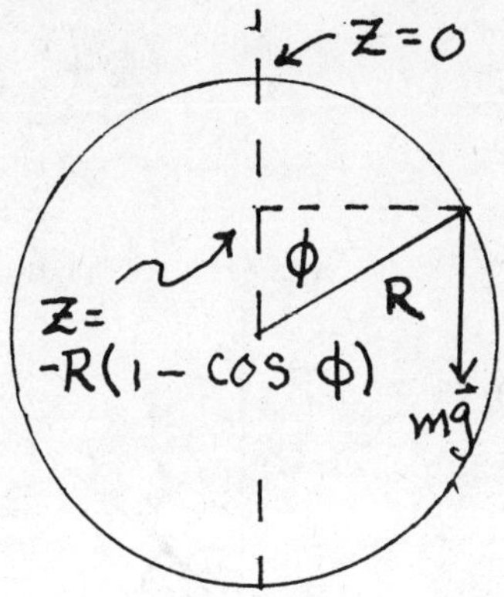

But, for circular motion

$$\tau = \frac{d\ell}{dt} = \frac{d}{dt}(mR^2\omega) = mR^2 \frac{d\omega}{dt} = mR^2\omega \frac{d\omega}{d\phi}$$

Equating the torque expressions,

$$mgR \sin \phi = mR^2\omega \frac{d\omega}{d\phi}$$

$$\int_0^{\phi} mgR \sin \phi \, d\phi = \int_{\omega_0}^{\omega} mR^2\omega \, d\omega$$

$$mgR(-\cos \phi + 1) = \frac{mR^2}{2}\omega_0^2 - \frac{mR^2}{2}\omega^2$$

So, $\omega^2 = \frac{2g}{R}(1 - \cos \phi) + \omega_0^2$

Using conservation of energy,

$$\frac{m\omega_0^2 R}{2} = \frac{m\omega^2 R}{2} - mgR(1 - \cos \phi)$$

Solving for ω^2 gives the same result as above.

10-12 (a) Angular momentum about the tube is conserved since, ignoring gravity, no external forces act on the system. Thus,

$$m\omega r^2 = m\omega_0 r_0^2$$

or $\omega = \omega_0 \frac{r_0^2}{r^2}$

(b) Since $m\omega r^2$ = constant,

$$\frac{d}{dt}(m\omega r^2) = 2m\omega r \frac{dr}{dt} + mr^2 \frac{d\omega}{dt} = 0$$

Solving for α,

$$\alpha = \frac{d\omega}{dt} = -\frac{2\omega}{r}\frac{dr}{dt} = -\frac{2\omega}{r}u = -\frac{2\omega_0 r_0^2 u}{r^3}$$

(c) Insert $\omega = u \frac{d\phi}{dr}$ into the expression in (a) and integrate:

$$\int_0^{\phi} d\phi = \frac{\omega_0 r_0^2}{u}\int_{r_0}^{r} \frac{dr}{r^2}$$

$$\phi = \frac{\omega_0 r_0^2}{u}\left(\frac{1}{r_0} - \frac{1}{r}\right)$$

Let $\beta = \frac{u}{\omega_0 r_0}$.

$$\beta\phi = 1 - \frac{r_0}{r}$$

$$r = \frac{r_0}{1 - \beta\phi}$$

(d) We assume that the string is pulled slowly enough that

$$r \frac{d\omega}{dt} << \omega^2 r$$

The acceleration is then given by $\omega^2 r$. Then

$$\begin{aligned} F &= m\omega^2 r \\ &= m\left(\omega_o^2 \frac{r_o^4}{r^4}\right) r \\ &= m\omega_o^2 r_o^4 \left(\frac{1-\beta\phi}{r_o}\right)^3 \\ &= m\omega_o^2 r_o \, (1-\beta\phi)^3 \end{aligned}$$

10-13

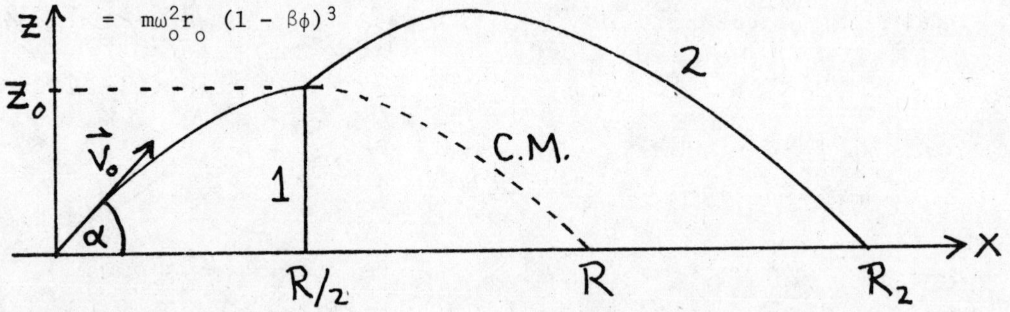

Conservation of momentum gives, for the instant after the explosion,

$$v_{2z} = -v_{1z} \equiv u$$

$$v_{1x} = 0 \qquad v_{2x} = 2v_o \cos\alpha$$

Letting T be the time at which the first fragment hits the ground,

$$Z_1(T) = Z_o - uT - \frac{1}{2} gT^2 = 0$$

$$Z_2(2T) = Z_o + u(2T) - \frac{1}{2} g(2T)^2 = 0$$

Subtracting the second expression from the first,

$$T = 2u/g$$

Clearly

$$\begin{aligned} R_2 &= \frac{R}{2} + (2v_o \cos\alpha)(2T) \\ &= \frac{R}{2} + \frac{8 v_o u}{g} \cos\alpha \end{aligned}$$

From ballistics, the maximum height of the trajectory is

$$Z_o = \frac{v_o^2 \sin^2\alpha}{2g}$$

Use $\quad Z_o - uT - \frac{1}{2} gT^2 = 0$

to get

$$u = \frac{1}{\sqrt{8}} v_o \sin\alpha$$

We now have

$$R_2 = \frac{R}{2} + 2\sqrt{2}\, \frac{v_o^2}{g} \sin\alpha \cos\alpha$$

but again from ballistics,

$$R = \frac{2 v_o^2}{g} \sin\alpha \cos\alpha$$

so our final result is

$$R_2 = \left(\frac{1}{2} + \sqrt{2}\right) R = 1.914 \, R$$

10-14 Radial forces contribute no torque so angular momentum is conserved.

$$d\ell = 0 = d(m\omega r^2) = 2m\omega r\,dr \quad mr^2 d\omega$$

or

$$\frac{d\omega}{\omega} = -2\,\frac{dr}{r}$$

Thus if the rod length is increased by 13 parts per million, the frequency decreases by 26 parts per million.

10-15 (a) $\frac{1}{2}\mu v_{12}^2 = \frac{1}{2}\kappa x^2$

So

$$v_{12} = \sqrt{\frac{\kappa}{\mu}}\,x = 1.61 \text{ m/s}$$

(b) 2 m/s

(c) $\vec{v}_2 - \vec{v}_1 = (v_{12}\cos 30°)\hat{i} + (v_{12}\sin 30°)\hat{j}$

$$m_1\vec{v}_1 + m_2\vec{v}_2 = (m_1 + m_2)\vec{v}_{CM}$$

$$= (m_1 + m_2)v_{CM}\hat{i}$$

Solving gives

$$v_1 = |1.441\,\hat{i} - 0.323\,\hat{j}| = 1.48 \text{ m/s}$$

$$v_2 = |2.839\,\hat{i} + 0.484\,\hat{j}| = 2.88 \text{ m/s}$$

10-16 (a) $\ell = (m_1 + m_2)R^2\omega$

$$\tau = |\vec{r} \times \vec{F}| = (m_1 - m_2)gR\cos\beta$$

$$\tau = \frac{d\ell}{dt} = (m_1 + m_2)R^2\,\frac{d\omega}{dt}$$

$$\frac{d\omega}{dt} = \frac{(m_1 - m_2)g}{(m_1 + m_2)R}\cos\beta$$

$$= \frac{(3 - 2)(9.8)}{(3 + 2)(0.75)}\cos\beta = 2.61\cos\beta$$

(b) Use $\frac{d\omega}{dt} = \frac{d\omega}{d\beta}\frac{d\beta}{dt} = \omega\,\frac{d\omega}{d\beta}$ to get

$$\int_0^\omega \omega\,d\omega = 2.61\int_0^\beta \cos\beta\,d\beta$$

or $\omega = 2.28\sqrt{\sin\beta}\ \text{s}^{-1}$

so $\ell = 6.43\sqrt{\sin\beta}\ \text{J}\cdot\text{s}$

(c) $\tau = 0,\ \ell = \ell_{max} = 6.43\ \text{J}\cdot\text{s}$

(d) $K = \frac{1}{2}(m_1 + m_2)R^2\omega^2 = 7.35\sin\beta$

$$U = -(m_1 - m_2)gR\sin\beta = 7.35\ \sin\beta$$

$$E = K + U = 0 \text{ for all } \beta.$$

10-17 (a) From the end of the rod holding the 2-kg mass,

$$h_{CM} = \frac{(3\text{ kg})(1.5\text{ m}) + (6\text{ kg})(0\text{ m})}{9\text{ kg}} = 0.5\text{ m}$$

$$V_{CM} = \frac{(4\text{ kg})(3\text{ m/s})}{(4 + 2 + 3)\text{kg}} = \frac{4}{3}\text{ m/s}$$

Its direction is parallel to the initial velocity of the 4 kg mass.

(b) Now consider the center of mass to be the origin. By conservation of angular momentum,

$$(4\text{ kg})(3\text{ m/s})(0.5\text{ m}) = (6\text{ kg})(0.5\text{ m})^2\omega + (3\text{ kg})(1\text{ m})^2\omega$$

or $\omega = \frac{4}{3}\text{ s}^{-1}$

(c) Initially,

$$K_i = \frac{1}{2}(4)(3)^2 = 18\text{ J}$$

After the collision,

$$K_f = \frac{1}{2}(9)\left(\frac{4}{3}\right)^2 + \frac{1}{2}(6)\left(\frac{4}{3}\cdot\frac{1}{2}\right)^2 + \frac{1}{2}(3)\left(\frac{4}{3}\cdot 1\right)^2$$

$$= 8 + \frac{4}{3} + \frac{8}{3} = 12\text{ J}$$

$$K_i - K_f = 6\text{ J}$$

10-18 (a) No. Since the string cannot stretch, the puck cannot move parallel to the string and away from the post. It cannot move toward the post and be parallel to the string as this would cause the tension in the string to instantaneously fall to zero.

(b) Since the tension is perpendicular to the direction of motion of the puck, no work is done. Thus, the kinetic energy and the velocity of the puck must remain constant.

(c) The force is not central. The tension of the string acts along a line from the puck to the outside of the post, not to the center of the post. Thus, angular momentum of the puck relative to O is not conserved.

(d) $\omega = \frac{v_o \cos\alpha}{r}$ where α is the angle between r and s.

But $\cos\alpha = \frac{s}{r}$ and $r = \sqrt{R^2 + s^2}$

So, $\omega = \frac{v_o s}{r^2} = \frac{v_o s}{s^2 + R^2}$

(e) $\ell = m\omega r^2 = mv_o s = mv_o(s_o - R\theta)$

(f) $s = s_o - R\theta$

$ds = -R\,d\theta$

$$\omega = \frac{d\theta}{dt} = -\frac{1}{R}\frac{ds}{dt} = \frac{v_o s}{s^2 + R^2}$$

So,

$$t = \int_{s_o}^{s} -\frac{1}{Rv_o}\frac{s^2 + R^2}{s}\,ds = \frac{R}{v_o}\left[\ln\frac{s_o}{s} + \frac{s_o^2 - s^2}{2R^2}\right]$$

10-19 (a)

$$\vec{r}_1 = (3t^2 + 7t)\hat{i} - 2t\hat{j}$$

$$\vec{v}_1 = (6t + 7)\hat{i} - 2\hat{j}$$

$$\vec{L}_1 = m_1\vec{r}_1 \times \vec{v}_1 = 2\begin{vmatrix} \hat{i} & \hat{j} & \hat{k} \\ (3t^2 + 7t) & (-2t) & 0 \\ (6t + 7) & (-2) & 0 \end{vmatrix}$$

$$= 2[(6t^2)\hat{k}] = 12t^2\hat{k}$$

$$\vec{r}_2 = 4t\hat{i} - 3\hat{k}$$

$$\vec{v}_2 = 4\hat{i}$$

$$\vec{L}_2 = -12\,\hat{j}$$

$$\vec{L} = \vec{L}_1 + \vec{L}_2 = -12\hat{j} + 12t^2\hat{k}$$

(b) $\vec{\tau} = \frac{d\vec{L}}{dt} = 24t\hat{k}$

(c) $\vec{P} = m_1\vec{v}_1 + m_2\vec{v}_2 = (12\,t + 26)\hat{i} - 4\hat{j}$

(d) $\vec{F} = \frac{d\vec{p}}{dt} = 12\hat{i}$

10-20 By conservation of angular momentum,

$$m_1v_1r_1 = m_2v_2r_2$$

$$v_2 = \frac{m_1}{m_2}v_1 = \frac{2}{3}(5\text{ m/s}) = \frac{10}{3}\text{ m/s}$$

$$T = \frac{2\pi r}{v} = \frac{2\pi(1.5)}{10/3}\text{ s} = 2.83\text{ s}$$

10-21 $m\omega_1 r_1^2 = m\omega_2 r_2^2$

$$\omega_2 = \left(\frac{r_1}{r_2}\right)^2\omega_1 = 27.78\text{ s}^{-1}$$

10-22 (a) The C.M. is 14.29 cm from the heavier particle. Thus,

$$\vec{L} = (m_1r_1^2 + m_2r_2^2)\vec{\omega}$$

$$L = 6.732\ \frac{\text{kg}\cdot\text{m}^2}{\text{s}}$$

(b) $\tau = \frac{dL}{dt} = \frac{6.732}{8} = 0.8415\ \frac{\text{kg}\cdot\text{m}^2}{\text{s}^2}$

10-23 $\vec{L} = \vec{R} \times \vec{P} + \sum_i \vec{r}_{Ci} \times \vec{p}_{Ci}$

$$\frac{d\vec{L}}{dt} = \frac{d}{dt}(\vec{R} \times \vec{P}) + \frac{d}{dt}\left(\sum_i \vec{r}_{Ci} \times \vec{p}_i\right) = 0$$

$$= \left(\frac{d\vec{R}}{dt} \times \vec{P}\right) + \left(\vec{R} \times \frac{d\vec{P}}{dt}\right) + \frac{d}{dt}\left(\sum_i \vec{r}_{Ci} \times \vec{p}_i\right) = 0$$

Now,

$$\frac{d\vec{R}}{dt} \times \vec{P} = \vec{V} \times M\vec{V} = 0$$

and if there are no external forces, $d\vec{P}/dt = 0$. Thus, the first two terms in the expression for $d\vec{L}/dt$ vanish, so that

$$\sum_i \vec{r}_{Ci} \times \vec{p}_i = \text{constant} \qquad \text{(a)}$$

Then, from the first expression for $d\vec{L}/dt$, we see that $d(\vec{R} \times \vec{P})/dt = 0$; hence,

$$\vec{R} \times \vec{P} = \text{constant} \qquad \text{(b)}$$

(a) The total angular momentum of the particle referred to the C.M. is constant.

(b) The angular momentum of the C.M. referred to 0 is constant.

V-1 Using 4-mm squares,

$$A = \left(\frac{16\ \text{mm}^2}{\text{square}}\right)\left(\frac{69\ \text{squares}}{\text{quadrant}}\right)(4\ \text{quadrants}) = 4416\ \text{mm}^2$$

$$= 44.16\ \text{cm}^2$$

Using 2-mm squares,

$$A = \left(\frac{4\ \text{mm}^2}{\text{square}}\right)\left(\frac{294\ \text{squares}}{\text{quadrant}}\right)(4\ \text{quadrants}) = 4704\ \text{mm}^2$$

$$= 47.04\ \text{cm}^2$$

To extrapolate, find an equation for the area of the circle as a linear function of the size of the small squares.

$$A(16) = 4416 = 16\,m + b$$

$$A(4) = 4704 = 4\,m + b$$

Solve to find b since $A(0) = b$. Multiplying the bottom expression by 4 and subtracting it from the top expression gives

$$-14400 = -3b$$

$$b = 4800\ \text{mm}^2 = 48.00\ \text{cm}^2$$

The true area is

$$A = \pi r^2 = 50.27\ \text{cm}^2$$

V-2 $A \cong \left(\frac{16\ \text{mm}^2}{\text{square}}\right)\left(\frac{77.5\ \text{squares}}{\text{quadrant}}\right)(4\ \text{quadrants}) = 49.6\ \text{cm}^2$

V-3 (a) $A = \int_1^3 (3 + x^2 - 2x)dx = 3x + \frac{x^3}{3} - x^2 \Big|_1^3 = 6\frac{2}{3}$

(b) $A = \int_0^{\pi/2} (5 \sin x - 2x)dx = -5 \cos x - x^2 \Big|_0^{\pi/2} = -\frac{\pi^2}{4} + 5$

V-4 First, find the points of intersection.

$$2x - y - 4 = \frac{y^2}{2} - y - 4 = 0$$

$$y = 4,\ -2$$

$$A = \int_{-2}^{4}\int_{y^2/4}^{(y+4)/2} dx\, dy = \int_{-2}^{4} \left(\frac{y+4}{2} - \frac{y^2}{4}\right)dy$$

$$= \frac{y^2}{4} + 2y - \frac{y^3}{12}\Big|_{-2}^{4} = 9$$

V-5 $\int_0^{\pi/4} (\cos x - \sin x)dx = \sin x + \cos x \Big|_0^{\pi/4} = \sqrt{2} - 1$

V-6 First, find the points of intersection.

$$x^2 + 4x - 12 = 0$$

$$x = 2, -6$$

The corresponding non-imaginary y-values are

$$y = \pm 2\sqrt{2}$$

$$A_s = \int_{-2\sqrt{2}}^{2\sqrt{2}} \left(\sqrt{12 - y^2} - \frac{y^2}{4}\right)dy = 2\int_0^{2\sqrt{2}} \left(\sqrt{12 - y^2} - \frac{y^2}{4}\right)dy$$

Make the following substitution:

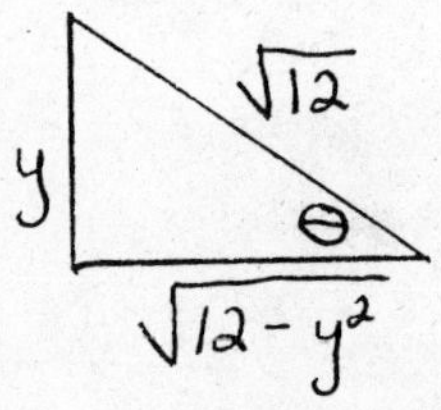

$$\sqrt{12 - y^2} = \sqrt{12} \cos \theta$$

$$\frac{-y\,dy}{\sqrt{12 - y^2}} = -\sqrt{12} \sin \theta \, d\theta$$

$$dy = \sqrt{12} \cos \theta \, d\theta$$

So,

$$A_1 = 24 \int_0^{\cos^{-1} 1/\sqrt{3}} \cos^2 \theta \, d\theta - 2\int_0^{2\sqrt{2}} \frac{y^2}{4} dy$$

$$= 24\left(\frac{\theta}{2} + \frac{\sin 2\theta}{4}\right)\Bigg|_0^{\cos^{-1} 1/\sqrt{3}} - \frac{y^3}{6}\Bigg|_0^{2\sqrt{2}}$$

$$= 11.46 + 5.66 - 3.77 = 13.35$$

$$A_2 = \pi r^2 - A_1 = 24.35$$

$$\frac{A_2}{A_1} = \frac{24.35}{13.35} = 1.824$$

V-7 $$A = 2\int_0^{\pi}\int_0^{a(1 - \cos \phi)} \rho \, d\rho \, d\phi = a^2\int_0^{\pi} (1 - \cos \phi)^2 d\phi$$

$$= a^2\int_0^{\pi} (1 - 2 \cos \phi + \cos^2\phi)d\phi$$

$$= a^2\left(\phi - 2 \sin \phi + \frac{\phi}{2} + \frac{\sin 2\phi}{4}\right)\Bigg|_0^{\pi} = \frac{3\pi}{2} a^2$$

V-8 First, find the points of intersection.

$$R = 2R \cos \phi$$

$$\phi = \cos^{-1} \frac{1}{2} = \pm \frac{\pi}{3}$$

$$A = \int_{-\pi/3}^{\pi/3}\int_R^{2 R \cos \phi} r \, dr \, d\phi = \frac{R^2}{2} \int_{-\pi/3}^{\pi/3} (4 \cos^2\phi - 1)d\phi$$

$$= \frac{R^2}{2} \left[4\left(\frac{\phi}{2} + \frac{\sin 2\phi}{4}\right) - \phi\right]\Bigg|_{-\pi/3}^{\pi/3}$$

$$= R^2 \left[\frac{2\pi}{3} + \frac{\sqrt{3}}{2} - \frac{\pi}{3}\right] = R^2\left(\frac{\pi}{3} + \frac{\sqrt{3}}{2}\right)$$

V-9 Let the side of length a contain the x-axis and the side of length b contain the y-axis.

$$M_x = \int_0^b\int_0^a y \, dx \, dy = \frac{ab^2}{2}$$

$$M_y = \int_0^b \int_0^a x \, dx \, dy = \frac{a^2 b}{2}$$

$$I_x = \int_0^b \int_0^a y^2 \, dx \, dy = \frac{ab^3}{3}$$

$$I_y = \int_0^b \int_0^a x^2 \, dx \, dy = \frac{a^3 b}{3}$$

V-10 $dA = dx\,dy = \rho\,d\rho\,d\phi$

$x = \rho \cos \phi$

$y = \rho \sin \phi$

$$M_x = \iint_S y \, dy \, dx = \iint_S (\rho \sin \phi)(\rho \, d\rho \, d\phi)$$

$$= \iint_S \rho^2 \sin \phi \, d\rho \, d\phi$$

$$M_y = \iint_S x \, dy \, dx = \iint_S \rho^2 \cos \phi \, d\rho \, d\phi$$

V-11 $$I_x = \iint y^2 \, dy \, dx = \int_0^{2\pi} \int_0^R (\rho \sin \phi)^2 \rho \, d\rho \, d\phi$$

$$= \frac{R^4}{4} \int_0^{2\pi} \cos^2 \phi \, d\phi = \frac{\pi R^4}{4}$$

V-12 $$\int_0^h \int_0^{2\pi} \int_0^R \rho d\rho \, d\phi \, dz = \pi R^2 h$$

V-13 $$A = \int_0^{2\pi} \int_0^{\beta} R^2 \sin \theta \, d\theta \, d\phi$$

$$= -R^2(\cos \beta - 1) \int_0^{2\pi} d\phi$$

$$= 2\pi R^2 (1 - \cos \beta)$$

Chapter 11

11-1 (a) $\tau = FR = 3.0 \text{ N}\cdot\text{m}$

(b) $\alpha = \frac{\tau}{I} = \frac{3.0}{\frac{1}{2}MR^2} = 5.56 \text{ rad/s}^2$

(c) $\omega = \alpha t = (5.56)(3) = 16.67 \text{ rad/s}$

11-2 (a) $\omega = 10 \text{ rpm} = \frac{(10)2\pi}{60}\frac{\text{rad}}{\text{s}} = 1.047 \frac{\text{rad}}{\text{s}}$

$\alpha = \frac{\omega}{t} = 0.209 \frac{\text{rad}}{\text{s}^2}$

(b) $\tau = I\alpha = \frac{2}{5}MR^2\alpha = \frac{2}{5}(200)(0.20)^2(0.209) = 0.669 \text{ N}\cdot\text{m}$

11-3 (a) $\omega = \omega_o + \alpha t$

$\omega = 0 \qquad \omega_o = \frac{(33.3)2\pi}{60}\frac{\text{rad}}{\text{s}}$

$\alpha = -\frac{\omega_o}{t} = -3.32 \times 10^{-2} \text{ rad/s}^2$

(b) $\theta = \omega_o t + \frac{1}{2}\alpha t^2$

$= (3.49)(105) - \frac{1}{2}(3.32 \times 10^{-2})(105)^2$

$= 183 \text{ rad} = 29.2 \text{ rev}$

(c) $I = \frac{1}{2}MR^2 = \frac{1}{2}(3.5)(0.15)^2 = 3.94 \times 10^{-2} \text{ kg}\cdot\text{m}^2$

$\tau = I\alpha = (3.94 \times 10^{-2})(-3.32 \times 10^{-2})$

$= -1.31 \times 10^{-3} \text{ N}\cdot\text{m}$

11-4 Notice that $2y = R\theta$, giving $\ddot{y} = a = \frac{1}{2}R\alpha$

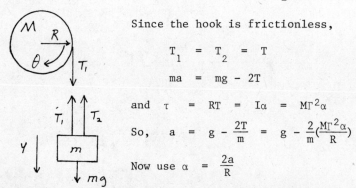

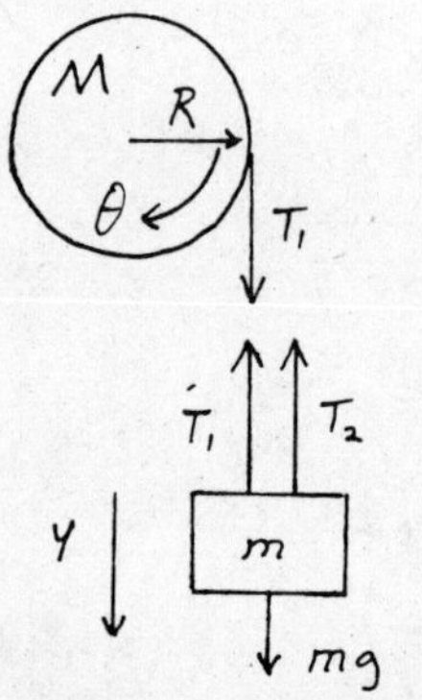

Since the hook is frictionless,

$T_1 = T_2 = T$

$ma = mg - 2T$

and $\tau = RT = I\alpha = M\Gamma^2\alpha$

So, $a = g - \frac{2T}{m} = g - \frac{2}{m}(\frac{M\Gamma^2\alpha}{R})$

Now use $\alpha = \frac{2a}{R}$

$a[1 + 4(\frac{M}{m})(\frac{\Gamma}{R})^2] = g$

$a = g[1 + 4(\frac{M}{m})(\frac{\Gamma}{R})^2]^{-1}$

$= 9.80[1 + 4(\frac{6}{24})(\frac{4}{8})^2]^{-1}$

$$a = 9.80\left(\frac{4}{5}\right) = 7.84 \text{ m/s}^2$$

11-5 $I = \frac{1}{12} M(a^2 + b^2)$

$$\Delta I = \frac{\partial I}{\partial a} a + \frac{\partial I}{\partial b} \Delta b$$

$$\Delta I = \frac{1}{6} M(a\Delta a + b\Delta b)$$

$$L = I\omega = \text{constant}$$

$$\omega \Delta I + I \Delta\omega = 0$$

$$\frac{\Delta\omega}{\omega} = -\frac{\Delta I}{I} = -2\left(\frac{a\Delta a + b\Delta b}{a^2 + b^2}\right)$$

Now use $\Delta a = \eta a$ and $\Delta b = \eta b$ to get $\frac{\Delta\omega}{\omega} = -2\eta = -2 \times 10^{-5}$

11-6

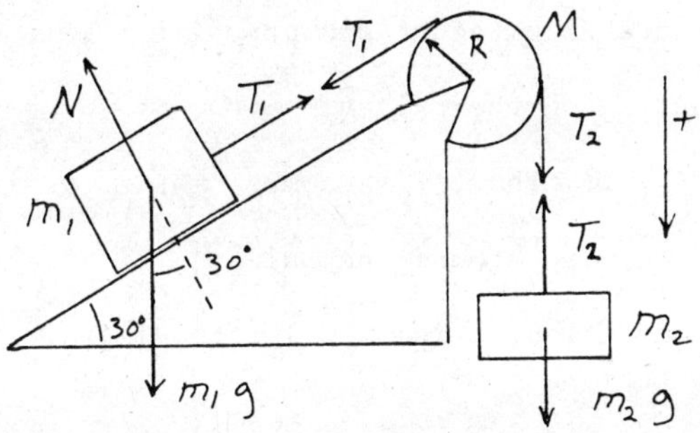

$m_1 = 5$ kg

$m_2 = 4$ kg

$R = 0.03$ m

$\Gamma = 0.02$ m

$M = 0.50$ kg

(1) $T_1 - m_1 g \sin 30° = m_1 a$

(2) $m_2 g - T_2 = m_2 a$

(3) $(T_2 - T_1)R = I\alpha = I\frac{a}{R} = M\Gamma^2 \frac{a}{R}$

(a) Add (1) and (2), then replace $(T_1 - T_2)$ using (3).

$$a = g\frac{m_2 - m_1 \sin 30°}{m_1 + m_2 + M\left(\frac{\Gamma}{R}\right)^2} = 1.59 \text{ m/s}^2$$

(b) $T_1 = m_1(a + g \sin 30°)$

$= 5(1.59 + 4.9) = 32.5$ N

$T_2 = m_2(g - a)$

$= 4(9.8 - 1.59) = 32.8$ N

They are not the same because some torque is needed to speed the rotation of the pulley.

(c)

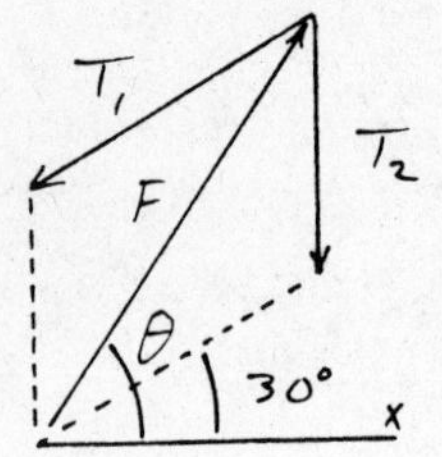

$$F_x = T_1 \cos 30° = 28.15 \text{ N}$$

$$F_y = T_2 + T_1 \sin 30° = 49.05 \text{ N}$$

$$F = \sqrt{F_y^2 + F_x^2} = 56.5 \text{ N}$$

$$\theta = \tan^{-1}\left(\frac{F_y}{F_x}\right) = 60.1°$$

11-7

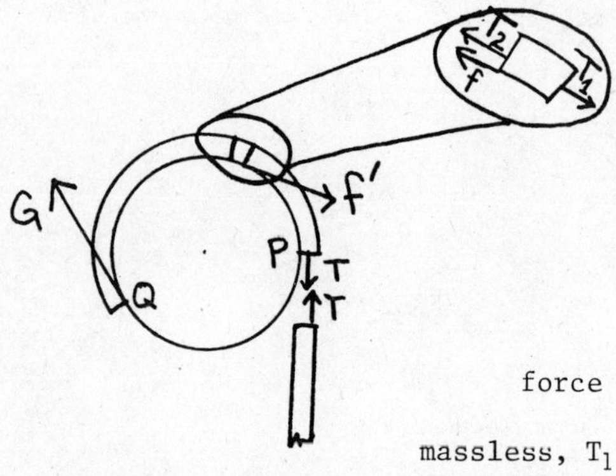

Imagine dividing the rope at point P. Each small segment of the upper length will have two opposite tensions and a frictional force $\vec{f}$. From Newton's third law, there is an equal and opposite frictional force $\vec{f}'$ acting on the pulley. Since the rope is massless, $T_1 = T_2 + f$. Thus, T_1 becomes smaller as it is considered to act at points going counterclockwise around the pulley, If any tension remains at point Q, it is exerted at the connection; this is $\vec{G}$ in the diagram. Thus, the torque on the disc is given by

$$\tau = \Sigma FR = R(G + \Sigma f') = R(G + \Sigma f) = RT$$

Had the pulley not been circular we could not treat the problem as if the tension T acts at one point, since the radius would not be constant. A noncircular pulley would have required more elaborate analysis.

11-8 $W = \tau\theta$

$$\frac{dW}{dt} = \tau\omega = P$$

$$\tau = \frac{P}{\omega} = \frac{(3000)(746)}{(2500)\left(\frac{2\pi}{60}\right)} = 8550 \text{ N}\cdot\text{m}$$

11-9 (a) First, find the linear acceleration of the string.

$$0.7 \text{ m} = \frac{1}{2}at^2 = \frac{1}{2}a(2.0 \text{ s})^2$$

$$a = \frac{0.7 \text{ m}}{2.0 \text{ s}^2} = 0.35 \text{ m/s}^2$$

$$\alpha = \frac{a}{r} = 35 \text{ s}^{-2}$$

(b) $\omega = \omega_o + \alpha t = (35 \text{ rad/s}^2)(2.0 \text{ s}) = 70 \text{ rad/s}$

(c) $\alpha = \frac{\tau}{I} = \frac{RF}{MR^2/2} = \frac{2F}{MR}$

$$F = \frac{MR\alpha}{2} = \frac{(4.0 \text{ kg})(0.01 \text{ m})(35 \text{ s}^{-2})}{2} = 0.70 \text{ N}$$

11-10 $T = 8.64 \times 10^4$ s

$R = 6.35 \times 10^6$ m

$M = 5.98 \times 10^{24}$ kg

$$K = \frac{1}{2} I\omega^2 = \frac{1}{5} MR^2\omega^2 = 2.55 \times 10^{29} \text{ J}$$

The consumption time is

$$t = \frac{\Delta K}{P} = \frac{I\omega\Delta\omega}{P} = \frac{I\omega^2}{P}\frac{\Delta\omega}{\omega}$$

$$= \left(\frac{60 \text{ s}}{8.64 \times 10^4 \text{ s}}\right)\left(\frac{2}{5} MR^2\right)\frac{\omega^2}{P}$$

$$= 1.77 \times 10^6 \text{ y}$$

11-11 $I_1 = 5 + 2 + 2(3)(0.8)^2 = 10.84 \text{ kg} \cdot \text{m}^2$

$I_2 = 5 + 2 + 2(3)(0.2)^2 = 7.24 \text{ kg} \cdot \text{m}^2$

$\omega_1 = \frac{2}{3}\pi$ rad/s

By conservation of momentum,

$$I_1\omega_1 = I_2\omega_2$$

$$\omega_2 = \omega_1\frac{I_1}{I_2} = 3.14 \frac{\text{rad}}{\text{s}}$$

$$\Delta K = \frac{1}{2} I_2\omega_2^2 - \frac{1}{2} I_1\omega_1^2$$

$$= \frac{1}{2}[7.24(3.14)^2 - 10.84(\frac{2\pi}{3})^2] = 11.9 \text{ J}$$

11-12 The work done as the blocks move is

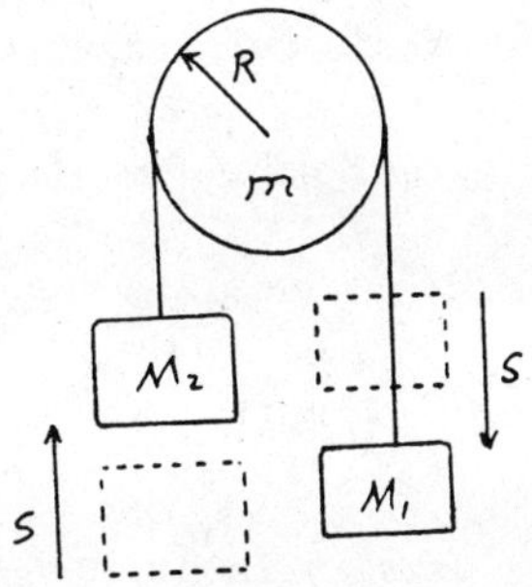

$$W = (M_1 - M_2)gs$$

The change in kinetic energy is

$$\Delta K = \frac{1}{2}(M_1 + M_2)v^2 + \frac{1}{2} I \omega^2$$

$$= \frac{1}{2}(M_1 + M_2)v^2 + \frac{1}{2}(\frac{1}{2} mR^2)(\frac{v}{R})^2$$

Equating the two,

$$(M_1 - M_2)gs = \frac{1}{2}(M_1 + M_2 + \frac{m}{2})v^2$$

Taking the time derivative of each side,

$$(M_1 - M_2)gv = (M_1 + M_2 + \frac{m}{2})va$$

So,

$$a = g \frac{M_1 - M_2}{M_1 + M_2 + \frac{m}{2}}$$

11-13

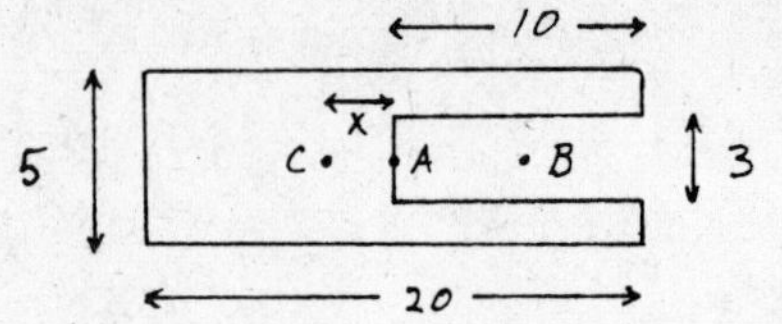

A = CM of solid rod

B = CM of hole

C = Resulting CM

Locate the coordinate origin at point C.

$$M_A x + M_B(x + 5) = 0$$

Since part B is a hole let M_B be negative

$$20(5)^2 x - 10(3)^2(x + 5) = 0$$

$$x = 1.098 \text{ cm}$$

11-14 The center of mass of the first segment is at (0, 2.5,0) with mass $\frac{M}{4}$. The second segment has its center of mass at (2.5, 5, 0) with mass $\frac{M}{4}$. The third segment's center of mass is at (5, 5, 5) with mass $\frac{M}{2}$.

$$X = \frac{\frac{M}{4}(0) + \frac{M}{4}(2.5) + \frac{M}{2}(5)}{M} = \frac{5}{4}(2.5) = 3.125 \text{ cm}$$

$$Y = \frac{\frac{M}{4}(2.5) + \frac{M}{4}(5) + \frac{M}{2}(5)}{M} = \frac{7}{4}(2.5) = 4.375 \text{ cm}$$

$$Z = \frac{\frac{M}{4}(0) + \frac{M}{4}(0) + \frac{M}{2}(5)}{M} = 2.500 \text{ cm}$$

11-15 The center of mass must lie along AA'. Let the origin of the coordinate system be at C. Treat the problem as two masses, one a disc of radius R and the other a disc of radius a with negative mass.

$$M\vec{R} = M_1\vec{R}_1 + M_2\vec{R}_2$$

$$0 = (\pi R^2)x - (\pi a^2)(x + b)$$

$$x = b\frac{a^2}{(R^2 - a^2)}$$

11-16 The mass density is $\sigma = \frac{M}{ab}$.

$$I_x = \int_{y_0}^{y_0+b}\int_{x_0}^{x_0+a} y^2\sigma\, dx\, dy = \frac{M}{b}\int_{y_0}^{y_0+b} y^2 dy$$

$$= \frac{M}{3b}[(y_0 + b)^3 - y_0^3]$$

By symmetry,

$$I_y = \frac{M}{3a}[(x_0 + a)^3 - x_0^3].$$

$$I_z = I_x + I_y = \frac{M}{3ab}[b(x_0 + a)^3 - bx_0^3 + a(y_0 + b)^3 - ay_0^3]$$

11-17

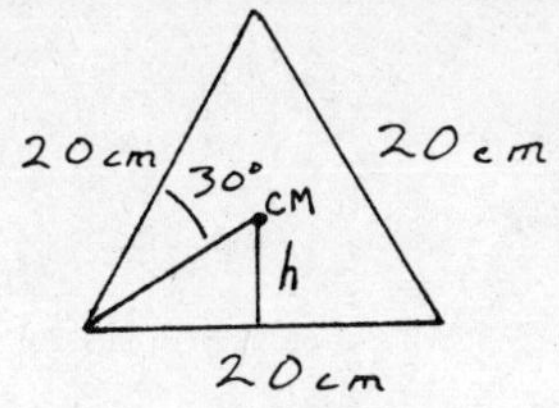

$$h = 10 \tan 30°$$

$$= 5.77 \text{ cm}$$

11-18 Divide the hemisphere into thin discs of thickness dz, each disc having its center of mass located on the z-axis. The center of mass of the hemisphere will then be located along the z-axis.

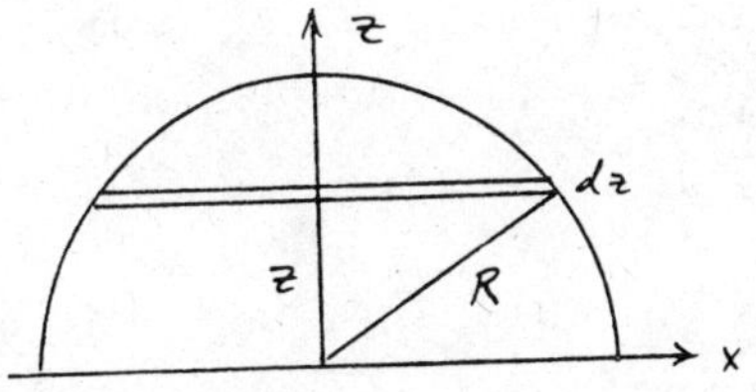

$$\rho_0 = \frac{M}{2/3\ \pi R^3}$$

$$Z = \frac{1}{M}\int_0^R \rho_0 \pi (R^2 - z^2) z dz = \frac{\rho_0 \pi}{M}\frac{R^4}{4} = \frac{3}{8}R$$

11-19 $I = \frac{M}{\ell}\int_{-\frac{\ell}{2}}^{\frac{\ell}{2}} x^2 dx = \frac{M}{2\ell}[(\frac{\ell}{2})^3 + (\frac{\ell}{2})^3] = \frac{1}{12} M\ell^2$

11-20

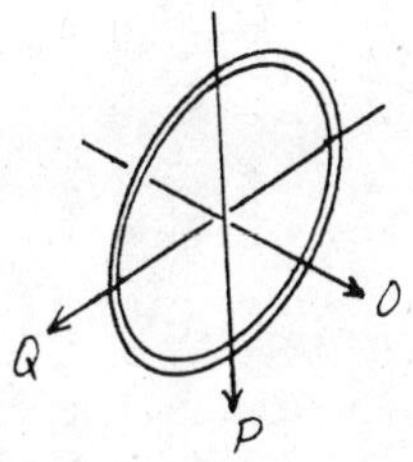

$$I_0 = MR^2 = 2\pi R^3 \lambda$$

$$I_0 = I_P + I_Q$$

By symmetry, $I_P = I_Q$

$$I_P = \frac{1}{2} I_0 = \pi R^3 \lambda$$

11-21 Divide the cylinder into thin cylinders of thickness dR and sum over R.

$$I = \rho_0 \int_{R_1}^{R_2} (2\pi \ell R) R^2 dR$$

$$I = \frac{2\pi\rho_0 \ell}{4} (R_2^4 - R_1^4)$$

$$M = \rho_0 \pi \ell (R_2^2 - R_1^2)$$

$$I = \frac{M}{2}(R_2^2 + R_1^2)$$

11-22 The rotational inertia for a thin disk is

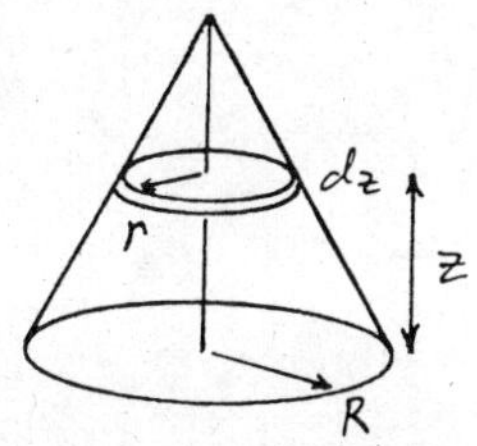

$$dI = \frac{1}{2} MR^2 = \frac{\rho_0}{2}(\pi r^2) r^2$$

So,

$$I_0 = \frac{\rho_0}{2}\int_0^h (\pi r^2) r^2 dz$$

$$r = \frac{(h - z)R}{h}$$

$$I_o = \frac{\rho_o \pi R^4}{2h^4} \int_o^h (h - z)^4 dz = \frac{\rho_o \pi R^4 h}{10} = \rho_o V \Gamma^2$$

Since $V = \frac{1}{3} \pi R^2 h$

$$\frac{\rho_o \pi R^4 h}{10} = \frac{1}{3} \rho_o \pi R^2 h \Gamma^2$$

$$\Gamma = R\sqrt{\frac{3}{10}}$$

11-23

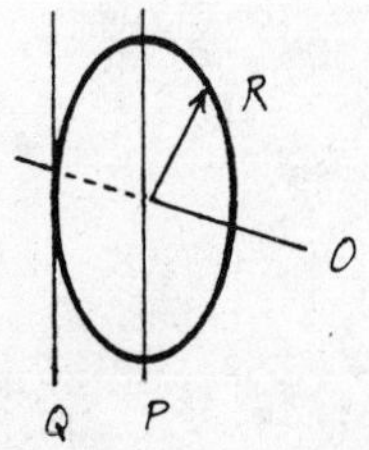

$$I_Q = I_P + MR^2$$

$$I_P = \frac{1}{2} I_o = \frac{1}{2} MR^2$$

$$I_Q = \frac{3}{2} MR^2$$

11-24

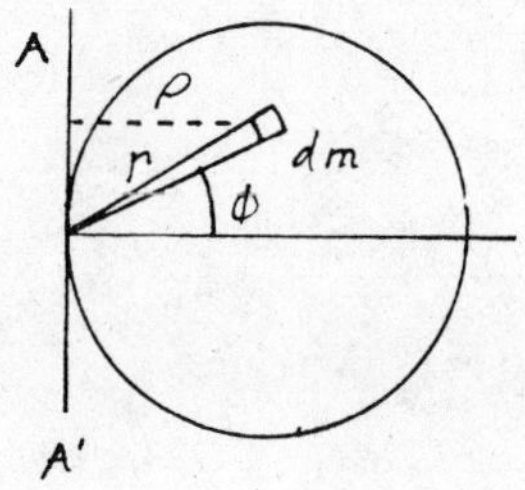

$$dm = \lambda r \, dr \, d\theta$$

$$\rho = r \cos \phi$$

$$r = 2R \cos \phi$$

$$I_A = \int_M \rho^2 dm$$

$$= \int_{-\pi/2}^{\pi/2} d\phi \int_o^{2R \cos \phi} r^2 \cos^2\phi \; \lambda r \, dr$$

$$= \int_{-\pi/2}^{\pi/2} d\phi \; \lambda \cos^2\phi \; \frac{1}{4}(2R \cos \phi)^4$$

$$= 4\lambda R^4 \int_{-\pi/2}^{\pi/2} \cos^6\phi \, d\phi$$

$$= 4\lambda R^4 \left(\frac{\cos^5\phi \;\; \sin \phi}{6} \Bigg|_{-\pi/2}^{\pi/2} + \frac{5}{6} \int_{-\pi/2}^{\pi/2} \cos^4 \phi \, d\phi \right)$$

$$= \frac{20 \; \lambda R^4}{6} \int_{-\pi/2}^{\pi/2} \cos^4 \phi \, d\phi = \frac{20 \; \lambda R^4}{6} \left(\frac{\cos^3 \phi \; \sin \phi}{4} \Bigg|_{-\pi/2}^{\pi/2} + \frac{3}{4} \int_{-\pi/2}^{\pi/2} \cos^2 \phi \, d\phi \right)$$

$$= \frac{5\lambda \; R^4}{2} \left(\frac{\phi}{2} + \frac{\sin 2\phi}{4} \right) \Bigg|_{-\pi/2}^{\pi/2} = \frac{5\lambda R^4 \pi}{4} = \frac{5}{4} MR^2$$

11-25

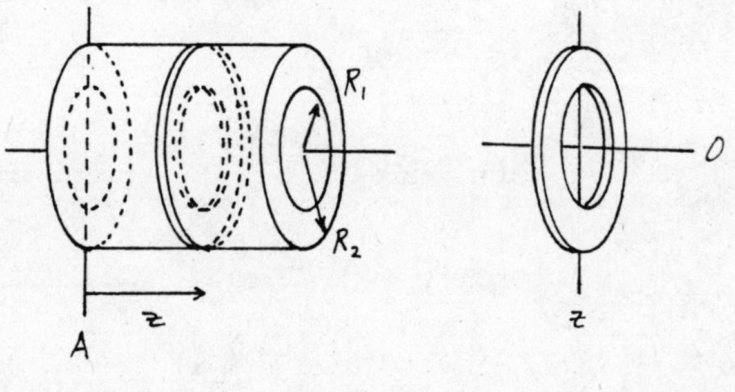

$$dI_z = \frac{1}{2} dI_o = \frac{1}{2}\left(\frac{dm}{2}(R_1^2 + R_2^2)\right)$$

$$= \frac{1}{4}(R_1^2 + R_2^2)dm$$

$$dI_A = dI_z + z^2 dm$$

$$I_A = \frac{1}{4}(R_1^2 + R_2^2)\int_M dm + \int_0^{\ell} z^2 \frac{M\,dz}{\ell}$$

$$= \frac{1}{4}(R_1^2 + R_2^2)M + \frac{1}{3}M\ell^2$$

11-26

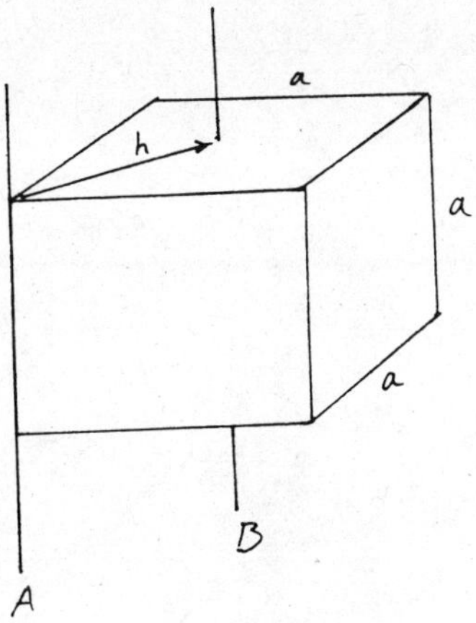

$$I_B = \frac{1}{12}M(a^2 + a^2) = \frac{1}{6}Ma^2$$

$$I_A = I_B + Mh^2 = \frac{1}{6}Ma^2 + M\frac{a^2}{2} = \frac{2}{3}Ma^2$$

11-27

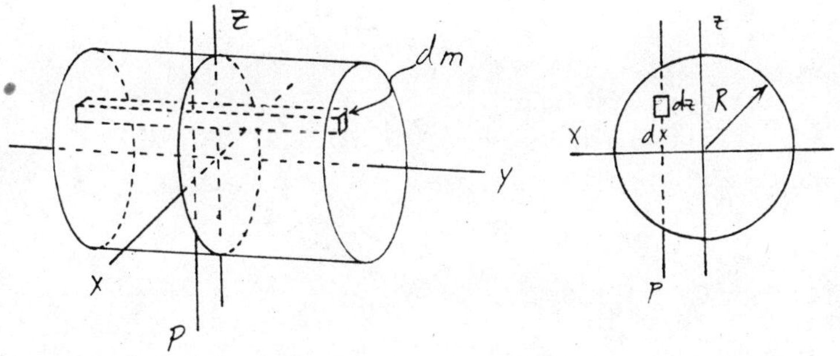

$$dI_z = dI_p + x^2 dm$$

$$dm = \frac{M}{\pi R^2}dx\ dz = \frac{M}{\pi R^2}\rho d\rho\ d\theta$$

$$x^2 = \rho^2 \cos^2\theta$$

From Table 16-1(a),

$$dI_p = \frac{1}{12}\ell^2 dm$$

$$I_z = \frac{1}{12}\ell^2\int_M dm + \frac{M}{\pi R^2}\int_0^{2\pi}\int_0^R (\rho^2 \cos^2\theta)\rho\, d\rho\ d\theta$$

$$= \frac{1}{12}M\ell^2 + \frac{1}{4}MR^2$$

11-28 $I = \frac{1}{2} MR^2 = \frac{1}{2}(1200)(0.9)^2$

$= 486\ \text{kg} \cdot \text{m}^2$

(a) $K = \frac{1}{2} I\omega^2 = \frac{1}{2}(486)(\frac{2\pi}{60} \times 3500)^2$

$= 3.26 \times 10^7\ \text{J}$

(b) $30\ \text{hp} = 2.24 \times 10^4\ \text{W}$

$t = \frac{3.26 \times 10^7\ \text{J}}{2.24 \times 10^4\ \text{W}} = 24.3\ \text{min}$

11-29 $I_1\omega_1 = I_2\omega_2$

$Mr_1^2\omega_1 = Mr_2^2\omega_2$

$\frac{T_1}{T_2} = \frac{\omega_2}{\omega_1} = \frac{r_1^2}{r_2^2} = \frac{\frac{2}{5}(30)^2}{0.09\ R_s^2} = 8.26 \times 10^{-9}$

$T_1 = (24.7)(8.26 \times 10^{-9}) = 2.04 \times 10^{-7}\ \text{days}$

$= 17.6\ \text{ms}$

11-30 Angular momentum is conserved. Rotational kinetic energy is not conserved.

(a) $L = I_A\omega_A + I_B\omega_B = (I_A + I_B)\omega$

$\omega = \frac{I_A\omega_A + I_B\omega_B}{I_A + I_B} = \frac{(7)^2 200 + (5)^2 800}{(7)^2 + (5)^2}$

$= 403\ \text{rpm}$

(b) $\Delta K = \frac{1}{2} I_A\omega_A^2 + \frac{1}{2} I_B\omega_B^2 - \frac{1}{2}(I_A + I_B)\omega^2$

$= \frac{10}{2}[(0.07 \times 200)^2 + (0.05 \times 800)^2$

$-((0.07)^2 + (0.05)^2)(403)^2](\frac{2\pi}{60})^2$

$= 32.6\ \text{J}$

(c) $W_F = \Delta K = \int_0^t \tau\omega\, dt = \int_0^{\Delta\theta} \tau\, d\theta$

$\tau = \frac{\Delta K}{\Delta\theta} = \frac{32.6}{\pi} = 10.4\ \text{N} \cdot \text{m}$

11-31 (a) $\frac{v^2}{R} = \omega^2 R = g$

$\omega = \sqrt{\frac{g}{R}} = 0.50\ \text{s}^{-1}$

(b) $\alpha = \frac{\omega}{t} = \frac{1}{t}\sqrt{\frac{g}{R}}$

$\tau = 2RF = I\alpha$

$$F = \frac{I\alpha}{2R} = \frac{(7.84 \times 10^9\ \text{kg}\cdot\text{m}^2)(5 \times 10^{-4}\ \text{s}^{-2})}{2(39.2\ \text{m})} = 5.00 \times 10^4\ \text{N}$$

(c) By conservation of angular momentum,

$$I\omega = I'\omega'$$

$$\frac{\omega}{\omega'} = \frac{I'}{I} = \frac{I - 2\ MR^2}{I} = 1 - \frac{2(255\ \text{tonne})(39.2\ \text{m})^2}{7.84 \times 10^6\ \text{tonne}\cdot\text{m}^2}$$

$$= 0.90$$

11-32 Let be the length of the shorter edge.

$$Mg\ \frac{\ell}{2} = \frac{1}{2}\ I\omega^2 = \frac{1}{2}\ (\frac{M\ell^2}{3})\omega^2$$

$$\omega = \sqrt{\frac{3g}{\ell}} = 7.67\ \text{rad/s}$$

11-33 Since $\Sigma\vec{F} = 0,\ F = Mg$

(a) $\tau = MgR = I\alpha = \frac{1}{2}\ MR^2\alpha$ $\qquad\alpha = 2g/R$

(b) $W = \frac{1}{2}\ I\omega^2 = \frac{1}{2}\ I(\alpha t)^2$

$$= \frac{1}{4}\ MR^2(\frac{4g^2t^2}{R^2}) = Mg^2t^2$$

$$= 144\ \text{J}$$

(c) $W = F\ell = Mg\ell = Mg^2t^2$

$$\ell = gt^2 = 2.45\ \text{m}$$

11-34

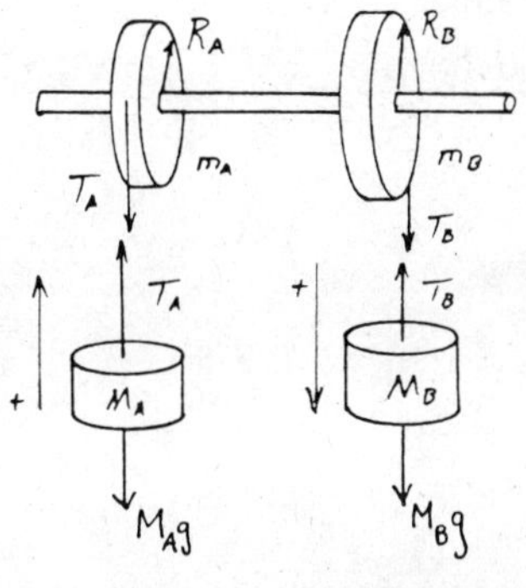

$$T_A = M_A(g + a_A)$$

$$T_B = M_B(g - a_B)$$

$$\frac{a_A}{R_A} = \alpha = \frac{a_B}{R_B}$$

Since $\Sigma\vec{\tau} = I\vec{\alpha}$,

$$T_BR_B - T_AR_A = \frac{1}{2}(m_BR_B^2 + m_AR_A^2)\alpha$$

$$\alpha = \frac{2(T_BR_B - T_AR_A)}{(m_BR_B^2 + m_AR_A^2)}$$

$$= \frac{2[R_BM_B(g - a_B) - R_AM_A(g + a_A)]}{m_BR_B^2 + m_AR_A^2}$$

$$= 2g\ \frac{(R_BM_B - R_AM_A)}{(m_BR_B^2 + m_AR_A^2)} - 2\alpha\ \frac{(R_B^2M_B + R_A^2M_A)}{(m_BR_B^2 + m_AR_A^2)}$$

$$= g\ \frac{R_BM_B - R_AM_A}{R_B^2(M_B + \frac{m_B}{2}) + R_A^2(M_A + \frac{m_A}{2})} = 1.49\ \text{rad/s}^2$$

$a_A = R_A\alpha = 0.119 \text{ m/s}^2$

$a_B = R_B\alpha = 0.149 \text{ m/s}^2$

11-35

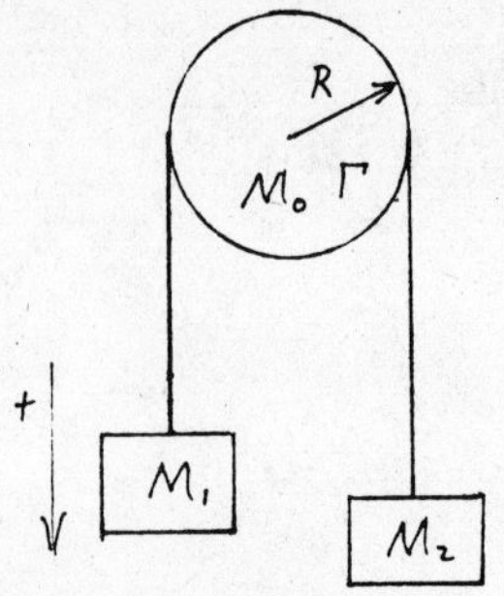

$M_o = 0.20 \text{ kg} \quad R = 7.5 \text{ cm}$

$\Gamma = 2.5 \text{ cm} \quad M_1 = 3.000 \text{ kg}$

$M_2 = 2.950 \text{ kg}$

$\tau_f = 1.5 \times 10^{-4} \text{ N m}$

(a) Initial Energy $= (M_1 - M_2)gs$

Final Energy $= \frac{1}{2}(M_1 + M_2)v^2 + \frac{1}{2} I\omega^2 + \tau_f\phi$

$$(M_1 - M_2)gs = \frac{1}{2}(M_1 + M_2)v^2 + \frac{1}{2} I\omega^2 + \tau_f\phi$$

(b) Taking the time derivative of the above equation,

$$(M_1 - M_2)gv = (M_1 + M_2)va + I\omega\alpha + \tau_f\omega$$

Using $\omega R = v$, $\alpha R = a$, and $I = M_o\Gamma^2$,

$$a = \frac{(M_1 - M_2)g - \tau_f/R}{M_1 + M_2 + M_o(\frac{\Gamma}{R})^2} = 0.082 \text{ m/s}^2$$

$$a = \frac{A - \delta}{B + \varepsilon} = \frac{A(1 - \frac{\delta}{A})}{B(1 + \frac{\varepsilon}{B})} \cong \frac{A}{B}(1 - \frac{\delta}{A} - \frac{\varepsilon}{B})$$

$$\text{percentage due to pulley} = -\frac{\varepsilon}{B} = \frac{-0.0222}{5.950} = -0.37\%$$

$$\text{percentage due to friction} = -\frac{\delta}{A} = -\frac{0.0020}{0.49} = -0.41\%$$

(c) $s = \frac{1}{2} at$

$$t = \sqrt{\frac{2s}{a}} = 3.50 \text{ s}$$

(d) $\frac{2s}{a} = t^2$

Taking derivatives,

$$\frac{-2s}{a^2} da = 2t\, dt$$

$$\frac{-t^2}{a} da = 2t\, dt$$

$$\frac{da}{a} = \frac{-2\, dt}{t} = \pm \frac{2(0.08 \text{ s})}{3.50 \text{ s}} = 4.6\%$$

(e) The main limitation is in the observer's reaction time. Increasing the fall time would improve results.

11-36

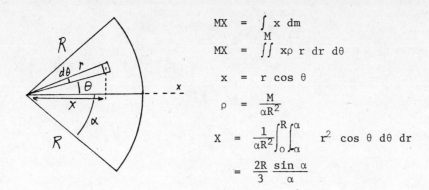

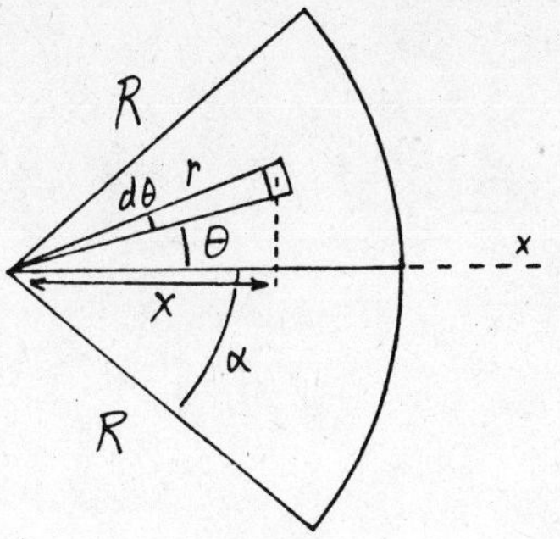

$$MX = \int_M x \, dm$$

$$MX = \iint x\rho \; r \, dr \, d\theta$$

$$x = r \cos \theta$$

$$\rho = \frac{M}{\alpha R^2}$$

$$X = \frac{1}{\alpha R^2}\int_0^R\int_{-\alpha}^{\alpha} r^2 \cos \theta \, d\theta \, dr$$

$$= \frac{2R}{3}\,\frac{\sin \alpha}{\alpha}$$

If $\alpha = \pi$, then $X = 0$ as expected since the wedge is now a disk.

If α approaches zero, $\frac{\sin \alpha}{\alpha} \to 1$. So, $\lim_{\alpha \to 0} X = \frac{2R}{3}$

11-37 $y = \sqrt{\frac{x}{a}}$ $\qquad y = \overline{\frac{x}{a}}$

$$A = 2\int_0^b y \, dx = \frac{2}{\sqrt{a}}\int_0^b \sqrt{x} \, dx = \frac{4}{3\sqrt{a}} b^{3/2}$$

$$AX = 2\int_0^b x \, y \, dx = \frac{2}{\sqrt{a}}\int_0^b x^{3/2} \, dx$$

$X = \frac{3}{5} b$, and $Y = 0$, obviously

11-38 $$M = \sigma_o \int_0^R\int_{-\pi}^{\pi} \cos (\phi/2)d\phi \; r \, dr$$

$$= \frac{\sigma_0 R^2}{2}(2 \sin \phi/2)\Big|_{-\pi}^{\pi} = 2\sigma_0 R^2$$

$$MX = \sigma_o \int_0^R\int_{-\pi}^{\pi} \cos(\phi/2) r \cos \phi \; r \, d\phi \, dr$$

$$= \frac{\sigma_0 R^3}{3}\int_{-\pi}^{\pi} \cos(\tfrac{\phi}{2}) \cos \phi \, d\phi$$

$$= 2\,\frac{\sigma_0 R^3}{3}\int_0^{\pi} \cos (\tfrac{\phi}{2})[1 - 2 \sin^2 \tfrac{\phi}{2}]d\phi$$

$$= \frac{2\sigma_0 R^3}{3}(2 \sin (\tfrac{\phi}{2}) - \tfrac{4}{3} \sin^3(\tfrac{\phi}{2}))\Big|_0^{\pi}$$

$$X = \frac{2}{9} R$$

$Y = 0$ from symmetry.

11-39 $$M = \int dm = \lambda_0\int_{-\ell/2}^{\ell/2} x^2 \, dx$$

$$= \frac{1}{12}\lambda_0 \ell^3$$

$$I_A = \int x^2 dm = 2\lambda_0\int_0^{\ell/2} x^4 dx = \frac{1}{80}\lambda_0 \ell^5 = \frac{3}{20} M\ell^2$$

11-40

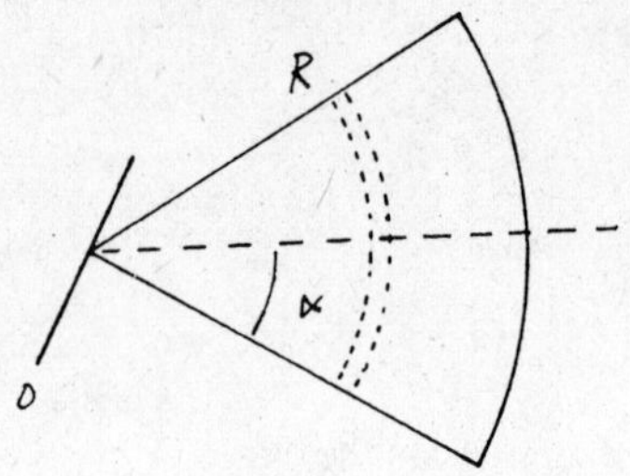

$$I_o = \sigma_o \int_0^R (2\alpha\, r) r^2 dr$$

$$= \frac{\sigma_o\ \alpha\ R^4}{2} = \frac{\alpha R^4}{2}\left(\frac{M}{\alpha R^2}\right)$$

$$= \frac{1}{2} MR^2 = M\Gamma^2$$

$$\Gamma = \frac{R}{\sqrt{2}}$$

11-41 $h = \ell \cos 30° = \frac{\sqrt{3}}{2}\ell$

$$y = \frac{1}{\sqrt{3}} x \qquad dm = 2\sigma_o\ y\, dx = \frac{2\sigma_o}{\sqrt{3}} x\ dx$$

$$dI_A = dI_o + x^2 dm = \frac{1}{12}(2\ y)^2 dm + x^2 dm$$

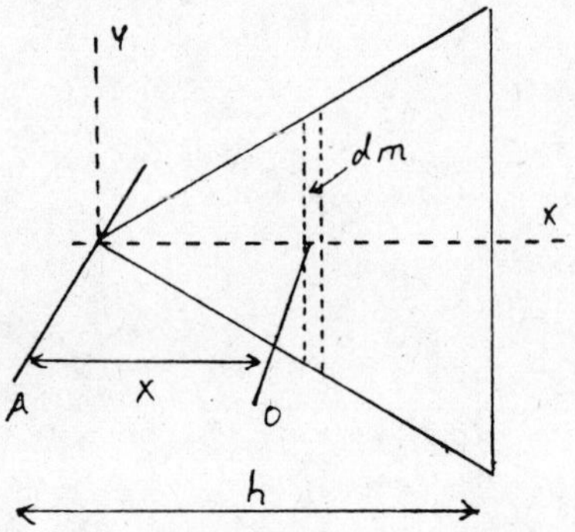

$$I_A = \frac{1}{3}\ \frac{2\ \sigma_o}{\sqrt{3}}\int_0^h xy^2 dx + \frac{2\sigma_o}{\sqrt{3}}\int_0^h x^3 dx$$

$$= \frac{5\ \sigma_o\ h^4}{9\sqrt{3}}$$

$$M = \sigma_o \int_0^h 2y\ dx = \frac{2\ \sigma_o\ h^2}{\sqrt{3}\ \ 2}$$

$$I_A = \frac{5}{9} Mh^2 = \frac{5}{12} M\ell^2 = 3.75 \times 10^{-3}\ \text{kg} \cdot \text{m}^2$$

11-42

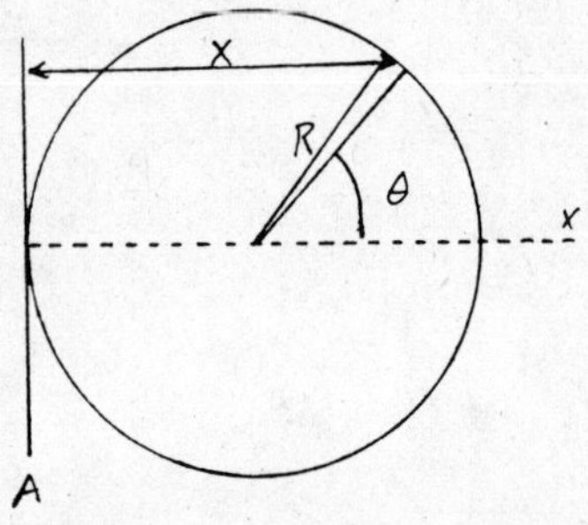

$$\lambda = \frac{M}{2\pi R}$$

$$dI_A = \lambda R\ x^2 d\theta$$

$$x^2 = R^2(1 + \cos\theta)^2$$

$$I_A = \lambda R^3 \int_0^{2\pi} (1 + \cos\theta)^2 d\theta$$

$$= \lambda R^3 \int_0^{2\pi} (1 + 2\cos\theta + \cos^2\theta) d\theta$$

$$= \frac{MR^2}{2\pi}\left[\theta + 2\sin\theta + \frac{\theta}{2} + \frac{1}{4}\sin 2\theta\right]_0^{2\pi}$$

$$= \frac{3}{2} MR^2$$

11-43 $dI_o = \frac{1}{2} r^2 dm$

$$dI_1 = \frac{1}{2} dI_o = \frac{1}{4} r^2 dm$$

$$dI_2 = dI_1 + z^2 dm$$

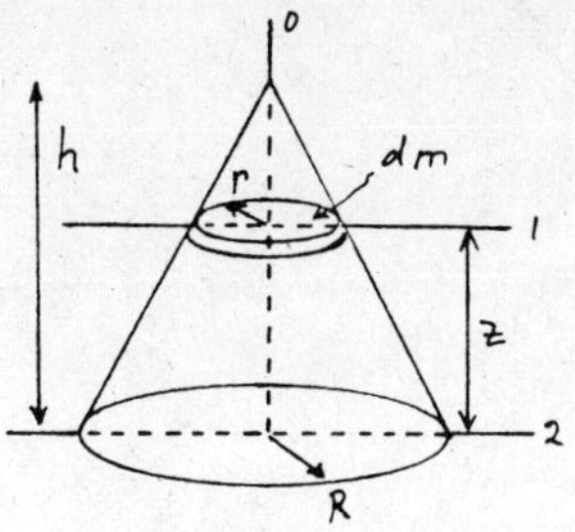

$$I_2 = \frac{1}{4}\int r^2 dm + \int z^2 dm$$

$$dm = \pi r^2 \rho_0 dz$$

$$\rho_0 = \frac{M}{\frac{1}{3}\pi R^2 h}$$

$$r = \frac{R}{h}(h - z)$$

$$I_2 = \frac{\pi R^4 \rho_o}{4h^4}\int_0^h (h-z)^4\, dz + \frac{\pi R^2 \rho_0}{h^2}\int_0^h z^2(h-z)^2 dz$$

$$= \frac{\pi R^4 \rho_0}{20 h^4}[(h-z)^5]_h^0 + \frac{\pi R^2 \rho_0}{h^2}\int_0^h (h^2z^2 - 2hz^3 + 2^4)dz$$

$$= \frac{\pi R^4 \rho_0 h}{20} + \pi R^2 \rho_0 h^3(\frac{1}{3} - \frac{1}{2} + \frac{1}{5})$$

$$= \pi R^2 h\rho_0(\frac{R^2}{20} + \frac{h^2}{30}) = \frac{3M}{10}(\frac{R^2}{2} + \frac{h^2}{3})$$

11-44

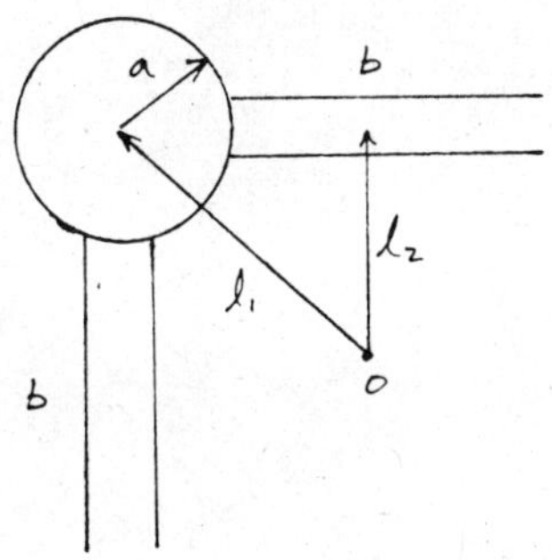

$$\ell_1 = \sqrt{2}\,(\frac{b}{2} + a) = 20\sqrt{2} \text{ cm}$$

$$\ell_2 = a + \frac{b}{2} = 20 \text{ cm}$$

For one sphere,

$$I_s = \frac{2}{3}Ma^2 + M\ell_1^2$$

For one rod,

$$I_r = \frac{1}{12}Mb^2 + M\ell_2^2$$

For the entire system,

$$I = 4(I_s + I_r) = 4M(\frac{2}{3}a^2 + \frac{1}{12}b^2 + \ell_1^2 + \ell_2^2) = 8M\Gamma^2$$

$$\Gamma = 25.5 \text{ cm}$$

11-45

r/R_s	δ	$\Delta r = r_i - r_{i-1}$	$(r/R_s)^2$	$(r/R_s)^2\delta\Delta r$	$(r/R_s)^4$	$(r/R_s)^4\delta\Delta r$
0	148	0	0	0	0	0
0.05	125	0.05	0.0025	0.0156	6.25×10^{-6}	0.000039
0.10	86	0.05	0.010	0.043	0.0001	0.00043
0.15	56	0.05	0.0225	0.063	0.0005	0.0014
0.20	36	0.05	0.04	0.072	0.0016	0.0029
0.30	12	0.10	0.09	0.108	0.0081	0.0097
0.40	4	0.10	0.16	0.064	0.0256	0.0102
0.60	0.5	0.20	0.36	0.036	0.130	0.013
0.80	0.1	0.20	0.64	0.012	0.41	0.0082

1.00 0 0.20 1.00 0 1.00 0

$$M = \Sigma\, 4\pi r_i^2\, \delta\, \Delta r$$

$$M\Gamma^2 = \Sigma (4\pi r_i^2\, \delta\, \Delta r)\frac{2}{3}\, r_i^2$$

$$\Gamma^2 = \frac{2}{3}\frac{\Sigma\, 4\pi r_i^4\, \delta\, \Delta r}{\Sigma\, 4\pi r_i^2\, \delta\, \Delta r} = \frac{2}{3}\frac{\Sigma\, r_i^4\, \delta\, \Delta r}{\Sigma\, r_i^2\, \delta\, \Delta r}$$

$$\Gamma^2 = 2R_s^2(0.000039 + 0.00043 + 0.0014 + 0.0029 + 0.0097 + 0.0102 + 0.013 + 0.0082)/$$
$$3(0.0156 + 0.043 + 0.063 + 0.072 + 0.108 + 0.064 + 0.036 + 0.012)$$

$$\Gamma = 0.22\ R_s$$

11-46 $I = \frac{Ma^2}{6} = \frac{(3\text{ kg})(0.20\text{ m})^2}{6} = 0.02\text{ kg}\cdot\text{m}^2$

(a) $\alpha = \frac{\tau}{I} = \frac{\beta t}{I} = \frac{(0.36)(2.5)}{(0.02)} = 45\text{ rad/s}^2$

(b) $\omega = \int_0^t \alpha\, dt = \int_0^t \frac{\beta}{I} t\, dt = \frac{\beta t^2}{2I} = 56.25\text{ rad/s}$

(c) $\phi = \int_0^t \omega\, dt = \int_0^t \frac{\beta t^2}{2I}\, dt = \frac{\beta}{2I}\frac{t^3}{3} = 46.88\text{ rad}$

(d) $L = I\omega = (0.02)(56.25) = 1.125\text{ kg}\cdot\text{m}^2/\text{s}$

12-1 $J = MV$

$$I\omega = MV(x - 0.5)$$

$$\frac{M\ell^2}{12}\omega = MV(x - 0.5)$$

The velocity of the end of the stick relative to the center of mass is

$$v_o = r\omega = (0.5)\omega = (0.5)12V(x - 0.5)$$

The velocity of the end of the stick relative to the surface is

$$v = 0 = -v_o + V$$

So

$$v_o = V = (0.5)12V(x - 0.5)$$

$$x = 0.67 \text{ m}$$

12-2 (a) By conservation of linear momentum,

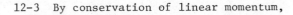

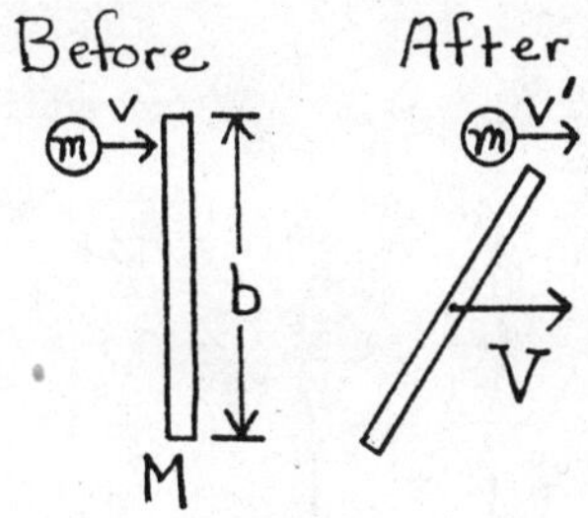

$$MV + mv' = mv$$

$$V = \frac{m}{M}(v - v') = 2.5 \text{ m/s}$$

(b) By conservation of angular momentum,

$$I\omega + \frac{mv'b}{2} = \frac{mvb}{2}$$

$$\frac{1}{12}Mb^2\omega = \frac{mvb}{4}$$

$$\omega = \frac{3m}{M}\frac{v}{b} = 10 \text{ rad/s}$$

(c) $$\frac{1}{2}mv^2 = \Delta E + \frac{1}{2}mv'^2 + \frac{1}{2}MV^2 + \frac{1}{2}I\omega^2$$

$$\Delta E = \frac{1}{2}m(v^2 - v'^2) - \frac{1}{2}MV^2 - \frac{1}{2}I\omega^2 = 2 \text{ J}$$

$$\frac{E}{mv^2/2} = \frac{2 \text{ J}}{16 \text{ J}} = 1/8$$

12-3 By conservation of linear momentum,

$$(M + m)V = mv$$

$$V = \frac{m}{M + m}v = \frac{0.02}{50.02}(400) = 0.16 \text{ m/s}$$

To find the new center of mass,

$$(M + m)x = \frac{mR}{2}$$

$$x = \frac{mR}{2(M + m)} = \frac{(0.02)(0.20)}{2(50.02)} = 4.0 \times 10^{-5} \text{ m} = 0.04 \text{ mm}$$

By conservation of angular momentum,

$$mv(\frac{R}{2} - x) = I\omega = [m(\frac{R}{2} - x)^2 + \frac{2}{5} MR^2 + Mx^2]\omega$$

$$\frac{mvR}{2} \cong \frac{2}{5} MR^2\omega$$

$$\omega \cong \frac{5}{4}\frac{mv}{MR} = 1.00 \text{ rad/s}$$

12-4 Using conservation of energy,

$$mg(h - R) = \frac{1}{2} mv^2 + \frac{1}{2} I_C\omega^2$$

$$= \frac{1}{2} (mv^2 + \frac{mv^2}{2}) = \frac{3mv^2}{4}$$

$$h = \frac{3v^2}{4g} + R = 0.46 \text{ m}$$

12-5 (1) $\alpha = a/r_2$

(2) $F - f = Ma$

(3) $r_2 f + Fr_1 = \Gamma^2 M\alpha = \Gamma^2 \frac{Ma}{r_2}$

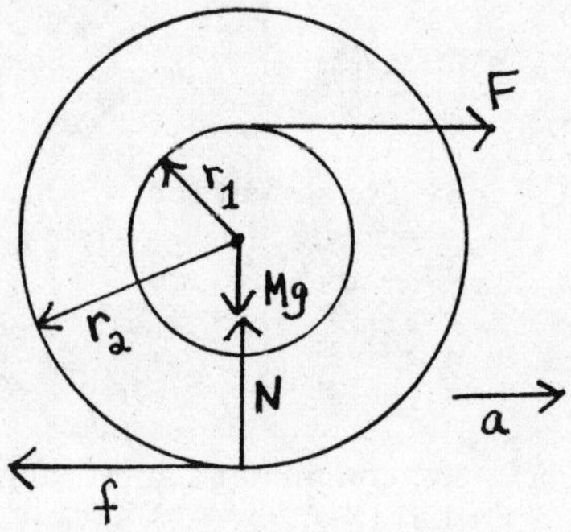

Dividing (3) by r_2 and adding to (2),

$$F(\frac{r_1}{r_2} + 1) = Ma(1 + \frac{\Gamma^2}{r_2^2})$$

$$a = \frac{F\left(\frac{r_1}{r_2} + 1\right)}{M\left(1 + \frac{\Gamma^2}{r_2^2}\right)} = \frac{5(\frac{8}{5})}{2(\frac{41}{25})} = 2.439 \text{ m/s}^2$$

$$f = F - Ma = 5 - 4.878 = 0.122 \text{ N}$$

$$f \le \mu_s N = \mu_s mg$$

$$\mu_s \ge \frac{f}{mg} = 0.0062$$

12-6 (1) $mg \sin\gamma - f = ma$

(2) $fR = I\alpha = mr^2 \frac{a}{R}$

When the object is placed on its flat end, (1) is the only applicable equation. Sliding begins when

$$f = mg \sin\gamma = \mu_s mg \cos\gamma$$

$$\mu_s = \tan\gamma = \tan 15° = 0.268$$

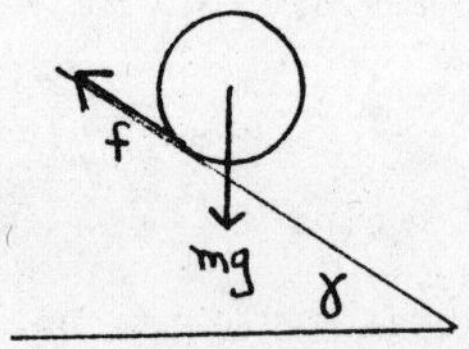

Multiplying (2) by $\frac{1}{R}$ and adding to (1) gives

$$a = \frac{g \sin\gamma}{1 + \frac{\Gamma^2}{R^2}}$$

Using (2),

$$f = m\Gamma^2 \frac{a}{R^2} = \frac{m\Gamma^2}{R^2} \frac{g \sin \gamma}{1 + \frac{\Gamma^2}{R^2}} = \frac{mg \sin \gamma}{1 + \frac{R^2}{\Gamma^2}}$$

$$\frac{R^2}{\Gamma^2} = \frac{mg \sin \gamma}{f} - 1 = \frac{mg \sin \gamma}{\mu_s mg \cos \gamma} - 1$$

$$= \frac{1}{\mu_s} \tan 30° - 1 = \frac{0.577}{0.268} = 1.155$$

$$\frac{R}{\Gamma} = 1.075$$

Since this value is close to unity, the object would possibly be a thin-walled cylinder.

12-7 (1) $I_1 = I_2 = I = \frac{1}{2} MR^2$

(2) $TR = I\alpha_1 = I\alpha_2 = I\alpha$

(3) $Mg - T = Ma_2$

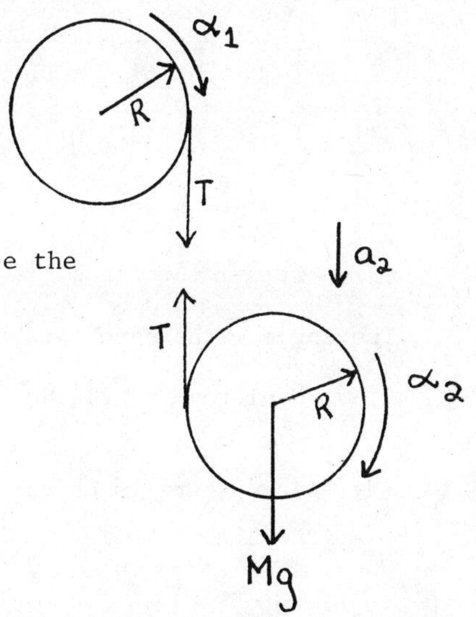

Divide (2) by R and add to (3).

$$Mg = \frac{I\alpha}{R} + Ma_2 = \frac{1}{2} MR\alpha + Ma_2$$

To find the relationship between α and a, let s be the distance the one disc has fallen.

$$s = \theta_1 R + \theta_2 R + \ell_o$$

$$a_2 = \frac{d^2s}{dt^2} = R(\alpha_1 + \alpha_2) = 2R\alpha$$

So,

$$g = \frac{1}{2} R\alpha + a_2 = \frac{a_2}{4} + a_2 = \frac{5}{4} a_2$$

$$a_2 = \frac{4}{5} g = 7.84 \text{ m/s}^2$$

$$\alpha = \frac{a}{2R} = 49.0 \text{ rad/s}^2$$

$$T = \frac{I\alpha}{R} = \frac{1}{2} MR\alpha = 15.7 \text{ N}$$

12-8

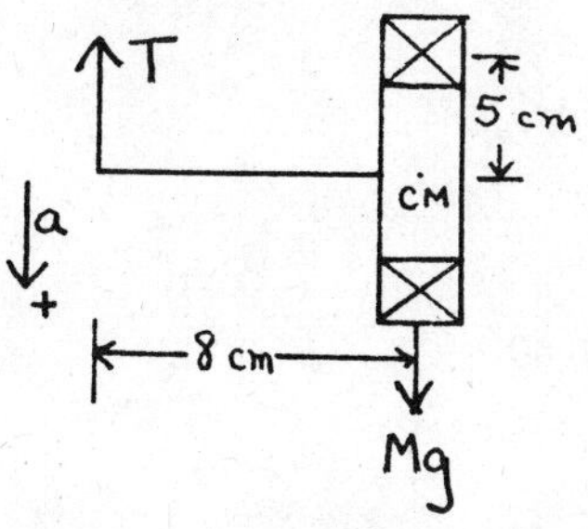

$$\frac{Mg}{3} = Mg - T$$

$$T = \frac{2}{3} Mg = 16.3 \text{ N}$$

$\tau_c = (16.3\text{ N})(0.08\text{ m}) = 1.30\text{ N}\cdot\text{m}$

$L_c = M'\Gamma^2\omega = 1.50\text{ kg}\cdot\text{m}^2/\text{s}$

Assuming that $\Omega << \omega$,

$\vec{\tau}_c = \vec{\Omega} \times \vec{L}_c$

$\Omega = \frac{\tau_c}{L_c} = 0.867\text{ rad/s}$

Since $\Omega << \omega$ the equation for $\vec{\tau}_c$ used above was appropriate.

12-9 $MV = 6\text{ N}\cdot\text{s}$

$V = \frac{6}{M} = 10.0\text{ m/s}$

$L = Jd = (6\text{ N}\cdot\text{s})(0.05\text{ m}) = 0.30\text{ kg}\cdot\text{m}^2/\text{s}$

$\omega = \frac{L}{I} = \frac{L}{\frac{1}{12}M\ell^2} = \frac{0.30}{\frac{1}{12}(0.6)(0.3)^2} \quad 66.67\text{ rad/s}$

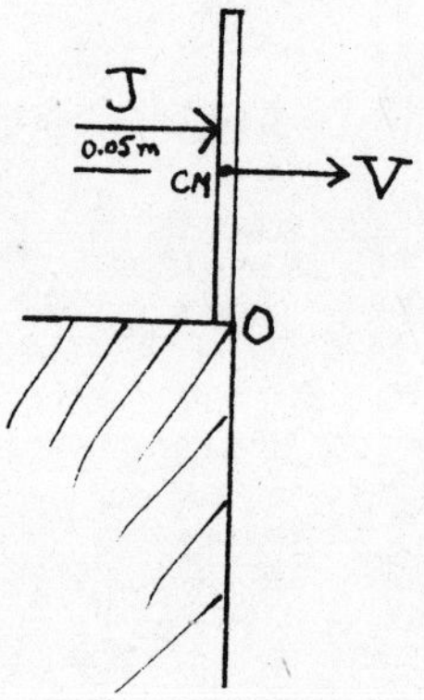

$X(1) = Vt = 10.0\text{ m}$

$Y(1) = Y_o - \frac{1}{2}gt^2 = 0.15 - \frac{1}{2}(9.8)(1)^2 = -4.75\text{ m}$

$\theta = \omega t = 66.67\text{ rad} = 10.61\text{ rev}$

The angle with respect to the vertical is

$\theta = 0.61\text{ rev} = 219.6°$

12-10 (a) Since there is no rotation about the center of mass,

$$\tau = 0 = f(0.5) + F(0.10) - Nx$$

$$F = \frac{Nx - 0.5\,f}{0.10} = \frac{mgx - (0.5)\mu_k\,mg}{0.10}$$

$$= 10\text{ mg }(x - 0.5\,\mu_k)$$

F will have its largest value when x has its largest value; the largest value of x is 0.25 m.

$F = 10\text{ mg}(0.25 - 0.20) = 147\text{ N}$

(b) $a = \frac{F - f}{m} = \frac{147 - (0.40)(30)(9.8)}{30} = 0.98\text{ m/s}^2$

(c) For incipient sliding,

$$F_{max} = f_{max} = \mu_s N = \mu_s\,mg = 0.60\text{ mg}$$

For incipient tipping and no sliding,

$$F = f < f_{max}$$

$$\Sigma\tau = F(0.1) + (0.5)f - Nx = 0$$

$$= 0.6\ F - Nx$$

$$F = \frac{mgx}{0.6}$$

$$F_{max} = \frac{0.25\ mg}{0.6} = 0.42\ mg$$

Since F_{max} is less for tipping than for sliding, the box will tip first.

12-11 For incipient sliding, the decelerating force is

$$f = f_{max} = \mu_s N = 0.60\ mg$$

For incipient tipping and no sliding,

$$\Sigma\tau = (0.60)f - Nx = 0$$

$$f = \frac{Nx}{0.6} = \frac{mgx}{0.6}$$

$$f_{max} = \frac{0.30\ mg}{0.60} = 0.50\ mg$$

The box will tip before it slides.

$$f = 0.5\ mg = ma_{max}$$

$$a_{max} = 4.9\ m/s^2$$

$$t = \frac{v}{a_{max}} = \frac{60\ km/h}{4.9\ m/s^2} = 3.40\ s$$

12-12

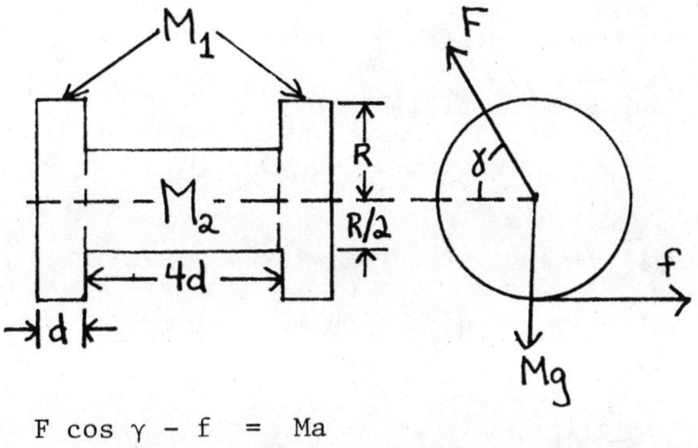

(1) $F\cos\gamma - f = Ma$

(2) $fR - \frac{FR}{2} = I_A\alpha = [\frac{1}{2} M_1 R^2 + \frac{1}{2} M (\frac{R}{2})^2]\frac{a}{R}$

$$M = M_1 + M_2 = (2d\pi R^2)\rho + [4d\pi(\frac{R}{2})^2]\rho = (3d\pi R^2)\rho$$

$$M_1 = \frac{2}{3} M$$

$$M_2 = \frac{1}{3} M$$

Rewriting (2),

$$I_A\alpha = [\frac{1}{3} MR^2 + \frac{1}{6} M(\frac{R}{2})^2]\frac{a}{R} = \frac{3}{8} MRa$$

So,

$$fR - \frac{FR}{2} = \frac{3}{8} MRa = \frac{3}{8}R(F \cos \gamma - f)$$

$$f = \frac{F(3 \cos \gamma + 4)}{11} \le \mu(Mg - F \sin \gamma)$$

Rewriting,

$$3 \cos \gamma + 4.4 \sin \gamma \le 4.8$$

Let $\phi = 34.3°$. The above equation can be written

$$\sin \phi \cos \gamma + \cos \phi \sin \gamma \le \frac{4.8}{5.325}$$

$$\sin (\phi + \gamma) \le 0.901$$

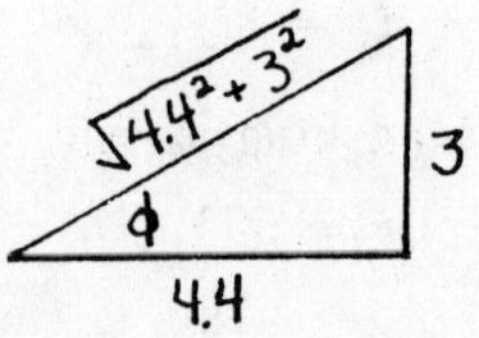

$$\phi + \gamma \le 64.3°$$

or

$$\phi + \gamma \ge 115.7°$$

So $\gamma \ge 81.4°$ or $\gamma \le 30.0°$

12-13 (a) Since the point of contact is fixed, no work is done.

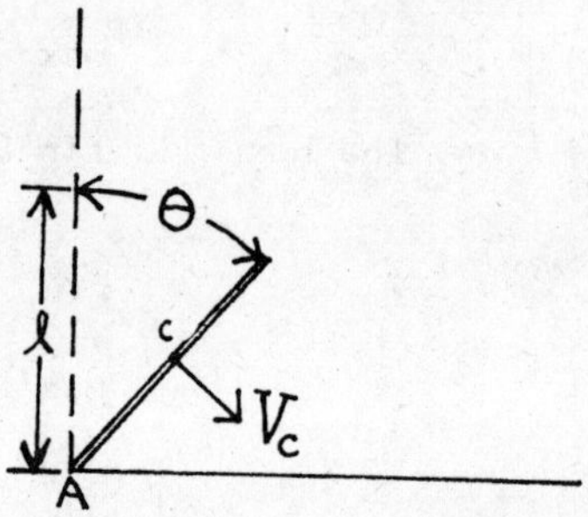

Using conservation of energy,

$$mg\frac{\ell}{2} = \frac{1}{2} I_A\omega^2 = \frac{1}{2} I_A\left(\frac{2v_c}{\ell}\right)^2$$

$$v_c = \sqrt{\frac{3g\ell}{4}} = 2.71 \text{ m/s}$$

(b) Since there are no external forces acting on the stick, the velocity of the center of mass must be straight down. The vertical velocity of the point of contact is zero.

$$v = 0 = \frac{-1}{2} \ell\omega + v_c \qquad \omega = -\frac{2v_c}{\ell}$$

Using conservation of energy, with no work done at the point of contact since there is no force in the direction of motion,

$$mg \frac{\ell}{2} = \frac{1}{2} mv_c^2 + \frac{1}{2} I_c\omega^2$$

$$= \frac{1}{2} mv_c^2 + \frac{1}{2}\left(\frac{m\ell^2}{12}\right)\left(\frac{4v_c^2}{\ell^2}\right)$$

Solving for v_c,

$$v_c = \sqrt{\frac{3g\ell}{4}} = 2.71 \text{ m/s}$$

(c) Using energy conservation for part (a),

$$\frac{mg\ell}{2} (1 - \cos \theta) = \frac{1}{2} I_A\omega^2 = \frac{1}{2}\left(\frac{m\ell^2}{3}\right)\omega^2 = \frac{2}{3} mv_c^2$$

Taking the time derivative of each side,

$$\frac{g\ell\omega}{2}\sin \theta = \frac{4}{3} v_c a_c = \frac{4}{3}\frac{\omega\ell}{2} a_c$$

At $\theta = \frac{\pi}{2}$,

$$a_c = \frac{3}{4} g$$

For part (b),

$$\frac{\ell}{2}(1 - \cos\theta)mg = \frac{1}{2} mv_c^2 + \frac{1}{2} I_c\omega^2$$

Since $v_c = \frac{\ell}{2} \omega \sin\theta$,

$$1 - \cos\theta = \frac{v_c^2}{g\ell}(1 + \frac{1}{3 \sin^2\theta})$$

Taking the time derivative,

$$\omega \sin\theta = \frac{v_c^2}{g\ell}(\frac{-2\omega \cos\theta}{3 \sin^3\theta}) + \frac{2a_c v_c}{g\ell}(1 + \frac{1}{3 \sin^2\theta})$$

Evaluating at $\theta = \frac{\pi}{2}$,

$$\omega = \frac{2a_c v_c}{g\ell}(\frac{4}{3}) = \frac{4a_c\omega}{3g}$$

$$a_c = \frac{3}{4} g$$

(d) For case (a), the centripal acceleration is due to the frictional force.

Thus,

$$f = ma = \frac{mv_c^2}{\ell/2} = 2.94 \text{ N}$$

The radial acceleration is

$$a_c = \frac{mg - N}{m} = \frac{3}{4} g$$

$$N = \frac{1}{4} mg = 0.49 \text{ N}$$

For case (b), the only reactive force is the normal force.

$$N = ma - mg = \frac{1}{4} mg = 0.49 \text{ N}$$

12-14 (a) Using conservation of energy,

$$mg\ \Delta h = \frac{1}{2} mv^2 + \frac{1}{2} I\omega^2$$

$$= \frac{1}{2} mv^2 + \frac{1}{2}(\frac{2}{5})\omega^2$$

$$= \frac{7}{10} mv^2$$

$$v = \sqrt{\frac{10g\ \Delta h}{7}} = \sqrt{\frac{10g\,[0.27-(0.20-0.01)]}{7}}$$

$$= 1.06 \text{ m/s}$$

(b) Since the marble rolls without slipping, the frictional force is zero.

$$ma = N + mg = \frac{mv^2}{r} = \frac{m}{r}(\frac{10\ g\ \Delta h}{7})$$

$$N = \frac{m}{r}\left(\frac{10\ g\ \Delta h}{7}\right) - mg$$

$$= \frac{10\ mg}{7}\left(\frac{0.08}{0.10 - 0.01}\right) - mg = 0.27\ mg = 0.040\ N$$

(c) When $N = 0$,

$$mg = \frac{mv^2}{r} = \frac{10\ mg\ \Delta h}{7r}$$

$$\Delta h = h - 0.19 = \frac{7r}{10}$$

$$h = \frac{7}{10}(0.09) + 0.19 = 0.253\ m = 25.3\ cm$$

12-15

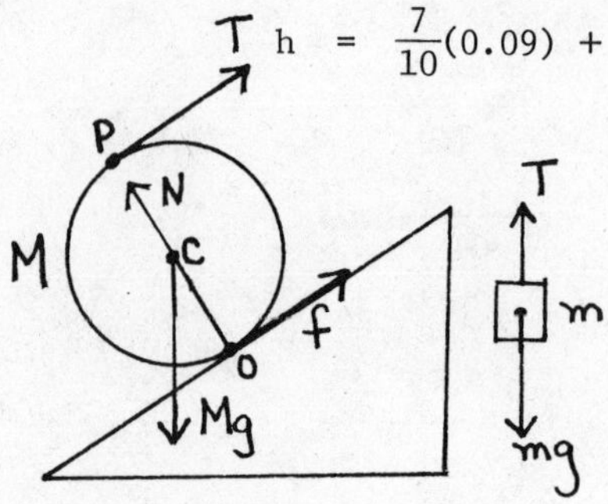

(1) $Mg \sin\theta - T - f = Ma_c$

(2) $(f - T)r = \frac{1}{2}Mr^2\alpha = \frac{1}{2}Mra_c$

(3) $T - mg = 2\ a_c m$

$a_o = 0$

$a_p = 2\ a_c = a_m$

Divide (2) by r. Multiply (3) by 2. Add (1), (2) and (3) together to get

$$Mg \sin\theta - 2\ mg = \frac{3}{2}Ma_c + 4\ ma_c$$

Solving for a_c,

$$a_c = \frac{M \sin\theta - 2m}{\frac{3}{2}M + 4m}\,g = 0.889\ m/s^2$$

$$a_m = 2a_c = 1.78\ m/s^2$$

12-16

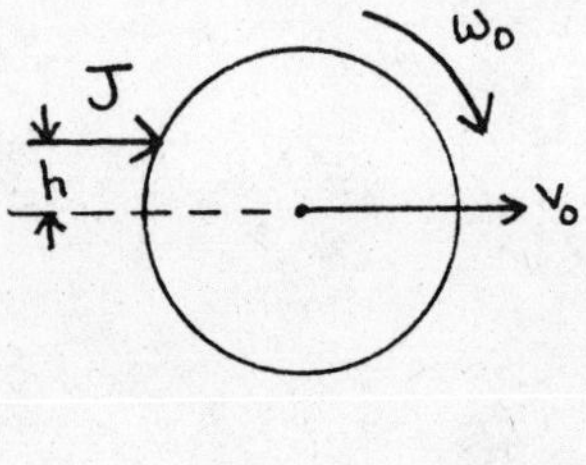

(1) $Jh = I_c\omega_o$

(2) $J = Mv_o$

If the ball rolls without slipping.

$R\omega_o = v_o$

So, $h = \frac{I_c\omega_o}{J} = \frac{I_c\omega_o}{Mv_o} = \frac{I_c}{MR} = \frac{2}{5}R$

12-17 Using equations (1) and (2) above, with $h = \frac{4}{5}R$,

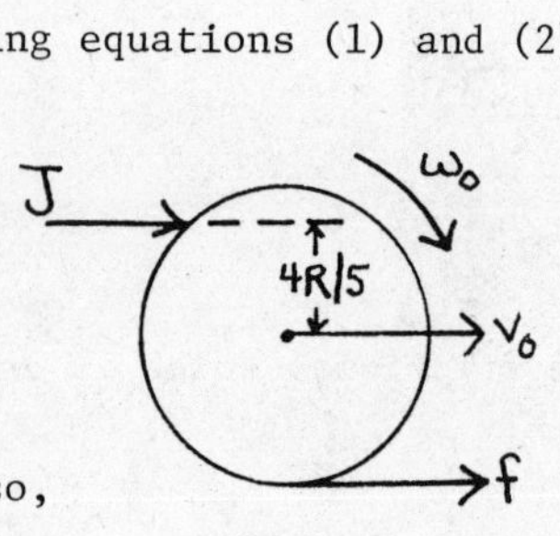

$$\frac{4}{5}RMv_o = I_c\omega_o = \frac{2}{5}MR^2\omega_o$$

$$v_o = \frac{1}{2}R\omega_o$$

Also,

$$(1)\quad \omega_1 = \omega_0 - \frac{R}{I_c}\int_0^T f\,dt = \omega_0 - \frac{5}{2MR}\int_0^T f\,dt$$

$$(2)\quad v_1 = v_0 + \frac{1}{M}\int_0^T f\,dt$$

Multiplying (1) by $\frac{2}{5}$ R and adding to (2) gives

$$\frac{2}{5} R\omega_1 + v_1 = \frac{2}{5} R\omega_0 + v_0$$

When the ball rolls without slipping,

$$R\omega_1 = v_1$$

So,

$$\frac{2}{5} v_1 + v_1 = \frac{2}{5}(2\ v_0) + v_0$$

$$v_1 = \frac{9}{7} v_0$$

12-18

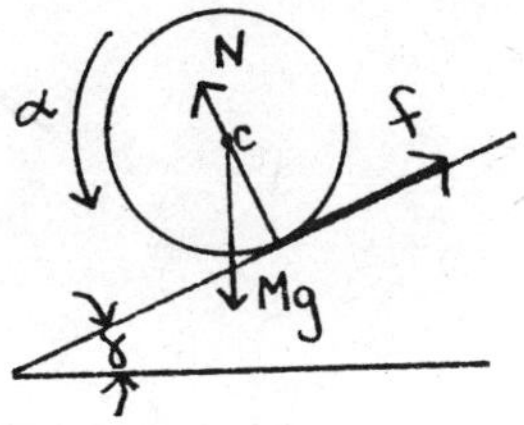

(a) $Mg \sin\gamma - f = Ma$ (1)

$Rf = I_c\alpha = \frac{2}{5} MRa$ (2)

Solving for a,

$$a = \frac{5}{7} g \sin\gamma$$

$$f = \frac{I_c\alpha}{R} = \frac{2}{5} Ma = \frac{2}{7} Mg \sin\gamma$$

$N = Mg \cos\gamma$ (3)

$$\mu_s \geq \frac{f}{N} = \frac{2}{7} \tan\gamma$$

(b) When slipping occurs,

$$f = \mu_k N = \mu_k Mg \cos\gamma$$

Equation (1) becomes

$Mg \sin\gamma - \mu_k Mg \cos\gamma = Ma$ (4)

Equation (2) becomes

$Rf = \mu_k MgR \cos\gamma = I_c\alpha = \frac{2}{5} MR^2\alpha$ (5)

Dividing (4) by (5),

$$\tan\gamma - \mu_k = \frac{5a\mu_k}{2R\alpha}$$

This gives

$$a = \frac{8}{5} R\alpha$$

$$Ma = Mg \sin \gamma - f = Mg \sin \gamma - \mu_k Mg \cos \gamma$$

$$a = g\left(\frac{\sqrt{2}}{2} - \frac{1}{5}\frac{\sqrt{2}}{2}\right) = 5.54 \text{ m/s}^2$$

$$\alpha = \frac{5a}{8R} = 69.2 \text{ rad/s}^2$$

(c) $$s = \frac{1}{2} at^2$$

$$t = \sqrt{\frac{2s}{a}} = 0.601 \text{ s}$$

(d) $$v = aT = 3.33 \text{ m/s}$$

$$\omega = \alpha T = 41.6 \text{ rad/s}$$

(e) $$K = \frac{1}{2} I_c\omega^2 + \frac{1}{2} Mv^2 = \frac{1}{2}\left(\frac{2}{5} MR^2\right)\omega^2 + \frac{1}{2} Mv^2$$

$$= 25.64 \text{ J}$$

(f) The displacement due to rolling is

$$s_r = R\theta = R\left(\frac{1}{2} \alpha t^2\right) = 0.625 \text{ m}$$

The displacement due to sliding is then

$$s_s = s - s_r = 1.00 - 0.625 = 0.375 \text{ m}$$

The force of friction is

$$f = \mu_k Mg \cos \gamma = 5.54 \text{ N}$$

$$W_f = s_s f = 2.08 \text{ J}$$

(g) $$\Delta PE = Mg\, s \sin \theta = 27.7 \text{ J}$$

$$K + W_f = 27.7 \text{ J}$$

12-19

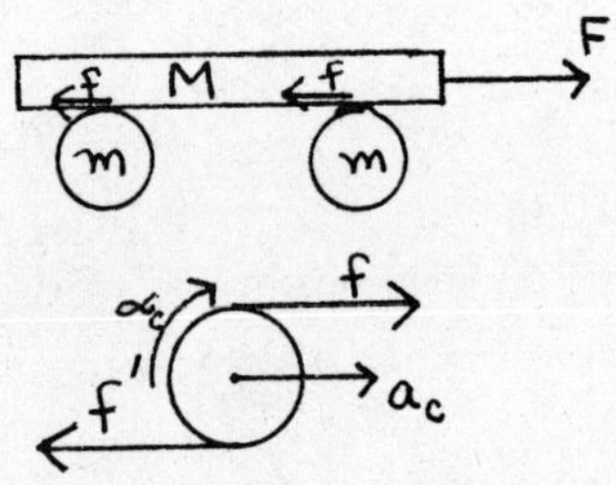

(1) $$F - 2f = Ma_p = 2Ma_c$$

(2) $$f - f' = ma_c$$

$$(f' + f)r = I_c\alpha = \frac{1}{2} mr^2\alpha$$

(3) $$f' + f = \frac{1}{2} ma_c$$

Adding (1), (2) and (3),

$$F = \left(2M + \frac{3}{2} m\right)a_c$$

$$a_c = \frac{F}{2M + \frac{3}{2} m} = 0.40 \text{ m/s}^2$$

$$a_p = 2a_c = 0.80 \text{ m/s}^2$$

From (1),

$$2f = F - 2Ma_c = 1.20\ \text{N}$$

$$f = 0.60\ \text{m/s}^2$$

From (2),

$$f' = f - ma_c = -0.20\ \text{N}$$

Chapter 13

13-1

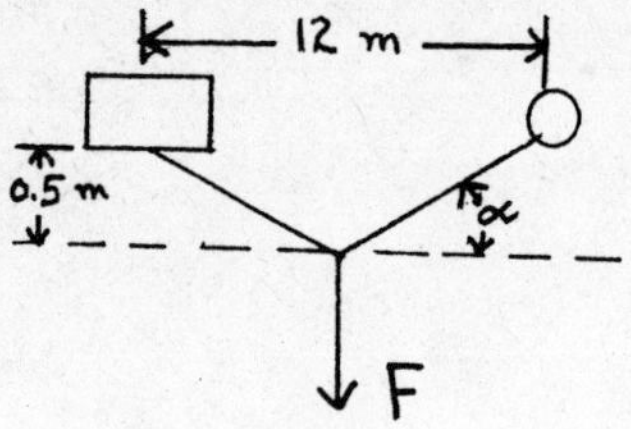

$\tan\alpha = \frac{0.5}{6}$

$\alpha = 4.76°$

$F = 2T\sin\alpha$

$T = \frac{F}{2\sin\alpha} = 3013\ \text{N}$

13-2 The torque about the front wheel is zero.

$0 = 1.2\ mg - 3(2\ F_r)$

Thus, the force at each rear wheel is

$F_r = 0.2\ mg = 2940\ \text{N}$

The force at each front wheel is then

$F_f = \frac{mg - 2F_r}{2} = 4410\ \text{N}$

13-3 The torque about the pivot is zero.

$0 = (0.13)mg - 0.05\ g(0.22)$

$m = 84.6\ \text{g}$

13-4

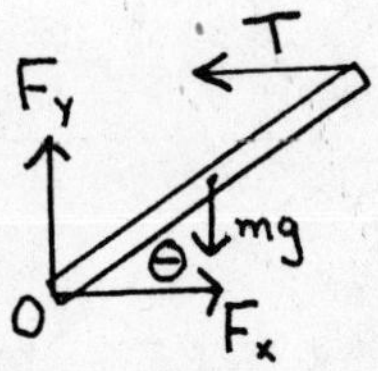

(a) The torque about O is zero.

$0 = \frac{mg\ell\cos\theta}{2} - T\ell\sin\theta$

$T = \frac{mg}{2\tan\theta} = 424\ \text{N}$

(b) $F_x = T = 424\ \text{N}$

$Fy = mg = 490\ \text{N}$

13-5

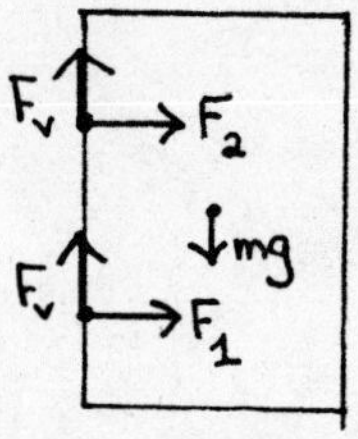

$2F_v = mg$

$F_v = \frac{mg}{2} = 122.5\ \text{N}$

$F_1 + F_2 = 0$

The torque about the upper hinge is zero.

$0 = (0.5)mg - (1.6)F_1$

$F_1 = 76.6\ \text{N}$

$F_2 = -76.6\ \text{N}$

13-6

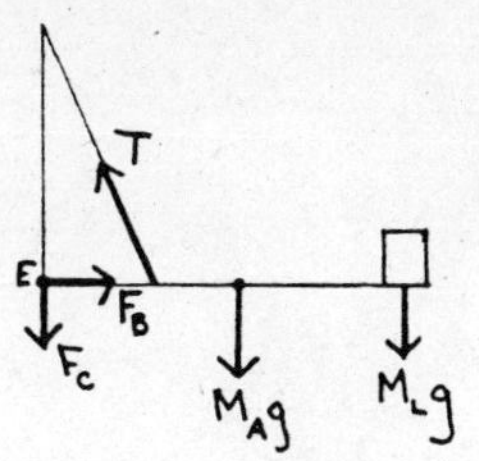

(a) $\Sigma\tau_E = 0 = 4T\cos 10° - 16\,M_A g - 36\,M_L g$

$$T = \frac{16\,M_A + 36\,M_L}{4\cos 10°}\,g = 1015\text{ N}$$

(b) $F_B = T\sin 10° = 176\text{ N}$

$F_c = T\cos 10° - (M_A + M_L)g = 872\text{ N}$

13-7

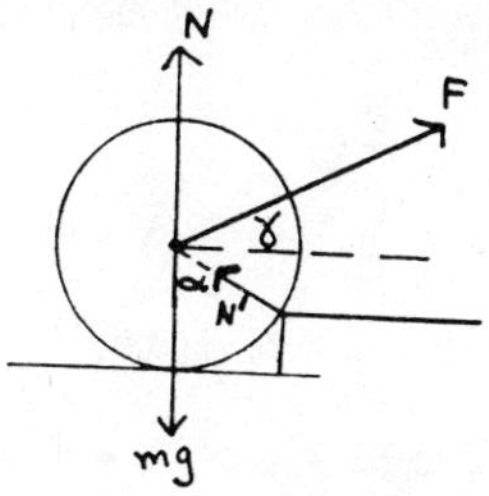

Solve the static equilibrium equations for N = 0.

$$\alpha = \cos^{-1}\frac{10}{20} = 60°$$

The torque about 0 is

$\tau = FR\sin(90 - \alpha + \gamma) - mgR\sin\alpha$

$\quad = FR\cos(\gamma - \alpha) - mgR\sin\alpha$

For static equilibrium, $\tau = 0$.

$FR\cos(\gamma - \alpha) = mgR\sin\alpha$

The force F is a minimum when $\gamma = 60°$.

$F = mg\sin\alpha = 849\text{ N}$

13-8

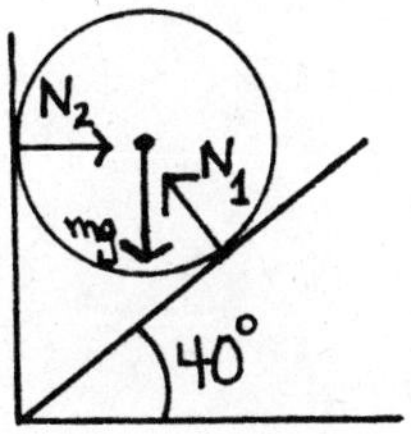

$N_1\sin 40° = N_2$

$N_1\cos 40° = mg$

$$N_1 = \frac{mg}{\cos 40°} = 154\text{ N}$$

$N_2 = N_1\sin 40° = 98\text{ N}$

13-9

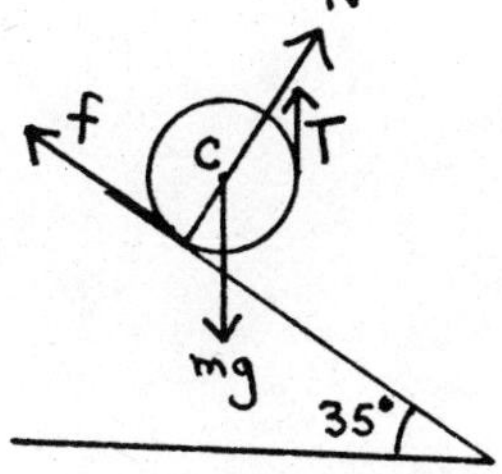

$f\cos 35° - N\sin 35° = 0$

$$\mu_s \geq \frac{f}{N} = \tan 35° = 0.70$$

13-10

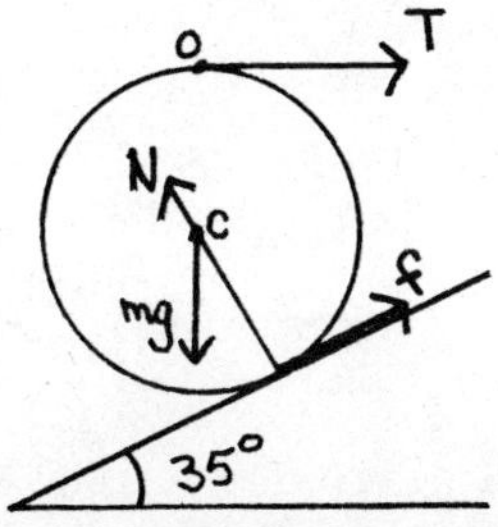

$\tau_c = 0 = fR - TR$

(1) $f = T$

(2) $N - mg\cos 35° - T\sin 35° = 0$

(3) $f + T\cos 35° - mg\sin 35° = 0$

Substitute (1) into (2) and (3). Multiply (3) by cot 35°. Substracting (3) from (2) gives

$$N - f(\sin 35° + \cos 35° + \frac{\cos^2 35°}{\sin 35°}) = 0$$

$$\mu \geq \frac{f}{N} = \frac{\sin 35°}{\sin^2 35° + \cos 35° + \cos^2 35°} = 0.315$$

This can be calculated directly by taking the torque about 0.

$$\tau_o = 0 = Rf \cos 35°(1 + \cos 35°) + Rf \sin^2 35° - RN \sin 35°$$

$$f(\cos 35° + \cos^2 35° + \sin^2 35°) = N \sin 35°$$

$$\mu \geq \frac{f}{N} = \frac{\sin 35°}{\sin^2 35° + \cos 35° + \cos^2 35°} = 0.315$$

13-11

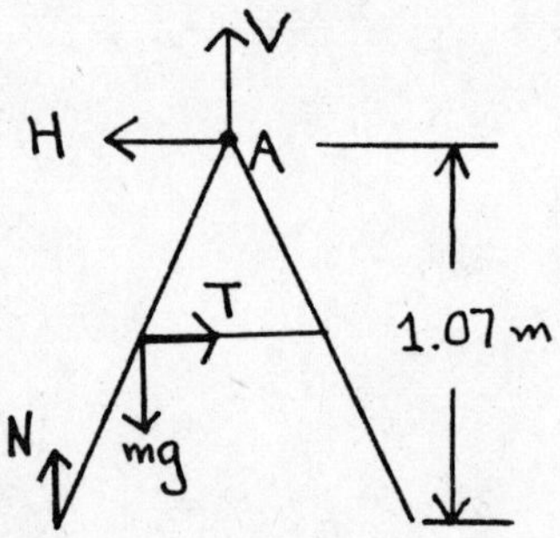

(a) For the entire assembly,

$$2N - 2\,mg = 0$$

For the left-hand plank,

$$\tau_A = 0 = mg(0.75) - N(1.5) + T(1.07)$$

$$mg(0.75) = T(1.07)$$

$$T = 0.70\,mg = 68.7\text{ N}$$

(b) $N + V - mg = 0$

$$V = mg - N = 0$$

$$T - H = 0$$

$$H = T = 68.7\text{ N}$$

13-12

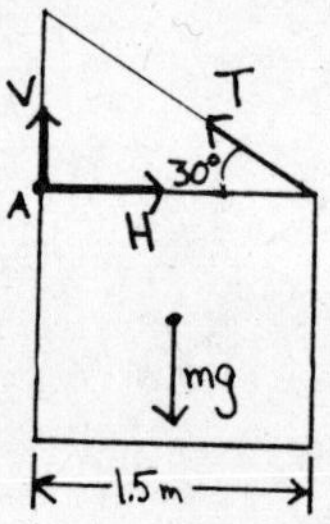

$$\Sigma\tau_A = 0 = 1.5\,T \sin 30° + 0.75\,mg$$

$$T = mg = 196\text{ N}$$

$$H = T \cos 30° = 169.7\text{ N}$$

$$V = mg - T \sin 30° = 98\text{ N}$$

13-13

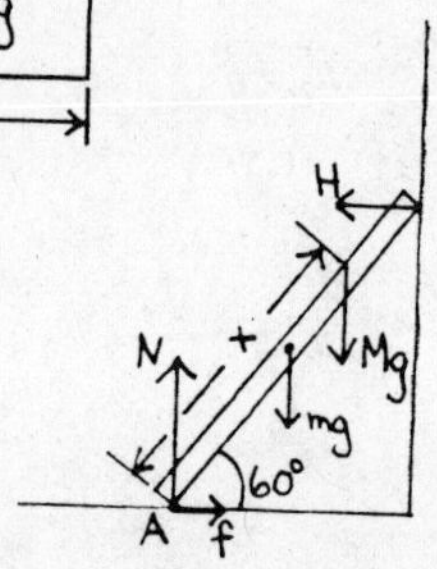

$$N = (M + m)g$$

$$H = f$$

$$H_{max} = f_{max} = \mu_s(m + M)g$$

$$\Sigma\tau_A = 0 = \frac{mgL}{2}\cos 60° + Mg\,x \cos 60° - HL \sin 60°$$

$$\frac{x}{L} = \frac{H \tan 60°}{Mg} - \frac{m}{2M} = \frac{\mu_s(m+M)\tan 60°}{M} - \frac{m}{2M}$$

$$= \frac{3}{2}\mu_s \tan 60° - \frac{1}{4} = 0.789$$

13-14 (1) $T \sin\gamma + mg - N = 0$

(2) $F - f - T\cos\gamma = 0$

(3) $\Sigma\tau_0 = F(L - \ell) - fL = 0$

Multiply (2) by tan γ and add to (1).

$$mg - N + (F - f)\tan\gamma = 0$$

$$\frac{f}{N} = \frac{f}{mg + (F - f)\tan\gamma}$$

Using (3),

$$f = \frac{F(L - \ell)}{L}$$

So,

$$\frac{f}{N} = \frac{F(1 - \frac{\ell}{L})}{mg + \frac{F\ell}{L}\tan\gamma}$$

Taking the limit as F goes to infinity,

$$\lim_{F\to\infty} \mu_s = \frac{f}{N} = \frac{1 - \frac{\ell}{L}}{\frac{\ell}{L}\tan\gamma} = \frac{\frac{L}{\ell} - 1}{\tan\gamma}$$

So, F can get infinitely large as long as

$$\frac{\ell}{L} = \frac{1}{1 + \mu_s \tan\gamma}$$

For $\gamma = 60°$ and $\mu_s = 0.40$,

$$\frac{\ell}{L} = \frac{1}{1 + (0.40)\tan 60°} = 0.591$$

13-15

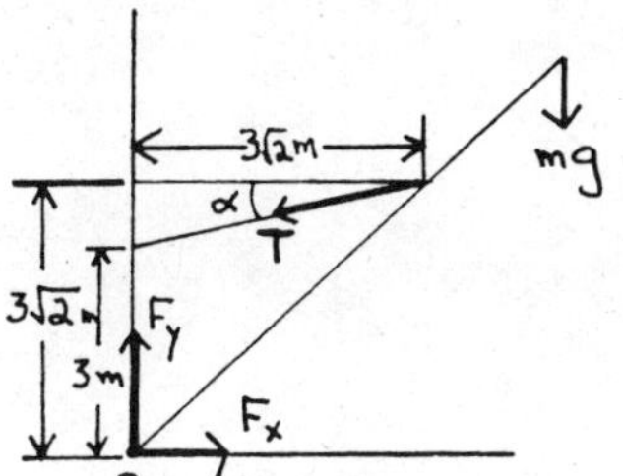

$$\alpha = \tan^{-1}\frac{3(\sqrt{2}-1)}{3\sqrt{2}} = 16.32°$$

(a) Taking the torque about P,

$$3\sqrt{2}T\cos\alpha - 12\,mg\cos 45° - 6T\sin\alpha\cos 45° = 0$$

$$T = \frac{12\,mg\cos 45°}{3\sqrt{2}\cos\alpha - 6\sin\alpha\cos 45°} = 5.78 \times 10^4\ \text{N}$$

(b) $F_y = mg + T\sin\alpha = 3.58 \times 10^4\ \text{N}$

$F_x = T\cos\alpha = 5.55 \times 10^4\ \text{N}$

(c)

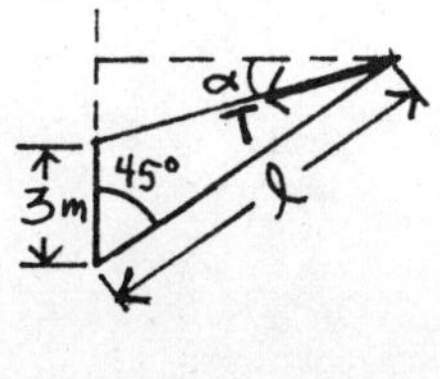

The torque about P due to the tension T is a constant equal in magnitude to the torque about P due to the load.

$$\tau_P = (3 \cos \alpha)T = \text{constant}$$

Taking the derivative with respect to the angle α,

$$(-3 \sin \alpha)T = 0$$

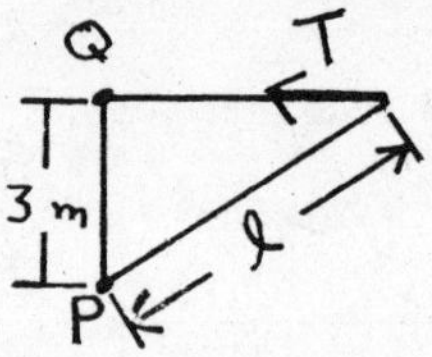

So, $\alpha = 0$

For $\alpha = 0$,

$$\ell = \frac{3}{\cos 45°} = 4.24 \text{ m}$$

13-16

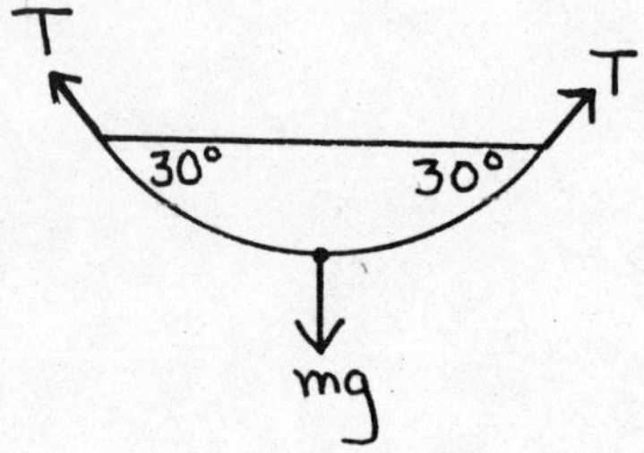

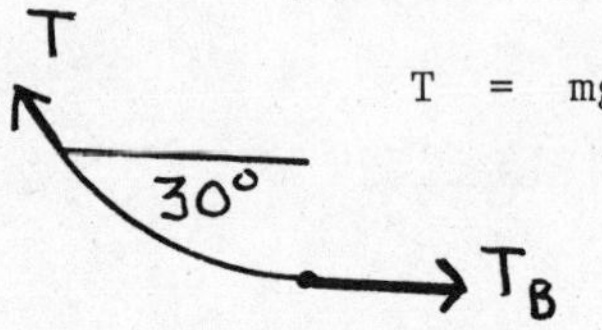

(a) $2T \sin 30° \; mg$

$$T = mg = 980 \text{ N}$$

(b) $T_B = T \cos 30° = 849 \text{ N}$

13-17

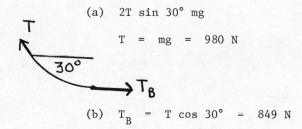

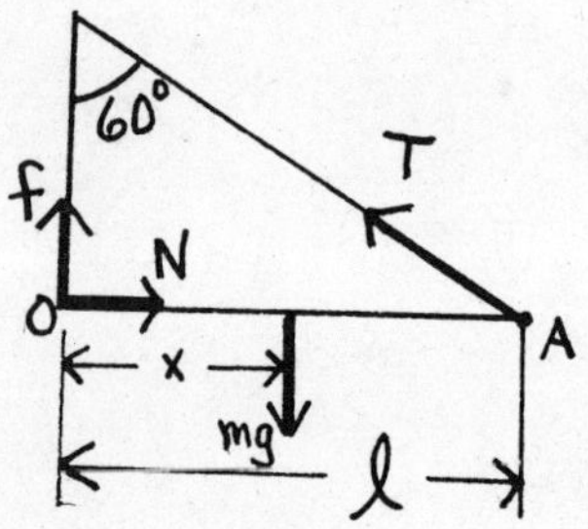

(1) $N = T \cos 30°$

$$\tau_O = 0 = xmg - T\ell \sin 30°$$

(2) $xmg = T\ell \sin 30°$

Dividing (2) by (1) gives

$$N = \frac{\sqrt{3}\, xmg}{\ell}$$

$$\tau_A = f\ell - mg(\ell - x) = 0$$

$$f = \frac{mg(\ell - x)}{\ell}$$

$$\frac{f}{N} = \frac{mg(\ell - x)}{\sqrt{3}\, xmg} = \frac{\ell - x}{\sqrt{3}\, x}$$

$$\frac{\ell - x}{x} \leq \sqrt{3}\, \mu_s$$

$$\frac{\ell}{x} \leq \sqrt{3}(0.20) + 1 = \frac{5 + \sqrt{3}}{5}$$

$$x \geq \frac{5}{5 + \sqrt{3}}$$

13-18 Referring to Ex. 13-6, for tipping to occur,

$$N_R = 0$$

Then, $N_F = Mg = 588 \text{ N}$

$$f_F = \mu_k N_F$$

$F = f_F$

$\Sigma\tau_o = 0 = (0.8\ m)N_F - (0.4\ m)Mg - (0.6\ m)F$

$= (0.8\ m)Mg - (0.4\ m)Mg - (0.6\ m)\mu_k Mg$

$= 0.4 - 0.6\ \mu_k$

$$\mu_k = \frac{0.4}{0.6} = \frac{2}{3}$$

13-19 Let the level of A and B define the point of zero potential energy.

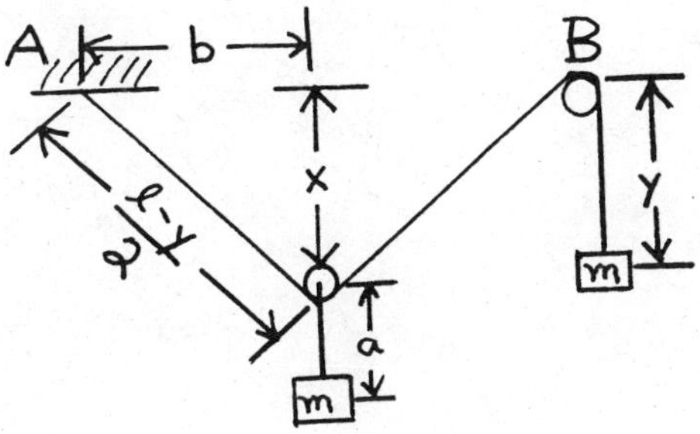

$U = -(a + x)mg - y\ mg$

$$x = \sqrt{\frac{\ell-y}{2}^2 - b^2}$$

$$U(y) = [-a + \sqrt{(\frac{\ell-y}{2})^2 - b^2}]mg - ymg$$

At equilibrium

$$\frac{dU}{dy} = 0 = \frac{(y-\ell)mg}{4\sqrt{(\frac{\ell-y}{2})^2 - b^2}} - mg$$

Solving for y,

$$y = \ell - \frac{4b}{\sqrt{3}}$$

13-20

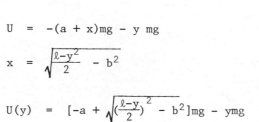

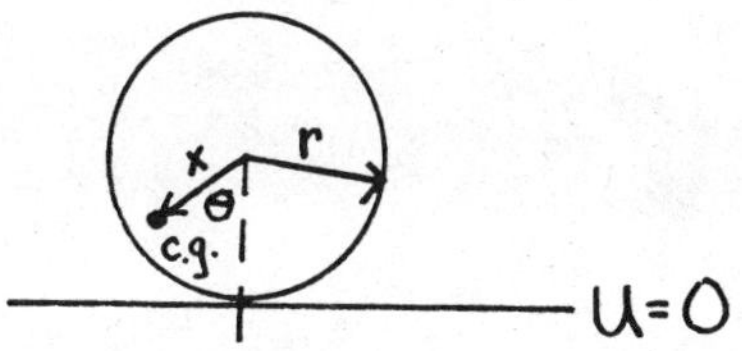

$U(\theta) = mg(r - x\cos\theta)$

At equilibrium,

$$\frac{dU}{d\theta} = 0 = mgx\sin\theta$$

$\theta = 0$

At $\theta = 0$, the center of mass is directly below the center of the sphere.

13-21 Originally, the center of mass is at a height

$$h = R + \frac{b}{2}$$

After rotating (rolling without slipping) through an angle α, the new center of mass height is

$$h' = (R + \frac{b}{2})\cos\alpha + R\,\alpha\sin\alpha$$

$$\alpha = \tan^{-1} \frac{0.15 - x_{min}}{0.3} = 26.6°$$

The block will slide before tipping. It slides at $\alpha = 21.8°$

(b) For $\mu_s = 0.50$, the angle for sliding is the same as that for tipping.

For $\mu_s = 0.60$, the angle for sliding is

$$\alpha = \tan^{-1} \mu_s = 31.0°$$

The block will tip before sliding.

The system will be in stable equilibrium if the new potential is greater than or equal to the old potential.

$$(R + \frac{b}{2}) \cos \alpha + R\alpha \sin \alpha \geq R + \frac{b}{2}$$

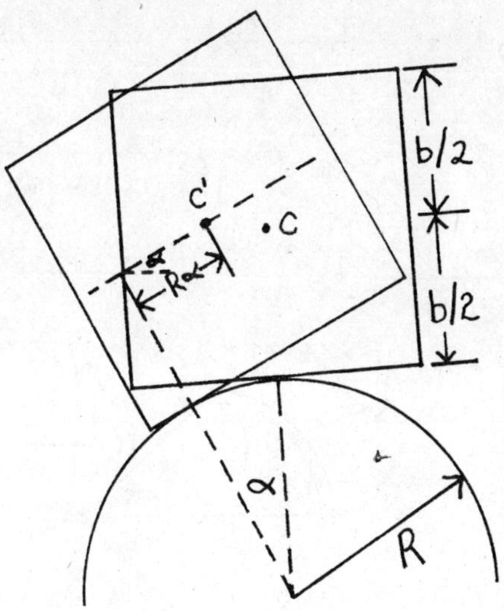

For α small,

$$\cos \alpha \cong 1 - \frac{\alpha^2}{2}$$

$$\sin \alpha \cong \alpha$$

So,

$$(R + \frac{b}{2})(1 - \frac{\alpha^2}{2}) + R\alpha^2 \geq R + \frac{b}{2}$$

$$(R + \frac{b}{2})(- \frac{\alpha^2}{2}) + R\alpha^2 \geq 0$$

$$\frac{R + \frac{b}{2}}{2} \leq R$$

$$R \geq \frac{b}{2}$$

13-22

$$\theta = \pi$$

$$T = T_0 e^{\mu_k \theta}$$

$$F = T_0 = Te^{-\mu\theta} = 78.4 \text{ N}$$

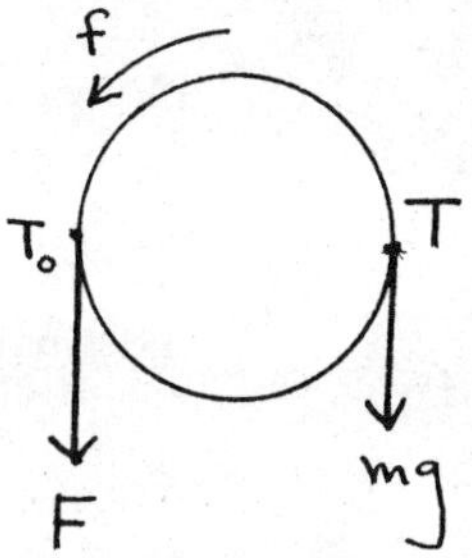

13-23

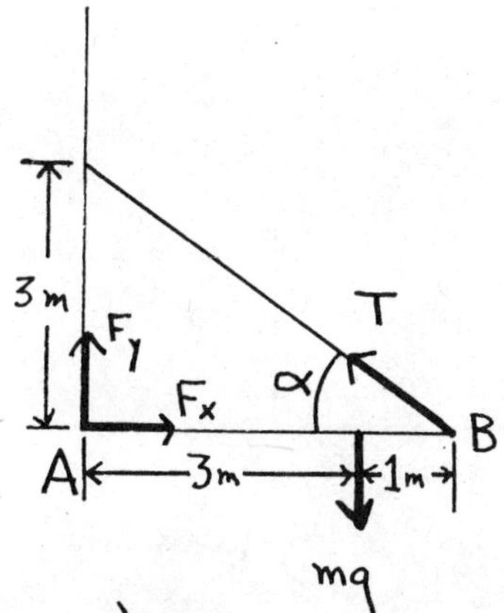

$$\alpha = \tan^{-1} \frac{3}{4} = 36.87°$$

$$\Sigma\tau_A = (4 \text{ m})T \sin \alpha - (3 \text{ m})mg$$

$$T = \frac{3 \text{ mg}}{4 \sin \alpha} = 250 \text{ N}$$

$$F_x = T \cos \alpha = 200 \text{ N}$$

$$F_y = mg - T \sin \alpha = 50 \text{ N}$$

13-24

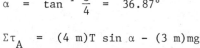

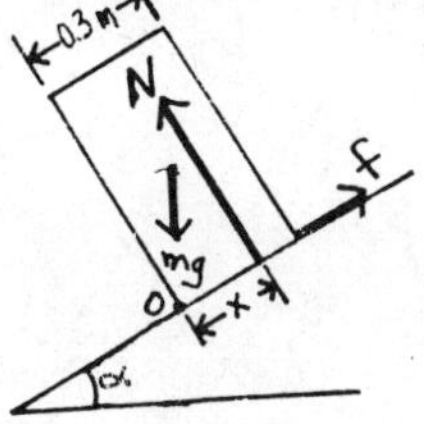

(a) For incipient sliding,

$$0 = f_{max} - mg \sin \alpha = \mu_s mg \cos \alpha - mg \sin \alpha$$

$$\alpha = \tan^{-1} \mu_s = 21.8°$$

For incipient tipping,

$$\Sigma\tau_0 = 0 = (0.15 \text{ m})mg \cos \alpha - (0.3 \text{ m})mg \sin \alpha - Nx$$

$$= 0.15 \text{ mg} \cos \alpha - 0.3 \text{ mg} \sin \alpha - mgx \cos \alpha$$

$$= (0.15 - x)\cos \alpha - 0.3 \sin \alpha$$

14-1 $a = \frac{F}{m} = \frac{Gm}{r^2} = \frac{(6.67 \times 10^{-11})(4 \times 10^7)}{10^4} = 2.67 \times 10^8 \text{ m/s}^2$

14-2 $G = \frac{Fr^2}{Mm} = \frac{(2.76 \times 10^{-8} \text{ N})(6 \times 0.0254 \text{ m})^2}{[\frac{4}{3}\pi(0.0254 \text{ m})^3][\frac{4}{3}\pi(4 \times 0.0254 \text{ m})^3](1.13 \times 10^4 \text{ kg/m}^3)^2}$

$= 1.665 \times 10^{-11} \text{ N} \cdot \text{m}^2/\text{kg}^2$

14-3

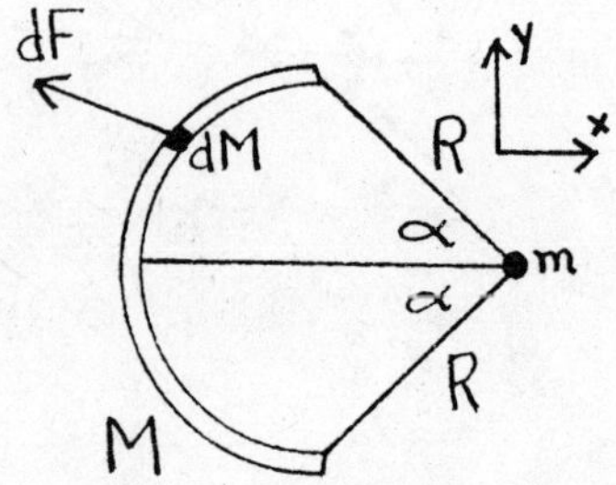

Since the vertical component of the gravitational force due to a piece of the wire dM in the upper half of the arc is cancelled by a corresponding piece of wire in the lower half of the arc, the total force on m is the sum of the horizontal forces.

$$F_m = \int dF_x = \int \cos\theta \, dF = \frac{Gm}{R^2}\int_{-\alpha}^{\alpha} \cos\theta \, dM$$

$$= \frac{Gm}{R^2}\int_{-\alpha}^{\alpha} (\cos\theta)\frac{M}{2\alpha}\, d\theta = \frac{GM\,m}{2\alpha R^2}\int_{-\alpha}^{\alpha} \cos\theta \, d\theta$$

$$= \frac{GMm}{R^2}\frac{\sin\alpha}{\alpha}$$

14-4 $F = 0 = \frac{GM_Em}{(L - x)^2} - \frac{GM_Mm}{x^2}$

$M_M(L - x)^2 = M_Ex^2$

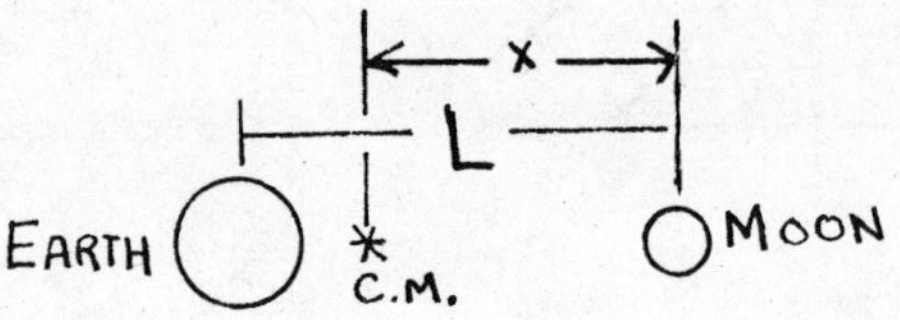

Solving for x,

$$x = L\frac{M_M \pm \sqrt{M_EM_M}}{M_M - M_E}$$

For $x > 0$,

$$x = L\frac{M_M - \sqrt{M_EM_M}}{M_M - M_E} = (3.84 \times 10^5 \text{ km})\frac{7.35 - \sqrt{(7.35)(597)}}{7.35 - 597}$$

$= 3.84 \times 10^4$ km from the moon's center

14-5

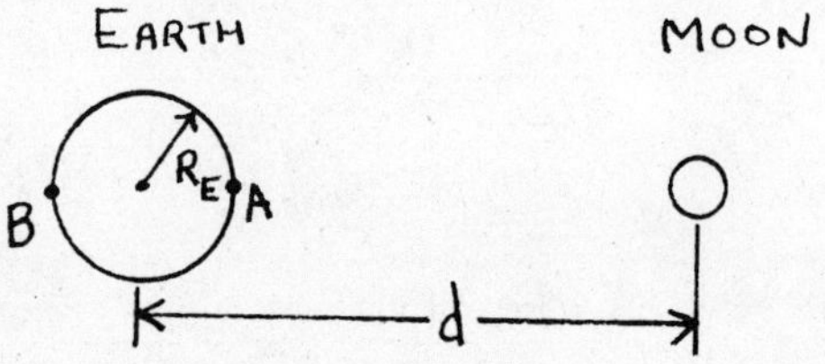

The acceleration of an object at the center of the earth due to the gravitational force of the moon is given by

$$a = \frac{GM_M}{d^2}$$

So,

$$\Delta a = \frac{-2GM_M}{d^3}\,\Delta d = \frac{-4GM_M}{d^3}\,R_E$$

and

$$g = \frac{GM_E}{R_E^2}$$

Thus,

$$\frac{\Delta g}{g} = \frac{\Delta a}{a} = \frac{-4M_M R_E^3}{M_E d^3} = -2.25 \times 10^{-7}$$

The negative sign indicates that the acceleration at A is greater than at B.

14-6 (a) $$F_M = \frac{GM_M m}{R_M^2}$$

$$\frac{F_M}{F_E} = \frac{M_M}{M_E}\left(\frac{R_E}{R_M}\right)^2 = 0.165$$

So,

$$F_M = 0.165\ g$$

(b) $$\frac{F_s}{F_E} = \frac{M_s}{M_E}\left(\frac{R_E}{R_s}\right)^2 = 27.9$$

$$F_s = 27.9\ g$$

14-7

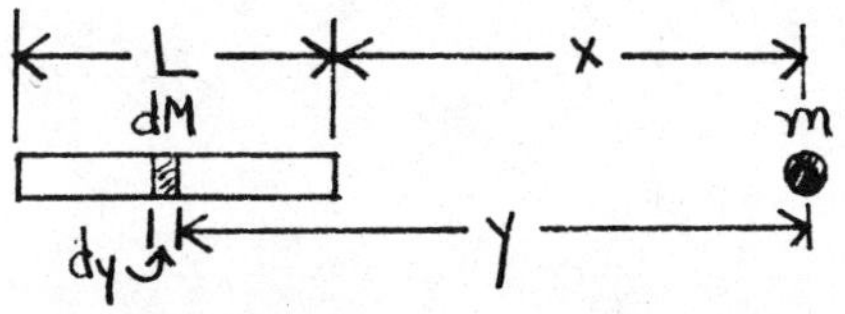

$$F = Gm\int_x^{x+L} \frac{dM}{y^2} = \frac{GmM}{L}\int_x^{x+L} \frac{dy}{y^2}$$

$$= \frac{GmM}{L}\left(-\frac{1}{y}\right)\Big|_x^{x+L} = \frac{GmM}{L}\left(\frac{1}{x} - \frac{1}{x+L}\right)$$

$$= \frac{GmM}{x(x+L)}$$

14-8

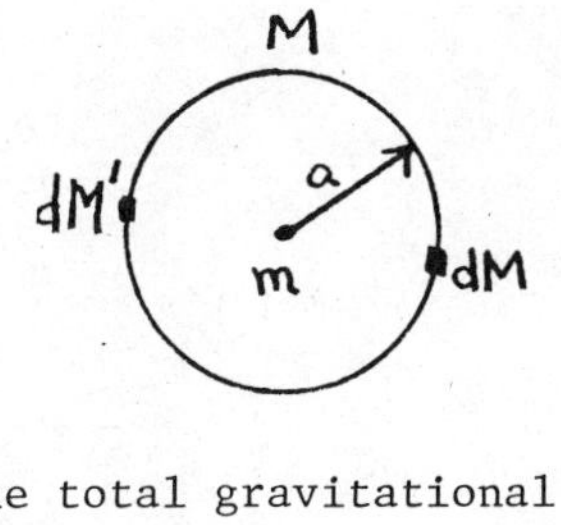

$$dM = \frac{Md\theta}{2\pi}$$

$$dU = -\frac{Gm\ dM}{a}$$

$$U = \frac{-GmM}{2\pi a}\int_0^{2\pi} d\theta = -\frac{GmM}{a}$$

The total gravitational force acting on m is zero since the force due to any dM is of magnitude equal to but oppositely directed from the force due to dM', with dM' located 180° from dM. The existence of a non-zero potential indicates that work was done to bring the system to its present configuration. The zero force acting on m indicates that m is in an equilibrium position.

14-9 Bringing M_1 into position,

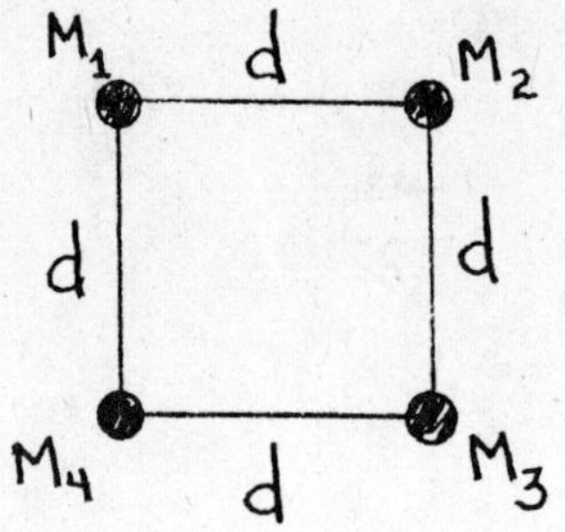

$$U_1 = 0$$

Bringing M_2,

$$U_2 = -\frac{GM^2}{d}$$

Bringing M_3,

$$U_3 = -\frac{GM^2}{d} - \frac{GM^2}{\sqrt{2}d}$$

Bringing M_4,

$$U_4 = -\frac{GM^2}{d} - \frac{GM^2}{d} - \frac{GM^2}{\sqrt{2}d}$$

$$U = U_1 + U_2 + U_3 + U_4 = -\frac{4GM^2}{d} - \frac{2GM^2}{\sqrt{2}\ d} = -\frac{GM^2}{d}(4 + \sqrt{2})$$

14-10 Let $d_{E,S}$ be the earth-to-sun distance.

$$E = K + U = \frac{1}{2}mv^2 - \frac{GM_Em}{R_E} - \frac{GM_Sm}{d_{E,S}} = 0$$

$$v = \sqrt{2G\left(\frac{M_E}{R_E} + \frac{M_S}{d_{E,S}}\right)}$$

$$= \sqrt{2(6.67 \times 10^{-11})\left(\frac{5.98 \times 10^{24}}{6.38 \times 10^6} + \frac{1.99 \times 10^{30}}{149.6 \times 10^9}\right)}$$

$$= 4.2 \times 10^4 \text{ m/s}$$

14-11 $W = GmM\left(\frac{1}{R_E} - \frac{1}{h + R_E}\right)$

$$\frac{mv^2}{h + R_E} = \frac{GMm}{(h + R_E)^2}$$

$$v^2 = \frac{GM}{h + R_E}$$

$$\frac{1}{2}mv^2 = \frac{1}{2}\frac{GMm}{h + R_E} = W = GMm\left(\frac{1}{R_E} - \frac{1}{h + R_E}\right)$$

$$\frac{1}{2(h + R_E)} = \frac{1}{R_E} - \frac{1}{h + R_E} = \frac{h}{R_E(h + R_E)}$$

$$h = R_E/2$$

14-12 (a) $\frac{1}{2}mv^2 + U_0 = U_R = \frac{-GM_Em}{R_E}$

The gravitational potential energy at the center of the earth is given by

$$U_0 = -Gm\int_0^{R_E}\frac{dM}{r} = -Gm\frac{M_E}{\frac{4}{3}\pi R_E^3}\int_0^{R_E}\frac{4\pi r^2 dr}{r}$$

$$= \frac{-3GM_Em}{2R_E}$$

So,

$$\frac{1}{2} mv^2 = \frac{3GM_E m}{2R_E} - \frac{GM_E m}{R_E} = \frac{GM_E m}{2R_E}$$

$$v = \sqrt{\frac{GM_E}{R_E}} = \sqrt{\frac{(6.67 \times 10^{-11})(5.97 \times 10^{24})}{6.37 \times 10^6}}$$

$$= 7.9 \text{ km/s}$$

(b) $$\frac{1}{2} mv^2 - \frac{3GM_E m}{2R_E} = 0$$

$$v = \sqrt{\frac{3GM_E}{R_E}} = 7.9\sqrt{3} \text{ km/s} = 13.7 \text{ km/s}$$

14-13 Consider the solid sphere to be made up of spherical shells of radius r' and thickness dr'. For $r \geq r'$, the potential due to each shell is, from Eq. 14-8,

$$dU = -\frac{Gm}{r} dM = \frac{-Gm}{r}(4\pi r'^2 \rho \, dr')$$

for $r \leq r'$, using Eq. 14-10,

$$dU = -Gm \frac{dM}{r'} = -GM(4\pi r' \rho \, dr')$$

When $r \geq R$, $r \geq r'$. Thus,

$$U(r \geq R) = \int_0^R \frac{-4\pi Gm\rho}{r} r'^2 \, dr'$$

$$= \frac{-4\pi Gm\rho R^3}{3r} = \frac{-GMm}{r}$$

$$F(r \geq R) = -\frac{dU}{dr} = \frac{GMm}{r^2}$$

Thus, the gravitational effect of the uniform solid sphere on a particle outside the sphere is the same as that of a pointlike object of the same mass as the sphere located at the center of the sphere.

For $r \leq R$,

$$U(r \leq R) = \int_0^r \frac{-4\pi Gm\rho}{r} r'^2 \, dr' + \int_r^R -4\pi Gm\rho r' \, dr'$$

$$= -4\pi Gm\rho \left(\frac{r^2}{3} + \frac{R^2}{2} - \frac{r^2}{2}\right)$$

$$= -4\pi Gm\rho \left(\frac{R^2}{2} - \frac{r^2}{6}\right)$$

$$F(r \leq R) = -\frac{dU}{dr} = 4\pi Gm\rho \left(\frac{r}{3}\right) = \frac{GMm}{R^3} r$$

$$F(0) = 0$$

$$U(0) = -4\pi Gm\rho \left(\frac{R^2}{2}\right) = -\frac{3}{2}\frac{GMm}{R}$$

$$F(R) = \frac{GMm}{R^2}$$

$$U(R) = -\frac{GMm}{R}$$

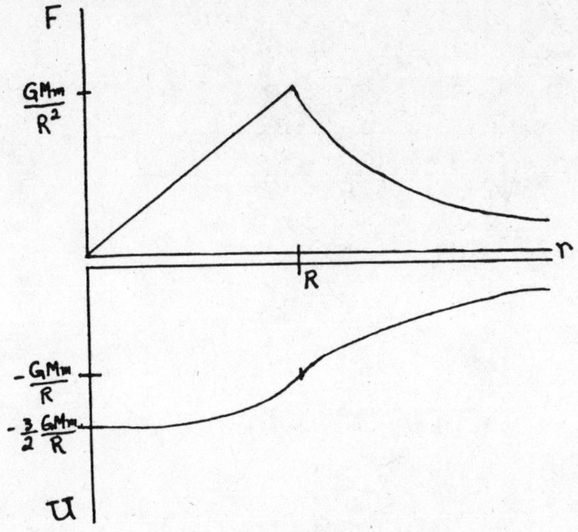

14-14

$$dM = \frac{M}{\pi R^2} 2\pi\rho \ d\rho = \frac{2M\rho \ d\rho}{R^2}$$

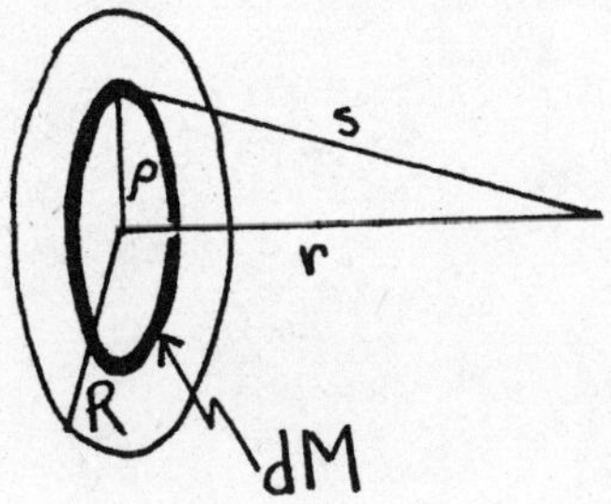

(a) $$dU = \frac{-Gm}{s} dM = \frac{-Gm}{\sqrt{\rho^2 + r^2}} \frac{2M\rho \ d\rho}{R^2}$$

$$U = \frac{-2GMm}{R^2} \int_0^R \frac{\rho}{\sqrt{\rho^2 + r^2}} d\rho = \frac{-2GMm}{R^2} \sqrt{\rho^2 + r^2} \Big|_0^R$$

$$= \frac{-2GMm}{R^2}(\sqrt{r^2 + R^2} - r)$$

(b) $$\vec{F}(r) = -\frac{\partial U}{\partial r} \hat{i} = \frac{2GMm}{R^2} \left(\frac{r}{\sqrt{r^2 + R^2}} - 1 \right)$$

(c) $$U(r) = -\frac{2GMm}{R^2}(r\sqrt{1 + \frac{R^2}{r^2}} - r)$$

$$= \frac{-2GMmr}{R^2}(1 + \frac{R^2}{2r^2} - \frac{R^4}{8r^4} + ... - 1)$$

For r >> R,

$$U(r) \cong \frac{-2GMmr}{R^2} (\frac{R^2}{2r^2}) = \frac{-GMm}{r}$$

$$F(r) = \frac{2GMm}{R^2} \left(\frac{r}{\sqrt{r^2 + R^2}} - 1 \right)$$

$$= \frac{2GMm}{R^2}[(1 + \frac{R^2}{r^2})^{-\frac{1}{2}} - 1]$$

$$= \frac{2GMm}{R^2}[1 - \frac{R^2}{2r^2} + \frac{3R^4}{8r^4} - ..... - 1]$$

For r >> R,

$$F(r) = \frac{2GMm}{R^2}\left(\frac{R^2}{2r^2}\right) = \frac{GMm}{r^2}$$

14-15 (a) Both m_1 and m_2 exert a force acting on the center of mass of M. Since forces add vectorally, the resultant force must act on the center of mass of M.

(b)

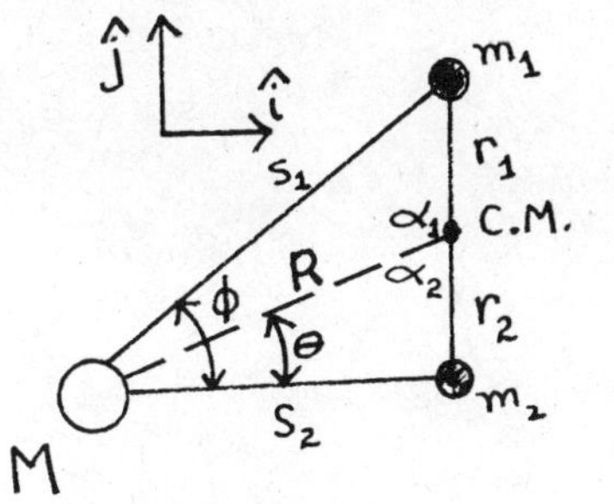

$$s_2^2 = R^2 + r_2^2 - 2r_2 R \cos \alpha_2$$

$$= 9 + 1 - 6 \cos \alpha_2$$

$$= 9 + 1 - 6 \cos 75° = 8.45 \text{ m}^2$$

$$s_2 = 2.91 \text{ m}$$

$$s_1^2 = R^2 + r_1^2 - 2r_1 R \cos \alpha_1$$

$$= 9 + 1 - 6 \cos 105° = 11.55 \text{ m}^2$$

$$s_1 = 3.40 \text{ m}$$

In units of G,

$$F_1 = \frac{m_1 M}{s_1^2} = 0.69$$

$$F_2 = \frac{m_2 M}{s_2^2} = 0.95$$

To find ϕ,

$$(r_1 + r_2)^2 = s_1^2 + s_2^2 - 2s_1 s_2 \cos \phi$$

$$\phi = \cos^{-1} \frac{s_1^2 + s_2^2 - (r_1 + r_2)^2}{2s_1 s_2} = 36°$$

The total force acting on M is given by

$$\vec{F} = \vec{F}_1 + \vec{F}_2 = (0.69 \cos 36°)\hat{i} + (0.69 \sin 36°)\hat{j} + 0.95\,\hat{i}$$

$$= 1.51\,\hat{i} + 0.41\,\hat{j}$$

Thus, the angle the total force makes with the horizontal is

$$\beta = \tan^{-1} \frac{0.41}{0.51} = 15.0°$$

The angle θ can be found as follows:

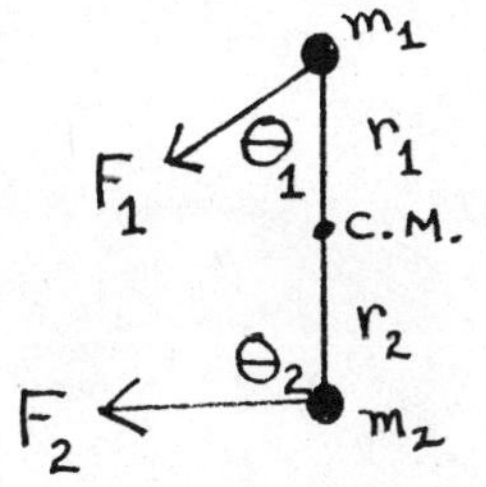

$$r_2^2 = R^2 + s_2^2 - 2s_2 R \cos \theta$$

$$\theta = \cos^{-1} \frac{R^2 + s_2^2 - r_2^2}{2s_2 R} = 19.6°$$

The force exerted by A on B acts below the center of mass of B.

To find the torque about the C.M., first find θ_1 and θ_2.

$$R^2 = r_2^2 + s_2^2 - 2r_2 s_2 \cos \theta_2$$

$$\theta_2 = \cos^{-1} \frac{r_2^2 + s_2^2 - R^2}{2r_2 s_2} = 85.6°$$

$$\theta_1 = \cos^{-1} \frac{r_1^2 + s_1^2 - R^2}{2r_1 s_1} = 58.5°$$

$$\tau = r_1F_1 \sin \theta_1 - r_2F_2 \sin \theta_2 = 0.36$$

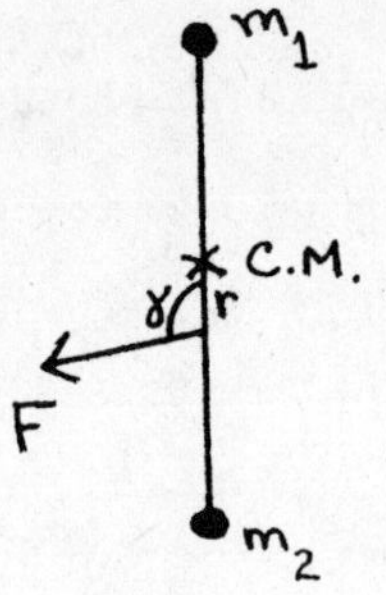

Alternately, the torque can be calculated as follows:

$$\tau = rF \sin \gamma$$

Using the law of sines,

$$r = R \frac{\sin(\theta - \beta)}{\sin \gamma}$$

$$\tau = RF \sin(\theta - \beta) = \sqrt{F_x^2 + F_y^2}\, R \sin(\theta - \beta) = 0.38$$

14-16 From 14-27,

$$a = \left(\frac{GM}{4\pi^2} T^2\right)^{1/3}$$

$$= \left[\frac{(6.67 \times 10^{-11})(5.97 \times 10^{24})(24 \times 3600)^2}{4\pi^2}\right]^{1/3}$$

$$= 4.22 \times 10^7 \text{ m}$$

14-17 To find the velocity of the satellite in its circular orbit, use

$$\frac{v_o^2}{r} = \frac{GM}{r^2}$$

$$v_o^2 = \frac{GM}{r}$$

To excape from earth's gravitational attraction,

$$\frac{1}{2} m(v^2 + v_o^2) - \frac{GMm}{r} = 0$$

$$v = \sqrt{\frac{2GM}{r} - v_o^2} = \sqrt{\frac{GM}{r}}$$

$$= \sqrt{\frac{GM}{R_E + 10^6 \text{ m}}} = 7.35 \times 10^3 \text{ m/s} = 7.35 \text{ km/s}$$

This is about 2/3 the escape velocity necessary at the earth's surface.

14-18 $a^3 = \frac{GM}{4\pi^2} T^2$

$$M = \frac{4\pi^2 a^3}{GT^2} = 1.89 \times 10^{27} \text{ kg}$$

$$\frac{M}{M_E} = \frac{1.89 \times 10^{27}}{5.98 \times 10^{24}} = 316$$

14-19 (a) $a = \frac{GM_m}{R_m^2} = \frac{(6.67 \times 10^{-11})(7.35 \times 10^{22})}{(1.74 \times 10^6)^2} = 1.62 \text{ m/s}^2$

(b) $\frac{v^2}{r} = a$

$$v = \sqrt{ar} = \sqrt{(1.62)(1.74 \times 10^6)} = 1.68 \times 10^3 \text{ m/s} = 1.68 \text{ km/s}$$

(c) $T = \frac{2\pi(1.74 \times 10^6 \text{ m})}{1.68 \times 10^3 \text{ m/s}} = 6.51 \times 10^3 \text{ s} = 108 \text{ minutes}$

14-20 The distance of the center of mass of the Sun-Jupiter system from the center of the sun is

$$x = \frac{m_J r_J}{m_S + m_J} \cong \frac{m_J a}{m_S + m_J} \quad \text{where a is the length of the semi-major axis}$$

$$= \frac{(318\ M_E)(5.20\ \text{A.U.})}{(1.99 \times 10^{30}\ \text{kg}) + (318\ M_E)}$$

$$= 7.44 \times 10^8 \text{ m}$$

This is approximately the radius of the sun.

14-21 $\frac{\delta e}{r} = 1 - e \cos \phi$

$\delta e = r - er \cos \phi$

Also,

$\delta = a/e - ae = a(\frac{1}{e} - e)$

$r = \sqrt{x'^2 + y^2}$

$r \cos \phi = -x'$

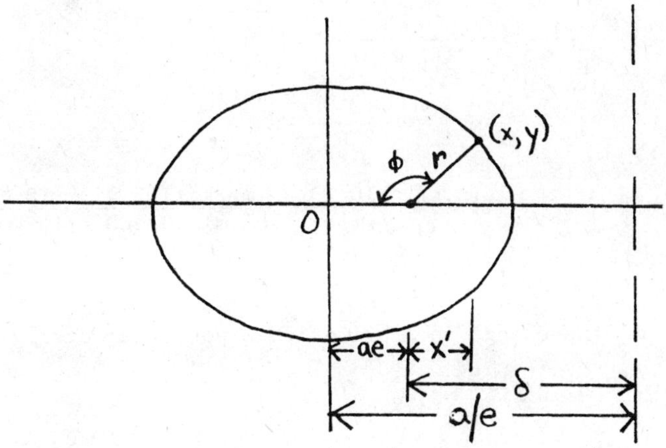

Thus,

$a(1 - e^2) = \sqrt{x'^2 + y^2} + ex'$

$x'^2 + y^2 = a^2(1 - e^2)^2 - 2aex'(1 - e^2) + e^2x'^2$

$x'^2(1 - e^2) + 2aex'(1 - e^2) = a^2(1 - e^2)^2 - y^2$

$x'^2 + 2aex' = a^2(1 - e^2) - \frac{y^2}{1 - e^2}$

Completing the square,

$(x' + ae)^2 - a^2e^2 = a^2(1 - e^2) - \frac{y^2}{1 - e^2}$

$x^2 = a^2 - \frac{y^2}{1 - e^2}$

$\frac{x^2}{a^2} + \frac{y^2}{a^2(1 - e^2)} = 1$

$\frac{x^2}{a^2} + \frac{y^2}{b^2} = 1$

14-22 Since the equation for elliptic motion in rectangular coordinates is

$\frac{x^2}{a^2} + \frac{y^2}{b^2} = 1$

the semi-major axis is

$a = \sqrt{25} = 5$

The semi-minor axis is

$$b = \sqrt{16} = 4$$

Since $b = a\sqrt{1 - e^2}$

$$e = \sqrt{1 - \frac{b^2}{a^2}} = 3/5$$

The foci are located at

$$x = \pm f = \pm ae = \pm(5)(\frac{3}{5}) = \pm 3$$

The directrices are located at

$$x = \pm d = \pm\frac{a}{e} = \pm(5)(\frac{5}{3}) = \pm 8\,\frac{1}{3}$$

The vertices are located at $(\pm a, 0) = (\pm 5, 0)$

14-23 $$\frac{v_i^2}{R_E + h} = \frac{GM_E}{(R_E + h)^2}$$

$$K_i = \frac{1}{2}mv_i^2 = \frac{1}{2}\frac{GM_E m}{R_E + h}$$

$$= \frac{1}{2}\frac{(6.67\ \ 10^{-11}\ \text{N}\cdot\text{m}^2/\text{kg}^2)(5.97 \times 10^{24}\ \text{kg})(500\ \text{kg})}{(6.38 \times 10^6\ \text{m}) + (0.500 \times 10^6\ \text{m})}$$

$$= 1.45 \times 10^{10}\ \text{J}$$

The change in gravitational potential energy is

$$\Delta U = \frac{GM_E m}{R_i} - \frac{GM_E m}{R_f}$$

$$= GM_E m\,(\frac{1}{R_i} - \frac{1}{R_f})$$

$$= (6.67 \times 10^{-11}\ \text{N}\cdot\text{m}^2/\text{kg}^2)(5.97 \times 10^{24}\ \text{kg})(500\ \text{kg})(-1.14 \times 10^{-8}\ \text{m}^{-1})$$

$$= -2.27 \times 10^9\ \text{J}$$

Also,

$$K_f = \frac{1}{2}mv_f^2 = \frac{1}{2}(500\ \text{kg})(2 \times 10^3\ \text{m/s})^2 = 1.00 \times 10^9\ \text{J}$$

The energy lost to friction is

$$E_f = K_i - K_f - \Delta U = (14.5 - 1.00 + 2.27) \times 10^9\ \text{J}$$

$$= 1.58 \times 10^{10}\ \text{J}$$

14-24 (a) $$T = \frac{2\pi r}{v} = \frac{2\pi(30{,}000 \times 9.46 \times 10^{15}\ \text{m})}{2.50 \times 10^5\ \text{m/s}} = 7.13 \times 10^{15}\ \text{s} = 2.3 \times 10^8\ \text{y}$$

$$= 0.23 \text{ billion years}$$

(b) $$M = \frac{4\pi^2 a^3}{GT^2} = \frac{4\pi^2(30{,}000 \times 9.46 \times 10^{15}\ \text{m})^3}{(6.67 \times 10^{-11}\ \text{N}\cdot\text{m}^2/\text{kg}^2)(7.13 \times 10^{15}\ \text{s})^2} = 2.66 \times 10^{41}\ \text{kg}$$

$$= 1.34 \times 10^{11} \text{ solar masses}$$

The number of stars is approximately 10^{11}.

14-25 The force acting on one particle due to another particle has magnitude

$$F = GM^2/h^2$$

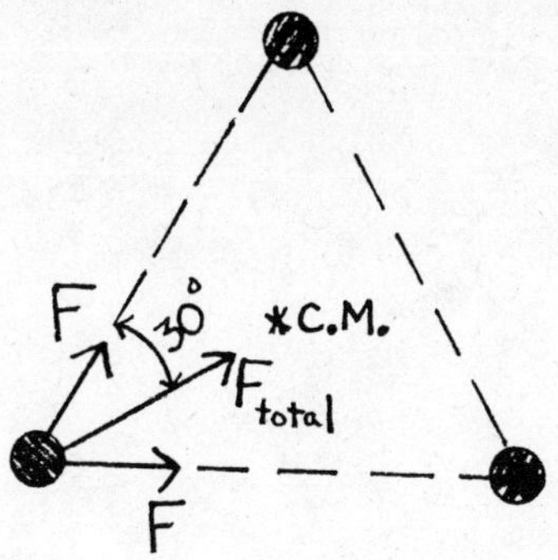

The total force acting on each particle is directed toward the center of mass and has magnitude

$$F_{total} = 2F \cos 30° = \sqrt{3}\, GM^2/h^2$$

The centripetal force is

$$F_c = MR\omega^2 = M\omega^2 h/(2 \cos 30°) = M\omega^2 h/\sqrt{3}$$

Equating the two,

$$M\omega^2 h/\sqrt{3} = \sqrt{3}\, GM^2/h^2$$

$$\omega = \sqrt{3GM/h^3}$$

14-26 (a) $a^3 = \dfrac{GM}{4\pi^2} T^2$

$$= (6.67 \times 10^{-11} \text{ N} \cdot \text{m}^2/\text{kg}^2)(1.99 \times 10^{30} \text{ kg})(88.0 \times 24 \times 3600 \text{ s})^2/4\pi^2$$

$$= 0.194 \times 10^{33} \text{ m}^3$$

$$a = 5.79 \times 10^{10} \text{ m} = 0.386 \text{ A.U.}$$

The velocity at the perihelion is 59.0 km/s and the radius at perihelion is r_{min}. From Example 19-4,

$$v_\phi(\text{perihelion}) = \frac{2\pi a^2\sqrt{1 - e^2}}{r_{min} T} = \frac{2\pi a^2\sqrt{1 - e^2}}{a(1 - e)T} = \frac{2\pi a}{T}\left(\frac{1 + e}{1 - e}\right)^{\frac{1}{2}}$$

$$\left(\frac{1 + e}{1 - e}\right)^{\frac{1}{2}} = \frac{Tv_\phi}{2\pi a} = \frac{(88.0 \times 24 \times 3600 \text{ s})(59.0 \times 10^3 \text{ m/s})}{2\ (5.79 \times 10^{10} \text{ m})} = 1.233$$

$$1 + e = (1.233)^2(1 - e) = 1.520 - 1.520e$$

$$e = 0.206$$

$$b = a(1 - e^2)^{\frac{1}{2}} = (5.79 \times 10^{10} \text{ m})\sqrt{1 - (0.206)^2} = 0.378 \text{ A.U.}$$

(b) $v\,(\text{perihelion}) - v_\phi(\text{aphelion}) = v_\phi(\text{perihelion}) \left[1 - \dfrac{1-e}{1+e}\right]$

$$= (59.0 \text{ km/s}) \left[1 - \frac{1-0.206}{1+0.206}\right]$$

$$= 20.2 \text{ km/s}$$

14-27 (a) $2a = (2 \times 1738) + 20 + 310 = 3806$ km

$$a = 1903 \text{ km}$$

$$r_{max} - r_{min} = a(1 + e) - a(1 - e) = 2ae$$

$$e = \frac{r_{max} - r_{min}}{2a} = \frac{(1738 + 310) - (1738 + 20)}{2(1903)} = 0.076$$

(b) $M = \frac{4\pi^2 a^3}{GT^2} = \frac{4\pi^2(1.903 \times 10^6 \text{ m})^3}{(6.67 \times 10^{-11} \text{ N}\cdot\text{m}^2/\text{kg}^2)(124.2 \times 60 \text{ s})^2} = 7.34 \times 10^{22} \text{ kg}$

$$\rho = \frac{7.34 \times 10^{22} \text{ kg}}{\frac{4}{3}\pi(1.738 \times 10^6 \text{ m})^3} = 3.34 \times 10^3 \text{ kg/m}^3 = 3.34 \text{ g/cm}^3$$

14-28 $M = \frac{1}{2} M_E = \frac{4}{3}\pi r^3 \rho = \frac{1}{2}(\frac{4}{3}\pi r_E^3 \rho)$

$$g = \frac{GM}{r^2} = \frac{1}{2}\frac{GM_E}{r^2} = \frac{1}{2}\frac{GM_E}{r_E^2/2^{2/3}} = \frac{1}{2^{1/3}} g_{earth} = 7.78 \text{ m/s}^2$$

14-29 (a) $g = \frac{C}{R_E^2}$

$$\Delta g \cong \frac{-2C}{R_E^3}\Delta R_E = \frac{-2C}{R_E^3} h$$

$$\frac{\Delta g}{g} \cong -\frac{2h}{R_E}$$

(b) At $h = 10^3$ m,

$$\frac{\Delta g}{g} \cong \frac{-2(10^3 \text{ m})}{6.378 \times 10^6 \text{ m}} = -3.14 \times 10^{-4}$$

Calculating exactly,

$$\frac{\Delta g}{g} = \left(\frac{1}{(R_E + h)^2} - \frac{1}{R_E^2}\right) \div \frac{1}{R_E^2} = \frac{R_E^2}{(R_E + h)^2} - 1$$

$$= \frac{(6.378 \times 10^6 \text{ m})^2}{(6.379 \times 10^6 \text{ m})^2} - 1 = -3.14 \times 10^{-4}$$

At $h = 10^6$ m,

$$\frac{\Delta g}{g} \cong \frac{-2(10^6 \text{ m})}{6.378 \times 10^6 \text{ m}} = -0.314$$

Calculating exactly,

$$\frac{\Delta g}{g} = \frac{(6.378 \times 10^6 \text{ m})^2}{(7.378 \times 10^6 \text{ m})^2} - 1 = -0.253$$

The approximation is not good when h is close to R_E.

14-30 (a) The force on a segment dm of rod one due to rod two is, from Problem 14-7,

$$dF = \frac{GM}{y(y + \ell)} dm$$

where y is the distance of the segment dm from the closer end of rod two. The total force on rod one is found by integrating over the length of the rod.

$$F = \int_x^{x+\ell} \frac{GM}{y(y + \ell)} dM = \int_x^{x+\ell} \frac{GM}{y(y + \ell)} \frac{M}{\ell} dy$$

$$= \frac{GM^2}{\ell} \int_x^{x+\ell} \frac{dy}{y(y+1)} = \frac{GM^2}{\ell^2} \int_x^{x+\ell} \left(\frac{1}{y} - \frac{1}{y+\ell}\right) dy$$

$$= \frac{GM^2}{\ell^2} \left(\ln \frac{x+\ell}{x} - \ln \frac{x+2\ell}{x+\ell}\right)$$

$$= \frac{GM^2}{\ell^2} \ln \frac{(x+\ell)^2}{x(x+2\ell)}$$

(b) $$F = \frac{GM^2}{\ell^2} \left[\ln\left(\frac{x+\ell}{x}\right) - \ln\left(\frac{x+2\ell}{x}\right)\right]$$

$$= \frac{GM^2}{\ell^2} \left[\ln(1 + \ell/x) - \ln\left(\frac{1+2\ell/x}{1+\ell/x}\right)\right]$$

$$= \frac{GM^2}{\ell^2} \left[2 \ln(1 + \ell/x) - \ln(1 + 2\ell/x)\right]$$

$$= \frac{GM^2}{\ell^2} \left[2\left(\frac{\ell}{x} - \frac{\ell^2}{2x^2} + \cdots\right) - \left(\frac{2\ell}{x} - \frac{2\ell^2}{x^2} + \cdots\right)\right]$$

$$\cong \frac{GM^2}{x^2}$$

14-31 (a) Let the particle be a distance x from the earth's center

$$F = -\frac{GmM_x}{x^2} = -\frac{Gm}{x^2}\frac{M_E x^3}{R_E^3} = -\frac{GmM_E}{R_E^3} x$$

In the form of Hooke's Law,

$$F = -\kappa x$$

where

$$\kappa = \frac{GmM_E}{R_E^3} = g \frac{m}{R_E}$$

(b) $$T = 2\pi\sqrt{m/\kappa} = 2\pi\sqrt{R_E/g} = 2\pi\sqrt{(6.38 \times 10^6 \text{ m})/(9.8 \text{ m/s}^2)}$$

$$= 5.07 \times 10^3 \text{ s} = 84.49 \text{ min}$$

The time to travel from one side of the earth to the other is half the period, which is 42.2 min.

14-32 (a) $$g = \frac{GM}{R_E^2}$$

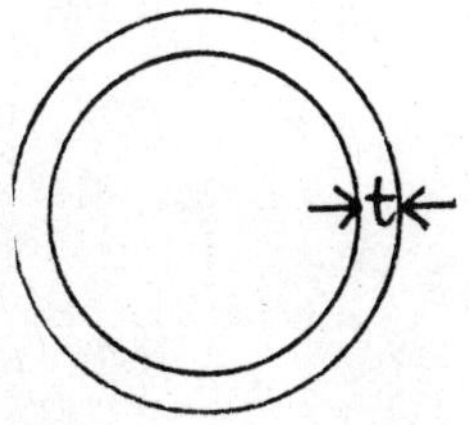

$$\Delta g = \frac{G\Delta M}{R_E^2} - \frac{2GM\,\Delta R_E}{R_E^3}$$

$$\frac{\Delta g}{g} = \frac{\Delta M}{M} - \frac{2\,\Delta R_E}{R_E}$$

$$= \frac{-4\pi R_E^2 t\,\rho_E}{M_E} - \frac{2t}{R_E}$$

$$= \frac{-4\pi(6.38 \times 10^8 \text{ cm})^2(15 \times 10^5 \text{ cm})(2.72 \text{ g/cm}^3)}{5.97 \times 10^{27} \text{ g}} - \frac{2(-15 \times 10^5 \text{ cm})}{6.38 \times 10^8 \text{ cm}}$$

$$= 1.21 \times 10^{-3}$$

(b) $\Delta g = \frac{G\Delta m}{r^2} = \frac{G}{r^2}(\frac{4}{3}\pi r^3 \Delta\rho)$

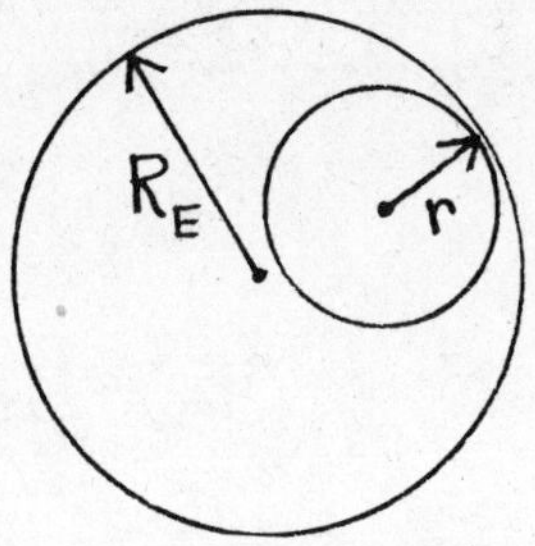

$$= \frac{4\pi rG}{3}(\rho - 2.72)$$

$$= \frac{4\pi rG}{3}\Delta\rho$$

$$\frac{\Delta g}{g} = \frac{4\pi R_E^2 r}{3 M_E}\Delta\rho$$

$$= \frac{4\pi(6.38 \times 10^8 \text{ cm})^2(2.5 \times 10^5 \text{ cm})}{3(5.97 \times 10^{27} \text{ g})}[(7.86 - 2.72)\text{g/cm}^3]$$

$$= 3.67 \times 10^{-4}$$

(c) $\frac{\Delta g}{g} = \frac{4\pi R_E^2 r}{3 M_E}\Delta\rho$

$$= \frac{4\pi(6.38 \times 10^8 \text{ cm})^2(2.5 \times 10^5 \text{ cm})}{3(5.97 \times 10^{27} \text{ g})}(-2.72 \text{ g/cm}^3)$$

$$= -1.94 \times 10^{-4}$$

14-33 The initial energy of the system is zero.

$$E = 0 = U + K = -\frac{GM^2}{R} + \frac{1}{2}MV^2 + \frac{1}{2}MV^2$$

$$MV^2 = \frac{GM^2}{R}$$

$$V = \sqrt{GM/R}$$

$$V_{rel} = V + V = 2\sqrt{GM/R}$$

14-34

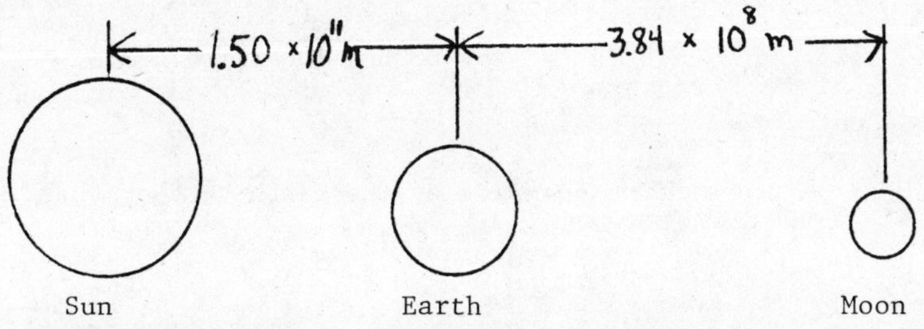

$M_s = 1.99 \times 10^{30}$ kg $\quad R_E = 6.38 \times 10^6$ m $\quad R_M = 1.74 \times 10^6$ m

$M_E = 5.97 \times 10^{24}$ kg $\quad M_M = 7.35 \times 10^{22}$ kg

The change in the potential energy of the system is the sum of the changes in potential energy of the 1-kg object with respect to each celestial body.

$$\Delta U_{sun} = -GmM_s\left[\frac{1}{(1.50 \times 10^{11} \text{ m})+(3.84 \times 10^8 \text{ m})-(1.74 \times 10^6 \text{ m})} - \frac{1}{(1.50 \times 10^{11} \text{ m})+(6.38 \times 10^6 \text{ m})}\right]$$

$$= 2.21 \times 10^6 \text{ J}$$

$$\Delta U_{earth} = -GmM_E \left[\frac{1}{(3.84 \times 10^8\ m)-(6.38 \times 10^6\ m)-(1.74 \times 10^6\ m)} - \frac{1}{(6.38 \times 10^6\ m)}\right]$$

$$= 6.14 \times 10^7\ J$$

$$\Delta U_{moon} = -GmM_M \left[\frac{1}{(1.74 \times 10^6\ m)} - \frac{1}{(3.84 \times 10^8\ m)-(6.38 \times 10^6\ m)}\right]$$

$$= -2.80 \times 10^6\ J$$

$$\Delta U = \Delta U_{sun} + \Delta U_{earth} + \Delta U_{moon} = 6.08 \times 10^7\ J$$

To find the escape velocity,

$$\frac{1}{2} mv^2 = \Delta U$$

$$v = \sqrt{\frac{2(6.08 \times 10^7\ J)}{1\ kg}} = 11.03\ km/s$$

This differs very little from v_E because the change in potential energy due to the moon and sun is quite small compared to that of the earth.

14-35 Twelve pairs are of the form

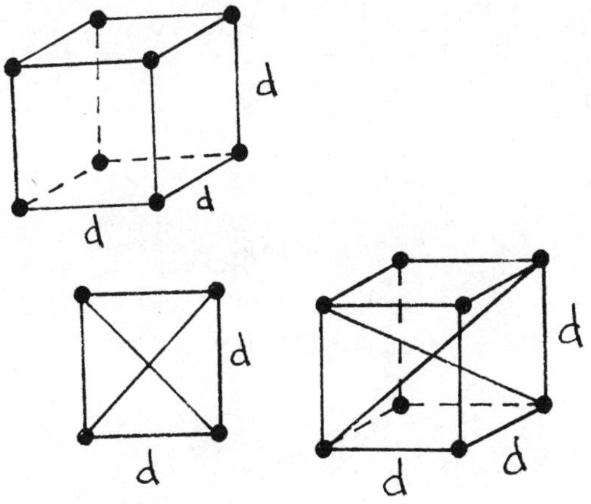

$$U = -\frac{GM^2}{d}$$

Twelve more pairs are of the form

$$U = -\frac{GM^2}{\sqrt{2}d}$$

The four diagonals of the cube each have gravitation potential energies of the form

$$U = \frac{-GM^2}{\sqrt{3}d}$$

$$U = -\frac{GM^2}{d}\left(12 + \frac{12}{\sqrt{2}} + \frac{4}{\sqrt{3}}\right)$$

$$= \frac{-(6.67 \times 10^{-11}\ N \cdot m^2/kg^2)(1.99 \times 10^{30}\ kg)^2}{9.46 \times 10^{15}\ m}(12 + 6\sqrt{2} + 4\sqrt{3}/3)$$

$$= -6.36 \times 10^{35}\ J$$

14-36 (a) Using Eq. 14-27,

$$a^3 = \frac{GM}{4\pi^2} T^2$$

$$= \frac{(6.67 \times 10^{-11}\ N \cdot m^2/kg^2)(1.99 \times 10^{30}\ kg)}{4\pi^2} (76.5\ y)^2$$

$$= \frac{(6.67 \times 10^{-11}\ N \cdot m^2/kg^2)(1.99 \times 10^{30}\ kg)}{4\pi^2} (76.5 \times 3.16 \times 10^7\ s)^2$$

$$= 19.65 \times 10^{36}\ m^3$$

$$a = 2.70 \times 10^{12}\ m = 18.0\ A.U.$$

$$b = a\sqrt{1 - e^2} = (18.0\ A.U.)\sqrt{1 - 0.967^2} = 4.6\ A.U.$$

(b) $r_{min} = a(1 - e) = (18\text{ A.U.})(1 - 0.967) = 0.59\text{ A.U.}$

$r_{max} = a(1 + e) = (18\text{ A.U.})(1 + 0.967) = 35.4\text{ A.U.}$

(c) From Example 14-4,

$$v_\phi(\text{perihelion}) = \frac{2\pi a}{T}\sqrt{\frac{1 + e}{1 - e}} = 54.2\text{ km/s}$$

$$v_\phi(\text{aphelion}) = \frac{2\pi a}{T}\sqrt{\frac{1 - e}{1 + e}} = 0.91\text{ km/s}$$

14-37 From Problem 14-16,

$$r_1 = 4.22 \times 10^7\text{ m}$$

For an elliptical trajectory,

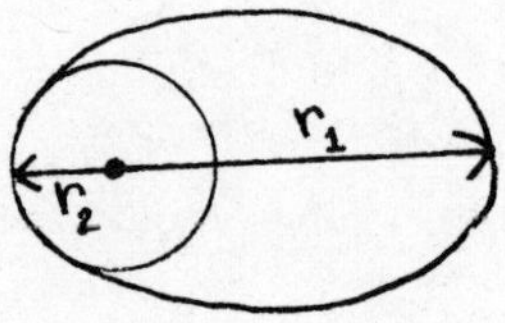

$$a = \frac{1}{2}(r_1 + r_2) = \frac{1}{2}[(4.22 \times 10^7\text{ m}) + (0.64 \times 10^7\text{ m})]$$

$$= 2.43 \times 10^7\text{ m}$$

For the rocket orbit to become elliptical instead of circular its velocity must be, from Example 14-5,

$$v_\phi(\text{apogee}) = \sqrt{\frac{a(1 - e^2)GM}{r_1^2}}$$

Since $r_1 r_2 = a^2(1 - e^2)$,

$$v_\phi(\text{apogee}) = \sqrt{\frac{r_2 GM}{ar_1}}$$

The initial velocity of the rocket is

$$v'_\phi = \sqrt{GM/r_1}$$

The necessary change in velocity is

$$\Delta v = \frac{GM}{r_1}\left(1 - \sqrt{\frac{r_2}{a}}\right)$$

$$= \sqrt{\frac{(6.67 \times 10^{-11}\text{ N}\cdot\text{m}^2/\text{kg}^2)(5.97 \times 10^{24}\text{ kg})}{4.22 \times 10^7\text{ m}}}\left(1 - \sqrt{\frac{0.64}{2.43}}\right)$$

$$= 1.50 \times 10^3\text{ m/s}$$

Since $v_r = 2800$ m/s,

$$\frac{M_f}{M_o} = e^{-1500/2800} = 0.585$$

$$M_o - M_f = M_o(1 - 0.585) = 0.415\, M_o = 415\text{ kg}$$

14-38 Since both stars have the same constant velocities, they are in circular orbits and have equal masses.

$$r_1 = r_2 = \frac{vT}{2\pi} = \frac{(40 \times 10^3\text{ m/s})(20 \times 24 \times 3600\text{ s})}{2\pi} = 1.10 \times 10^{10}\text{ m}$$

The separation of the stars is

$$d = r_1 + r_2 = 2.20 \times 10^{10} \text{ m}$$

From Eq. 14-27,

$$M = m_1 + m_2 = \frac{4\pi^2}{GT^2} a^3 = \frac{4\pi^2}{GT^2}(r_1 + r_2)^3$$

$$= \frac{4\pi^2(2.20 \times 10^{10} \text{ m})^3}{(6.67 \times 10^{-11} \text{ N}\cdot\text{m}^2/\text{kg}^2)(20 \times 24 \times 3600 \text{ s})^2} = 2.11 \times 10^{30} \text{ kg}$$

$$m_1 = m_2 = 1.06 \times 10^{30} \text{ kg}$$

14-39 Measuring from the diagram,

$$\frac{m_A}{m_B} \cong 2$$

$$T \cong 50 \text{ y}$$

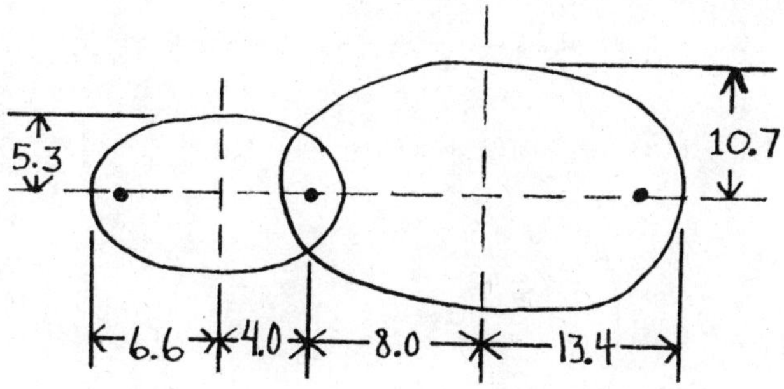

(a) $\frac{m_A}{m} = \frac{2m_B}{2m_B + m_B} = 0.67$

$$\frac{m_B}{m} = 0.33$$

From Eq. 14-27,

$$m = \frac{4\pi^2}{GT^2} a^3 = \frac{4\pi^2}{GT^2}\left(\frac{r_{min} + r_{max}}{2}\right)^3$$

$$= \frac{4\pi^2}{(6.67 \times 10^{-11} \text{ N}\cdot\text{m}^2/\text{kg}^2)(50 \times 365 \times 24 \times 3600 \text{ s})^2}\left(\frac{40 \text{ A.U.}}{2}\right)^3$$

$$= 6.43 \times 10^{30} \text{ kg}$$

$$m_A = 0.67(6.43 \times 10^{30} \text{ kg}) = 4.31 \times 10^{30} \text{ kg}$$

$$m_B = 0.33(6.43 \times 10^{30} \text{ kg}) = 2.12 \times 10^{30} \text{ kg}$$

(b) Since $r_{max} = a(1 + e)$,

$$e = \frac{r_{max} - a}{a} = \frac{32 - 20}{20} = 0.60$$

(c) $a_A = \frac{m_B}{m} a = 0.33(20 \text{ A.U.}) = 6.6 \text{ A.U.}$

$$b_A = \frac{m_B}{m} b = 0.33(20 \text{ A.U.})\sqrt{1 - e^2} = 5.3 \text{ A.U.}$$

$$a_B = \frac{m_A}{m} a = 0.67(20 \text{ A.U.}) = 13.4 \text{ A.U.}$$

$$b_B = \frac{m_A}{m} b = 0.67(20 \text{ A.U.})\sqrt{1 - e^2} = 10.7 \text{ A.U.}$$

(d) $ea_A = 0.60(6.6 \text{ A.U.}) = 4.0 \text{ A.U.}$

$$ea_B = 0.60(13.4 \text{ A.U.}) = 8.0 \text{ A.U.}$$

VI-1 $f(x) = e^x = e^o + e^x\Big|_{x=0}\cdot x + \frac{1}{2} e^x\Big|_{x=0}\cdot x^2 + \ldots + \frac{1}{n!} e^x \cdot x^n$

$= 1 + x + \frac{1}{2} x^2 + \frac{1}{3} x^3 + \ldots + \frac{x^n}{n!}$

VI-2 $f(x) = \cos\theta = (\cos 0) - \sin\theta\Big|_{\theta=0}\cdot\theta - \frac{1}{2}\cos\theta\Big|_{\theta=0}\cdot\theta^2 + \frac{1}{3}\sin\theta\Big|_{\theta=0}\cdot\theta^3$

$+ \frac{1}{4}\cos\theta\Big|_{\theta=0}\cdot\theta^4 - \frac{1}{5}\sin\theta\Big|_{\theta=0}\cdot\theta^5 - \frac{1}{6}\cos\theta\Big|_{\theta=0}\cdot\theta^6 + \ldots$

$= 1 - 0 - \frac{\theta^2}{2} + 0 + \frac{\theta^4}{4} - 0 - \frac{\theta^6}{6} + \ldots$

$= 1 - \frac{\theta^2}{2} + \frac{\theta^4}{4} - \frac{\theta^6}{6} + \ldots$

VI-3 $f(x) = \ln(1 + x) = \ln 1 + \frac{1}{1 + x}\Big|_{x=0}\cdot x - \frac{1}{(1 + x)^2}\Big|_{x=0}\cdot x^2 + \frac{1}{(1 + x)^3}\Big|_{x=0}\cdot x^3 + \ldots$

$= 0 + x - x^2 + x^3 + \ldots$

VI-4 $f(\theta)\Big|_{\theta=0} = e^{-\theta}\sin\theta\Big|_{\theta=0} = 0$

$$\frac{df}{d\theta}\Big|_{\theta=0} = e^{-\theta}(\cos\theta - \sin\theta)\Big|_{\theta=0} = 1$$

$$\frac{d^2f}{d\theta^2}\Big|_{\theta=0} = -2e^{-\theta}\cos\theta\Big|_{\theta=0} = -2$$

$$\frac{d^3f}{d\theta^3}\Big|_{\theta=0} = 2e^{-\theta}(\sin\theta + \cos\theta)\Big|_{\theta=0} = 2$$

$$\frac{d^4f}{d\theta^4}\Big|_{\theta=0} = -4\, e^{-\theta}\sin\theta\Big|_{\theta=0} = 0$$

$$\frac{d^5f}{d\theta^5}\Big|_{\theta=0} = -4e^{-\theta}(\cos\theta - \sin\theta)\Big|_{\theta=0} = -4$$

$f(\theta) = 0 + (1\cdot\theta) + \frac{1}{2!}(-2\cdot\theta^2) + \frac{1}{3!}(2\cdot\theta^3)$

$+ \frac{1}{4!}(0\cdot\theta^4) + \frac{1}{5!}(-4\cdot\theta^5) + \ldots$

$= \theta - \theta^2 + \frac{\theta^3}{3} - \frac{\theta^5}{30} + \ldots$

VI-5 $(1 + x)^{\frac{1}{2}} = 1 + \frac{1}{2}x + \frac{\frac{1}{2}(\frac{1}{2} - 1)}{2!}x^2 + \frac{\frac{1}{2}(\frac{1}{2} - 1)(\frac{1}{2} - 2)}{3!}x^3 + \ldots$

$= 1 + \frac{1}{2}x - \frac{1}{8}x^2 + \frac{1}{16}x^3 + \ldots$

$\sqrt{10} = (9 + 1)^{\frac{1}{2}} = 3(1 + \frac{1}{9})^{\frac{1}{2}} = 3(1 + \frac{1}{2}\cdot\frac{1}{9} - \frac{1}{8}\cdot\frac{1}{81} + \frac{1}{16}\cdot\frac{1}{729} + \ldots)$

$\cong 3.16229$

Using only the first two terms,

$$\sqrt{10} \cong 3(1 + \frac{1}{18}) = 3.16667$$

The percentage error using only the first two terms is

$$\frac{3.16667 - 3.16229}{3.16229} = 0.00139 = 0.139\%$$

VI-6 $$f(\theta) = f(0) + \left.\frac{df}{d\theta}\right|_{\theta=0} \cdot \theta + \frac{1}{2}\left.\frac{d^2f}{d\theta^2}\right|_{\theta=0} \cdot \theta^2 + \ldots \frac{1}{s!}\left.\frac{d^sf}{d\theta^s}\right|_{\theta=0} \cdot \theta^s + \ldots$$

$$= \theta - \frac{1}{3}\theta^3 + \frac{1}{5}\theta^5 + \ldots$$

$$= \sin\theta$$

15-1 (a) $U(\xi) = 2\xi^2 - 4\xi + 1$

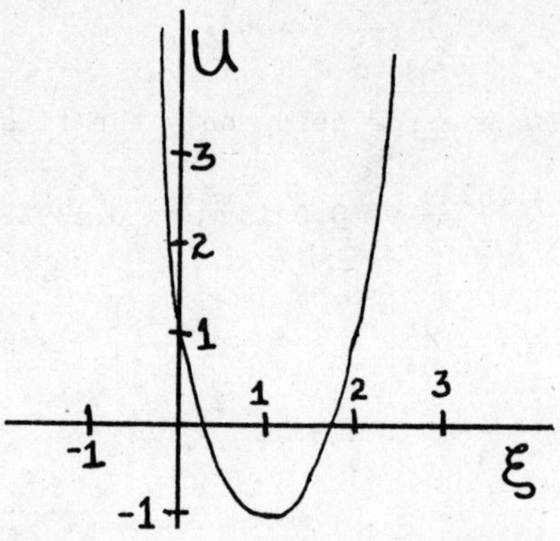

(b) $\frac{dU}{d\xi} = 4\xi - 4 = 0 \qquad \xi = 1$

$$\left.\frac{d^2U}{d\xi^2}\right|_{\xi=1} = 4 > 0$$

$\xi = 1$ is a point of stable equilibrium.

(c) $U(\xi) = 2(\xi^2 - 2\xi + 1) - 1 = 2(\xi - 1)^2 - 1$

Let $x = \xi - 1$

$U(x) = 2x^2 - 1$

$F(x) = \frac{-dU}{dx} = -4x$

$\kappa = 4$

(d) From Eq. 15-3,

$$U(\xi) = U(\xi = 1) + (\xi - 1)\left.\frac{dU(\xi)}{d\xi}\right|_{\xi=1} + \frac{1}{2}(\xi - 1)^2 \left.\frac{d^2U(\xi)}{d\xi^2}\right|_{\xi=1} + \ldots$$

$$= -1 + 0 + 2(\xi - 1)^2$$

Using Eq. 15-2,

$\kappa = 4$

15-2 (a) $U = mgh = mg(\ell - \ell\cos\theta) = mg\ell(1 - \cos\theta)$

(b) $$U(\theta) = U(0) + \theta\left.\frac{dU}{d\theta}\right|_{\theta=0} + \frac{1}{2}\theta^2 \left.\frac{d^2U}{d\theta^2}\right|_{\theta=0}$$

$$= 0 + \theta(mg\ell\sin\theta)\Big|_{\theta=0} + \frac{1}{2}\theta^2(mg\ell\cos\theta)\Big|_{\theta=0}$$

$$= \frac{1}{2}\theta^2 mg\ell$$

(c) $F(\theta) = -\frac{dU}{d\theta} = -\theta\, mg\ell$

$\kappa = mg\ell$

15-3 (a)

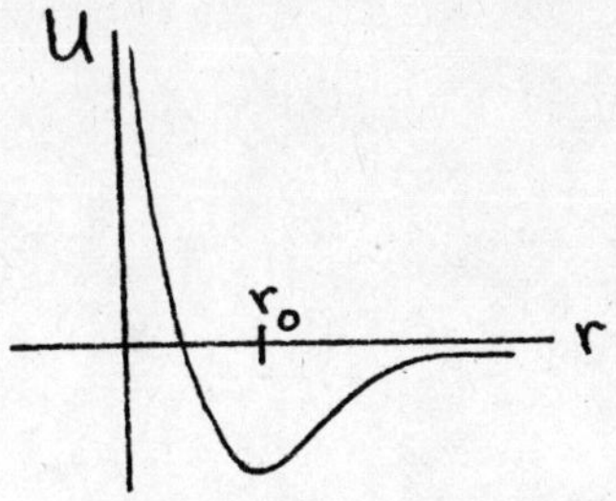

(b) $$\frac{dU}{dr} = \frac{U_o}{r_o}\left[-12\left(\frac{r_o}{r}\right)^{13} + 12\left(\frac{r_o}{r}\right)^{7}\right] = 0$$

$$r = r_o$$

$$\frac{d^2U}{dr^2} = \frac{U_o}{r_o^2}\left[156\left(\frac{r_o}{r}\right)^{14} - 84\left(\frac{r_o}{r}\right)^{8}\right]\Bigg|_{r=r_o} > 0$$

$r = r_o$ is a point of stable equilibrium.

(c) $$U(r) = U(r_o) + \frac{1}{2}(r - r_o)^2 \left.\frac{d^2U}{dr^2}\right|_{r=r_o}$$

$$= -U_o + \frac{1}{2}(r - r_o)^2\left[\frac{U_o}{r_o^2}(156 - 84)\right]$$

$$= -U_o + 36\,\frac{U_o(r - r_o)^2}{r_o^2}$$

Let the displacement from stable equilibrium be given by

$$x = r - r_o$$

Then,
$$U(x) = -U_o + 36\,\frac{U_o x^2}{r_o^2}$$

(d) $$F(x) = -\frac{dU}{dx} = -\frac{72\,U_o x}{r_o^2}$$

$$\kappa = 72\,\frac{U_o}{r_o^2}$$

15-4 (a) $$v_{max} = A\omega_o = (0.10\text{ m})(2\pi \times 3\text{ rad/s}) = 1.88\text{ m/s}$$

$$a_{max} = A\omega_o^2 = (0.10\text{ m})(2\pi \times 3\text{ rad/s})^2 = 35.5\text{ m/s}^2$$

(b) $$x = A\sin\phi$$

$$\phi = \sin^{-1}\frac{x}{A} = 53.1°$$

$$v = A\omega_o\cos\phi = 1.13\text{ m/s}$$

$$a = -A\omega_o^2\sin\phi = 28.4\text{ m/s}^2$$

(c) $$x = A\sin\omega_o t$$

$$\omega_o t = \sin^{-1}\frac{x}{A} = 0.643\text{ rad}$$

$$t = \frac{0.643}{6\pi} = 0.034\text{ s}$$

15-5 (a)

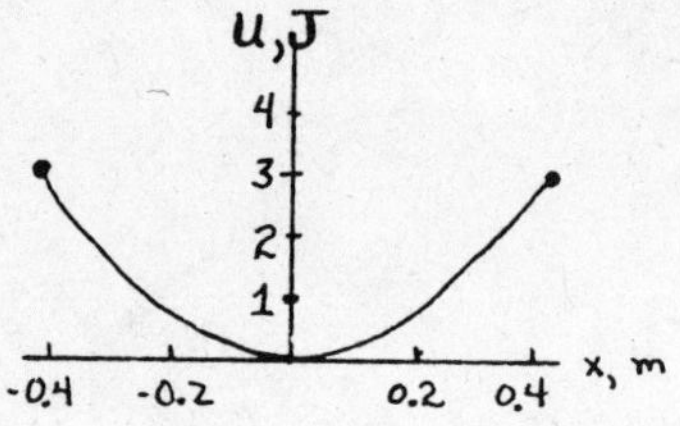

$U(x) = \frac{1}{2}\kappa x^2 = 20x^2\ \text{J} \cdot \text{m}^2$

(b) $E = K + U = 1.5 + 2 = 3.5\ \text{J}$

$E = \frac{1}{2}\kappa A^2$

$A = \sqrt{\frac{2E}{\kappa}} = \sqrt{\frac{7\ \text{J}}{40\ \text{N/m}}} = 0.42\ \text{m}$

(c) $E = \frac{1}{2}mv_{max}^2$

$v_{max} = \sqrt{\frac{2E}{m}} = \sqrt{\frac{7\ \text{J}}{0.50\ \text{kg}}} = 3.74\ \text{m/s}$

(d) $\frac{1}{2}\kappa x^2 = \frac{1}{2}E$

$x = \sqrt{\frac{E}{\kappa}} = 0.30\ \text{m}$

(e) $\omega_o = \sqrt{\frac{\kappa}{m}} = 8.94\ \text{rad/s}$

$T = \frac{2\pi}{\omega_o} = 0.702\ \text{s}$

(f) Initially,

$\frac{1}{2}\kappa x_o^2 = U$

$x_o = \sqrt{\frac{2U}{\kappa}} = \sqrt{\frac{2\times 2}{40}} = 0.316\ \text{m}$

$\phi_o = \sin^{-1}\frac{x_o}{A} = 48.8°, 131.2°$

For $v < 0$,

$\phi_o = 131.2° = 2.29\ \text{rad}$

(g) $x = A\sin(\omega t + \phi_o) = (0.42\ \text{m})\sin(8.94\ t + 2.29)$

15-6 $F = -\kappa x = m_1\frac{d^2x_1}{dt^2} = m_2\frac{d^2x_2}{dt^2}$

$x = x_1 + x_2$

$\frac{d^2x}{dt^2} = \frac{d^2x_1}{dt^2} + \frac{d^2x_2}{dt^2} = \frac{-\kappa x}{m_1} + \frac{-\kappa x}{m_2}$

$\frac{m_1m_2}{m_1 + m_2}\frac{d^2x}{dt^2} = -\kappa x$

$\omega = \sqrt{\frac{\kappa}{M}} = \sqrt{\frac{\kappa(m_1 + m_2)}{m_1m_2}}$

15-7 (a) $a_{max} = \omega_o^2 A$

$$\omega_o = \sqrt{\frac{a_{max}}{A}} = \sqrt{\frac{0.30\ m/s^2}{0.04\ m}} = 2.74\ rad/s$$

$$T = \frac{2\pi}{\omega_o} = 2.29\ s$$

(b) $v_{max} = \omega_o A = (2.74\ rad/s)(0.04\ m) = 0.11\ m/s$

(c) $x = (0.04\ m)\sin(2.74\ t + \frac{3\pi}{2})$

15-8 (a) $v_{max} = \omega_o A$

$$A = \frac{v_{max}}{\omega_o} = \frac{1.5\ m/s}{2.0\ rad/s} = 0.75\ m$$

(b) $x = (0.75\ m)\sin 2.0t$

15-9 Let the unprimed variables refer to the system before the spring is cut and the primed variables refer to the system after the string is cut. Since the spring is uniform, the force it exerts before and after being cut can be written as follows

$$F = -\kappa x = -\kappa' x' = -\kappa' f x$$

So,

$$\kappa' = \frac{\kappa}{f}$$

Likewise, the strength parameter for the piece of length $(1-f)\ell$ is

$$\kappa'' = \frac{\kappa}{1-f}$$

(b)

$$\omega_o' = \sqrt{\frac{\kappa'}{m'}} = \sqrt{\frac{\kappa/f}{(1-f)m}} = \omega_o \sqrt{\frac{1}{f(1-f)}}$$

$$\omega_o'' = \frac{\kappa''}{m''} = \sqrt{\frac{\kappa/(1-f)}{fm}} = \omega_o \sqrt{\frac{1}{f(1-f)}}$$

15-10 (a) Using the result of Example 15-5,

$$\omega_o^2 = \frac{\kappa}{m + \frac{m_s}{3}}$$

$$\frac{\omega_1^2}{\omega_2^2} = \frac{m_2 + m_s/3}{m_1 + m_s/3} = \frac{T_2^2}{T_1^2}$$

$$m_s = 0.060\ kg = 60\ g$$

(b) $\kappa = \left(m_1 + \frac{m_s}{3}\right)\omega_1^2 = \left(m_1 + \frac{m_s}{3}\right)\left(\frac{2\pi}{T_1}\right)^2 = 28.7\ N/m$

15-11 $\omega = \frac{2\pi}{T} = 2.09$ rad/s

$v_o = A\omega = (0.30\text{ m})(2.09\text{ rad/s}) = 0.63$ m/s

$x = (0.30\text{ m})\sin 2.09t$

15-12 $y = 60\sin(8t + \pi/8)$

(a) $R = 60$ m

$\theta_o = \phi_o = \frac{\pi}{8}$

$\omega = \omega_o = 8$ rad/s

(b) $y = 40 = 60\sin(8t + \pi/8) = 60\sin\theta$

$\theta = \sin^{-1}\frac{40}{60} = 41.8°, 138.2°$

For $v < 0$, $\frac{\pi}{2} < \theta < \pi$. So,

$\theta = 138.2° = 2.4$ rad

(c) $a = g = \frac{d^2y}{dt^2} = 3840\sin\theta$

$\theta = \sin^{-1}\frac{9.8}{3840} = 0.15°, 179.8°$

For $v > 0$ and $a < 0$, $0 < \theta < \frac{\pi}{2}$

$\theta = 0.15° = 0.003$ rad

(d) $\theta = 8t + \frac{\pi}{8}$

For part (b),

$8t + \frac{\pi}{8} = 2.4$

$t = 0.25$ s

For part (c),

$8t + \frac{\pi}{8} = 0.003$

Since $\pi/8 > 0.003$, this must be rewritten.

$8t + \frac{\pi}{8} = 2\pi + 0.003$

$t = 0.74$ s

15-13 $T = 2\pi\sqrt{\frac{I_o}{Mg\ell}} = 2\pi\sqrt{\frac{M\ell^2/3}{Mg\ell}} = 2\pi\sqrt{\frac{\ell}{3g}} = 1.47$ s

$\nu = \frac{1}{T} = 0.68$ Hz

$T' = 2\pi\sqrt{\frac{M\ell^2 + M\ell^2/3}{Mg\ell}} = 2\pi\sqrt{\frac{4\ell}{3g}} = 2T = 2.94$ s

$\nu' = \frac{1}{2}\nu = 0.34$ Hz

15-14 $J = \int F\,dt = mv_0$

$$v_0 = \frac{J}{m} = \frac{0.15\ \text{N}\cdot\text{s}}{0.20\ \text{kg}} = 0.75\ \text{m/s}$$

$$\theta_0 = \phi_0 = 0$$

The angular speed is

$$\eta = \frac{d\theta}{dt} = \omega\Theta \cos \omega t$$

(a) At $t = 0$,

$$\ell\eta = v_0 = \ell\omega\Theta$$

$$\Theta = \frac{v_0}{\ell\omega} = \frac{v_0}{\ell\sqrt{g/\ell}} = \frac{v_0}{\sqrt{g\ell}} = 0.189\ \text{rad}$$

(b) $\theta(t) = (0.189\ \text{rad}) \sin \sqrt{g/\ell}\ t = (0.189\ \text{rad}) \sin 2.47t$

15-15 (a) $T = 2\pi\sqrt{\dfrac{I_0}{\Gamma}}$

$$\Gamma = \frac{4\pi^2}{T^2} I_0 = \frac{4\pi^2}{T^2}\left(\frac{M\ell^2}{12}\right) = 5.14 \times 10^{-8}\ \text{N}\cdot\text{m}$$

(b) Setting the torque about the axis of the fiber due to the fiber equal in magnitude to the torque due to gravity,

$$\Gamma\phi = mg\frac{\ell}{2}$$

The vertical deflection is

$$\phi\frac{\ell}{2} = \frac{mg\ell^2}{4\Gamma} = 0.48\ \text{cm}$$

15-16

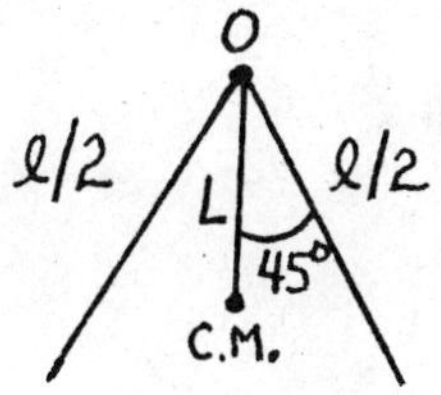

$$I_0 = 2\left[\frac{1}{3}\left(\frac{m}{2}\right)\left(\frac{\ell}{2}\right)^2\right] = \frac{m\ell^2}{12}$$

$$T = 2\pi\sqrt{\frac{I_0}{mgL}} = 2\pi\sqrt{\frac{I_0}{(mg\ell \cos 45°)/4}}$$

$$= 2\pi\sqrt{\frac{\sqrt{2}\ell}{3g}} = 0.689\ \text{s}$$

15-17 $I_0\alpha = -mgr \sin\theta \cong -mgr\,\theta$

$$I_0 = \frac{1}{2}mR^2 + mr^2$$

$$T = 2\pi\sqrt{\frac{I_0}{mgr}} = 2\pi\sqrt{\frac{R^2 + 2r^2}{2gr}}$$

For T to be a minimum,

$$\frac{dT}{dr} = 0 = \frac{\pi}{\sqrt{\dfrac{R^2 + 2r^2}{2gr}}}\left(\frac{2r^2 - R^2}{2gr^2}\right)$$

$$r = \frac{R}{\sqrt{2}}$$

$$T_{min} = 2\pi\sqrt{\frac{R^2 + R^2}{2gR/\sqrt{2}}} = 2\pi\sqrt{\frac{\sqrt{2}R}{g}} = 1.07 \text{ s}$$

15-18 $T = 2\pi\sqrt{\frac{I_o}{Mg\ell}}$

$$I_o = I_c + M\ell^2 = \frac{2}{5} MR^2 + M\ell^2$$

$$T = 2\pi\sqrt{\frac{\frac{2}{5} R^2 + \ell^2}{g\ell}} = 2\pi\sqrt{\frac{\ell}{g}} \cdot \sqrt{\frac{2R^2}{5\ell^2} + 1}$$

$$= T_o\sqrt{\frac{2R^2}{5\ell^2} + 1} \cong T_o[1 + \frac{1}{5}\frac{R^2}{\ell^2}] = T_o(1 + 0.00008)$$

$$T_o = 2\pi\sqrt{\frac{\ell}{g}} = 1.0035 \text{ s}$$

$$T = T_o(1 + 0.00008) = 1.0036 \text{ s}$$

15-19

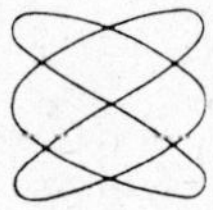

15-20 $x(t) = Ae^{-\beta t} \sin(\omega^o t + \phi_o)$

$$\frac{dx}{dt} = -A\beta e^{-\beta t} \sin(\omega^o t + \phi_o) + Ae^{-\beta t} \omega^o \cos(\omega^o t + \phi_o)$$

$$\frac{d^2x}{dt^2} = A\beta^2 e^{-\beta t} \sin(\omega^o t + \phi_o) - A\beta e^{-\beta t} \omega^o \cos(\omega^o t + \phi_o)$$

$$-A\beta e^{-\beta t} \omega^o \cos(\omega^o t + \phi_o) - Ae^{-\beta t}(\omega^o)^2 \sin(\omega^o t + \phi_o)$$

Dividing Eq. 15-31 by m,

$$\frac{d^2x}{dt^2} + \frac{b}{m}\frac{dx}{dt} + \frac{\kappa}{m} x = 0 = \frac{d^2x}{dt^2} + 2\beta \frac{dx}{dt} + \omega_o^2 x$$

$$\frac{d^2x}{dt^2} + 2\beta \frac{dx}{dt} + \omega_o^2 x = A\beta^2 e^{-\beta t} \sin(\omega^o t + \phi_o) - A\beta e^{-\beta t}\omega^o \cos(\omega^o t + \phi_o)$$

$$-A\beta e^{-\beta t}\omega^o \cos(\omega^o t + \phi_o) - Ae^{-\beta t}(\omega^o)^2 \sin(\omega^o t + \phi_o)$$

$$+ 2\beta[-A\beta e^{-\beta t} \sin(\omega^o t + \phi_o) + Ae^{-\beta t} \omega^o \cos(\omega^o t + \phi_o)]$$

$$+ \omega_o^2 Ae^{-\beta t} \sin(\omega^o t + \phi_o)$$

$$= (\beta^2 - (\omega^o)^2 - 2\beta^2 + \omega_o^2) \sin(\omega^o t + \phi_o) +$$

$$(-\beta\omega^o - \beta\omega^o + 2\beta\omega^o) \cos(\omega^o t + \phi_o)$$

$$= (-\beta^2 - \omega^{o2} + \omega_o^2) \sin(\omega^o t + \phi_o)$$

$$= [-\beta^2 - (\omega_o^2 - \beta^2) + \omega_o^2] \sin(\omega^o t + \phi_o)$$

$$= 0$$

15-21 (a) $(\omega^{0})^2 = \frac{\omega_o^2}{(1.25)^2} = \omega_o^2 - (\frac{b}{2m})^2$

$$b = 2m\omega_o\sqrt{1 - \frac{1}{(1.25)^2}} = 2m\sqrt{\frac{\kappa}{m}} \cdot \sqrt{1 - \frac{1}{(1.25)^2}} = 13.42 \text{ N·s/m}$$

(b) $\beta = \frac{b}{2m} = 13.42 \text{ s}^{-1}$

$$e^{-\beta t} = \frac{1}{3}$$

$$t = \frac{\ell n\ 3}{\beta} = 0.0819 \text{ s}$$

15-22 $x(t) = Ae^{-\beta t} \sin (\omega^{0}t + \phi_o)$

$$v(t) = \frac{dx}{dt} = -A\beta e^{-\beta t} \sin (\omega^{0}t + \phi_o) + A\omega^{0}e^{-\beta t} \cos (\omega^{0}t + \phi_o)$$

For the given parameters,

$$v(t) = [2e^{-t} \cos (2t + \phi_o) - e^{-t} \sin (2t + \phi_o)](1 \text{ m/s})$$

where t is in seconds. The graph for $\phi_o = 0$ is as follows:

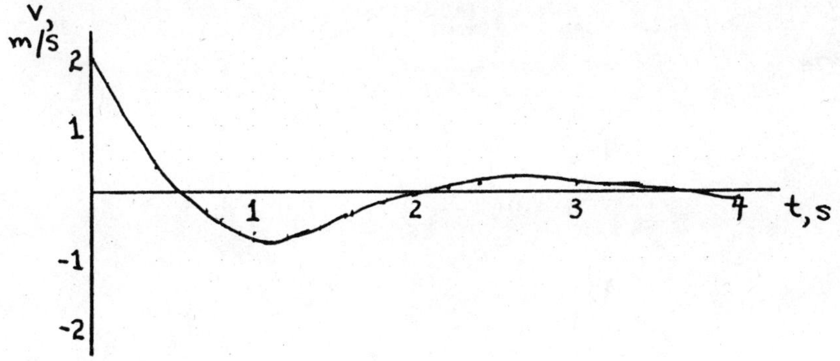

15-23 $$A = \frac{F_o/\kappa}{\sqrt{\frac{1}{Q^2}\frac{\omega^2}{\omega_o^2} + (\frac{\omega^2}{\omega_o^2} - 1)^2}}$$

$$D^2 = \frac{1}{Q^2}\frac{\omega^2}{\omega_o^2} + (\frac{\omega^2}{\omega_o^2} - 1)^2$$

$$\frac{dD^2}{d\omega} = \frac{2}{Q^2}\frac{\omega}{\omega_o^2} + \frac{4\omega}{\omega_o^2}(\frac{\omega^2}{\omega_o^2} - 1) = 0$$

$$\frac{1}{Q^2} + 2(\frac{\omega^2}{\omega_o^2} - 1) = 0$$

$$\frac{\omega^2}{\omega_o^2} = 1 - \frac{1}{2Q^2}$$

$$\omega^2 = \omega_R^2 = \omega_o^2(1 - \frac{1}{2Q^2})$$

To show that this is equivalent to Eq. 15-39,

$$Q = \frac{m\omega_o}{b}$$

So,

$$\omega^2 = \omega_o^2(1 - \frac{b^2}{2m^2\omega_o^2}) = \omega_o^2 - \frac{b^2}{2m^2}$$

$$\omega = \sqrt{\omega_o^2 - \frac{b^2}{2m^2}}$$

15-24 (a) $\omega_R = \sqrt{\omega_o^2 - \frac{b^2}{2m^2}} = \sqrt{\omega_o^2 - \frac{\omega_o^2}{2Q^2}} = \omega_o \sqrt{1 - \frac{1}{2Q^2}}$

$= \sqrt{\frac{\kappa}{m}} \cdot \sqrt{1 - \frac{1}{2Q^2}} = 19.72$ rad/s

(b) $A_R = \dfrac{F_o/\kappa}{\sqrt{\frac{1}{Q^2}\frac{\omega_R^2}{\omega_o^2} + \left(\frac{\omega_R^2}{\omega_o^2} - 1\right)^2}} = 0.318$ m

$\delta_R = \tan^{-1}\left[Q\left(\frac{\omega_R}{\omega_o} - \frac{\omega_o}{\omega_R}\right)\right] = 159.3° = 2.78$ rad

(c) $A_o = \dfrac{F_o/\kappa}{\sqrt{\frac{1}{Q^2}}} = \dfrac{QF_o}{\kappa} = 0.30$ m

$\tan \delta = 0$

$\delta_o = 0°, 180°$

(d) $A = \dfrac{F_o/\kappa}{\sqrt{\frac{1}{9Q^2} + (\frac{1}{9} - 1)^2}} = 0.218$ m

$\delta = \tan^{-1}\left[Q(\frac{1}{3} - 3)\right] = 104.0° = 1.82$ rad

(e) $A = \dfrac{F_o/\kappa}{\sqrt{\frac{9}{Q^2} + (9 - 1)^2}} = 0.024$ m

$\delta = \tan^{-1}\left[Q(3 - \frac{1}{3})\right] = 76.0° = 1.33$ rad

15-25 (a)

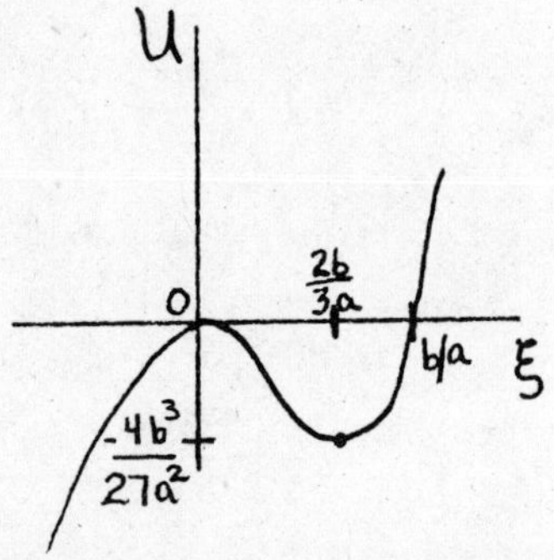

(b) $\dfrac{dU}{d\xi} = 3a\xi^2 - 2b\xi = 0$

$\xi = 0, \dfrac{2b}{3a}$

$\dfrac{d^2U}{d\xi^2} = 6a\xi - 2b$

For $\xi = 0$, $\frac{d^2U}{d\xi^2} < 0$. Thus, $\xi = 0$ is a point of unstable equilibrium. For $\xi = \frac{2b}{3a}$, $\frac{d^2U}{d\xi^2} > 0$. Thus, $\xi = \frac{2b}{3a}$ is a point of stable equilibrium.

(c) $U(\xi) \cong U(\frac{2b}{3a}) + \frac{1}{2}(\xi - \frac{2b}{3a})^2 (6a\xi - 2b)\Big|_{\xi = \frac{2b}{3a}}$

$= -\dfrac{4b^2}{27a^2} + b(\xi - \dfrac{2b}{3a})^2$

(d) $U(x) = \frac{-4b^2}{27a^2} + bx^2$

$F(x) = \frac{-dU}{dx} = -2bx$

$\kappa = 2b$

15-26 (a)

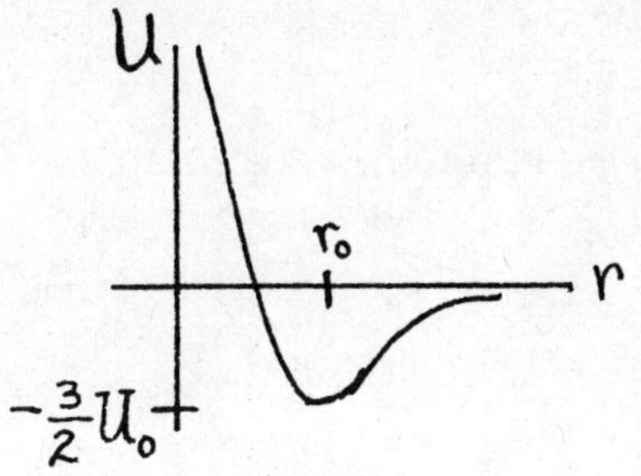

(b) $\frac{dU}{dr} = \frac{U_o}{r_o}\left[-5\left(\frac{r_o}{r}\right)^6 + 2C\left(\frac{r_o}{r}\right)^3\right]$

$= \frac{U_o}{r_o}(-5 + 2C) = 0$

$C = \frac{5}{2}$

$\frac{d^2U}{dr^2} = \frac{U_o}{r_o^2}\left[30\left(\frac{r_o}{r}\right)^7 - 15\left(\frac{r_o}{r}\right)^4\right]$

Since $\frac{d^2U}{dr^2} > 0$ for $r = r_o$, this is a point of stable equilibrium.

(c) $U(r) = U(r_o) + \frac{1}{2}(r - r_o)^2 \left.\frac{d^2U}{dr^2}\right|_{r=r_o}$

$= U_o(1 - \frac{5}{2}) + \frac{1}{2}(r - r_o)^2\left(\frac{15U_o}{r_o^2}\right)$

$= -\frac{3}{2}U_o + \frac{15U_o}{2r_o^2}(r - r_o)^2$

(d) $U(x) = -\frac{3}{2}U_o + \frac{15U_o}{2r_o^2}x^2$

$F(x) = \frac{-dU}{dx} = -\frac{15U_o}{r_o^2}x$

$\kappa = \frac{15U_o}{r_o^2}$

15-27

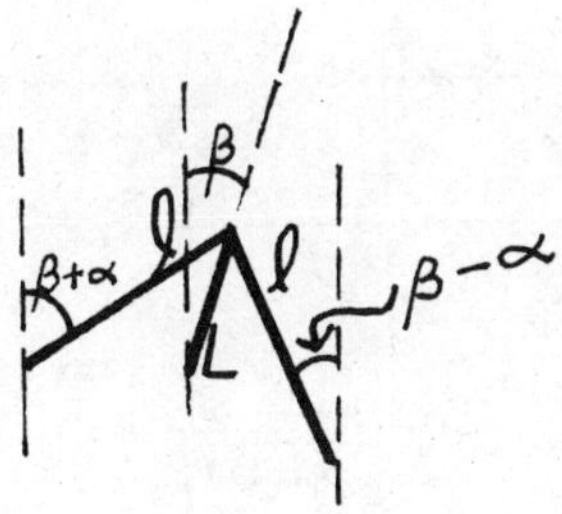

(a) $U(\beta) = mg[(L\cos\beta - \ell\cos(\alpha + \beta)) + (L\cos\beta - \ell\cos(\alpha - \beta))]$

Using the following identity,

$\cos(\alpha \pm \beta) = \cos\alpha\cos\beta \mp \sin\alpha\sin\beta$

the potential is

$$U(\beta) = mg[2L \cos\beta - 2\ell \cos\alpha \cos\beta]$$

$$= 2\ mg \cos\beta(L-\ell \cos\alpha)$$

(b) $$\frac{dU}{d\beta} = 2mg \sin\beta(\ell \cos\alpha - L) = 0$$

Equilibrium occurs at $\beta = 0$. For the equilibrium to be stable,

$$\frac{d^2U}{d\beta^2} > 0$$

$$\left.\frac{d^2U}{d\beta^2}\right|_{\beta=0} = 2\ mg(\ell \cos\alpha - L) > 0$$

So,

$$\ell \cos\alpha > L$$

(c) Since

$$U(\beta) = U(0) + \frac{1}{2}\beta^2 \left.\frac{d^2U}{d\beta^2}\right|_{\beta=0}$$

and

$$U(\beta) = -\int F\ d\beta = \frac{1}{2}\kappa\beta^2 + \text{constant}$$

the strength parameter is

$$\kappa = \left.\frac{d^2U}{d\beta^2}\right|_{\beta=0} = 2\ mg(\ell \cos\alpha - L)$$

The moment of inertia about the pivot point is

$$I_o = 2m(\ell^2 + L^2 - 2\ell L \cos\alpha)$$

Thus,

$$T = \frac{2\pi}{\omega} = 2\pi\sqrt{\frac{I_o}{\kappa}} = 2\pi\sqrt{\frac{\ell^2 + L^2 - 2\ell L \cos\alpha}{g(\ell \cos\alpha - L)}}$$

$$= 2\pi\sqrt{\frac{9L^2 + L^2 - 3L^2}{g(\frac{3}{2}L - L)}} = 2\pi\sqrt{\frac{14\ L}{g}}$$

15-28 $$\frac{1}{2}\kappa x^2 + \frac{1}{2}mv^2 = E$$

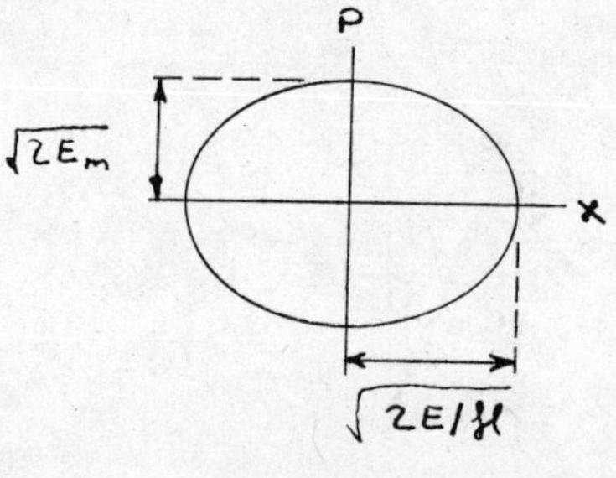

Using $p = mv$,

$$\frac{1}{2}\kappa x^2 + \frac{1}{2}\frac{p^2}{m} = E$$

$$\frac{\kappa x^2}{2E} + \frac{p^2}{2Em} = 1$$

$$\frac{p^2}{2Em} + \frac{x^2}{2E/\kappa} = 1$$

$$\text{Area} = \pi ab = 2\pi E\ \sqrt{m/\kappa} = 2\pi E/\omega_o$$

15-29 $$\frac{d^2x}{dt} = A\omega^2$$

$$f_{max} = \mu_s N = \mu_s mg = mA\omega^2$$

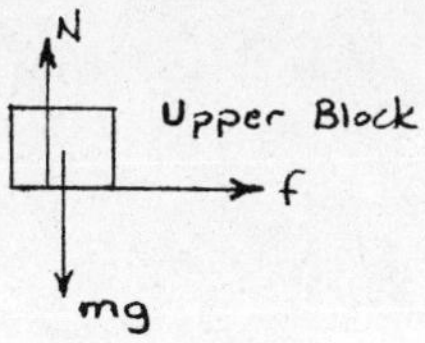

$$A = \frac{\mu_s g}{\omega^2} = 0.066 \text{ m}$$

15-30 (a) $$I = \int r^2 \, dM = \int_0^{\ell/4} \frac{r^2 M}{\ell} dr + \int_0^{3\ell/4} \frac{r^2 M}{\ell} dr$$

$$= \frac{M}{3\ell}\left[\left(\frac{\ell}{4}\right)^3 + \left(\frac{3\ell}{4}\right)^3\right] = \frac{M}{3\ell}\left(\frac{28\ell^3}{64}\right) = \frac{7M\ell^2}{48}$$

(b) $$\tau = \frac{-Mg\ell}{4} \sin\theta = I_o \alpha$$

$$\alpha = -\frac{Mg\ell}{4I_o} \sin\theta = -\frac{12g}{7\ell} \sin\theta$$

For small oscillations,

$$\alpha = -\frac{12g}{7\ell} \theta$$

(c) $$T = 2\pi\sqrt{\frac{I_o}{Mg\ell/4}} = 2\pi\sqrt{\frac{7\ell}{12g}} = 1.68 \text{ s}$$

15-31 If $a_{max} > g$, the block will separate from the plate.

$$a_{max} = g = A\omega_o^2$$

$$A = \frac{g}{\omega_o^2} = \frac{gT^2}{4\pi^2} = 0.36 \text{ m}$$

15-32 (a) $$\omega_o = \sqrt{\frac{\kappa}{m}} = 20 \text{ rad/s}$$

(b)

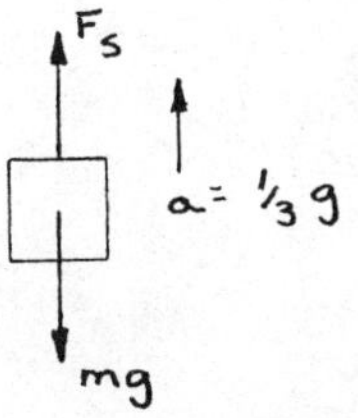

$$F_s - mg = \frac{1}{3} mg$$

$$F_s = \frac{4}{3} mg = 6.53 \text{ N}$$

$$x_s = \frac{F_s}{\kappa} = 3.26 \text{ cm}$$

(c) When the acceleration of the car is zero, the new equilibrium position can be found as follows:

$$F'_s = mg = 4.90 \text{ N}$$

$$x'_s = \frac{F'_s}{\kappa} = 2.45 \text{ cm}$$

Thus,

$$x_o = x'_s - x_s = -0.81 \text{ cm}$$

Since $v_o = 0$, and using Eq. 15-17c,

$$A^2 = x_o^2$$

$$A = 0.81 \text{ cm}$$

Since $v_o = 0$ and the upward direction is positive,

$$\phi_o = \frac{3\pi}{2}$$

15-33 (a) $T = \frac{2\pi}{\omega_o} = 2\sqrt{\frac{m + M}{\kappa}} = 0.397 \text{ s}$

(b) The velocity of the mass of putty on impact is given by

$$v = -\sqrt{2gh} = -1.40 \text{ m/s}$$

Conservation of momentum can be used to determine the initial velocity of the SHM.

$$v_o = \frac{mv}{M + m} = -0.525 \text{ m/s}$$

The initial position of the block relative to its new equilibrium position is

$$x_o = \frac{F_f}{\kappa} - \frac{F_i}{\kappa} = \frac{(M + m)g}{\kappa} - \frac{Mg}{\kappa} = \frac{mg}{\kappa} = 0.0147 \text{ m}$$

The angular frequency is

$$\omega_o = \sqrt{\frac{\kappa}{M + m}} = 15.81 \text{ rad/s}$$

Using Eq. 15-17b,

$$\phi_o = \tan^{-1}\frac{\omega_o x_o}{v_o} = \tan^{-1}\frac{x_o}{v_o}\sqrt{\frac{\kappa}{m + M}} = 156.1° = 2.72 \text{ rad}$$

Using Eq. 15-17c,

$$A = \sqrt{x_o + (\frac{v_o}{\omega_o})^2} = 0.0363 \text{ m}$$

The equation of motion measured from the new equilibrium position is

$$x(t) = (0.0363 \text{ m}) \sin (15.81 t + 2.72)$$

(c) $E = \frac{1}{2}\kappa A^2 = 0.132 \text{ J}$

15-34 $T = 2\pi\sqrt{\frac{\ell}{g}}$ $\qquad \frac{dT}{dg} = \frac{-\pi}{g}\sqrt{\frac{\ell}{g}}$

$$\frac{dT}{T} = -\frac{1}{2}\frac{dg}{g}$$

However, we know

$$g = \frac{GM_E}{R_E^2}$$

and

$$dg = -2\, GM_E \frac{dR}{R_E^3}$$

So,

$$\frac{dg}{g} = -2\frac{dR}{R_E}$$

This gives

$$\frac{dT}{T} = \frac{dR}{R_E}$$

The height above the earth is then

$$h = dR = R_E \frac{dT}{T} = (6.378 \times 10^3 \text{ km}) \frac{60 - 59.914}{60}$$

$$= 9.142 \text{ km}$$

15-35 Using Eq. 15-18,

The tangential acceleration of the bug is given by

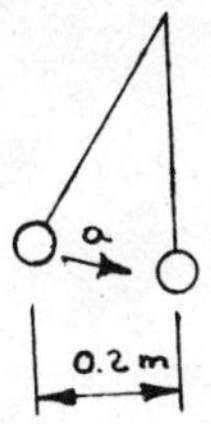

$$a = \ell \frac{d^2\theta}{dt^2} = g \sin\theta$$

The tangential force exerted by the bob on the bug is given by

$$F_t = ma - mg \sin\theta = 0$$

The radial force exerted by the bob on the bug is given by

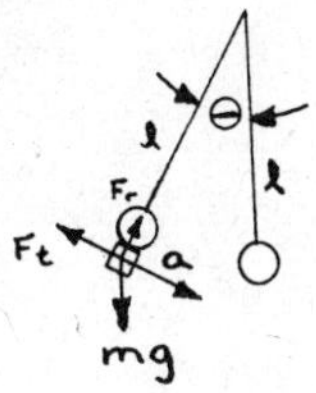

$$F_r = mg \cos\theta$$

This force, equal in magnitude to the holding force exerted by the bug, is a minimum when θ is a maximum.

$$F_r = mg \cos\left(\sin^{-1}\frac{0.2 \text{ m}}{2 \text{ m}}\right) = 9.75 \times 10^{-4} \text{ N}$$

15-36 $U(r) = \frac{-GMm}{r} = \frac{-GMm}{\sqrt{R^2 + x^2}}$

$$= -\frac{GMm}{R}\left(1 + \frac{x^2}{R^2}\right)^{-\frac{1}{2}}$$

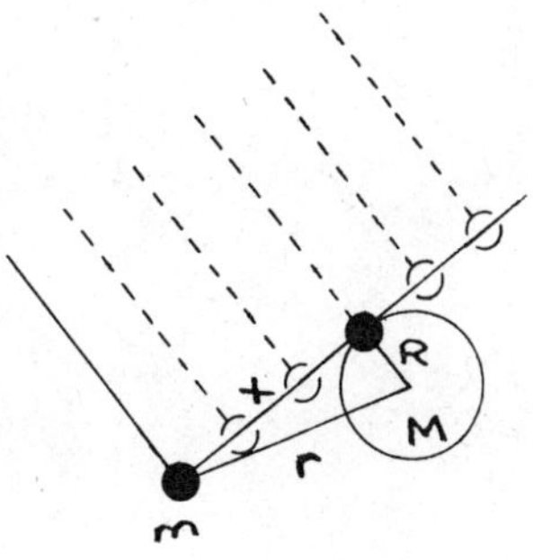

$$U(x) \cong \frac{-GMm}{R}\left(1 - \frac{x^2}{2R^2}\right)$$

$$= \frac{-GMm}{R} + \frac{1}{2}\frac{GMm}{R^3}x^2$$

So,

$$U(x) = U_o + \frac{1}{2}\kappa x^2$$

where

$$\kappa = \frac{GMm}{R^3}$$

$$T = \frac{2\pi}{\omega} = 2\pi\sqrt{\frac{m}{\kappa}} = 2\pi\sqrt{\frac{R^3}{GM}}$$

$$= 2\pi\sqrt{\frac{R}{g}} = 5.06 \times 10^3 \text{ s} = 84.48 \text{ min}$$

For a satellite in circular orbit,

$$a = \frac{v^2}{R} = \omega^2 R = \frac{GM}{R^2}$$

$$T = \frac{2\pi}{\omega} = 2\pi\sqrt{\frac{R^3}{GM}} = 84.48 \text{ min}$$

15-37 $$F_x = \frac{-GmM}{r^2}\left(\frac{r^3}{R_E^3}\right)\sin\theta$$

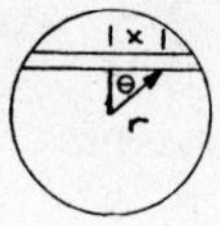

$$= \frac{-GmMr}{R_E^3}\,\frac{x}{r}$$

$$= \frac{-GmM}{R_E^3}\,x$$

$$T = \frac{2\pi}{\omega} = 2\pi\sqrt{\frac{m}{\kappa}} = 2\pi\sqrt{\frac{R_E^3}{GM}} = 2\pi\sqrt{\frac{RE}{G}}$$

$$= 84.48 \text{ min}$$

15-38 (a) $$F = F_1 + F_2 = (\kappa_1 + \kappa_2)x$$

$$\kappa = \kappa_1 + \kappa_2$$

(b) $$x = \frac{F}{\kappa_1} + \frac{F}{\kappa_2} = \frac{F}{\kappa}$$

$$\kappa = \frac{\kappa_1\kappa_2}{\kappa_1 + \kappa_2}$$

(c) $$F = F_1 + F_2 = (\kappa_1 + \kappa_2)x$$

$$\kappa = \kappa_1 + \kappa_2$$

15-39 $$\tau_o \cong -\kappa\ell^2\theta - mgL\,\theta$$

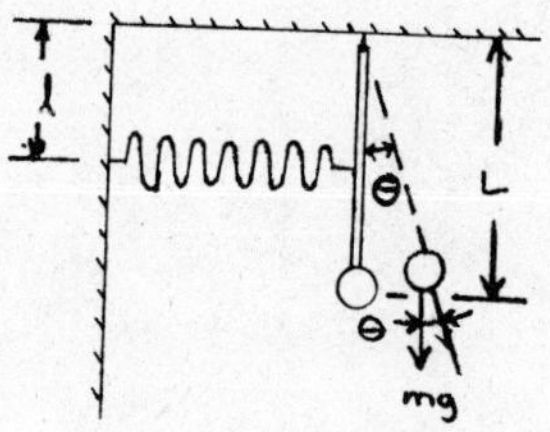

$$I_o\alpha \cong -(\kappa\ell^2 + mgL)\theta = -\kappa\theta$$

$$I_o = mL^2$$

$$\omega_o^2 = \frac{\kappa}{I_o} = \frac{\kappa\ell^2 + mgL}{mL^2}$$

$$T = \frac{2\pi}{\omega^o} = 2\pi\left(\frac{\kappa}{m}\frac{\ell^2}{L^2} + \frac{g}{L}\right)^{\frac{1}{2}}$$

15-40 (a) Initially, $$x = A\sin\omega t$$

$$v = A\ \cos\omega t$$

Using conservation of momentum,

$$(m + M)v' = Mv = MA\omega\cos\phi_1$$

The new total energy of the system is

$$E' = \frac{1}{2}\kappa A'^2 = \frac{1}{2}(m + M)v'^2 + \frac{1}{2}\kappa x'^2$$

$$= \frac{1}{2}\,\frac{M^2A^2\omega^2\cos^2\phi_1}{m + M} + \frac{1}{2}\kappa A^2\sin^2\phi_1$$

$$= \frac{1}{2}\kappa A^2\left[\frac{M}{m + M}\cos^2\phi_1 + \sin^2\phi_1\right]$$

$$= \frac{1}{2}\kappa A^2\left[\frac{M\cos^2\phi_1 + (m+M)\sin^2\phi_1}{m+M}\right]$$

So,

$$A' = A\sqrt{\frac{M + m\sin^2\phi_1}{M+m}}$$

(b) $$\omega'^2 = \frac{\kappa}{m+M} = \omega^2\frac{M}{m+M}$$

$$\omega' = \omega\sqrt{\frac{M}{m+M}}$$

15-41 (a) Taking the right-hand direction to be positive,

$$\phi_0 = \pi - \frac{\pi}{4} = \frac{3\pi}{4}$$

$$\omega = 2\pi(\frac{300}{60}) = 10\pi \text{ rad/s}$$

$$x(t) = (15 \text{ cm}) \sin(10\pi t + \frac{3\pi}{4})$$

(b) $$a_m = \omega^2 A = (100\,\pi^2)(0.15) = 148.0 \text{ m/s}^2$$

$$F_m = ma_m = 740.2 \text{ N}$$

15-42 $$F(x) = -\frac{GMm}{x^2}\frac{x^3}{R_E^3} = -\frac{GMm}{R_E^3}x = -\frac{gm}{R_E}x$$

So,

$$\kappa = \frac{gm}{R_E}$$

(a) $$T = \frac{2\pi}{\omega} = 2\pi\sqrt{\frac{m}{\kappa}} = 2\pi\sqrt{\frac{RE}{g}} = 84.5 \text{ min}$$

(b) $$v = \omega R_E = 2\pi R_E/T = 7.9 \times 10^3 \text{ m/s}$$

(c) Near the earth's surface,

$$a = \frac{v^2}{R_E} = g$$

So,

$$v = \sqrt{gR_E} = \frac{2\pi\, R_E}{2\pi\sqrt{R_E/g}} = \frac{2\pi R_E}{T}$$

This is the same velocity as in part (b) so the orbit is stable.

15-43 (a) From the dimensions given,

$\ell' > a$ and $\ell' > b$

Thus, using the property of conjugate circles of oscillation,

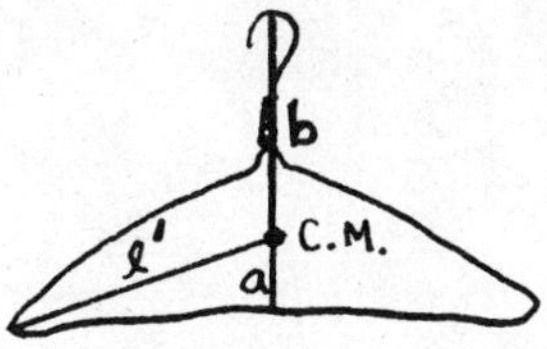

$$a = b = \frac{20.5 \text{ cm}}{2} = 10.25 \text{ cm}$$

Then, $$\ell' = \sqrt{a^2 + (\frac{41.0 \text{ cm}}{2})^2} = 22.9 \text{ cm}$$

(b) $T = 2\pi\sqrt{\frac{L}{g}} = 2\pi\sqrt{\frac{\ell' + \ell}{g}} = 2\pi\sqrt{\frac{\ell' + b}{g}} = 1.16\ s$

15-44 $\tau_o \cong - mg\ell\theta$

$I_o = \frac{mL^2}{12} + m\ell^2$

Since $I_o\alpha = \tau_o$,

$(\frac{L^2}{12} + \ell^2)\frac{d^2\theta}{dt^2} = -g\ell\theta$

Since $\frac{d^2\theta}{dt^2} = -\frac{\kappa}{m}\theta$,

$$\omega^2 = \frac{g\ell}{\frac{L^2}{12} + \ell^2}$$

$T^2 = \frac{4\pi^2}{g\ell}(\frac{L^2}{12} + \ell^2)$

$\frac{d(T^2)}{d\ell} = 0 = \frac{8\pi^2}{g} - \frac{4\pi^2}{g\ell^2}(\frac{L^2}{12} + \ell^2)$

The solution for ℓ is then

$$\ell = \frac{L}{\sqrt{12}}$$

15-45 (a) $\omega^o = \frac{2\pi}{T} = 20.94\ \text{rad/s}$

$b = \frac{mg}{v_\infty} = 0.261\ \text{kg/s}$

From Eq. 15-33b,

$\frac{\kappa}{m} = (\omega^o)^2 + (\frac{b}{2m})^2$

$\kappa = m[(\omega^o)^2 + (\frac{b}{2m})^2] = 877\ \text{N/m}$

(b) $\beta = \frac{b}{2m} = 0.0653\ s^{-1}$

$e^{-\beta t} = 0.5$

$t = \frac{\ell n\ 2}{\beta} = 10.6\ s$

(c) $\frac{E}{E_o} = \frac{\frac{1}{2}\kappa A^2}{\frac{1}{2}\kappa A_o^2} = \frac{(\frac{1}{2}A_o)^2}{A_o^2} = \frac{1}{4}$

$E_o - E = E_o - \frac{1}{4}E_o = \frac{3}{4}E_o$

(d) $\omega_R = \sqrt{\omega_o^2 - \frac{b^2}{2m^2}} \cong \omega_o = \sqrt{\frac{\kappa}{m}} = 20.94\ \text{rad/s}$

(e) $A = \frac{F_o/\kappa}{\sqrt{\frac{1}{Q^2}\frac{\omega^2}{\omega_o^2} + (\frac{\omega^2}{\omega_o^2} - 1)^2}}$

Since $\omega_R < \omega_o$, $\omega << \omega_o$.

This means,

$$A = F_o/\kappa = \frac{F_o}{877 \text{ N/m}}$$

(f) $\mathcal{Q} = \frac{\omega_o m}{b} = 160.5$

(g) $A_m \cong \mathcal{Q}F_o/\kappa = \frac{160.5\ F_o}{877 \text{ N/m}}$

(h) $F = \kappa A_m = \mathcal{Q}F_o = 160.5\ F_o$

15-46 $E = \frac{1}{2} mv^2 + \frac{1}{2} \kappa x^2$

$$\frac{dE}{dt} = mv \frac{d^2x}{dt^2} + \kappa x \frac{dx}{dt} = v(m \frac{d^2x}{dt^2} + \kappa_x)$$

Using Eq. 15-31,

$$m \frac{d^2x}{dt^2} + \kappa x = -bv$$

So,

$$\frac{dE}{dt} = -bv^2 = vF_v$$

Relating this to power,

$$P_v = \frac{dW_v}{dt} = F_v v = -bv^2$$

15-47 (a) $x(t) = A \sin(\omega t - \delta)$

$v(t) = A\omega \cos(\omega t - \delta)$

$P(t) = v\ F(t) = A\omega F_o \cos(\omega t - \delta)\cos \omega t$

$= A\omega F_o(\cos \omega t \cos \delta + \sin \omega t \sin \delta)\cos \omega t$

$= A\omega F_o(\cos^2 \omega t \cos \delta + \cos \omega t \sin \omega t \sin \delta\ )$

(b) $\overline{P} = \frac{1}{T} \int_0^T A\omega F_o (\cos^2 \omega t \cos \delta + \cos \omega t \sin \omega t \sin \delta)dt$

$= \frac{A\omega F_o}{T}[\cos \delta \int_0^T \cos^2 \omega t\ dt + \sin \delta \int_0^T \cos \omega t \sin \omega t\ dt]$

$= \frac{A\omega F_o}{2T} \cos \delta$

$= \frac{1}{2} v_m F_o \cos \delta$

16-1 $F = s_nA = (5.57 \times 10^8 \text{ Pa})\pi(4 \times 10^{-3})^2 = 2.80 \times 10^4 \text{ N}$

16-2 $m(g + a) = \frac{s_nA}{f}$

$$a = \frac{s_nA}{mf} - g = \frac{(2.41 \times 10^8 \text{ Pa})\pi(6 \times 10^{-3} \text{ m})^2}{(800 \text{ kg})(2.5)} - (9.81 \text{ m/s}^2)$$

$$= 3.82 \text{ m/s}^2$$

16-3 Using Eq. 16-14,

$$Y = \frac{F_n \ell}{A\,\Delta\ell} = \frac{(5.0 \text{ kg})(9.81 \text{ m/s}^2)(3 \text{ m})}{\pi(0.5 \times 10^{-3})^2(1.5 \times 10^{-3})} = 1.25 \times 10^9 \text{ Pa}$$

16-4 The stretched length of the wire is given by

$$\ell = \ell_0 + \Delta\ell = \ell_0 + \frac{F_n\ell_0}{AY} = \ell_0(1 + \frac{\Delta mg}{Ay})$$

$$\frac{\ell}{\ell_0} = 1 + \frac{\Delta mg}{AY} = 1 + \frac{(9 \text{ kg})(9.8 \text{ m/s}^2)}{\pi(4.06 \times 10^{-4} \text{ m})^2(12.4 \times 10^{10} \text{ Pa})}$$

$$= 1.0014$$

$$\frac{T}{T_0} = \sqrt{\frac{\ell}{\ell_0}} = 1.0007$$

$$\Delta T = T_0(\frac{T}{T_0} - 1) = 0.0007\, T_0$$

16-5 $\Delta\ell_{Cu} = \frac{F\ell}{AY} = \frac{(100 \text{ kg})(9.81 \text{ m/s}^2)(1 \text{ m})}{\pi(0.25 \times 10^{-3} \text{ m})^2(12.4 \times 10^{10} \text{ Pa})} = 4.03 \text{ cm}$

$$\Delta\ell_W = 6.00 - 4.03 = 1.97 \text{ cm}$$

$$r_w = \sqrt{\frac{F\ell}{\pi Y\Delta\ell_w}} = \sqrt{\frac{(981 \text{ N})(1 \text{ m})}{\pi(35.5 \times 10^{10} \text{ Pa})(1.97 \times 10^{-2} \text{ m})}} = 0.211 \text{ mm}$$

16-6 Using Eq. 16-21,

$$\frac{\Delta V}{V} = \frac{-p}{\beta} = \frac{-4 \times 10^7 \text{ Pa}}{3.1 \times 10^{10} \text{ Pa}} = -1.29 \times 10^{-3}$$

Also,

$$V = \frac{4}{3}\pi r^3$$

$$\frac{\Delta V}{V} = 3\,\frac{\Delta r}{r}$$

Thus,

$$\Delta r = \frac{1}{3}(\frac{\Delta V}{V})r = \frac{1}{3}(-1.29 \times 10^{-3})(10 \text{ cm}) = 4.3 \times 10^{-3} \text{ cm}$$

16-7 The stress on the exterior face is given by

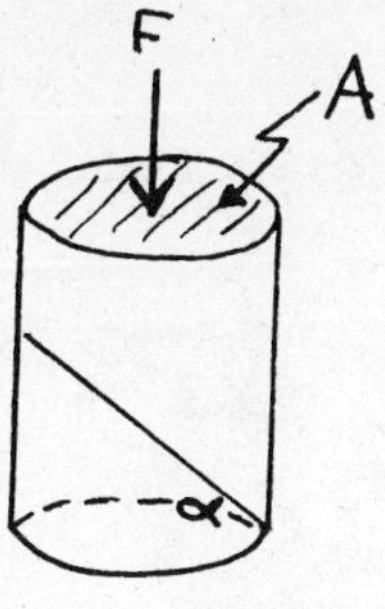

$$s = \frac{F}{A}$$

The normal stress on the family of planes oriented at an angle α with respect to the horizontal is

$$s_n \le s \cos^2 \alpha$$

$$s \ge \frac{s_n}{\cos^2 \alpha}$$

$$s_{min} = \frac{s_n}{(\cos^2 \alpha)_{max}} = \frac{s_n}{1} = s_n = 4.6 \times 10^7 \text{ Pa}$$

The tangential stress on the family of planes oriented at an angle α with respect to the horizontal is

$$s_t \le \frac{1}{2} s \sin 2\alpha$$

$$s \ge \frac{2 s_t}{\sin 2\alpha}$$

$$s_{min} = \frac{2 s_t}{(\sin 2\alpha)_{max}} = \frac{2 s_t}{1} = 2(1.16 \times 10^7 \text{ Pa}) = 2.32 \times 10^7 \text{ Pa}$$

The block will fail when

$$F = sA = (2.32 \times 10^7 \text{ Pa})\pi(0.30 \text{ m})^2 = 6.56 \times 10^6 \text{ N}$$

This occurs along the plane with the orientation

$$\alpha = \frac{1}{2} \sin^{-1} 1 = 45°$$

16-8 $F = 2\pi r^2 s_t = 2\pi(5 \times 10^{-3} \text{ m})^2(3.08 \times 10^8 \text{ Pa}) = 4.84 \times 10^4 \text{ N}$

$$s_n = k \frac{F}{(w - 2r)d} = 2.0 \frac{4.84 \times 10^4 \text{ N}}{(0.08 \text{ m} - 0.02 \text{ m})d} = 4.45 \times 10^8 \text{ Pa}$$

$d = 3.63$ mm

16-9 $\frac{\Delta\rho}{\rho} = -\frac{\Delta V}{V} = \frac{p}{\beta}$

$$\Delta\rho = \frac{p}{\beta} \rho = \frac{(2 \times 10^8 \text{ Pa})}{(7.8 \times 10^{10} \text{ Pa})} (2.70 \times 10^3 \text{ kg/m}^3)$$

$$= 6.92 \text{ kg/m}^3$$

$$\frac{\Delta\rho}{\rho} = 0.26\%$$

16-10 Using Eq. 16-22,

$$\phi = \frac{2\ell\tau}{\pi\mu R^4} = \frac{2(0.05 \text{ m})(100 \text{ N} \cdot \text{m})}{\pi(8.0 \times 10^{10} \text{ Pa})(5.0 \times 10^{-3} \text{ m})^4} = 0.0637 \text{ rad}$$

$$\gamma_R = \frac{R}{\ell} \phi = 0.0064 \text{ rad}$$

16-11 (a) $\tau = \int_{R_1}^{R_2} \frac{2\pi\mu r^3\phi}{\ell}\,dr = \frac{\pi\mu}{2\ell}(R_2^4 - R_1^4)\phi$

(b) Since the mass per unit length is the same,

$$\pi R^2 = \pi(R_2^2 - R_1^2)$$

Also,

$$\frac{\phi_R}{\phi_{12}} = \frac{R_2^4 - R_1^4}{R^4} = \frac{R_2^4 - R_1^4}{(R_2^2 - R_1^2)^2} = \frac{R_2^2 + R_1^2}{R_2^2 - R_1^2}$$

16-12 (a) From Ex. 16-3,

$$y = \frac{-Mg\ell^3}{3Y\mathcal{J}} = \frac{-4Mg\ell^3}{Ywh^3} = 7.53 \text{ cm}$$

(b) $s_n = \frac{Y\xi}{R} = Y\xi\,\frac{d^2y}{dx^2}$

$$\frac{d^2y}{dx^2} = -\frac{Mg}{Y\mathcal{J}}(\ell - x)$$

s_n is a maximum for $x = 0$, $\xi = \frac{h}{2}$

so, the tensile stress is a maximum at the top surface of the beam at the supported end.

(c) $s_n = \frac{Yh}{2}\frac{Mg\ell}{Y\mathcal{J}} = \frac{6\,Mg\ell}{wh^2} = 7.06 \times 10^8 \text{ Pa}$

16-13 (a) $\tau = \frac{-(\ell - x)^2}{2\,\ell}\,Mg = Y\mathcal{J}\,\frac{d^2y}{dx^2}$

$$\frac{d^2y}{dx^2} = \frac{-(\ell - x)^2}{2\,\ell}\,\frac{Mg}{Y\mathcal{J}}$$

$$\frac{dy}{dx} = \frac{(\ell - x)^3}{6\,\ell}\,\frac{Mg}{Y\mathcal{J}} + C_1$$

For $x = 0$, $\frac{dy}{dx} = 0$.

Thus,

$$C_1 = -\frac{\ell^2 Mg}{6Y\mathcal{J}}$$

$$\frac{dy}{dx} = \frac{(\ell - x)^3}{6\,\ell}\,\frac{Mg}{Y\mathcal{J}} - \frac{\ell^2 Mg}{6Y\mathcal{J}}$$

$$y = -\frac{-(\ell - x)^4}{24\,\ell}\,\frac{Mg}{Y\mathcal{J}} - \frac{\ell^2 Mgx}{6Y\mathcal{J}} + C_2$$

For $x = 0$, $y = 0$.

Thus,

$$C_2 = \frac{\ell^3 Mg}{24Y\mathcal{J}}$$

$$y = -\frac{(\ell - x)^4}{24\,\ell}\,\frac{Mg}{Y\mathcal{J}} - \frac{\ell^2 Mgx}{6Y\mathcal{J}} + \frac{\ell^3 Mg}{24Y\mathcal{J}}$$

For $x = \ell$,

$$y = -\frac{Mg\ell^3}{8Y\mathcal{I}}$$

(b) $$y = -\frac{g\rho wh\ell^4}{8Y\mathcal{I}} = -\frac{3g\rho wh\ell^4}{2Ywh^3} = -\frac{3(9.81\ \text{m/s}^2)(7.86 \times 10^3\ \text{kg/m}^3)(0.80\ \text{m})^4}{2(20 \times 10^{10}\ \text{Pa})(0.02\ \text{m})^2}$$

$$= 0.59\ \text{mm}$$

16-14 The decrease in the length of the rod due to the mass m acting on dx is given by

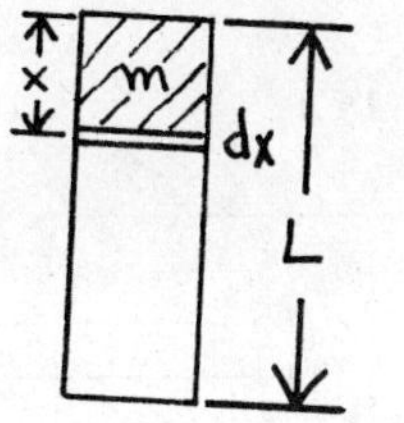

$$dy = \frac{-mg}{YA}\,dx = \frac{-Mgx}{YA\ell}\,dx$$

$$\Delta\ell = \int dy = \int_0^{\ell} -\frac{Mgx}{YA\ell}\,dx =$$

$$= \frac{-Mg\ell}{2YA} = \frac{-\rho g\ell^2}{2Y}$$

$$= \frac{-(7.90 \times 10^3\ \text{kg/m}^3)(9.81\ \text{m/s}^2)(6.0\ \text{m})^2}{2(20.0 \times 10^{10}\ \text{Pa})}$$

$$= -6.97 \times 10^{-6}\ \text{m} = -0.007\ \text{mm}$$

The change in length is the same as that produced by a load equal to half the weight of the rod.

16-15 (a) The additional force on each wire is given by

$$F_A = F_B = \frac{1}{2}Mg = (40\ \text{kg})(9.8\ \text{m/s}^2) = 392\ \text{N}$$

$$\Delta\ell_A = \frac{\ell F_A}{Y_A A_A} = \frac{(1\ \text{m})(392\ \text{N})}{(12.4 \times 10^{10}\ \text{Pa})\pi(0.8 \times 10^{-3}\ \text{m})^2} = 1.57\ \text{mm}$$

$$\Delta\ell_B = \frac{\ell F_B}{Y_B A_B} = \frac{(1\ \text{m})(392\ \text{N})}{(9.0 \times 10^{10}\ \text{Pa})\pi(0.5 \times 10^{-3}\ \text{m})^2} = 5.55\ \text{mm}$$

$$\alpha \cong \frac{5.55 - 1.57}{2 \times 10^3} = 0.00199\ \text{rad} = 0.114°$$

(b) For equal extension,

$$F_A + F_B = (80\ \text{kg})(9.8\ \text{m/s}^2) = 784\ \text{N}$$

$$\Delta\ell_A = \Delta\ell_B = \frac{\ell F_B}{Y_B A_B} = \frac{\ell(784 - F_B)}{Y_A A_A}$$

$$F_B = 784\left(\frac{Y_B A_B}{Y_A A_A + Y_B A_B}\right) = 173.2\ \text{N}$$

$$\frac{x}{2\ \text{m}} = \frac{F_B}{784\ \text{N}}$$

$$x = 0.44\ \text{m}$$

16-16 (a) $\gamma \cong \frac{.5}{5} = \frac{1}{10}$

$$\frac{F}{A} = \mu\gamma$$

$$\mu = \frac{F}{A\gamma} = \frac{4.0 \text{ N}}{(0.05 \text{ m})^2/10} = 1.6 \times 10^4 \text{ Pa}$$

(b) Using Eq. 16-17,

$$Y = 2\mu(1 + \sigma) = 3\mu = 4.8 \times 10^4 \text{ Pa}$$

16-17 (a) Using Eq. 16-15,

$$\kappa = \frac{YA}{\ell} = \frac{(7.0 \times 10^{10} \text{ Pa})\pi(0.5 \times 10^{-3} \text{ m})^2}{2 \text{ m}} = 2.75 \times 10^4 \text{ N/m}$$

$$T = 2\pi\sqrt{\frac{m}{\kappa}} = 2\pi\sqrt{\frac{1.2 \text{ kg}}{2.75 \times 10^4 \text{ N/m}}} = 0.0415 \text{ s}$$

(b) $\frac{Mg}{A} = s_n$

$$M = \frac{A s_n}{g} = \frac{\pi(0.5 \times 10^{-3} \text{ m})^2(1.2 \times 10^8 \text{ Pa})}{9.8 \text{ m/s}^2} = 9.62 \text{ kg}$$

(c) The amplitude must be half of the extension caused by the 9.62-kg load. Since stress is proportional to strain below the proportional limit,

$$M = \frac{1}{2}(9.62 \text{ kg}) = 4.81 \text{ kg}$$

(d) The elongation with the 1.2-kg load is given by

$$\Delta\ell = \frac{Mg\ell}{YA} = 4.28 \times 10^{-4} \text{ m}$$

Since the wire cannot sustain compression, this is the maximum amplitude for SHM.

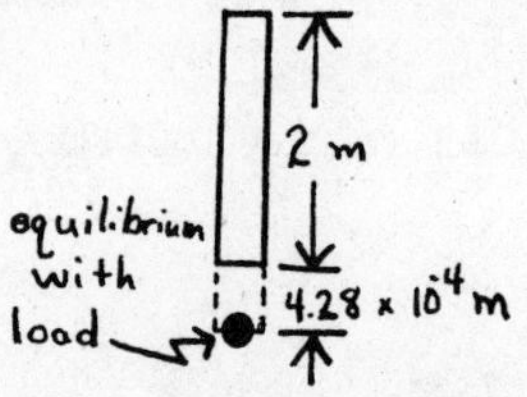

16-18 (a) $Ma = T - Mg$

$$T = M(a + g) = (2000 \text{ kg})(11.4 \text{ m/s}^2) = 2.28 \times 10^4 \text{ N}$$

$$\frac{s_n}{3} = \frac{T}{A} = \frac{T}{\pi d^2/4}$$

$$d = \sqrt{\frac{12 T}{\pi s_n}} = \sqrt{\frac{12(2.28 \times 10^4 \text{ N})}{(2.41 \times 10^8 \text{ Pa})}} = 1.9 \text{ cm}$$

(b) $s_n = \frac{M(a + g)}{\pi d^2/4}$

$$a = \frac{\pi d^2 s_n}{4 M} - g = 3.9 \text{ m/s}^2$$

(c) $\Delta \ell = \frac{T\ell}{YA} = \frac{(1600 \text{ kg})(11.4 \text{ m/s}^2)(50 \text{ m})}{(20.0 \times 10^{10} \text{ Pa})\pi(0.95 \times 10^{-2} \text{ m})^2} = 1.6 \text{ cm}$

16-19 $\tau = \frac{P}{\omega} = \frac{500 \times 10^3 \text{ W}}{(2\pi \times 200 \text{ rpm})(60 \text{ s/min})^{-1}} = 2.39 \times 10^4 \text{ N} \cdot \text{m}$

Using Eq. 16-22,

$$\phi = \frac{2\ell\tau}{\pi\mu R^4} = \frac{2(4.6 \text{ m})(2.39 \times 10^4 \text{ N} \cdot \text{m})}{\pi(8.1 \times 10^{10} \text{ Pa})(0.04 \text{ m})^4} = 0.338 \text{ rad}$$

16-20 $T = 2\pi\sqrt{I/\Gamma}$

$$\Gamma = \frac{4\pi^2}{T^2} I = \frac{4\pi^2}{T^2}\left(\frac{MR^2}{2}\right)$$

$$= \frac{4\pi^2(0.100 \text{ kg})(0.05 \text{ m})^2}{2(300 \text{ s})^2} = 5.48 \times 10^{-8} \text{ N} \cdot \text{m}$$

Using Eq. 16-23,

$$R^4 = \frac{d}{2^4} = \frac{2\ell\Gamma}{\pi\mu} = \frac{2(1 \text{ m})(5.48 \times 10^{-8} \text{ N} \cdot \text{m})}{\pi(3.0 \times 10^{10} \text{ Pa})} = 1.16 \times 10^{-18} \text{ m}^4$$

$d = 0.066 \text{ mm}$

16-21 $y = \frac{-Mg\ell^3}{3Y\mathcal{I}}$

$$F = Mg = \frac{-3Y\mathcal{I}}{\ell^3} y$$

$$\ddot{y} = \frac{F}{M} = \frac{-3Y\mathcal{I}}{M\ell^3} y$$

$$\omega = \sqrt{\frac{3Y\mathcal{I}}{M\ell^3}}$$

$$T = \frac{2\pi}{\omega} = 2\pi\sqrt{\frac{M\ell^3}{3Y\mathcal{I}}}$$

$$= 2\pi\sqrt{\frac{4M\ell^3}{Ywh^3}} = 2\pi\sqrt{\frac{4(65 \text{ kg})(3 \text{ m})^3}{(1.4 \times 10^{10} \text{ Pa})(0.35 \text{ m})(0.04 \text{ m})^3}} = 0.94 \text{ s}$$

16-22 (a) Since the vertical deflection is inversely proportional to $\mathcal{I}$, the orientation used should be that having the largest $\mathcal{I}$. This corresponds to the orientation shown in the diagram.

(b) $y = -\frac{Mg\ell^3}{3Y\mathcal{I}} = \frac{-(2000 \text{ kg})(9.8 \text{ m/s}^2)(2 \text{ m})^3}{3(20 \times 10^{10} \text{ Pa})(295 \times 10^{-8} \text{ m}^4)} = -8.86 \text{ cm}$

(c) $y = \frac{-Mg\ell^3}{3Y\mathcal{I}} = \frac{-Mg\ell^3}{3Y(wh^3/12)} = \frac{-4Mg\ell^3}{Yw^4}$

$$w^4 = \frac{-4Mg\ell^4}{Yy} = \frac{-4(2000 \text{ kg})(9.8 \text{ m/s}^2)(2 \text{ m})^3}{(20 \times 10^{10} \text{ Pa})(0.089 \text{ m})} = 0.35 \times 10^{-4} \text{ m}$$

$w = 7.7 \times 10^{-2} \text{ m} = 7.7 \text{ cm}$

(d) $\frac{\text{Area (I beam)}}{\text{Area(square beam)}} = \frac{19.68 \text{ cm}^2}{7.7^2 \text{ cm}^2} = 0.33$

It takes $\frac{1}{3}$ the amount of steel to make an I beam as it takes to make a square beam.

17-1 $F = pA = (0.2\ \text{atm})(4.5 \times 22\ \text{m}^2) = 2.0 \times 10^6\ \text{N}$

17-2 $F = mg = pA$

$\frac{m}{A} = \frac{p}{g} = \frac{1.013 \times 10^5\ \text{N/m}^2}{9.80\ \text{m/s}^2} = 1.03 \times 10^4\ \text{kg/m}^2$

$A = 4\pi R_E^2 = 4\pi(6.38 \times 10^6\ \text{m})^2$

$m = 5.27 \times 10^{18}\ \text{kg}$

17-3 $p = \frac{F_x}{A_x} = \frac{20\ \text{N}}{\frac{1}{2}(0.1\ \text{m})^2} = 4000\ \text{Pa}$

By symmetry, all components of $\vec{F}$ are equal.

$\vec{F} = F_x(\hat{i} + \hat{j} + \hat{k}) = (20\ \text{N})(\hat{i} + \hat{j} + \hat{k})$

17-4 $dA = R^2 \sin\theta\ d\theta\ d\phi$

$dF_x = -p\ dA \sin\theta \sin\phi$

$F_x = -pR^2\int_0^\pi\int_0^\pi \sin^2\theta \sin\phi\ d\theta\ d\phi$

$= -\frac{pR^2\pi}{2}\int_0^\pi \sin\phi\ d\phi$

$= -\pi R^2 p$

17-5 $F = p_x A_x = (1.013 \times 10^5\ \text{Pa})[\pi(0.4\ \text{m})^2] = 5.09 \times 10^4\ \text{N}$

17-6 (a) $p - p_o = -\rho_o g\Delta y = -(1.29\ \text{kg/m}^3)(9.81\ \text{m/s}^2)(412.4\ \text{m}) = 5.22 \times 10^3\ \text{Pa}$

(b) $p - p_o = p_o(1 - e^{-\alpha y}) = p_o(1 - e^{-g\rho_o y/p_o}) = 0.669\ \text{atm}$

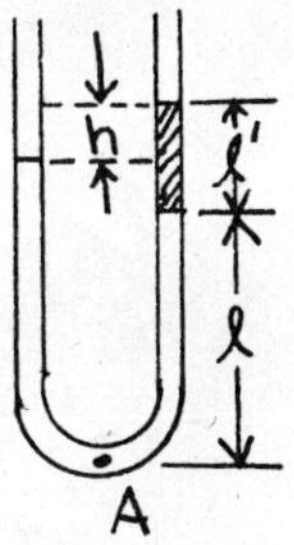

17-7 $p = p_o + \rho gz = (1.013 \times 10^5\ \text{Pa}) + (1030\ \text{kg/m}^3)(9.81\ \text{m/s}^2)(40\ \text{m})$

$= 5.05 \times 10^5\ \text{Pa} = 4.99\ \text{atm}$

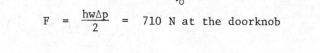

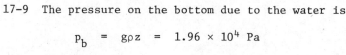

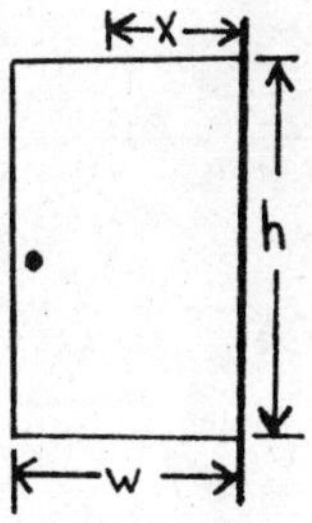

17-8 $\tau_H = Fw = (\Delta p)h\int_0^w x\ dx = (\Delta p)h\frac{w^2}{2}$

$F = \frac{hw\Delta p}{2} = 710\ \text{N}$ at the doorknob

17-9 The pressure on the bottom due to the water is

$p_b = g\rho z = 1.96 \times 10^4\ \text{Pa}$

So, $F_b = p_b A = 5.87 \times 10^6\ \text{N}$

On the sides,

$$F_s = (30\text{ m})\int_0^h g\rho z\, dz = (30\text{ m})\, g\rho\frac{h^2}{2} = 5.89 \times 10^5 \text{ N}$$

On the ends,

$$F_e = (10\text{ m})\int_0^h g\rho z\, dz = 1.96 \times 10^5 \text{ N}$$

17-10 $A = \frac{F}{p} = \frac{mg}{p} = \frac{(80\text{ kg})(9.81\text{ m/s}^2)}{(1.013 \times 10^5\text{ Pa})} = 77.4\text{ cm}^2$

17-11 $p_{load} = \frac{F_{load}}{A_{load}} = p_{pump} = \frac{f}{A_{pump}}$

$$f = F_{load}\frac{A_{pump}}{A_{load}} = F_{load}\frac{d^2_{pump}}{d^2_{load}} = (5000)(9.8)\left(\frac{2}{30}\right)^2 = 218\text{ N}$$

From Eq. 17-6,

$$\frac{\Delta V}{V} = -\lambda\Delta p = -\frac{\Delta\rho}{\rho}$$

$$\frac{\Delta\rho}{\rho} = \lambda\Delta p = (40 \times 10^{-11})\left(\frac{218}{\pi \times 0.01^2}\right) = 2.78 \times 10^{-4}$$

17-12 (a) Assuming the fluid does not compress,

$$M = \frac{F}{g} = \frac{W/h}{g} = 1.02 \times 10^4\text{ kg}$$

(b) $p = \frac{F}{A} = \frac{Mg}{\pi r^2} = 4.97 \times 10^6\text{ N/m}^2$

17-13 $$\tau_A = gL\left[\int_0^2 (h-y)y\, dy - \int_0^{1.25} (h'-y)y\, dy\right]$$

$$= \rho gL\left[\left(\frac{hy^2}{2} - \frac{y^3}{3}\right)\Big|_0^2 - \left(\frac{h'y^2}{2} - \frac{y^3}{3}\right)\Big|_0^{1.25}\right]$$

$$= 5.89 \times 10^4\text{ N}\cdot\text{m}$$

17-14 Using Eq. 17-13,

$$\Delta p_0 = \rho g\Delta h = -2.67 \times 10^3\text{ Pa}$$

$$p = p_0 + \Delta p_0 = (1.013 - 0.027) \times 10^5\text{ Pa} = 0.986 \times 10^5\text{ Pa}$$

17-15 $p_0 = \rho gh$

$$h = \frac{p_0}{\rho g} = \frac{1.013 \times 10^5\text{ Pa}}{(0.984 \times 10^3\text{ kg/m}^3)(9.81\text{ m/s}^2)} = 10.5\text{ m}$$

17-16 $p_0 = \rho gh_1 = p + \rho gh_2$

$$p = \rho g(h_1 - h_2) = 3.50 \times 10^3\text{ Pa}$$

In units of torr,

$$p = h_1 - h_2 = 26.3\text{ mm Hg} = 26.3\text{ torr}$$

The gauge pressure is given by

$$p - p_0 = -\rho gh_2$$

In torr,

$$p - p_o = -h_2 = -734.2 \text{ torr}$$

17-17 $p_A = p_o + (\ell + \ell' - h)\rho_w g = p_o + \ell'\rho_k g + \ell\rho_w g$

Solving for h,

$$h = \ell'(1 - \frac{\rho_k}{\rho_w}) = 1.08 \text{ cm}$$

17-18 $p_A - p_o = (0.35\ \rho_{Hg} + 0.70\ \rho_{oil})g$

$$= 5.24 \times 10^4 \text{ Pa}$$

17-19 $p_A = p_B - 0.18\rho_w g - 0.08\rho_k g + 0.36\rho_w g$

$$p_A - p_B = 0.18\rho_w g - 0.08\rho_k g = 1.12 \times 10^3 \text{ Pa}$$

17-20 $F_g = (m + \rho_s V)g$

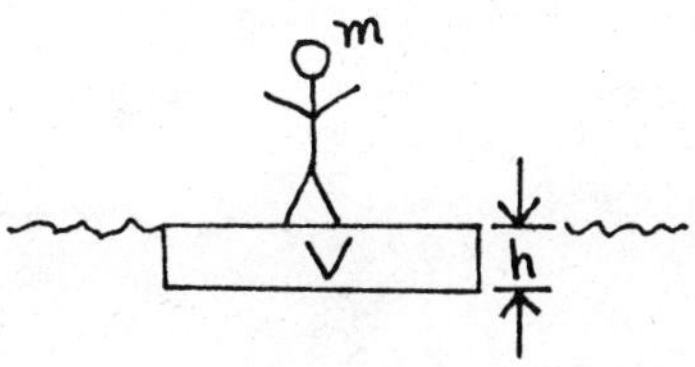

$F_b = \rho_w Vg$

Since $F_b = F_g$,

$$m + \rho_s V = \rho_w V$$

$$V = Ah = \frac{m}{\rho_w - \rho_s}$$

$$A = \frac{m}{(\rho_w - \rho_s)h} = 1.07 \text{ m}^2$$

17-21 The mass of the immersed crown is

$$M' = M - \frac{M\rho_w}{\rho_{crown}}$$

$$\rho_{crown} = \frac{M}{M-M'}\rho_w = 19.2 \times 10^4 \text{ kg/m}^3$$

The crown is essentially pure gold.

17-22 $F_b = F_g$

$$\rho g V = \rho g(\frac{2}{3}\pi r^3) = mg$$

$$m = \frac{2}{3}\pi r^3 \rho = 0.611 \text{ kg}$$

17-23 $F_b = \rho_w g\ V = \rho_w g\ \pi r^2 z = F_g = mg$

$$z = \frac{m}{\rho_w \pi r^2} = \frac{0.20 \text{ kg}}{(10^3 \text{ kg/m}^3)\pi(0.02 \text{ m})^2} = 15.9 \text{ cm}$$

17-24 On scale B the new reading is

$$B = m - \rho_w V = m - \frac{m}{\rho_{A\ell}}\rho_w = 0.38 \text{ kg}$$

Since the sum of the scale readings must remain unchanged,

$$A = C = 0.40 + \frac{1}{2}(0.22) = 0.51 \text{ kg}$$

17-25 The buoyant force exerted by the water on the copper block is

$$F_b = \rho_w g V_c = \rho_w g(\frac{m}{\rho_c}) = 0.27 \text{ N}$$

Since the block exerts a downward reaction force on the water, the scale reads

$$W = (0.30 \text{ kg})(9.8 \text{ m/s}^2) + 0.27 \text{ N} = 3.21 \text{ N}$$

17-26 $p_b = p_a - \rho g h$ (1)

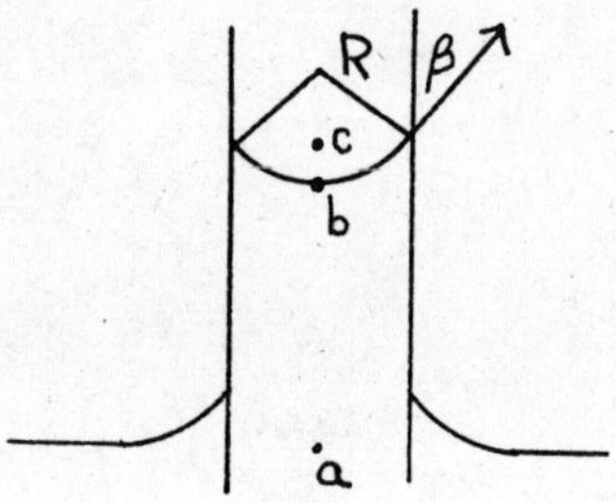

Using Eq. 17-16,

$p_c - p_b = \frac{2\gamma}{R}$ (2)

Also,

$p_c = p_a$ (3)

Adding (1) to (2) and subtracting (3),

$$0 = \frac{2\gamma}{R} - \rho g h$$

$$h = \frac{2\gamma}{\rho g R} = \frac{2\gamma \cos \beta}{\rho g r}$$

17-27 Cutting the bubble in half, the vertical force acting on the top half of the bubble is

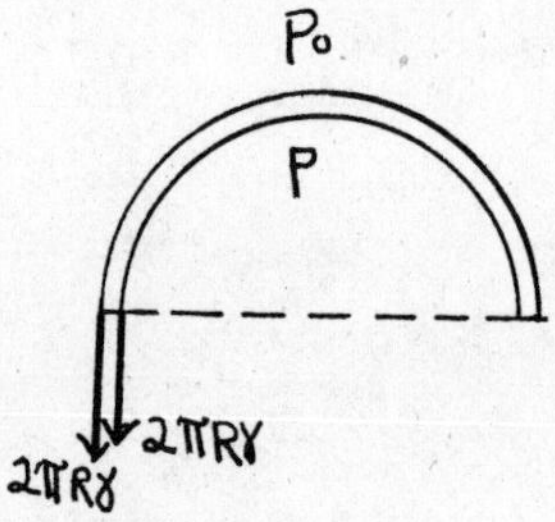

$$F_p = (p - p_o)\pi R^2$$

Since the bubble is not collapsing, this force must be just balanced by the force due to the surface tension of the lower half of the bubble. Since the bubble is two-sided, the force is

$$F_s = 2(2\pi R\gamma)$$

Equating the two forces,

$$4\pi R\gamma = (p - p_o)\pi R^2$$

Thus,

$$p - p_0 = \frac{4\gamma}{R} = \frac{120 \times 10^{-3}\ \text{N/m}}{0.015\ \text{m}} = 8.0\ \text{Pa} = 7.9 \times 10^{-5}\ \text{atm}$$

17-28 Using Eq. 17-20,

$$\gamma = \frac{\rho g h R}{2 \cos \beta} = 59 \times 10^{-3}\ \text{N/m}$$

17-29 $h = \frac{V}{A} = \frac{M}{\rho A} = 2.45 \times 10^{-3}\ \text{cm}$

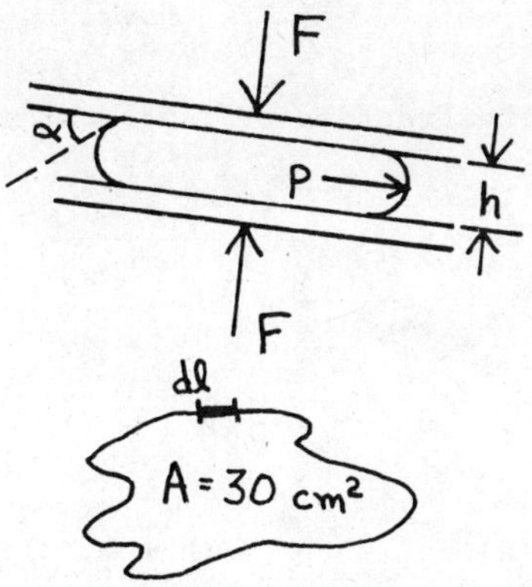

In the horizontal direction, the pressure difference between the mercury and the air times a unit of surface area is equal to the surface tension times a unit length of solid-liquid interface.

$$(\Delta p) t\, d\ell = \frac{F}{A} t\, d\ell = 2\gamma \cos \alpha\, d\ell$$

$$F = \frac{2\gamma A}{t} \cos \alpha = \frac{2\gamma A}{t} \cos (180° - \beta)$$

$$= \frac{2(487 \times 10^{-3}\ \text{N/m})(30 \times 10^{-4}\ \text{m}^2)}{2.45 \times 10^{-5}\ \text{m}} \cos 32°$$

$$= 101.1\ \text{N}$$

17-30 (a) $2\ell\ \gamma \cos \beta = \rho g h \ell d$

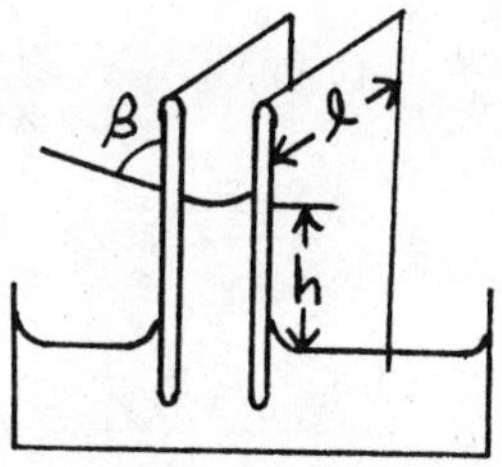

$$h = \frac{2\gamma \cos \beta}{\rho g d}$$

(b) $p - p_0 = (\frac{1}{R_1} + \frac{1}{R_2})$

Taking R_2 to be infinity,

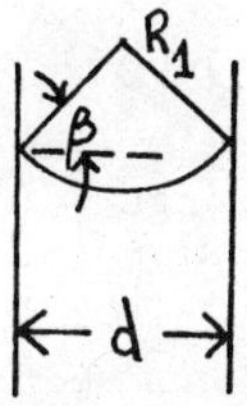

$$\Delta p = \frac{\gamma}{R_1} = \frac{2\gamma \cos \beta}{d}$$

Also,

$$\Delta p = \rho g h$$

Equating the two,

$$h = \frac{2\gamma \cos \beta}{\rho g d}$$

17-31 $p(y) = p_0 e^{-\alpha y}$

For small αy,

$$p = p_0(1 - \alpha y + \frac{\alpha^2 y^2}{2} + \cdots)$$

$$\cong p_0(1 - \alpha y) = p_0(1 - \frac{g\rho_0}{p_0} y) = p_0 - g\rho_0 y$$

17-32 $p_o = \rho g h$

$$h = \frac{p_o}{\rho g} = \frac{1.013 \times 10^5 \text{ Pa}}{(9.81 \text{ m/s}^2)(1.29 \text{ kg/m}^3)} = 8.00 \text{ km}$$

17-33 The mass of a column of air with cross sectional area A and height y is

$$M = A\int_0^y \rho(y)dy = A\rho_o\int_0^y \frac{p(y)}{p_o} dy$$

$$= A\rho_o\int_0^y e^{-\alpha y} dy$$

The fraction of the total mass of the column that lies below y is given by

$$f = \frac{A\rho_o\int_0^y e^{-\alpha y} dy}{A\rho_o\int_0^\infty e^{-\alpha y} dy} = \frac{e^{-\alpha y}\Big|_0^y}{e^{-\alpha y}\Big|_0^\infty} = 1 - e^{-\alpha y}$$

$$y = -\frac{1}{\alpha} \ln(1 - f) = 18.4 \text{ km}$$

17-34 (a) $p = p_o + (h + 1.20)\rho_w g$

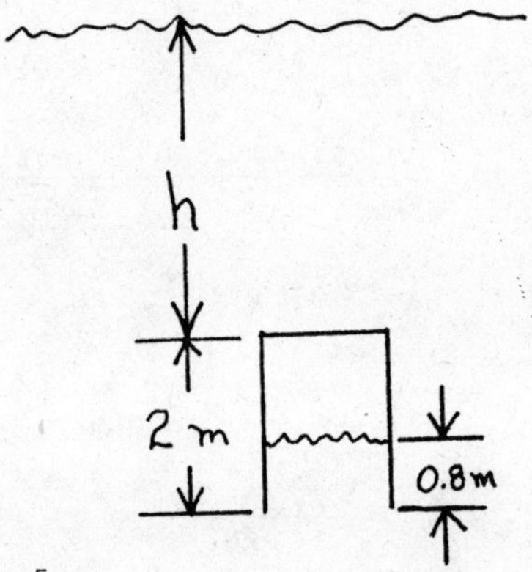

Also,

$$\frac{p}{p_o} = \frac{V}{V_o} = \frac{A(2 \text{ m})}{A(1.2 \text{ m})}$$

Thus,

$$h = \frac{p-p_o}{\rho_w g} - 1.20$$

$$= (\frac{2}{1.2} - 1)\frac{p_o}{\rho_w g} - 1.20 = 5.48 \text{ m}$$

(b) $p = p_o + (h + 2)\rho_w g = (1.013 + 0.76) \times 10^5 \text{ Pa}$

$$= 1.77 \times 10^5 \text{ Pa} = 1.75 \, p_o$$

17-35 (a) $dF_x = \rho g L(h - y)dy$

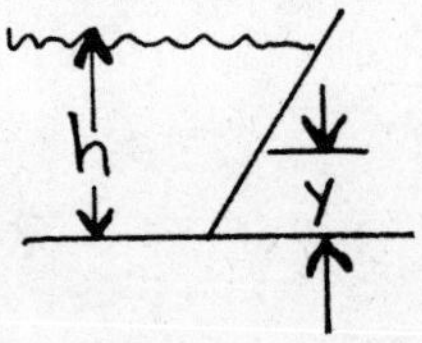

$$F_x = \int_0^h dF_x = \rho g L \int_0^h (h - y)dy$$

$$= \frac{\rho g L h^2}{2}$$

$$\tau_H = \rho g L\int_0^h y(h - y)dy = \frac{\rho g L h^3}{6}$$

The line of action is given by

$$\bar{y} = \frac{\tau_H}{F_x} = \frac{h}{3}$$

(b) $F_y = \rho g L\int_0^{h \tan \alpha} (h - y)dx = \rho g L \int_0^{h \tan \alpha} (h - \frac{x}{\tan \alpha})dx$

$$= \frac{\rho g L h^2}{2} \tan \alpha$$

$$\tau_H = \rho g L \int_0^{h \tan \alpha} x(h - \frac{x}{\tan\alpha})dx = \frac{\rho g L h^3}{6}\tan^2 \alpha$$

$$\bar{x} = \frac{\tau_H}{F_y} = \frac{h}{3} \tan \alpha$$

17-36 $p_v + \frac{\rho_w}{\rho_{Hg}} (1.20 \text{ cm}) + 74.08 = 75.92$

$$p_v = 75.92 - 74.08 - 0.09 = 1.75 \text{ cm Hg}$$

17-37 $\frac{4}{3} \pi R^3 \rho_w g = 4\pi R^2 t \rho_s g$

$$t = \frac{R}{3} \frac{\rho_w}{\rho_s} = \frac{10.0 \text{ cm}}{3} \frac{1.00}{7.86} = 0.424 \text{ cm}$$

17-38 $m = \rho(V_o + \pi r^2 x)$

$$\rho_w(V_o + \pi r^2 x_w) = \rho_a(V_o + \pi r^2 x_a)$$

$$V_o = \frac{\pi r^2(\rho_a x_a - \rho_w x_w)}{\rho_w - \rho_a} = 18.74 \text{ cm}^3$$

$$m = \rho_w(V_o + \pi r^2 x_w) = \frac{998 \text{ kg}}{\text{m}^3}[18.74 \text{ cm}^3 + \pi(8.5 \text{ mm})^2]$$

$$= \frac{998 \text{ kg}}{\text{m}^3}[(18.74 \text{ cm}^3) + \pi(\frac{8.5 \text{ mm}}{2})^2 (2.00 \text{ cm})]$$

$$= 19.8 \text{ g}$$

17-39 $M_{a\ell} g - \frac{M_{a\ell}}{\rho_{a\ell}} \rho_o g = M_B g - \frac{M_B}{\rho_B} \rho_o g$

$$M_{a\ell} - M_B = \frac{M_{a\ell}}{\rho_{a\ell}} \rho_o - \frac{M_B}{\rho_B} \rho_o \cong M_B \rho_o (\frac{1}{\rho_{a\ell}} - \frac{1}{\rho_B})$$

$$\frac{M_{a\ell} - M_B}{M_B} \cong \rho_o (\frac{1}{\rho_{a\ell}} - \frac{1}{\rho_B}) = 0.00032 = 0.032\%$$

17-40 (a) $p = p_o e^{-\alpha y}$

$$y = -\frac{1}{\alpha} \ln \frac{p}{p_o} = 4358 \text{ m}$$

(b) $F_g = F_b$

$$mg = \frac{p}{p_o}(\rho_{air} - \rho_{He})Vg$$

$$V = \frac{m}{\rho_{air} - \rho_{He}} \frac{p_o}{p} = 1.857 \times 10^3 \text{ m}^3$$

(c) $m = \frac{p}{p_o}(\rho_{air} - \rho_{H_2})V = 1295 \text{ kg}$

17-41 Let V be the volume of the cork.

$$m = \rho_c V$$

$$M_w = \rho_w V$$

Summing the forces in the vertical direction,

$$0 = M_w g - mg - T \cos \alpha$$

In the horizontal direction,

$$ma_c = M_w a_c - T \sin \alpha$$

Solving for α ,

$$\alpha = \tan^{-1} \frac{a_c}{g} = \frac{R\omega^2}{g}$$

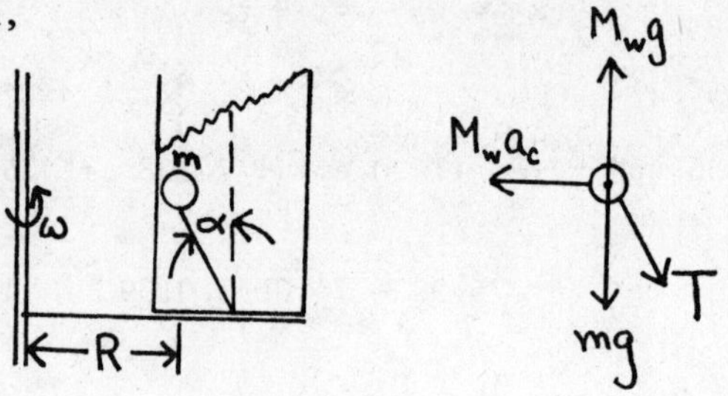

The water surface also tips inward since $a_c \propto R$. For a beaker radius much less than R the surface is approximately flat.

17-42 When the tube is depressed a depth z below its equilibrium position the restoring force is given by

$$F_r = -\rho_w g \pi r^2 z = m \frac{d^2z}{dt^2}$$

An equation of this form describes simple harmonic motion.

$$\omega^2 = \frac{\pi r^2 \rho_w g}{m}$$

$$\omega = \sqrt{\frac{\pi r^2 \rho_w g}{m}} = 7.85 \text{ s}^{-1}$$

$$\nu = \frac{\omega}{2\pi} = 1.25 \text{ Hz}$$

17-43 (1) $2\pi r_1 \gamma \cos \beta = \pi r_1^2 \rho g h_1$

(2) $2\pi (r_2 + r_3) \gamma \cos \beta = \pi (r_3^2 - r_2^2) \rho g h_2$

Solving for r_1,

$$r_1 = \frac{h_2}{h_1} \frac{r_3^2 - r_2^2}{r_2 + r_3} = 0.75 \text{ mm}$$

$$t = r_2 - r_1 = 1.25 \text{ mm}$$

Using (1),

$$\gamma = \frac{r_1 \rho g h_1}{2 \cos \beta} = 72.0 \times 10^{-3} \text{ N/m}$$

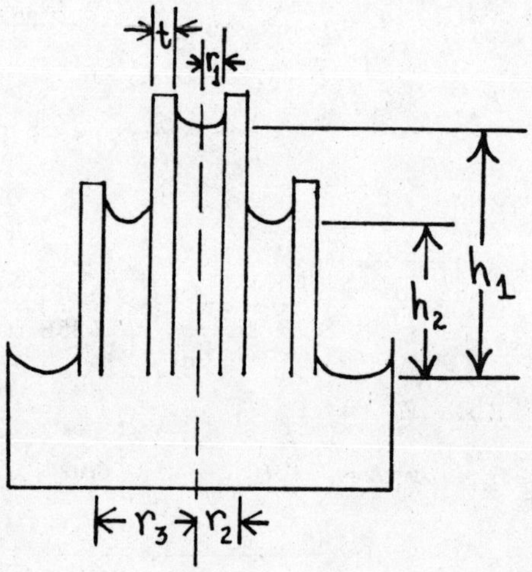

17-44 The change in surface energy is found as follows:

$$A = \frac{V}{h}$$

$$\Delta A = -\frac{V}{h^2} \Delta h$$

$$\Delta W = \gamma \Delta A = -\frac{V\gamma}{h^2} \Delta h$$

The work done by the surface tension is given by

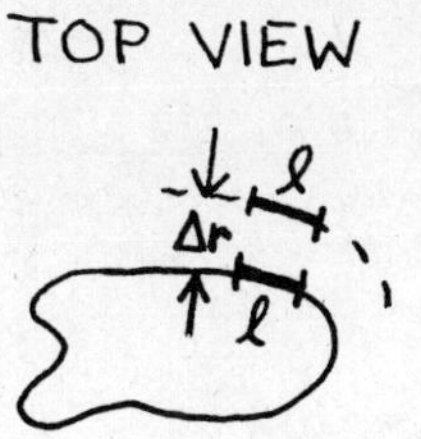

$$\Delta W = \int_0^L \gamma \ell \cos\alpha \, \Delta r = \gamma \cos\alpha \, \Delta A$$

$$= \frac{-V\gamma \cos\alpha}{h^2} \Delta h$$

The change is gravitational energy is

$$\Delta E = -mg \frac{\Delta h}{2} = -\frac{V\gamma}{h^2} \Delta h(1 + \cos\alpha)$$

Solving for h,

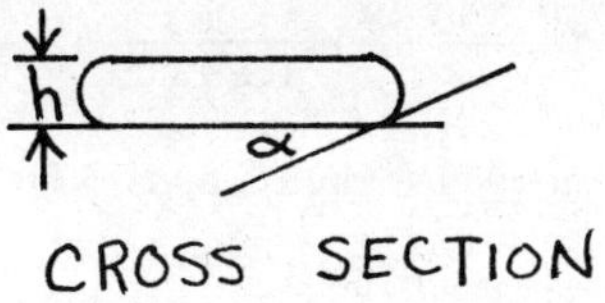

$$h = \sqrt{\frac{2V\gamma}{mg}(1 + \cos\alpha)}$$

$$= \sqrt{\frac{2\gamma}{\rho g}(1 + \cos\alpha)}$$

$$= \sqrt{\frac{2\gamma}{\rho g}(1 - \cos\beta)}$$

For mercury on glass,

$$h = \sqrt{\frac{2(472 \times 10^{-3} \text{ N/m})}{(13.6 \times 10^3 \text{ kg/m}^3)(9.81 \text{ m/s}^2)}(1 - \cos 148°)}$$

$$= 3.62 \text{ mm}$$

17-45

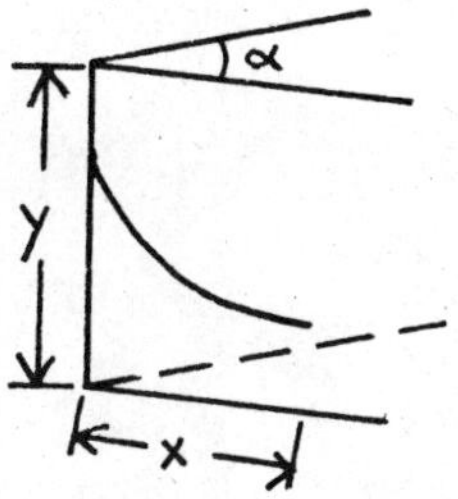

Front view

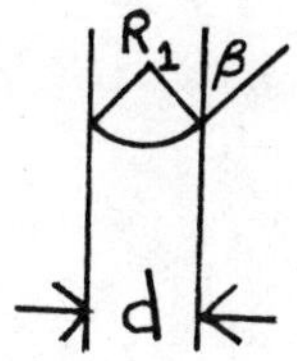

Side view

Using Eq. 17-17 with $R_2 = \infty$,

$$p - p_o = \gamma(\frac{1}{R_1}) = \gamma(\frac{2\cos\beta}{d}) \cong \gamma(\frac{2\cos\beta}{x\alpha})$$

Also,

$$p - p_o = \rho g y$$

Thus,

$$\rho g y = (\frac{2\cos\beta}{x\alpha})$$

$$xy = \frac{2\gamma \cos\beta}{\rho g \alpha} = \text{constant}$$

17-46 $Mg + \pi r^2 t \rho_{a\ell} g = 2\pi r\gamma + \pi r^2(t + d)\rho_w g$

$$M = \frac{2\pi r\gamma}{g} + \pi r^2[t(\rho_w - \rho_{a\ell}) + d\rho_w]$$

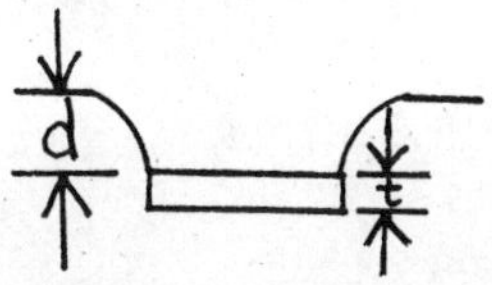

$$= (4.66 \times 10^{-4} \text{ kg}) + (-0.63 \times 10^{-4} \text{ kg})$$

$$= 0.40 \text{ g}$$

18-1 (a) $\overline{v} = \dfrac{0.01(2.0 \times 10^{-4}\ m^3/s)}{(0.1 \times 10^{-3}\ m)(3 \times 10^{-3}\ m)} = 6.67\ m/s$

(b) $\overline{v} = \dfrac{0.99(2.0 \times 10^{-4}\ m^3/s)}{\pi(0.015\ m)^2} = 0.280\ m/s$

18-2 $q = \vec{v}_1 \cdot \Delta\vec{A}_1 = \dfrac{10\ cm}{h}(2.0\ m^2)\cos 20° = 104\ cm^3/s$

18-3 $\overline{v} = (\dfrac{0.150\ kg}{s})(\dfrac{m^3}{1.293\ kg})(\dfrac{1}{\pi(0.05\ m)^2}) = 14.8\ m/s$

Since this is much less than 150 m/s, the flow can be considered incompressible.

18-4 As in Example 18-1,

$$v = \sqrt{2gh}$$

$$\frac{dm}{dt} = q\rho = Av\rho = A\rho\sqrt{2gh}$$

$$= (0.0001\ m^2)(10^3\ kg/m^3)\sqrt{2(9.8\ m/s^2)(2\ m)} = 0.626\ kg/s$$

18-5 $\frac{1}{2}gt^2 = 0.7\ m$

$$t = \sqrt{\frac{1.4\ m}{9.8\ m/s^2}} = 0.38\ s$$

$$vt = 2.5\ m$$

$$v = \frac{2.5\ m}{0.38\ s} = 6.6\ m/s$$

$$h = \frac{v^2}{2g} = \frac{(6.6\ m/s)^2}{2(9.8\ m/s^2)} = 2.22\ m$$

18-6 $A_2 = \dfrac{A_1}{2}$

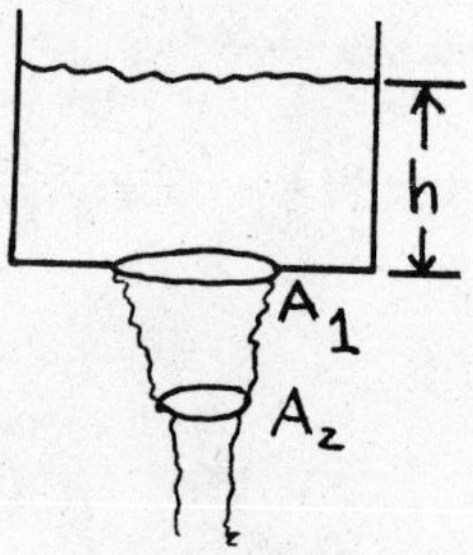

$$v_1 = \sqrt{2gh}$$

$$v_2 = \frac{v_1A_1}{A_2} = 2v_1$$

$$\rho gy + \frac{1}{2}\rho v_1^2 = \frac{1}{2}\rho v_2^2$$

$$y = \frac{v_2^2 - v_1^2}{2g} = \frac{3v_1^2}{2g} = \frac{3(2gh)}{2g} = 3h = 3(2.5\ m) = 7.5\ m$$

18-7 (a) $p_o + \rho gh_1 + 0 = p_o + 0 + \frac{1}{2}\rho v_3^2$

$$v_3 = \sqrt{2gh_1}$$

$$q = A_3v_3 = \frac{\pi d^2}{4}\sqrt{2gh_1} = \pi d^2\sqrt{gh_1}$$

(b) $p + \rho gh_2 + \frac{1}{2}\rho v_2^2 = p_o + 0 + \frac{1}{2}\rho v_3^2$

Since $v_2 = v_3$,

$$p = p_0 - \rho g h_2$$

Since $p \geq 0$,

$$h \leq \frac{p_0}{\rho g} = \frac{(1.013 \times 10^5 \text{ Pa})}{(10^3 \text{ kg/m}^3)(9.81 \text{ m/s}^2)} = 10.3 \text{ m}$$

The maximum value of h_2 occurs for $h_1 = 0$.

$$h_2 = 10.3 \text{ m}$$

18-8 As in Example 18-2,

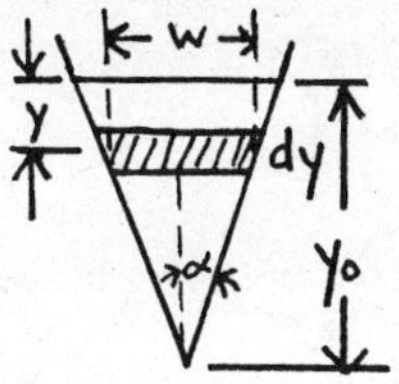

$$v = \sqrt{2gy}$$

$$w = 2(y_0 - y)\tan\alpha$$

$$dq = vw\,dy$$

$$q = 2\sqrt{2g}\tan\alpha\int_0^{y_0}\frac{1}{2}(y_0 y^{\frac{1}{2}} - y^{3/2})\,dy$$

$$= 2\sqrt{2g}\tan\alpha\left(\frac{2}{3}y_0^{5/2} - \frac{2}{5}y_0^{5/2}\right)$$

$$= \frac{8}{15}\sqrt{2gy_0^5}\tan\alpha$$

18-9 $z = (100 \text{ m})\sin 2° = 3.49 \text{ m}$

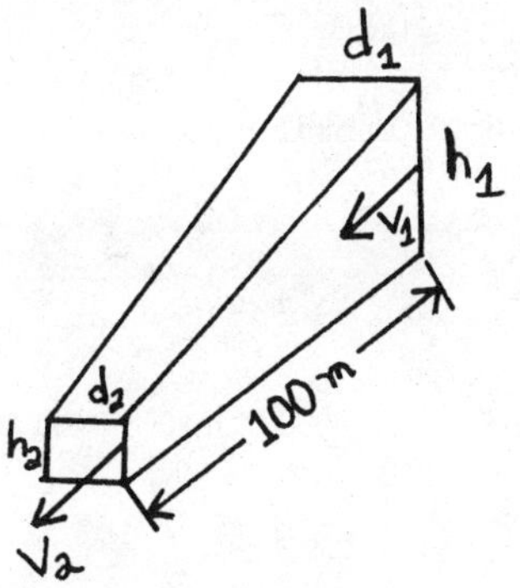

Applying Bernoulli's Equation,

$$\frac{1}{2}\rho\overline{v}_1^2 + \rho g z = \frac{1}{2}\rho\overline{v}_2^2$$

Also

$$\overline{v}_1 d_1 h_1 \cos 2° = \overline{v}_2 d_2 h_2 \cos 2°$$

$$\overline{v}_1^2 - \overline{v}_2^2 = v_1^2\left[1 - \left(\frac{d_1 h_1}{d_2 h_2}\right)^2\right] = -2gz$$

$$\frac{h_1 d_1}{h_2 d_2} = \sqrt{1 + \frac{2gz}{v_1^2}} = 2.30$$

$$h_2 = \frac{h_1 d_1}{2.30\,d_2} = 3.91 \text{ cm}$$

18-10 $p + \rho g h_1 = p_0 + \rho' g h_2$

$$p = p_0 + g(\rho' h_2 - \rho h_1)$$

18-11 (a) $\overline{v} = \frac{q}{A} = \frac{2 \text{ m}^3/\text{min}}{\pi(0.05 \text{ m})^2}\left(\frac{\text{min}}{60 \text{ s}}\right) = 4.24 \text{ m/s}$

(b) $\overline{v}' = \frac{q}{A'} = \frac{2 \text{ m}^3/\text{min}}{\pi(0.03 \text{ m})^2}\left(\frac{\text{min}}{60 \text{ s}}\right) = 11.79 \text{ m/s}$

(c) $p + \frac{1}{2}\rho\overline{v}^2 = p' + \frac{1}{2}\rho\overline{v}'^2$

$$p - p' = \frac{1}{2}\rho(\overline{v}'^2 - \overline{v}^2) = \frac{1}{2}(10^3 \text{ kg/m}^3)(11.79^2 - 4.24^2)\text{m}^2/\text{s}^2$$

$$= 6.05 \times 10^4 \text{ Pa} = 0.587 \text{ atm}$$

18-12 $p_1 + \rho g z_1 = p_2 + \rho g z_2 + \rho' g h$

$$p_1 - p_2 = \rho g h(\frac{\rho'}{\rho} - 1)$$

Also,

$$\frac{1}{2}\bar{v}_1^2 + \frac{p_1}{\rho} = \frac{1}{2}\bar{v}_2^2 + \frac{p_2}{\rho}$$

and

$$\bar{v}_1 A_1 = \bar{v}_2 A_2$$

Thus,

$$q = \bar{v}_2 A_2 = A_2 \sqrt{\frac{2(p_1 - p_2)}{\rho\left(1 - \frac{A_2^2}{A_1^2}\right)}} = A_2 \sqrt{\frac{2gh(\frac{\rho'}{\rho} - 1)}{1 - \frac{A_2^2}{A_1^2}}}$$

The energy equation must be used as it applies to non-steady, turbulent flow whereas Bernoulli's Equaiton is applicable only to steady flows.

18-13 Using Eq. 18-8,

$$W_{in} + 0 + (\frac{p_1}{\rho_1} - \frac{p_2}{\rho_2}) = (u_2 - u_1) + g(h_2 - h_1) + 0$$

$$u_2 - u_1 = W_{in} - g(h_2 - h_1)$$

Since the above applies to a unit mass, it must be multiplied by $\frac{dm}{dt}$ to find the rate of frictional heating.

$$\frac{dm}{dt}(u_2 - u_1) = P - g(h_2 - h_1)\frac{dm}{dt}$$

$$= 10\text{ hp}(\frac{746\text{ W}}{\text{hp}}) - (9.8\text{ m/s}^2)(30\text{ m})(20\text{ kg/s})$$

$$= 1.58 \times 10^3\text{ W}$$

$$\frac{u_2 - u_1}{L}\frac{dm}{dt} = \frac{1.58 \times 10^3\text{ W}}{200\text{ m}} = 7.9\text{ W/m}$$

$$\frac{1.58 \times 10^3\text{ W}}{7.46 \times 10^3\text{ W}} = 0.21 = 21\%$$

18-14 $p_1 + \frac{v_1^2}{2} + \rho g y_1 = p_2 + \rho g y_1$

Also,

$$p_a = p_3 + \rho' g h + \rho g y_1$$

and

$$p_b = p_2 + \rho g(h + y_1)$$

Since $p_a = p_b$,

$$p_2 - p_3 = gh(\rho' - \rho) \cong p_2 - p_1 = \rho\frac{v_1^2}{2}$$

$$v_1 = \sqrt{2gh(\frac{\rho'}{\rho} - 1)}$$

The Prandtl tube is self-contained.

18-15 Using Eq. 18-13,

$$\frac{q_B}{q_A} = \left(\frac{R_B}{R_A}\right)^4 \left(\frac{L_A}{L_B}\right) = 2.53$$

18-16 Using Eq. 18-13,

$$p_A - p_B = \frac{8q\eta L}{\pi R^4}$$

$$= \frac{8(2.0 \times 10^{-2}\ \text{m}^3/\text{s})(1.0 \times 10^{-3}\ \text{Pa} \cdot \text{s})(20\ \text{m})}{\pi(9.5 \times 10^{-3}\ \text{m})^4}$$

$$= 1.25 \times 10^5\ \text{Pa} = 1.24\ \text{atm}$$

18-17 Using Eq. 18-18,

$$v = \frac{N_R \eta}{\rho \ell}$$

$$v_{air} = \frac{100(1.81 \times 10^{-5}\ \text{Pa} \cdot \text{s})}{(1.29\ \text{kg/m}^3)(0.10\ \text{m})} = 1.40 \times 10^{-2}\ \text{m/s}$$

$$v_{water} = \frac{100(1.00 \times 10^{-3}\ \text{Pa} \cdot \text{s})}{(1.00 \times 10^3\ \text{kg/m}^3)(0.10\ \text{m})} = 1.00 \times 10^{-3}\ \text{m/s}$$

$$v_{glycerine} = \frac{100(1.49\ \text{Pa} \cdot \text{s})}{(1.26 \times 10^3\ \text{kg/m}^3)(0.10\ \text{m})} = 1.18\ \text{m/s}$$

18-18 (a) $(\frac{1}{2} \rho v^2)\ell^2 C_D = mg$

$$v = \sqrt{\frac{2\ mg}{\rho \ell^2 C_D}} = \sqrt{\frac{2(80\ \text{kg})(9.8\ \text{m/s}^2)}{(10^3\ \text{kg/m}^3)(0.6\ \text{m}^2)(0.5)}} = 2.29\ \text{m/s}$$

(b) $$(\frac{1}{2} \rho v^2)\pi(r^2)C_D = (\frac{1}{2} \rho v^2)\pi\left(\frac{3V}{4\pi}\right)^{2/3} C_D$$

$$= (\frac{1}{2} \rho v^2)\pi\left(\frac{3m}{4\pi\rho}\right)^{2/3} C_D = mg$$

$$v^2 = \frac{2\ mg}{\rho \pi C_D}\left(\frac{3m}{4\pi\rho}\right)^{-2/3} = \frac{2(9.8\ \text{m/s}^2)(80\ \text{kg})}{(10^3\ \text{kg/m}^3)\pi(0.5)}\left[\frac{240\ \text{kg}}{4\pi(10^3\ \text{kg/m}^3)}\right]^{-2/3}$$

$$v = \sqrt{13.97}\ \text{m/s} = 3.74\ \text{m/s}$$

18-19 $[Y] = \frac{F}{A} = MLT^{-2}L^{-2} = ML^{-1}T^{-2}$

$[\rho] = ML^{-3}$

$[c] = LT^{-1}$

$LT^{-1} = M^{\alpha}L^{-\alpha}T^{-2\alpha}M^{\beta}L^{-3\beta}$

$-\alpha - 3\beta = 1$

$\alpha + \beta = 0$

$-2\alpha = -1$

Thus,

$$\alpha = \frac{1}{2} \quad \beta = -\frac{1}{2}$$

$$c = \sqrt{\frac{Y}{\rho}}$$

18-20 Equating Froude numbers for the ship and the model,

$$\frac{v_1}{\sqrt{\ell_1 g}} = \frac{v_2}{\sqrt{\ell_2 g}}$$

$$v_2 = v_1\sqrt{\frac{\ell_2}{\ell_1}} = (0.517 \text{ m/s})\sqrt{\frac{2 \text{ m}}{200 \text{ m}}} = 0.0517 \text{ m/s} = 1.5 \text{ knots}$$

18-21 Equating Reynolds numbers,

$$\frac{\rho_A v_A \ell_A}{\eta_A} = \frac{\rho_w v_w \ell_w}{\eta_w}$$

$$v_w = \frac{\rho_A}{\rho_w}\left(\frac{\ell_A}{\ell_w}\right)\left(\frac{\eta_w}{\eta_A}\right)v_A$$

$$= \left(\frac{1.29}{1.00 \times 10^3}\right)\left(\frac{1}{15}\right)\left(\frac{1.002}{1.81 \times 10^{-2}}\right)(90 \text{ km/h}) = 0.43 \text{ km/h} = 11.9 \text{ cm/s}$$

18-22 $\frac{\ell_1}{\ell_2} = \frac{1}{10}$

$$\frac{\rho_1}{\rho_2} = \frac{13.6 \times 10^3 \text{ kg/m}^3}{1.00 \times 10^3 \text{ kg/m}^3} = 13.6$$

$$\frac{\eta_1}{\eta_2} = \frac{1.45 \times 10^{-3} \text{ Pa} \cdot \text{s}}{0.653 \times 10^{-3} \text{ Pa} \cdot \text{s}} = 2.22$$

Equating the Froude numbers,

$$\frac{g_1}{g_2} = \frac{v_1^2}{v_2^2} \cdot \frac{\ell_2}{\ell_1}$$

Equating the Euler numbers,

$$\frac{\Delta p_1}{\Delta p_2} = \frac{\rho_1}{\rho_2} \cdot \frac{v_1^2}{v_2^2}$$

Equating the Reynolds numbers,

$$\frac{v_1}{v_2} = \frac{\eta_1}{\eta_2} \cdot \frac{\rho_2}{\rho_1} \cdot \frac{\ell_2}{\ell_1}$$

The ratio of pressure differences is

$$\frac{\Delta p_1}{\Delta p_2} = \frac{\rho_2}{\rho_1}\left(\frac{\eta_1}{\eta_2}\right)^2\left(\frac{\ell_2}{\ell_1}\right)^2 = 36.2$$

$$\Delta p_2 = \frac{\Delta p_1}{36.2} = 0.011 \text{ atm}$$

The ratio of effective g values is

$$\frac{g_1}{g_2} = \left(\frac{\eta_1}{\eta_2}\right)^2\left(\frac{\rho_2}{\rho_1}\right)^2\left(\frac{\ell_2}{\ell_1}\right)^3 = 26.6$$

$$g_2 = \frac{g_1}{26.6} = \frac{9.81 \text{ m/s}^2}{26.6} = 0.369 \text{ m/s}^2$$

18-23 (a) Using Eq. 18-13,

$$q = \pi R^2\overline{v} = \frac{\pi R^4}{8\eta L}(p_A - p_B)$$

Using Eq. 18-12,

$$p_A - p_B = \frac{4\eta L\, v_o}{R^2}$$

Thus,

$$\overline{v} = \frac{R^2}{8\eta L}\left(\frac{4\eta L\, v_o}{R^2}\right) = \frac{1}{2}\, v_o$$

(b) $$N_R = \frac{\rho v_o \ell}{2\eta} = \frac{(930\ \text{kg/m}^3)(21.9\ \text{m/s})(0.01\ \text{m})}{2(0.12\ \text{Pa}\cdot\text{s})}$$

$$= 849$$

Since this is much less than 2000, the flow is actually laminar.

18-24 $$ma = -bv + \frac{4}{3}\pi R^3 g(\rho-\rho_o)$$

At terminal velocity there is no acceleration.

$$0 = -bv_\infty + \frac{4}{3}\pi R^3\, g(\rho-\rho_o)$$

$$v_\infty = \frac{4\pi R^3}{3b}\, g(\rho-\rho_o)$$

Using Stokes' Law,

$$b = 6\pi\eta R$$

Thus,

$$v_\infty = \frac{2}{9}\frac{R^2 g}{\eta}(\rho-\rho_o)$$

18-25 (a) $$v = \frac{N_R\eta}{2\rho_o R} = \frac{2}{9}\frac{R^2 g}{\eta}(\rho-\rho_o)$$

$$R^3 = \frac{9}{4}\frac{N_R\eta^2}{g\rho_o(\rho-\rho_o)}$$

$$= \frac{9}{4}\frac{(0.5)(1.81\times10^{-5}\ \text{Pa}\cdot\text{s})^2}{(9.8\ \text{m/s}^2)(1.20\ \text{kg/m}^3)(1000 - 1.20\ \text{kg/m}^3)}$$

$$= 3.14\times10^{-14}\ \text{m}^3 = 31.4\times10^{-15}\ \text{m}^3$$

$$R = 3.15\times10^{-5}\ \text{m} = 0.032\ \text{mm}$$

(b) $$R^3 = \frac{9}{4}\frac{(0.5)(1.49\ \text{Pa}\cdot\text{s})^2}{(9.8\ \text{m/s}^2)(1.26\times10^3\ \text{kg/m}^3)(1.49\times10^3\ \text{kg/m}^3)}$$

$$= 1.36\times10^{-7}\ \text{m}^3 = 136\times10^{-9}\ \text{m}^3$$

$$R = 5.14\times10^{-3}\ \text{m} = 5.14\ \text{mm}$$

(c) For part (a),

$$v_\infty = \frac{N_R\eta}{2\rho_o R} = \frac{(0.5)(1.81\times10^{-5}\ \text{Pa}\cdot\text{s})}{2(1.20\ \text{kg/m}^3)(0.032\times10^{-3}\ \text{m})} = 0.12\ \text{m/s}$$

For part (b),

$$v_\infty = \frac{(0.5)(1.49\ \text{Pa}\cdot\text{s})}{2(1.26\times10^3\ \text{kg/m}\ )(5.14\times10^{-3}\ \text{m})} = 5.8\ \text{cm/s}$$

18-26 (a) Using Eq. 18-18,

$$v = \frac{N_R\eta}{\rho d} = \frac{(3.0\times10^6)(1.81\times10^{-5}\ \text{Pa}\cdot\text{s})}{(0.80\ \text{kg/m}^3)(0.25\ \text{m})}$$

$$= 272\ \text{m/s}$$

(b) $C_D = 0.11$

$C_L = 1.25$

$$\frac{F_D}{\ell} = C_D R\rho v^2 = (0.11)(0.125\ \text{m})(0.80\ \text{kg/m}^3)(272\ \text{m/s})^2$$

$$= 814\ \text{N/m}$$

$$\frac{F_\ell}{\ell} = C_L R\rho v^2 = \frac{C_L}{C_D}\left(\frac{F_D}{\ell}\right) = 9.25\times10^3\ \text{N/m}$$

(c) $C_L = 1.0$

$$M = \frac{F_\ell}{g} = \frac{C_L R\rho v^2\ell}{g}$$

$$= \frac{(1.0)(0.125\ \text{m})(0.80\ \text{kg/m}^3)(272\ \text{m/s})^2(10\ \text{m})}{9.8\ \text{m/s}^2}$$

$$= 7.55\times10^3\ \text{kg}$$

(d) $$p_u - p_\ell = \frac{1}{2}\rho(\overline{v}_\ell^2 - \overline{v}_u^2) = \rho(\overline{v}_\ell - \overline{v}_u)\left(\frac{\overline{v}_\ell + \overline{v}_u}{2}\right)$$

$$= \rho\Delta v\ \overline{v}$$

Also,

$$p_u - p_\ell = \frac{F_\ell}{\ell s}$$

Equating the two expressions for the pressure difference,

$$\Delta v = \frac{F_\ell/\ell}{\rho\overline{v}\ s} + \frac{9.25\times10^3\ \text{N/m}}{(0.80\ \text{kg/m}^3)(272\ \text{m/s})(1.3\ \text{m})} = 32.7\ \text{m/s}$$

18-27 (a) $$N_R = \frac{\rho d\overline{v}}{\eta} = \frac{\rho d}{\eta}\frac{q}{A}$$

$$= \frac{(10^3\ \text{kg/m}^3)(0.10\ \text{m})}{(1.00\times10^{-3}\ \text{Pa}\cdot\text{s})}\frac{(10^{-2}\ \text{m}^3/\text{s})}{\pi(0.05\ \text{m})^2} = 1.27\times10^5$$

(b) $$p_A - p_B = \frac{F}{\pi R^2} = \frac{qR}{\pi R^2} = \frac{q}{\pi R^2}\left(\frac{\eta L}{\pi R^2}\right)[0.209(N_R)^{3/4}]$$

$$= \frac{(10^{-2}\ \text{m}^3/\text{s})(10^{-3}\ \text{Pa}\cdot\text{s})(20\ \text{m})}{\pi^2(0.05\ \text{m})^4}(0.209)(1.27\times10^5)^{3/4}$$

$$= 4.56\times10^3\ \text{Pa}$$

18-28 $v = \sqrt{2gx}$

$\frac{1}{2} gt^2 = (h - x)$

$y^2 = v^2t^2 = \frac{2(h-x)}{g} v^2 = 4x(h - x)$

$\frac{d^2(y^2)}{dx^2} = 0 = 4h - 8x$

$x = \frac{h}{2}$

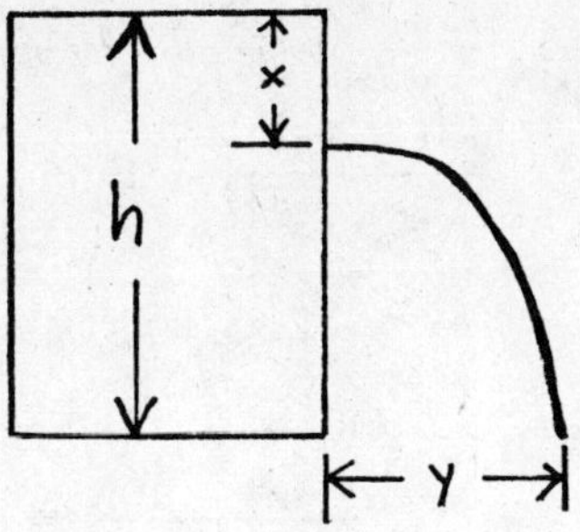

18-29 $v_1A_1 = v_2A_2$

$v_2 = v_1 \frac{A_1}{A_2} = (20 \text{ m/s}) \frac{6\text{m}^2}{1.5 \text{ m}^2} = 80 \text{ m/s}$

$p_1 + \frac{1}{2} \rho_1 v_1^2 = p_2 + \frac{1}{2} \rho_2 v_2^2$

Assuming density remains constant,

$\rho_2 = \rho_1 \frac{p_2}{p_1} \cong \rho_1$

$p_2 = p_1 + \frac{1}{2} \rho_1 (v_1^2 - v_2^2)$

$= (1.01 \times 10^5 \text{ Pa}) + \frac{1}{2}(1.20 \text{ kg/m}^3)(-6000 \text{ m}^2/\text{s}^2)$

$= 0.974 \times 10^5 \text{ Pa}$

To justify the constant density assumption,

$\frac{p_1 - p_2}{p_1} = \frac{1.01 - 0.974}{1.01} = 3.6\%$

The result can be iterated by using the calculated value of p to obtain a better value for ρ_2.

$\rho_2 = \rho_1 \frac{0.974}{1.01} = 0.96 \rho_1$

$p_2 = p_1 + \frac{1}{2} \rho_1 (v_1^2 - 0.96 v_2^2)$

$= 0.976 \times 10^5 \text{ Pa}$

18-30 (a) $p + \frac{1}{2} \rho \overline{V}^2 = p_0 + \frac{1}{2} \rho \overline{v}^2$

Also,

$\pi R^2 \overline{V} = \pi r^2 \overline{v} = \pi (\frac{d}{2})^2 \overline{v}$

Thus,

$p - p_0 = \frac{1}{2} \rho (\overline{v}^2 - \overline{V}^2) = \frac{1}{2} \rho \overline{v}^2 (1 - \frac{d^4}{4R^4})$

$\overline{v} = \sqrt{\frac{8(p - p_0)R^4}{(4R^4 - d^4)\rho}}$

The thrust is given by

$$F = \bar{v}\frac{dm}{dt} = \bar{v}^2 \rho A = \bar{v}^2 \rho\, \pi r^2 = \bar{v}^2 \rho\pi (\frac{d}{2})^2$$

$$= \frac{2(p - p_o)R^4}{(4R^4 - d^4)} \pi d^2$$

(b) $F = Mg = [(25\ \text{kg})+(10^3\text{kg/m}^3)\pi(0.30\ \text{m})^2(1\ \text{m})](9.8\ \text{m/s}^2)$

$= 3016\ \text{N}$

$$p - p_o = \frac{F(4R^4 - d^4)}{2\pi R^4\ d^2} = \frac{(3016\ \text{N})(0.032\ \text{m}^4)}{2\pi(0.0081\ \text{m}^4)(0.0016\ \text{m}^2)} = 1.19 \times 10^6\ \text{Pa}$$

$$p = p_o + (11.9 \times 10^5\ \text{Pa}) = 12.9 \times 10^5\ \text{Pa} = 12.7\ \text{atm}$$

18-31 $p_1 + \frac{1}{2}\rho\bar{v}_1^2 = p_2 + \frac{1}{2}\rho\bar{v}_2^2 + L(\sin\alpha)\rho g$

Also,

$$p_1 + y_1\rho g = p_2 + y_2\rho g + \rho' gh$$

$$p_1 - p_2 = (y_2 - y_1)\rho g + \rho' gh = (L\sin\alpha - h)\rho g + \rho' gh$$

Using the first equation,

$$p_1 - p_2 = \frac{1}{2}\rho(\bar{v}_2^2 - \bar{v}_1^2) + L(\sin\alpha)\rho g$$

Equating the two expressions for the pressure difference,

$$\frac{1}{2}(\bar{v}_2^2 - \bar{v}_1^2) = \frac{\rho'}{\rho} gh - hg$$

$$\frac{1}{2}\bar{v}_2^2\left(1 - \frac{A_2^2}{A_1^2}\right) = gh(\frac{\rho'}{\rho} - 1)$$

So,

$$q = \bar{v}_2 A_2 = A_2 \left(\frac{2gh(\frac{\rho'}{\rho} - 1)}{1 - (A_2/A_1)^2}\right)^{\frac{1}{2}}$$

This is the same expression as that in Example 18-3. Notice that h is independent of L and α. It is the <u>vertical</u> difference in manometer heights.

18-32 Using Eq. 18-8,

$$W_{in} = g(h_2 - h_1) = gh$$

Since this applies to a unit mass, it must be multiplied by $\frac{dm}{dt}$ to obtain power.

$$P = W_{in}(\frac{dm}{dt}) = gh(\frac{dm}{dt}) = gh\rho q$$

$$h = \frac{P}{g\rho q} = \frac{(3.0 \times 746\ \text{W})}{(9.8\ \text{m/s}^2)(10^3\ \text{kg/m}^3)(2 \times 10^{-2}\ \text{m}^3/\text{s})} = 11.4\ \text{m}$$

18-33 Using Eq. 18-13,

$$\eta = \frac{\pi R^4(p_A - p_B)}{8qL} = \frac{\pi(0.70 \times 10^{-3}\ \text{m})^4[(1.01 \times 10^5\ \text{Pa})/20]}{8[(292 \times 10^{-6}\ \text{m}^3)/600\ \text{s}](1.50\ \text{m})}$$

$= 0.652 \times 10^{-3}$ Pa · s

The liquid is water.

18-34 (a) $c = \lambda^{\alpha} g^{\beta}$

$LT^{-1} = L^{\alpha} L^{\beta} T^{-2\beta}$

$\alpha + \beta = 1$

$-2\beta = -1$

Thus,

$\alpha = \beta = \frac{1}{2}$

$c \propto \sqrt{\lambda g}$

(b) $c = \lambda^{\alpha} \rho^{\beta} \gamma^{\gamma}$

$LT^{-1} = L^{\alpha} M^{\beta} L^{-3\beta} M^{\gamma} T^{-2\gamma}$

$\beta + \gamma = 0$

$-2\gamma = -1$

$\alpha - 3\beta = 1$

Thus,

$\gamma = \frac{1}{2}$

$\beta = \alpha = \frac{-1}{2}$

$c \propto \sqrt{\frac{\gamma}{\rho\lambda}}$

(c) $c = h^{\alpha} g^{\beta}$

$LT^{-1} = L^{\alpha} L^{\beta} T^{-2\beta}$

$\alpha + \beta = 1$

$-2\beta = -1$

Thus,

$\beta = \alpha = \frac{1}{2}$

$c \propto \sqrt{hg}$

(d) For part (a), $h \gg \lambda$.

$\tanh \frac{2\pi h}{\lambda} \cong 1$

For ordinary wavelengths, $\frac{2\pi\lambda}{\rho\lambda}$ is small.

Thus,

$c \cong \sqrt{\frac{g\lambda}{2\pi}}$

For part (b), $h \gg \lambda$.

For small wavelengths, $\frac{g\lambda}{2\pi}$ is small.

Thus,

$$c \cong \sqrt{\frac{2\pi\gamma}{\rho\lambda}}$$

For part (c), $\lambda \gg h$.

For long wavelengths, $\frac{2\pi\gamma}{\rho\lambda}$ is small.

Thus,

$$c \cong \sqrt{hg}$$

18-35 (a) $N_R = \frac{\rho v d}{\eta} = \frac{(10^3\ \text{kg/m}^3)(10^{-2}\ \text{m/s})(10^{-2}\ \text{m})}{10^{-3}\ \text{Pa}\cdot\text{s}} = 100$

(b) From Figure 18-13,

$$C_D = 1.5$$

$$F_D = (\tfrac{1}{2}\rho v^2)(\ell d)C_D$$

$$\frac{F_D}{\ell} = \tfrac{1}{2}(10^3\ \text{kg/m}^3)(10^{-2}\ \text{m/s})^2(10^{-2}\ \text{m})(1.5)$$

$$= 7.5 \times 10^{-4}\ \text{N/m}$$

19-1 Since $v_{light} >> v_{sound}$,

$$d \cong (334 \text{ m/s})(16.2 \text{ s}) = 5.41 \text{ km}$$

19-2 Let d be the distance the stone drops.

$$t = \frac{d}{v_s} + \sqrt{\frac{2d}{g}}$$

$$d + \sqrt{\frac{2}{g}}\, v_s \sqrt{d} - v_s t = 0$$

$$\sqrt{d} = -\frac{1}{2}\left(\sqrt{\frac{2}{g}}\, v_s \pm \sqrt{\frac{2v_s^2}{g} + 4v_s t}\right)$$

$$= (-1.50.9 \pm \sqrt{36394})/2$$

Choose the positive root so that $\sqrt{d} > 0$.

$$\sqrt{d} = 19.9 \text{ m}^{\frac{1}{2}}$$

$$d = 397 \text{ m}$$

If the speed of sound is ignored,

$$t = \sqrt{\frac{2d'}{g}}$$

$$d' = \frac{1}{2} gt^2 = 510 \text{ m}$$

The percentage error is given by

$$\frac{d' - d}{d} = 0.28 = 28\%$$

19-3

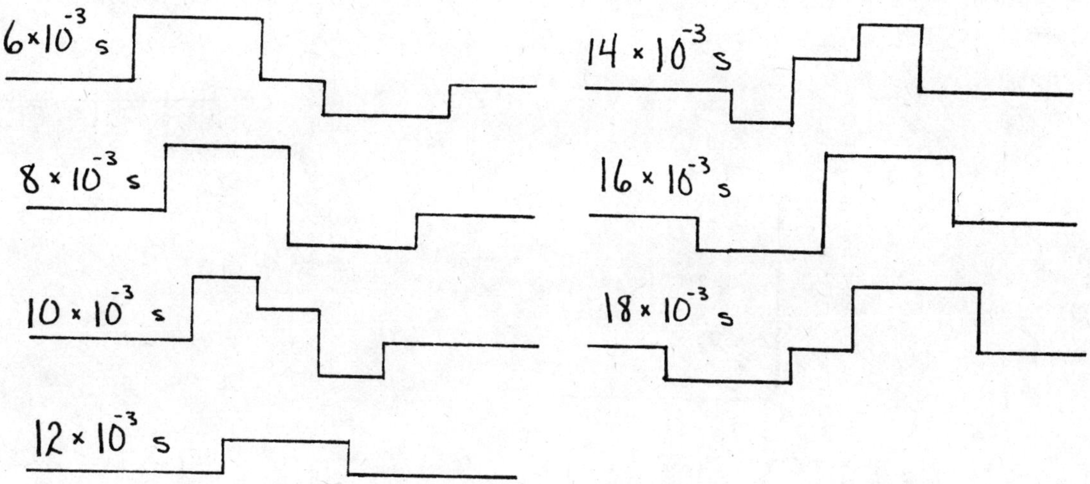

19-4 (a) $y(x, 0) = Ae^{-\sigma x^2}$

x, m	y, m
0	0.030
±0.2	0.020
±0.4	0.006
±0.6	0.0008

To find FWHM,

$$y(x, 0) = \frac{1}{2} A = Ae^{-\sigma x^2}$$

$$\sigma x^2 = \ln 2$$

$$x = \pm\sqrt{\frac{\ln 2}{\sigma}} = \pm 0.26 \text{ m}$$

$$\text{FWHM} = 2x = 0.52 \text{ m}$$

(b) $y(0, t) = Ae^{-\sigma(vt)^2}$

$$= Ae^{-100\sigma t^2}$$

For $t = \frac{1}{10}x$, the y-values are equal.

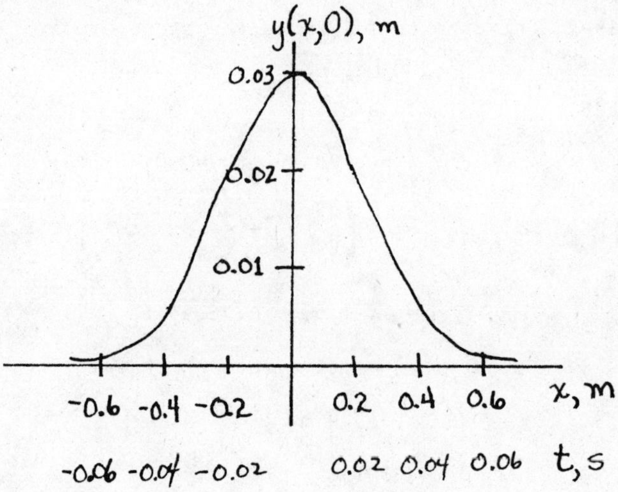

19-5 (a) The maximum height occurs when $x = vt$.

$$\frac{1}{2}\left(\frac{0.1\ C^3}{C^2}\right) = \frac{0.1\ C^3}{C^2 + [x - v(t + \Delta t)]^2} = \frac{0.1\ C}{C^2 + v^2(\Delta t^2)}$$

$$v = \frac{C}{\Delta t} = \frac{0.04 \text{ m}}{2 \times 10^{-3} \text{ s}} = 20 \text{ m/s}$$

(b) $\text{FWHM} = 2x(\eta, 0) = 2x(0.2 \text{ cm}, 0) = 2(4 \text{ cm}) = 8 \text{ cm}$

(c)

x,cm	η(x, 0),cm
0	0.40
±2	0.32
±4	0.20
±6	0.12
±8	0.08
±10	0.06

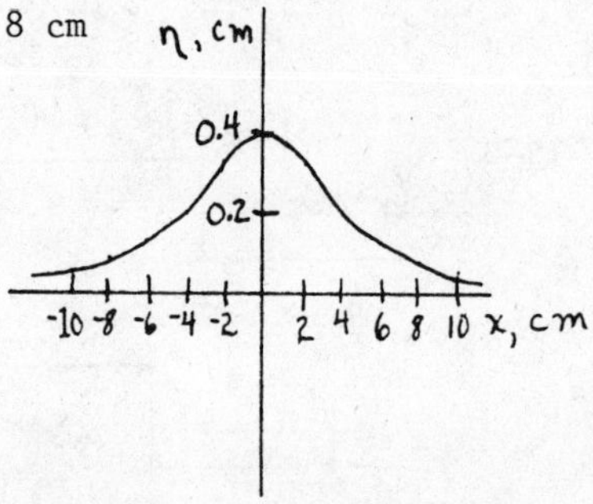

19-6 (a)

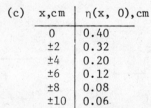

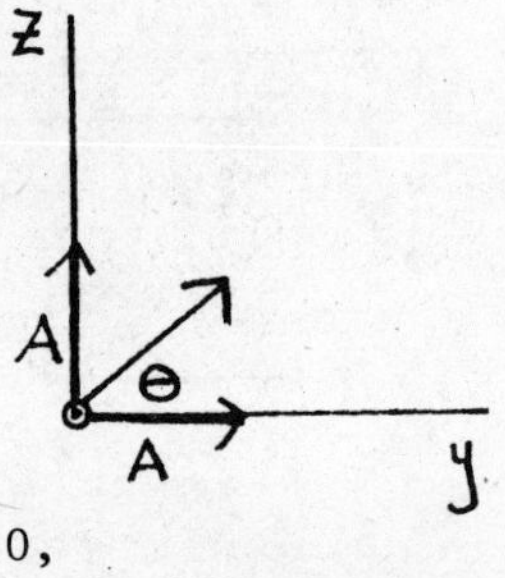

$$\tan\theta = \frac{A}{A} = 1$$

$$\theta = 45°$$

(b) At $x = 0$,

$$y(0, t) = A\sin(kvt - \phi_0) = A\sin\alpha$$

$$z(0, t) = A\sin(kvt - \phi_0 + \frac{\pi}{2}) = A\sin(\alpha + \frac{\pi}{2}) = A\cos\alpha$$

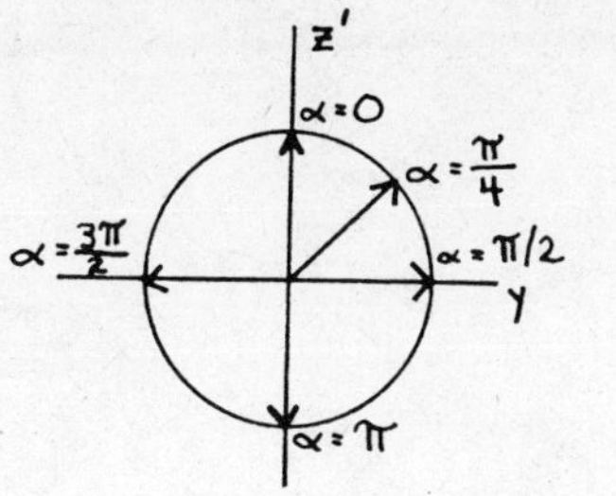

The height of the wave remains constant at A. The polarization of the wave moves in a circle about the x-axis.

19-7 $v = \sqrt{\frac{F}{\mu}} = \sqrt{\frac{F\ell}{m}} = \sqrt{\frac{(750\ \text{N})(5\ \text{m})}{(0.060\ \text{kg})}} = 250\ \text{m/s}$

19-8 $v = \frac{4450\ \text{km}}{9.5\ \text{h}} = 468\ \text{km/h}$

$\bar{d} = \frac{v^2}{g} = \left(\frac{468\ \text{km}}{3600\ \text{s}}\right)^2 \frac{10^3}{9.8\ \text{km/s}^2} = 1.72\ \text{km}$

19-9 $F(x) = \frac{x}{L} Mg = \mu x g$

Using Eq. 19-9,

$v = \sqrt{\frac{F}{\mu}} = \sqrt{xg}$

19-10 Let the string have mass M and radius R.

To find the tension in the undisturbed string,

$2T \sin \frac{d\theta}{2} = \frac{M}{2\pi} d\theta \frac{v_o^2}{R}$

For $d\theta$ small,

$2T \frac{d\theta}{2} = \frac{M}{2\pi} d\theta \frac{v_o^2}{R}$

$T = \frac{M}{2\pi R} v_o^2 = \mu v_o^2$

Using Eq. 19-9,

$v_{wave} = \sqrt{\frac{T}{\mu}} = v_o$

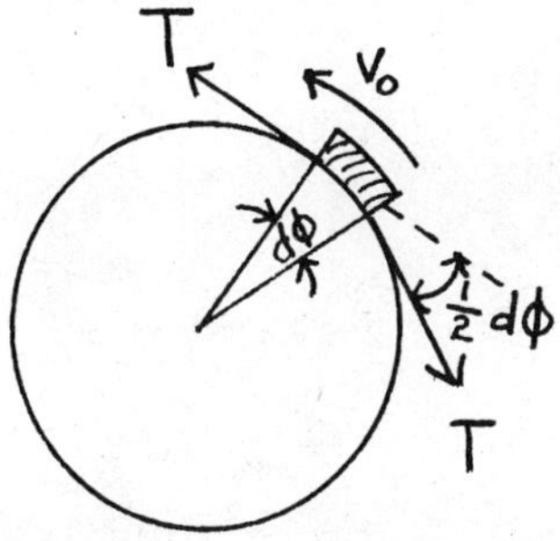

19-11 (a) For $u > L$ and $u < -L$,

$y = 0$

For $-L \le u \le 0$,

$y = (L + u)\tan\theta = \frac{h}{L}(L + u)$

For $0 \le u \le L$,

$y = (L - u)\tan\theta = \frac{h}{L}(L - u)$

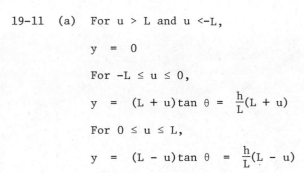

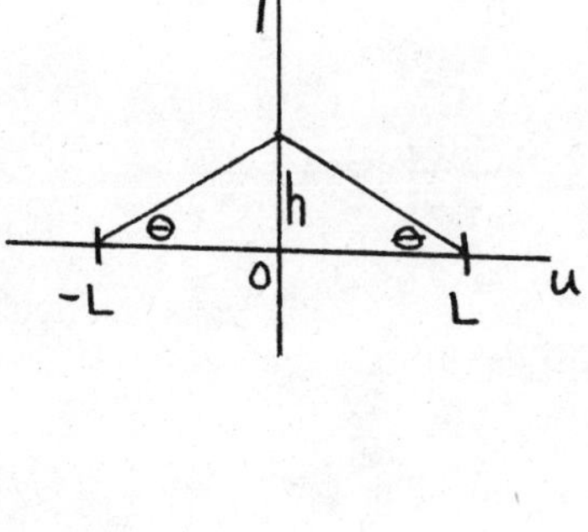

(b) $P = -F\left(\frac{\partial y}{\partial x}\right)\left(\frac{\partial y}{\partial t}\right)$

For $u > L$ and $u < -L$,

$P = -F(0)(0) = 0$

For $-L \le u \le 0$,

$$P = -F\left(\frac{\partial y}{\partial u}\right)^2\left(\frac{\partial u}{\partial x}\right)\left(\frac{\partial u}{\partial y}\right) = -F\left(\frac{h}{L}\right)^2(1)(-v) = Fv(h/L)^2$$

For $0 \le u \le L$,

$$P = -F\left(-\frac{h}{L}\right)^2(1)(-v) = Fv(h/L)^2$$

(c) $$E = \int_{-\infty}^{\infty} P(0, t)dt = \int_{-L/v}^{L/v} Fv(h/L)^2\, dt = 2Fh^2/L$$

$$= 2(v^2\mu)h^2/L = \frac{2(50\text{ m/s})^2(0.08\text{ kg/m})(1.5\times 10^{-2}\text{ m})^2}{(0.10\text{ m})}$$

$$= 0.90\text{ J}$$

19-12 (a) Let $u = 10\pi t - 3\pi x + \frac{\pi}{4}$

$$\frac{du}{dt} = 10\pi - 3\pi\frac{dx}{dt} = 0$$

$$\frac{dx}{dt} = \frac{10}{3}\text{ m/s}$$

The velocity is in the positive x-direction.

(b) $y(0.10, 0) = (0.35\text{ m})\sin(-0.3\pi + \frac{\pi}{4}) = -0.055\text{ m} = -5.5\text{ cm}$

(c) $k = \frac{2\pi}{\lambda} = 3\pi$

$\lambda = \frac{2}{3}\text{ m}$

$\omega = 2\pi\nu = 10\pi$

$\nu = 5\text{ Hz}$

(d) $$v_y = \frac{dy}{dt} = (0.35)(20\pi)\cos(20\pi t - 3\pi x + \frac{\pi}{4})$$

$$v_{y,max} = (20\pi)(0.35) = 22.0\text{ m/s}$$

19-13 (a) $y(0, 0) = A\cos\phi_0 = 0.02\text{ m}$

$$\left.\frac{dy}{dt}\right|_{0,0} = -A\omega\sin\phi_0 = -2.0\text{ m/s}$$

Also,

$$\omega = \frac{2\pi}{T} = \frac{2\pi}{0.025\text{ s}}$$

$$A^2 = (0.02\text{ m})^2 + \left(\frac{2.0\text{ m/s}}{2\pi/0.025\text{ s}}\right)^2$$

$$A = 0.022\text{ m} = 2.2\text{ cm}$$

(b) $\phi_0 = \tan^{-1}\frac{0.025/\pi}{0.02} = 21.7° = 0.379\text{ rad}$

(c) $v_{y,max} = A\omega = 2\pi(2.2\text{ cm})/0.025\text{ s} = 550\text{ cm/s}$

(d) $\lambda = v_x T = (30\text{ m/s})(0.025\text{ s}) = 0.75\text{ m}$

$k = 2\pi/\lambda = 2\pi/0.75 = 8\pi/3$

$$\omega = 2\pi/0.025 = 80\pi$$

$$y(x, t) = (2.2\text{ cm})\cos(80\pi t + \frac{8\pi}{3}x + 0.379)$$

19-14 (a) $y(0,t) = 2A\cos(\omega t - \phi_0)$

$$v_y = \frac{dy}{dt} = -2A\omega\sin(\omega t - \phi_0)$$

$$\phi_0 = \sin^{-1}\frac{v_y(t=0)}{-2A\omega} = \sin^{-1}\frac{\frac{1}{2}v_{y,max}}{-2A\omega}$$

$$= \sin^{-1}\frac{\frac{1}{2}(-2A\omega)}{-2A\omega} = \sin^{-1}\tfrac{1}{2} = \frac{\pi}{6}, \frac{5\pi}{6}$$

For v increasing,

$$\phi_0 = \frac{5\pi}{6}$$

Thus,

$$y(0,t) = 2A\cos(\omega t - \frac{5\pi}{6})$$

(b) $v = \sqrt{\frac{F}{\mu}} = \sqrt{\frac{400\text{ N}}{0.05\text{ kg/m}}} = 89.4\text{ m/s}$

$$\lambda = \frac{v}{\nu} = \frac{89.4\text{ m/s}}{100\text{ s}^{-1}} = 89.4\text{ cm}$$

(c) $A = \frac{30\text{ m/s}}{\omega} = \frac{30\text{ m/s}}{2\pi(100\text{ s}^{-1})} = 4.77\text{ cm}$

(d) $k = \frac{2\pi}{\lambda} = 7.03\text{ m}^{-1}$

In SI units,

$$y^+(x,t) = (0.0477\text{ m})\cos(7.03x - 200\pi t + \frac{5\pi}{6})$$

$$y^-(x,t) = (0.0477\text{ m})\cos(7.03x + 200\pi t - \frac{5\pi}{6})$$

19-15 $\overline{P} = 2\pi^2 FA^2\nu/\lambda = 2\pi^2\mu v^2 A^2\nu/\lambda$

$$= 2\pi^2\mu v A^2\nu^2$$

$$= 2\pi^2(0.075\text{ kg/m})(\frac{10}{3}\text{ m/s})(0.35\text{ m})^2(5\text{ s}^{-1})^2$$

$$= 15.1\text{ W}$$

$$E = \overline{P}T = \frac{15.1\text{ W}}{5\text{ s}^{-1}} = 3.0\text{ J}$$

19-16 (a) $E = \overline{P}T = 2\pi^2 FA^2\nu T/\lambda = 2\pi^2 FA^2/\lambda$

$$= 2\pi^2\mu v^2 A^2/\lambda = 2\pi^2\mu\lambda\nu^2 A^2$$

$$= \frac{\mu\lambda}{2}(4\pi^2\nu^2 A^2) = \frac{\mu\nu}{2}v^2_{\perp,max}$$

$$\lambda = \frac{2E}{\mu v^2_{\perp,max}} = \frac{2(4\text{ J})}{(0.10\text{ kg/m})(15.0\text{ m/s})^2} = 0.36\text{ m}$$

(b) $F = \mu v^2 = (0.10\ \text{kg/m})(30.0\ \text{m/s})^2 = 90\ \text{N}$

(c) $\nu = \dfrac{v}{\lambda} = \dfrac{30.0\ \text{m/s}}{0.36\ \text{m}} = 83\ \text{Hz}$

(d) $A = \dfrac{v_{\perp,max}}{\omega} = \dfrac{15.0\ \text{m/s}}{2\pi(84.3\ \text{s}^{-1})} = 2.83\ \text{cm}$

19-17 $v = \sqrt{\dfrac{Y}{\rho}} = \sqrt{\dfrac{16.7 \times 10^{10}\ \text{N/m}}{21.4 \times 10^3\ \text{kg/m}}} = 2.79 \times 10^3\ \text{m/s}$

19-18 $Y = v^2\rho = \dfrac{L^2}{t^2}\rho$

$dY = -2\dfrac{L^2}{t^3}\rho dt$

$\left|\dfrac{dY}{Y}\right| = \left|\dfrac{-2\ dt}{t}\right| = \dfrac{2v\ dt}{L}$

$L = \dfrac{2v\ dt}{|dY/Y|} = \dfrac{2(3.75 \times 10^3\ \text{m/s})(120 \times 10^{-6}\ \text{s})}{0.01}$

$= 90.0\ \text{m}$

19-19 The wave equation for waves in a slinky is derived just as it is for waves in a solid rod. Thus, for a slinky of cross-sectional area S,

$$v = \sqrt{\frac{Y}{\rho}} = \sqrt{\frac{1}{\rho}}\sqrt{\frac{\text{stress}}{\text{strain}}} = \sqrt{\frac{\ell S}{M}}\sqrt{\frac{F/S}{\Delta\ell/\ell}}$$

$$= \sqrt{\frac{\ell S}{M}}\sqrt{\frac{\kappa\ \Delta\ell/S}{\Delta\ell/\ell}} = \ell\sqrt{\frac{\kappa}{M}} = (2\ \text{m})\sqrt{\frac{6\ \text{N/m}}{0.5\ \text{kg}}} = 6.9\ \text{m/s}$$

19-20 (a) Using Eq. 19-27,

$$\frac{A_t}{A_i} = \frac{2\sqrt{\mu_1}}{\sqrt{\mu_1} + \sqrt{\mu_2}} = \frac{2\sqrt{\mu_1}}{\sqrt{\mu_1} + \sqrt{2\mu_1}} = \frac{2}{1 + \sqrt{2}}$$

$$A_t = \frac{2}{1 + \sqrt{2}} A_i = 8.3\ \text{cm}$$

The phase is the same as that of the incident wave.

(b) $$\frac{A_r}{A_i} = \frac{\sqrt{\mu_1} - \sqrt{\mu_2}}{\sqrt{\mu_1} + \sqrt{\mu_2}} = \frac{1 - \sqrt{2}}{1 + \sqrt{2}}$$

$$A_r = \frac{1 - \sqrt{2}}{1 + \sqrt{2}} A_i = -1.7\ \text{cm}$$

The relative phase difference is π.

(c) $$\frac{\overline{P}_r}{\overline{P}_i} = \frac{A_r^2}{A_i^2} = \left(\frac{1 - \sqrt{2}}{1 + \sqrt{2}}\right)^2$$

$$\overline{P}_r = \left(\frac{1 - \sqrt{2}}{1 + \sqrt{2}}\right)^2 \overline{P}_i = 2.9\ \text{W}$$

19-21 $$\frac{A_t}{A_i} = \frac{2\sqrt{\mu}}{\sqrt{\mu} + \sqrt{2\mu}} = \frac{2}{1 + \sqrt{2}}$$

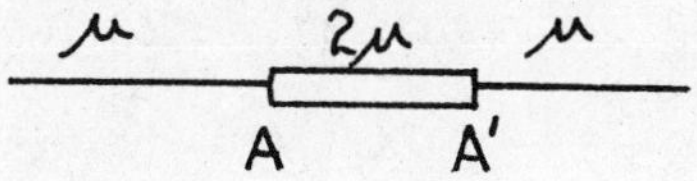

$$\frac{A'_t}{A'_i} = \frac{2\sqrt{2\mu}}{\sqrt{2\mu} + \sqrt{\mu}} = \frac{2\sqrt{2}}{1 + \sqrt{2}}$$

$$A'_t = \frac{2\sqrt{2}}{1 + \sqrt{2}} A'_i = \frac{2\sqrt{2}}{1 + \sqrt{2}} A_t = \frac{2\sqrt{2}}{1 + \sqrt{2}} \left(\frac{2}{1 + \sqrt{2}}\right) A_i = 0.97\ A_i$$

It is necessary to specify the "first" pulse because the reflected wave at A' will reflect at A and return to A'.

19-22 The velocity of the pulse in the lighter string is

$$v' = \sqrt{\frac{F}{\mu}}$$

In the heavier string,

$$v = \sqrt{\frac{F}{3\mu}}$$

Thus,

$$v' = \sqrt{3}\ v$$

The travel time for a round-trip pulse on the lighter string is

$$t = 2\left(\frac{\ell}{v'}\right) = 2\left(\frac{\ell}{\sqrt{3}\ v}\right) = \frac{2}{\sqrt{3}}\left(\frac{2\ \text{m}}{20\ \text{m/s}}\right) = 0.115\ \text{s}$$

The distance between reflected pulses is then

$$d = vt = (20\ \text{m/s})(0.115\ \text{s}) = 2.3\ \text{m}$$

For a pulse moving from left to right,

$$\frac{A_r}{A_i} = \frac{\sqrt{3\mu} - \sqrt{\mu}}{\sqrt{3\mu} + \sqrt{\mu}} = 0.268$$

$$\frac{A_t}{A_i} = \frac{2\sqrt{3\mu}}{\sqrt{3\mu} + \sqrt{\mu}} = 1.268$$

For a pulse moving from right to left,

$$\frac{A'_r}{A_i} = \frac{\sqrt{\mu} - \sqrt{3\mu}}{\sqrt{\mu} + \sqrt{3\mu}} = -0.268$$

$$\frac{A'_t}{A_i} = \frac{2\sqrt{\mu}}{\sqrt{\mu} + \sqrt{3\mu}} = 0.732$$

The first four pulses reflected back have amplitudes as follows:

$$A_0 = 0.268\ A$$

$A_1 \;=\; 0.732(-1)(1.268\text{ A}) \;=\; -0.928\text{ A}$

$A_2 \;=\; 0.732(-1)(-0.268)(-1)(1.268\text{ A}) \;=\; -0.249\text{ A}$

$A_3 \;=\; 0.732(-1)(-0.268)^2(-1)^2(1.268\text{ A}) \;=\; -0.067\text{ A}$

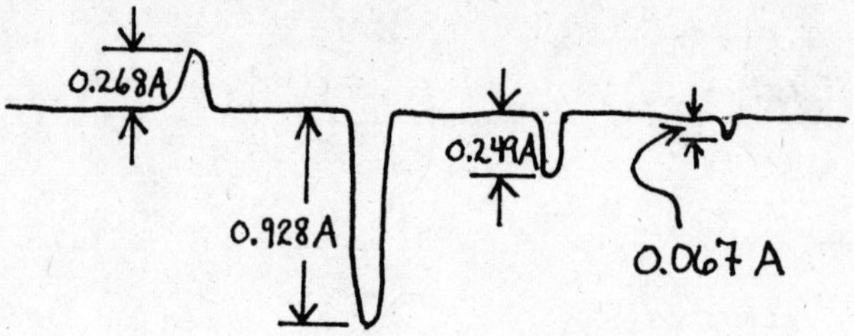

For n > 0,

$$A_n \;=\; A_1(0.268)^{n-1}$$

The total energy reflected is then, using a constant of proportionality k,

$$E \;=\; k\sum_{n=0}^{\infty}A_n^2 \;=\; k(A_0^2 + \sum_{n=1}^{\infty}A_n^2) \;=\; k[A_0^2 + A_1^2\sum_{n=1}^{\infty}(0.268)^{2(n-1)}]$$

$$=\; kA^2[(0.268)^2 + \frac{(0.928)^2}{1-(0.268)^2}] \;=\; kA^2$$

19-23 $\lambda_1 \;=\; \frac{3}{2}\lambda_2$

$\nu_1 \;=\; \nu_2$

Since $v \;=\; \lambda\nu$,

$v_1^2 \;=\; \frac{9}{4}v_2^2$

Thus,

$F_1 \;=\; \frac{9}{4}F_2$

19-24 $v \;=\; \sqrt{\frac{F}{\mu}} \;=\; \sqrt{\frac{FL}{m}} \;=\; \sqrt{\frac{(90\text{ N})(0.35\text{ m})}{(0.41\times 10^{-3}\text{ kg})}} \;=\; 277\text{ m/s}$

$\nu_n \;=\; \frac{v}{\lambda_n} \;=\; \frac{v}{2L/n}$

$\nu_1 \;=\; \frac{277\text{ m/s}}{2(0.35\text{ m})} \;=\; 396\text{ Hz}$

$\nu_2 \;=\; 2\nu_1 \;=\; 792\text{ Hz}$

$\nu_3 \;=\; 3\nu_1 \;=\; 1188\text{ Hz}$

19-25 (a) $\frac{\nu_A'}{\nu_A} \;=\; \frac{v'}{v} \;=\; \sqrt{\frac{F'}{F}}$

$F' \;=\; \left(\frac{\nu_A'}{\nu_A}\right)^2 F \;=\; 4F$

(b) $\nu \;=\; \frac{v}{2L}$

$$\frac{\nu'_C}{\nu_A} = \frac{L_A}{L_C}$$

$$L_C = \frac{\nu_A}{\nu'_C} L_C = \frac{440 \text{ Hz}}{2(261.6 \text{ Hz})}(0.70 \text{ m}) = 58.9 \text{ cm}$$

$$\Delta L = L_A - L_C = 11.1 \text{ cm}$$

19-26

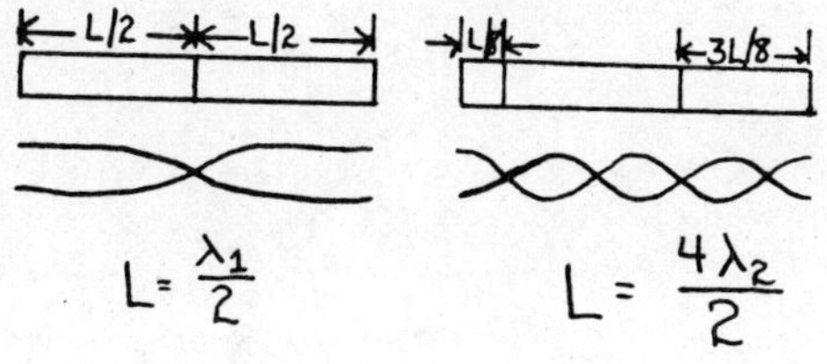

The clamp can be placed a distance of either L/8 or 3L/8 from either end.

19-27 $v_1 = \sqrt{\frac{F}{\mu_1}} \qquad v_2 = \sqrt{\frac{F}{\mu_2}}$

$$\lambda_1 = \frac{2}{3} L_1 \qquad \lambda_2 = L_2$$

$$\nu_1 = \frac{v_1}{\lambda_1} = \nu_2 = \frac{v_2}{\lambda_2}$$

$$\frac{3}{2L_1}\sqrt{\frac{F}{\mu_1}} = \frac{1}{L_2}\sqrt{\frac{F}{\mu_2}}$$

$$\frac{L_2}{L_1} = \frac{2}{3}\sqrt{\frac{\mu_1}{\mu_2}} = \frac{2}{3}\sqrt{\frac{4}{5}} = 0.596$$

19-28 $\mu = 0.10 \text{ kg/m} \qquad F = (50 \text{ kg})(9.8 \text{ m/s}^2) = 490 \text{ N}$

$$\lambda = 4L = 4 \text{ m}$$

(a) $T = \frac{\lambda}{v} = \lambda\sqrt{\frac{\mu}{F}} = (4 \text{ m})\sqrt{\frac{0.10 \text{ kg/m}}{490 \text{ N}}} = 0.057 \text{ s}$

(b) $A = \frac{v_\perp}{\omega} = \frac{v_\perp T}{2\pi} = \frac{(15 \text{ m/s})(0.057 \text{ s})}{2\pi} = 0.136 \text{ m}$

(c) Using Eq. 19-30,

$$y(x,t) = A \sin kx \sin(\omega t - \phi_0)$$

$$y(L, \tfrac{T}{4}) = A = A \sin kL \sin(\tfrac{\omega T}{4} - \phi_0)$$

$$= A \sin \frac{2\pi L}{\lambda} \sin(\tfrac{\omega T}{4} - \phi_0)$$

$$= A \sin \frac{\pi}{2} \sin(\tfrac{\omega T}{4} - \phi_0)$$

Solving for ϕ_0,

$$\frac{\omega T}{4} - \phi_0 = \frac{\pi}{2}$$

$$\phi_0 = \frac{\omega T}{4} - \frac{\omega}{2} = \frac{\omega}{4}(\frac{2\pi}{\omega}) - \frac{\pi}{2} = 0$$

Thus,

$$y(x,t) = A \sin kx \sin \omega t$$

$$= A \sin(\pi x/2L) \sin \omega t$$

(d) Using Eq. 19-36,

$$E = \frac{1}{2}\mu \int_0^L [(\frac{\partial y}{\partial t})^2 + v^2(\frac{\partial y}{\partial x})^2]dx$$

$$\frac{\partial y}{\partial t} = A\omega \sin(\pi x/2L)\cos \omega t$$

$$\frac{\partial y}{\partial x} = \frac{\pi A}{2L} \cos(\pi x/2L)\sin \omega t$$

Thus,

$$E = \frac{1}{2} \mu A^2\omega^2 \int_0^L [\sin^2(\pi x/2L)\cos^2\omega t + \cos^2(\pi x/2L)\sin^2 \omega t]dx$$

$$= \frac{1}{4} \mu A^2\omega^2 L = \frac{1}{4}(0.10 \text{ kg/m})(0.136 \text{ m})^2[2\pi/(0.057 \text{ s})]^2(1 \text{ m})$$

$$= 5.62 \text{ J}$$

(e) $T_p = 2\pi\sqrt{\frac{\ell}{g}} = 2\pi\sqrt{\frac{1 \text{ m}}{9.8 \text{ m/s}^2}} = 2.01 \text{ s}$

$$\frac{T_p}{T} = \frac{2.01}{0.057} = 35.3$$

Since the pendulum period is far from resonance, it is reasonable to assume that the hanging ball remains stationary.

19-29 $y(x,0) = f(x) = \begin{cases} \frac{4h}{L} x & 0 \le x \le \frac{1}{4} L \\ \frac{4h}{3L} (L-x) & \frac{1}{4} L \le x \le L \end{cases}$

$$B_n = \frac{2}{L}\int_0^{\frac{1}{4}L} \frac{4h}{L} x \sin(\frac{n\pi}{L} x)dx + \frac{2}{L}\int_{\frac{1}{4}L}^{L} \frac{4h}{3L}(L-x)\sin(\frac{n\pi}{L} x)dx$$

$$= \frac{32 h}{3n^2\pi^2}\sin \frac{n\pi}{4}$$

$$y(x,t) \cong \sum_{n=1}^{4} \frac{32 h}{3n^2\pi^2} \sin \frac{n\pi}{4} \sin\left(\frac{n\pi}{L} x\right)\cos\left(\frac{n\pi v}{L} t\right)$$

$$= \frac{32h}{3\pi^2}\left[\frac{\sqrt{2}}{2} \sin\left(\frac{\pi}{L} x\right) \cos\left(\frac{\pi v}{L} t\right) + \frac{1}{4} \sin\left(\frac{2\pi}{L} x\right)\cos\left(\frac{2\pi v}{L} t\right) - \frac{1}{9\sqrt{2}} \sin\left(\frac{3\pi}{L} x\right) \cos\left(\frac{3\pi v}{L} t\right)\right.$$

$$\left. + \frac{1}{25\sqrt{2}} \sin\left(\frac{5\pi}{L} x\right)\cos\left(\frac{5\pi v}{L} t\right)\right]$$

The expansion resembles that of Example 19-10 because both expressions describe a triangular wave. The significant second harmonic is present because the peak of the triangle is off-center.

19-30 $y(x,0) = f(x) = \begin{cases} \frac{4h}{L}(\frac{L}{4} - x) & 0 \le x \le \frac{1}{2} L \\ \frac{4h}{L}(x - \frac{3L}{4}) & \frac{1}{2} L \le x \le L \end{cases}$

From the shape of the curve it can be seen that

$$B_n = 0$$

The Fourier coefficients are

$$A_m = \frac{2}{L}\int_0^{\frac{1}{2}L} \frac{4h}{L}\left(\frac{L}{4} - x\right) \cos\left(\frac{2\pi m}{L} x\right)dx + \frac{2}{L}\int_{\frac{1}{2}L}^{L} \frac{4L}{L}\left(x - \frac{3L}{4}\right)\cos\left(\frac{2\pi m}{L} x\right)dx$$

$$= \frac{4\ h}{\pi^2 m^2}(1 - \cos \pi m)$$

$$= \begin{cases} 0 & \text{m even} \\ \dfrac{8\ h}{\pi^2 m^2} & \text{m odd} \end{cases}$$

$$y(x,t) = \sum_{m=1}^{\infty} \frac{8h}{\pi^2 m^2} \cos \frac{2\pi m}{L}(x - vt) \qquad \text{m odd}$$

$$= \frac{8h}{\pi^2} \sum_{m=1}^{\infty} \frac{1}{m^2} \cos \frac{2\pi m}{L}(x - vt) \qquad \text{m odd}$$

19-31 Using Eq. 19-44,

$$y(x,t) = \sum_n B_n \sin\left(\frac{n\pi}{L} x\right) \cos \omega_n t$$

From Eq. 19-40b,

$$B_n = \frac{2}{L}\int_0^L f(x) \sin\left(\frac{\pi n}{L} x\right)dx$$

For $\varepsilon << L$, $f(x) \cong h$.

$$B_n = \frac{2h}{L}\int_0^L \sin\left(\frac{\pi n}{L} x\right)dx$$

$$= -\frac{2h}{n\pi} \cos\left(\frac{\pi n}{L} x\right)\Big|_0^L$$

$$= \frac{2h}{n\pi}(1 - \cos \pi n)$$

$$= \begin{cases} 0 & \text{n even} \\ \dfrac{4h}{n\pi} & \text{n odd} \end{cases}$$

Thus,

$$y(x,t) = \frac{4h}{\pi} \sum_n \frac{1}{n} \sin\left(\frac{n\pi}{L} x\right)\cos\left(\frac{n\pi v}{L} t\right) \qquad \text{n odd}$$

19-32 $y(\frac{1}{2} L,0) = 1 = \frac{4}{\pi} \sum_{n=1}^{\infty} \frac{1}{n} \sin \frac{\pi n}{2} \qquad$ n odd

$$= \frac{4}{\pi}\left(1 - \frac{1}{3} + \frac{1}{5} - \frac{1}{7} + \frac{1}{9} - \frac{1}{11} + \cdots\right)$$

$$\pi = 4\left(1 - \frac{1}{3} + \frac{1}{5} - \frac{1}{7} + \frac{1}{9} - \frac{1}{11} + \cdots\right)$$

When evaluated to the term 1/31,

$$\pi \cong 3.079$$

For alternating series, the error is approximately equal to the last term. Thus, to get an error of 0.001, n must equal approximately 10^3.

19-33 $e^{-\alpha t} = e^{-\frac{v\beta}{2}(vt)} = e^{-\frac{v\beta}{2}x}$

For x = 2.08 m

$$\frac{1}{2} = e^{-\frac{v\beta}{2}(2.08\text{ m})} = e^{-\frac{\beta}{2}(200\text{ m/s})(2.08\text{ m})}$$

$$\beta = \frac{2\ \ell n\ 2}{(200\text{ m/s})(2.08\text{ m})} = 0.0033\text{ m}^{-2}\cdot\text{s}$$

$$\alpha = \frac{v^2\beta}{2} = \frac{(200\text{ m/s})^2(0.0033\text{ m}^{-2}\cdot\text{s})}{2} = 66\text{ s}^{-1}$$

19-34 Using Eq. 19-45,

$$v_g = v - \lambda\frac{dv}{d\lambda} = v_o - \frac{\lambda}{2}\sqrt{\frac{g}{2\pi\lambda_o}} = \sqrt{\frac{g\lambda_o}{2\pi}} - \frac{1}{2}\sqrt{\frac{g\lambda_o}{2\pi}} = \frac{1}{2}v_o$$

19-35 (a) $\frac{\partial y}{\partial x} = -2\sigma(x - vt)Ae^{-\sigma(x - vt)^2}$

Evaluated at t = 0,

$$\frac{\partial y}{\partial x} = -2\sigma xAe^{-\sigma x^2} = -2(10\text{ m}^{-2})(0.03\text{ m})$$
$$= -2(10)(3 \times 10^{-2})xe^{-10x^2}$$

This is small everywhere.

$$\frac{\partial^2 y}{\partial x^2} = 4\sigma^2(x - vt)^2Ae^{-\sigma(x - vt)^2} - 2\sigma Ae^{-\sigma(x - vt)^2}$$

Evaluated at t = 0, x = 0,

$$\frac{\partial^2 y}{\partial x^2} = -2\sigma A$$

$$R = \frac{-1}{2\sigma A} = \frac{-1}{2(10\text{ m}^{-2})(0.03\text{ m})} = -1.67\text{ m}$$

(b) $a_c = \frac{v^2}{R} = \frac{(10\text{ m/s})^2}{1.67\text{ m}} = 59.9\text{ m/s}^2 = 6.11\text{ g}$

(c) $F_c = ma_c = (0.010\text{ kg})(59.9\text{ m/s}^2) = 0.60\text{ N}$

(d) $F_s = \mu v^2 = (0.50\text{ kg/m})(10\text{ m/s})^2 = 50\text{ N}$

$F_c = \frac{0.60}{50}F_s = 0.012\ F_s$

19-36

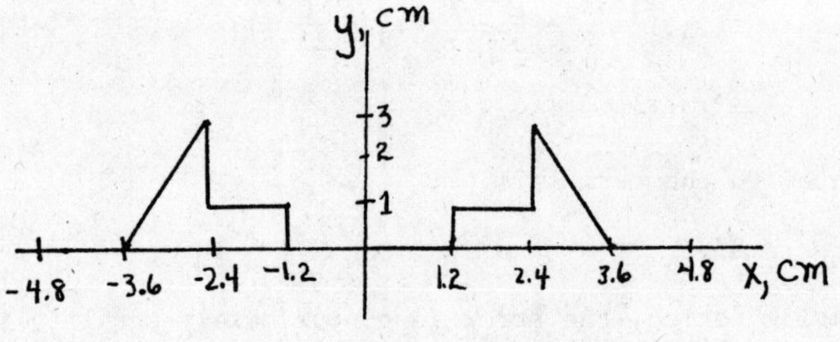

19-37 $v = \sqrt{\frac{F}{\mu}} = \sqrt{\frac{As}{\rho A}} = \sqrt{\frac{s}{\rho}}$

$$v_{max} = \sqrt{\frac{2.7 \times 10^9 \text{ Pa}}{7.86 \times 10^3 \text{ kg/m}^3}} = 586 \text{ m/s}$$

19-38 $F = \mu v^2 = (\rho \frac{\pi d^2}{4})v^2 = (8.93 \times 10^3 \text{ kg/m}^3)(1.5 \times 10^{-3} \text{ m})^2\pi(200 \text{ m/s})^2/4$

$= 631$ N

19-39
$$\begin{aligned} y(x,t) &= 0.20 \sin(80\pi t - 4\pi x) \\ &= 0.20 \cos[4\pi(x - 20t) + \frac{\pi}{2}] \\ &= 0.20 \cos(4\pi x - 80\pi t + \frac{\pi}{2}) \\ &= 0.20 \cos[2\pi(\frac{x}{1/2} - \frac{t}{1/40}) + \frac{\pi}{2}] \end{aligned}$$

$v = 20$ m/s, $\lambda = 0.50$ m, $\omega = 80\pi$ rad/s, $\nu = 40$ Hz,

$k = 4\pi \text{ m}^{-1}$, $T = 0.025$ s, $\phi_o = \frac{\pi}{2}$

19-40 Comparing Eq. 19-18 with Eq. 19-2, it can be seen that the longitudinal displacement from equilibrium of the rod material can be written as follows:

$$\xi(x,t) = \Xi \cos(kx - \omega t + \phi_o)$$

$$\frac{\partial\xi}{\partial t} = \Xi\omega \sin(kx - \omega t + \phi_o)$$

$$\frac{\partial\xi}{\partial x} = -\Xi k \sin(kx - \omega t + \phi_o)$$

$$v_{particle} = \frac{\partial\xi}{dt} = -\frac{\omega}{k}\frac{\partial\xi}{\partial x} = -\frac{\omega}{k}\frac{s}{Y} = -v\frac{s}{Y}$$

19-41 (a) For $h >> \lambda$, $\tanh \frac{2\pi h}{\lambda} \cong 1$.

$$v \cong \sqrt{\frac{g\lambda}{2\pi} + \frac{2\pi\gamma}{\rho\lambda}}$$

Since $\lambda >> \sqrt{\gamma/\rho g}$,

$$\frac{g\lambda}{2\pi} >> \frac{2\pi\lambda}{\rho\lambda}$$

Thus,

$$v \cong \sqrt{\frac{g\lambda}{2\pi}}$$

(b) For $h << \lambda$, $\tanh \frac{2\pi h}{\lambda} \cong \frac{2\pi h}{\lambda}$. Also, as in part (a), the second term in parentheses is much smaller than the first term.

$$v \cong \sqrt{\frac{g\lambda}{2\pi}(\frac{2\pi h}{\lambda})} = \sqrt{gh}$$

(c) Once again, $\tanh \frac{2\pi h}{\lambda} \cong 1$. Since $\lambda << \sqrt{\gamma/\rho g}$, the first term in parentheses is much smaller than the second

$$v \cong \sqrt{\frac{2\pi\gamma}{\rho\lambda}}(1) = \sqrt{\frac{2\pi\gamma}{\rho\lambda}}$$

19-42 Since $h >> \lambda$,

$$v = \sqrt{\frac{g\lambda}{2\pi} + \frac{2\pi\gamma}{\rho\lambda}}$$

$$= \sqrt{\frac{(9.8)(0.03)}{2\pi} + \frac{2\pi(7.6 \times 10^{-2})}{(10^3)(0.03)}}$$

$$= 0.25 \text{ m/s} = 25 \text{ cm/s}$$

19-43 $\lambda >> \sqrt{\gamma/\rho g}$

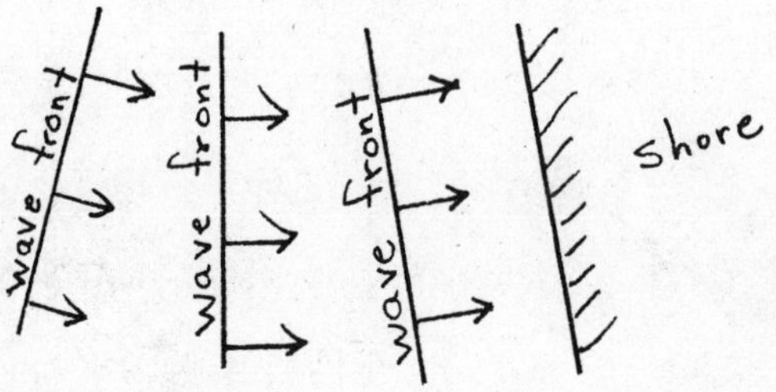

Far from shore, $h >> \lambda$.

$$v_\infty \cong \sqrt{\frac{g\lambda}{2\pi}} = 1.77 \text{ m/s}$$

At $h = 0.5$ m,

$$v_{0.5} = v\sqrt{\tanh\frac{2\pi h}{\lambda}} = (1.77 \text{ m/s})\sqrt{\tanh\frac{\pi}{2}} = 1.70 \text{ m/s}$$

At $h = 0.1$ m,

$$v_{0.1} = (1.77 \text{ m/s})\sqrt{\tanh\frac{\pi}{10}} = 0.98 \text{ m/s}$$

$$T_\infty = \frac{\lambda}{v_\infty} = 1.13 \text{ s}$$

$$T_{0.5} = \frac{\lambda}{v_{0.5}} = 1.18 \text{ s}$$

$$T_{0.1} = \frac{\lambda}{v_{0.1}} = 2.04 \text{ s}$$

19-44 $A_r^2 = \left(\frac{k_1 - k_2}{k_1 + k_2}\right)^2 A_i^2$

$$A_t^2 = \frac{4k_1^2}{(k_1 + k_2)}A_i^2$$

$$\bar{P}_r + \bar{P}_t = \frac{1}{2} F\omega A_i^2 \left[k_1 \frac{(k_1 - k_2)^2}{(k_1 + k_2)^2} + k_2 \frac{4k_1^2}{(k_1 + k_2)^2}\right]$$

$$= \frac{1}{2} F\omega A_i^2 k_1 = \bar{P}_i$$

19-45 $\mu = \frac{4.3 \times 10^{-3} \text{ kg}}{0.70 \text{ m}} = 6.14 \times 10^{-3} \text{ kg/m}$

$$v = \lambda\nu = 2(0.70 \text{ m})(261.6 \text{ Hz}) = 366 \text{ m/s}$$

$$F = \mu v^2 = (6.14 \times 10^{-3}\ \text{kg/m})(366\ \text{m/s})^2 = 822\ \text{N}$$

19-46 $\lambda_G = 2(0.35\ \text{m}) = \dfrac{v}{\nu_G}$

$$\lambda_A = 2L_A = \frac{v}{\nu_A}$$

$$L_G - L_A = L_G - \frac{\nu_G}{\nu_A} L_G = L_G\left(1 - \frac{\nu_G}{\nu_A}\right) = (0.35\ \text{m})\left(1 - \frac{392}{440}\right) = 0.038\ \text{m}$$

The string must be shortened by 3.8 cm.

$$L_A = \frac{v}{2\nu_A} = \frac{1}{2\nu_A}\sqrt{\frac{F}{\mu}}$$

$$dL_A = \frac{dF}{4\nu_A\sqrt{F\mu}}$$

$$\frac{dL_A}{L_A} = \frac{1}{2}\frac{dF}{F}$$

$$\frac{dF}{F} = 2\frac{dL_A}{L_A} = 2\,\frac{0.6\ \text{cm}}{(35 - 3.8)\text{cm}} = 3.9\%$$

19-47 From symmetry there will be no cosine terms. Thus,

$$A_n = 0$$

Since the average value of the function is zero,

$$A_o = 0$$

$$\eta(u) = \begin{cases} \dfrac{2hu}{L} & 0 \le \eta \le L/2 \\ \dfrac{2h}{L}(u - L) & L/2 \le \eta \le L \end{cases}$$

$$B_n = \frac{4h}{L^2}\left[\int_o^{L/2} u \sin\left(\frac{2\pi n}{L}u\right)du + \int_{L/2}^{L} (u - L)\sin\left(\frac{2\pi n}{L}u\right)du\right]$$

$$= \frac{4h}{L^2}\left[\int_o^{L} u \sin\left(\frac{2\pi n}{L}u\right)du - \int_{L/2}^{L} L\sin\left(\frac{2\pi n}{L}u\right)du\right]$$

$$= \frac{4h}{L^2}\left\{\left(\frac{L^2}{4\pi^2 n^2}\sin\left(\frac{2\pi n}{L}u\right) - \frac{uL}{2\pi n}\cos\left(\frac{2\rho n}{L}u\right)\right)\Bigg]_o^L + \frac{L^2}{2\pi n}\cos\left(\frac{2\pi n}{L}u\right)\Bigg]_{L/2}^{L}\right\}$$

$$= -\frac{2h}{\pi n}\cos n\pi = \begin{cases} -\dfrac{2h}{n\pi} & n \text{ even} \\ \dfrac{2h}{n\pi} & n \text{ odd} \end{cases}$$

$$\eta(x,t) = \eta(u) = \sum_{n=1}^{\infty}\frac{2h}{n\pi}(-1)^{n+1}\sin\left[\frac{2\pi n}{L}(x - vt)\right]$$

$$= \frac{2h}{\pi}\sum_{n=1}^{\infty}\frac{(-1)^{n+1}}{n}\sin\left[\frac{2\pi n}{L}u\right]$$

19-48 (a) For $h >> \lambda$,

$$v^2 = \left(\frac{g\lambda}{2\pi} + \frac{2\pi\gamma}{\rho\lambda}\right)\tanh \frac{2\pi h}{\lambda} \cong \frac{g\lambda}{2\pi} + \frac{2\pi\gamma}{\rho\lambda}$$

$$\frac{d(v^2)}{d\lambda} = \frac{g}{2\pi} - \frac{2\pi\gamma}{\rho\lambda_o^2} = 0$$

$$\lambda_o = 2\sqrt{\frac{\gamma}{\rho g}}$$

(b) $$\lambda_o = 2\pi\sqrt{\frac{7.6 \times 10^{-2}\ \text{N/m}}{(10^3\ \text{kg/m}^3)(9.8\ \text{m/s}^2)}} = 1.75 \times 10^{-2}\text{m} = 1.75\ \text{cm}$$

(c) $$v_g = v - \lambda\frac{dv}{d\lambda} = v = \sqrt{\frac{g\lambda_o}{2\pi} + \frac{2\pi\gamma}{\rho\lambda_o}} = \left(\sqrt{\frac{g\gamma}{\rho}} + \frac{g\gamma}{\rho}\right)^{\frac{1}{2}} = \left(2\sqrt{\frac{g\lambda}{\rho}}\right)^{\frac{1}{2}}$$

$$= \left(2\sqrt{\frac{(9.8\ \text{m/s}^2)(7.6 \times 10^{-2}\ \text{N/m})}{10^3\ \text{kg/m}^3}}\right)^{\frac{1}{2}} = 0.23\ \text{m/s}$$

Since $v_g = v$, dispersion does not occur.

20-1 $\dfrac{v_{38}}{v_{-28}} = \dfrac{\sqrt{273 + 38}}{\sqrt{273 - 28}} = 1.13$

20-2 $v = \sqrt{\dfrac{\gamma p_o}{\rho_o}} = \sqrt{\dfrac{1.67(1.013 \times 10^5 \text{ Pa})}{0.1786 \text{ kg/m}^3}} = 973 \text{ m/s}$

20-3 $\Delta t = L\left(\dfrac{1}{v_{air}} - \dfrac{1}{v_{Cu}}\right) = L\,\dfrac{v_{Cu} - v_{air}}{v_{air}\, v_{Cu}}$

$$L = \frac{v_{air}\, v_{Cu}}{v_{Cu} - v_{air}}\,\Delta t = \frac{(331.5 \text{ m/s})(3.75 \times 10^3 \text{ m/s})}{(3750 - 331.5)\text{m/s}}\,(6.4 \times 10^{-3} \text{ s})$$

$$= 2.33 \text{ m}$$

20-4 $v_\ell = \sqrt{Y/\rho} = \sqrt{(12.4 \times 10^{10} \text{ Pa})/(8.93 \times 10^3 \text{ kg/m}^3)} = 3726 \text{ m/s}$

$$v'_\ell = \sqrt{\frac{\beta + \frac{4}{3}\mu}{\rho}} = \sqrt{\frac{(13.1 + \frac{4}{3} \times 4.5) \times 10^{10} \text{ Pa}}{8.93 \times 10^3 \text{ kg/m}^3}} = 4625 \text{ m/s}$$

$$v'_t = \sqrt{\mu/\rho} = \sqrt{(4.5 \times 10^{10} \text{ Pa})/(8.93 \times 10^3 \text{ kg/m}^3)} = 2245 \text{ m/s}$$

20-5 (a) Using Eq. 20-9,

$$I = \frac{1}{2} A^2 k\beta_a \omega = \frac{1}{2}\frac{(\delta p_o)^2}{k\beta_a}\omega = \frac{1}{2}\frac{(\delta p_o)^2}{v\rho_o}$$

Thus,

$$\delta p_o = \sqrt{2(10^{-2} \text{ W/m}^2)(331.5 \text{ m/s})(1.29 \text{ kg/m}^3)} = 2.92 \text{ N/m}^2$$

(b) $\delta p_o = Ak\beta_a = Akv^2\rho_o = 2\pi A \dfrac{v^2}{\lambda}\rho_o$

$$A = \frac{\lambda\, \delta p_o}{2\pi v^2 \rho_o} = \frac{(0.80 \text{ m})(2.92 \text{ N/m}^2)}{2\pi(331.5 \text{ m/s})^2(1.29 \text{ kg/m}^3)} = 2.62 \times 10^{-6} \text{ m}$$

20-6 (a) $\nu_o = \dfrac{v}{2\pi}\sqrt{\dfrac{\pi r^2}{V(\ell + 8r/3\pi)}}$

$$= \frac{331.5 \text{ m/s}}{2\pi}\sqrt{\frac{\pi(2.5 \times 10^{-2} \text{ m})^2}{(2 \times 10^{-3} \text{ m}^3)(4.0 + 2.1)(10^{-2} \text{ m})}} = 211.7 \text{ Hz}$$

(b) $V_c = \dfrac{\nu_o^2}{\nu_c^2} V = \left(\dfrac{211.7}{261.6}\right)^2 (2 \times 10^{-3} \text{ m}^3) = 1.31 \times 10^{-3} \text{ m}^3$

$$V_{liquid} = V - V_c = (2.00 - 1.31) \times 10^{-3} \text{ m}^3 = 0.69 \times 10^{-3} \text{ m}^3 = 0.69 \text{ liters}$$

20-7 (a) $\nu_o = \dfrac{v}{\lambda} = \dfrac{v}{4L} = \dfrac{331.5 \text{ m/s}}{4(4.88 \text{ m})} = 17.0 \text{ Hz}$

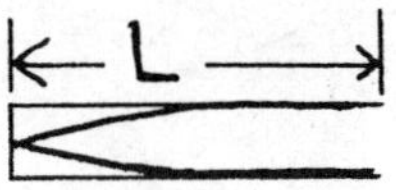

(b) $\nu_o = \dfrac{v}{\lambda} = \dfrac{v}{2L} = 34.0 \text{ Hz}$

(c) For the closed pipe,

$$\nu = \frac{v(20° \text{ C})}{v(0° \text{ C})}\,\nu_o = \sqrt{1 + 20/273}\;\nu_o = 17.6 \text{ Hz}$$

For the open pipe,

$$\nu = \sqrt{1 + 20/273}\,\nu_0 = 35.2 \text{ Hz}$$

20-8 $L_{open} = \frac{\lambda}{2} = \frac{v}{2\nu} = \frac{331.5 \text{ m/s}}{2(261.6 \text{ Hz})}$

$= 0.634$

$L_{closed} = \frac{3\lambda}{4} = \frac{3v}{4\nu} = \frac{3(331.5 \text{ m/s})}{4(261.6 \text{ Hz})}$

$= 0.950 \text{ m}$

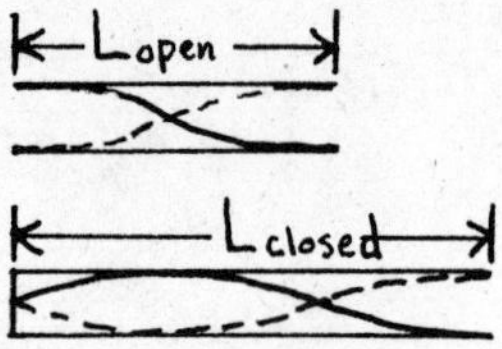

20-9 $L = \frac{1}{2}\lambda = (96.6 - 58.0)\text{cm} = 38.6 \text{ cm}$

$v = \lambda\nu = 2(0.386 \text{ m})(440 \text{ Hz}) = 339.7 \text{ m/s}$

$\left(\frac{v}{v_0}\right)^2 = 1 + (T/273)$

$T = 273[(339.7/331.5)^2 - 1] = 13.7° \text{ C}$

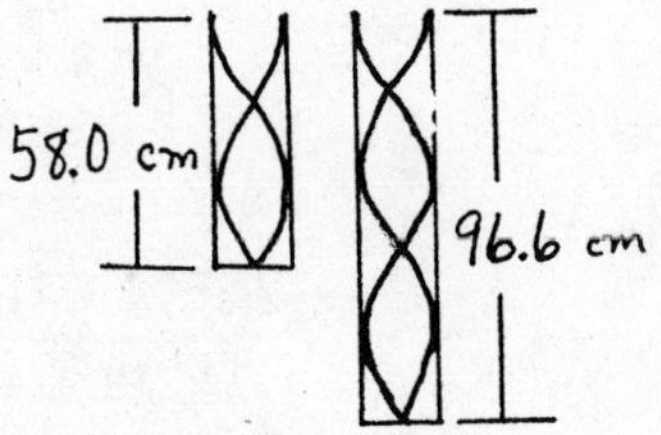

20-10 $\nu = \frac{v}{\lambda} = v_0\sqrt{1 + T/273}$

$\nu = \frac{v}{\lambda} = \frac{v_0\sqrt{1 + T/273}}{kL(1 + \alpha T)} = \nu_0 \frac{\sqrt{1 + T/273}}{1 + \alpha T}$

$\cong \nu_0\left(1 + \frac{T}{2(273)}\right)(1 - \alpha T)$

$\cong \nu_0\left(1 + \frac{T}{2(273)} - \alpha T\right)$

For $\alpha = \frac{1}{2(273)} = 1.83 \times 10^{-3}$ the scheme will work. This value for α is too large for the design to be practical.

20-11 (a) In ξ and $v_g\left(= \frac{\partial \xi}{\partial t}\right)$, nodes occur when

$x = \frac{nL}{2} \qquad n = 0, 1, 2, 3....$

Antinodes occur when

$x = \frac{nL}{4} \qquad n = 1, 3, 5....$

In δp and $\delta \rho$, which are both proportional to $\frac{\partial \xi}{\partial x}$, nodes occur when

$x = \frac{nL}{4} \qquad n = 1, 3, 5....$

Antinodes occur when

$x = \frac{nL}{2} \qquad n = 0, 1, 2, 3....$

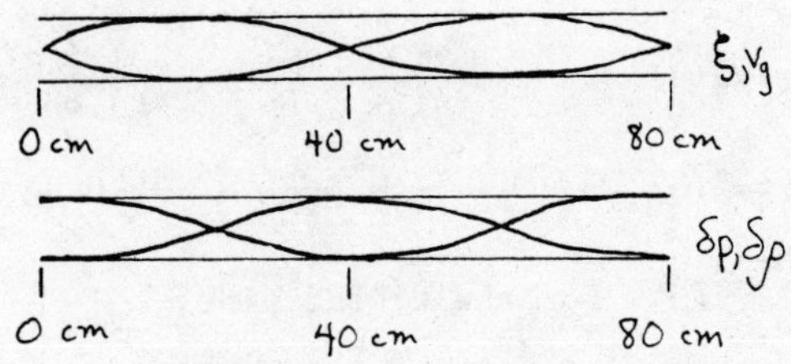

(b) Since there are nodes at both ends in ξ and v_g, the pipe is closed at both ends.

(c) $\lambda = L = 80$ cm

$\nu = \frac{v}{\lambda} = 414.4$ Hz

The fundamental frequency for this pipe, since it is closed on both ends, is

$\nu_o = \frac{\nu}{2} = 207.2$ Hz

(d) $$\delta p(x,t) = -v^2\rho_o \frac{\partial\xi}{\partial x} = -v^2\rho_o A \frac{2\pi}{L} \cos \frac{2\pi}{L} \sin(\frac{2\pi v}{L} t + \frac{\pi}{4})$$

$$\delta p(\frac{L}{2}, 0) = (2 \times 10^{-3})p_o = v^2\rho_o A \frac{2\pi}{L}(\frac{\sqrt{2}}{2})$$

$$A = \frac{(2 \times 10^{-3})p_o L}{\sqrt{2}\,\pi v^2\rho_o} = \frac{(2 \times 10^{-3})(1.013 \times 10^5 \text{ Pa})(0.80 \text{ m})}{\sqrt{2}\,\pi(331.5 \text{ m/s})^2(1.29 \text{ kg/m}^3)} = 0.257 \text{ mm}$$

(e) $$K(t) = \frac{1}{2}\rho_o S\int_0^L (\frac{\partial\xi}{\partial t})^2 dx$$

$$= \frac{1}{2}\rho_o SA^2 \frac{4\pi^2 v^2}{L^2} \cos^2(\frac{2\pi v}{L}t + \frac{\pi}{4})\int_0^L \sin^2 \frac{2\pi}{L} x\, dx$$

$$= \frac{\rho_o SA^2\pi^2 v^2}{L} \cos^2(\frac{2\pi v}{L} t + \frac{\pi}{4})$$

$$U(t) = \frac{1}{2}\rho_o Sv^2\int_0^L (\frac{\partial\xi}{\partial x})^2 dx$$

$$= \frac{1}{2}\rho_o Sv^2A^2 \frac{4\pi^2}{L^2} \sin^2(\frac{2\pi v}{L} t + \frac{\pi}{4})\int_0^L \cos^2 \frac{2\pi}{L} x\, dx$$

$$= \frac{\rho_o SA^2\pi^2 v^2}{L} \sin^2(\frac{2\pi v}{L} t + \frac{\pi}{4})$$

$$E = \rho_o SA^2\pi^2 v^2/L$$

$$= \frac{(1.29 \text{ kg/m}^3)(25 \times 10^{-4} \text{ m}^2)(0.257 \times 10^{-3} \text{ m})^2\pi^2(331.5 \text{ m/s})^2}{0.80 \text{ m}}$$

$$= 288.8\ \mu\text{J}$$

20-12 Using Eq. 20-13b,

$$\overline{P} = 2\pi B^2 v\rho_o\omega^2 = 2\pi A^2 r^2 v\rho_o\omega^2$$

$$= 2\pi(0.002 \text{ m})^2(0.15 \text{ m})^2(343.4 \text{ m/s})(1.20 \text{ kg/m}^3)(2\pi \times 1200 \text{ s}^{-1})^2$$

$$= 13.25 \text{ kW}$$

20-13 $$A_2 = \sqrt{\frac{P_2}{P_1}}\frac{r_1}{r_2} A_1 = \frac{10\sqrt{2}}{18}(1.5 \text{ mm}) = 1.18 \text{ mm}$$

20-14 From Example 20-5,

$$\nu_{n_x n_y n_z} = \frac{1}{2} v[(\frac{n_x}{L_x})^2 + (\frac{n_y}{L_y})^2 + (\frac{n_z}{L_z})^2]^{1/2}$$

$$= \frac{343.4}{2} [(\frac{n_x}{8})^2 + (\frac{n_y}{6})^2 + (\frac{n_z}{3})^2]^{1/2}$$

ν_{100} = 21.5 Hz		ν_{002} = 114.5 Hz	
ν_{010} = 28.6 Hz		ν_{210} = 51.6 Hz	
ν_{001} = 57.2 Hz		ν_{201} = 71.5 Hz	
ν_{110} = 35.8 Hz		ν_{120} = 61.1 Hz	
ν_{101} = 61.1 Hz		ν_{021} = 80.9 Hz	
ν_{011} = 64.0 Hz		ν_{211} = 77.1 Hz	
ν_{111} = 67.5 Hz		ν_{220} = 71.5 Hz	
ν_{200} = 42.9 Hz		ν_{300} = 64.4 Hz	
ν_{020} = 57.2 Hz		ν_{030} = 85.8 Hz	
		ν_{310} = 70.5 Hz	

The twelve lowest frequencies are, in Hz, 21.5, 28.6, 35.8, 42.9, 51.6, 57.2, 57.2, 61.1, 61.1, 64.0, 64.4, and 67.5.

20-15 $\nu_1 = \nu_2 \pm \nu_B = 440 \pm \frac{1}{0.200} = 435 \text{ or } 445 \text{ Hz}$

Since $\nu_1 \propto \sqrt{\kappa/m_1}$, ν_1 decreases. Since the beat frequency, ν_B, also decreases, the + sign is the correct choice in the above equation.

$\nu_1 = 445 \text{ Hz}$

20-16 $\nu = \frac{v}{\lambda} = \frac{1}{\lambda}\sqrt{F/\mu}$

$\Delta\nu = \frac{1}{2\lambda}\sqrt{1/\mu F}\,\Delta F$

$\frac{\Delta\nu}{\nu} = \frac{\Delta F}{2F}$

$\frac{\Delta F}{F} = 2\frac{\Delta\nu}{\nu} = \frac{2(4.1)}{261.6} = 3.1 \times 10^{-2} = 3.1\%$

20-17 $\lambda = \frac{331.5 \text{ m/s}}{3000 \text{ Hz}} = 0.11 \text{ m}$

(a) For maxima,

$d \sin\theta = n\lambda \qquad n = 0, 1, 2....$

$\theta_1 = \sin^{-1}\frac{\lambda}{d} = \sin^{-1}\frac{0.11 \text{ m}}{0.50 \text{ m}} = 12.7°$

$\theta_2 = \sin^{-1}\frac{2(0.11 \text{ m})}{0.50 \text{ m}} = 26.1°$

(b) For minima,

$d \sin\theta' = (n + \tfrac{1}{2})\lambda \qquad n = 0, 1, 2....$

$\theta'_0 = \sin^{-1}\frac{\lambda}{2d} = \sin^{-1}\frac{0.11 \text{ m}}{2(0.50 \text{ m})} = 6.3°$

$$\theta_1' = \sin^{-1}\frac{3\lambda}{2d} = \frac{3(0.11\text{ m})}{2(0.50\text{ m})} = 19.3°$$

(c) You should add amplitudes.

$$A = A_1 + A_2 = 2A_1$$

$$I \propto A^2 = 4A_1^2$$

$$I = 4I_1 \cong 4\left(\frac{0.5\text{ W}}{4\pi(5.0\text{ m})^2}\right) = 6.37 \times 10^{-3}\text{ W/m}^2$$

20-18 (a) $r_1^2 = x^2 + (y + 2\lambda)^2$

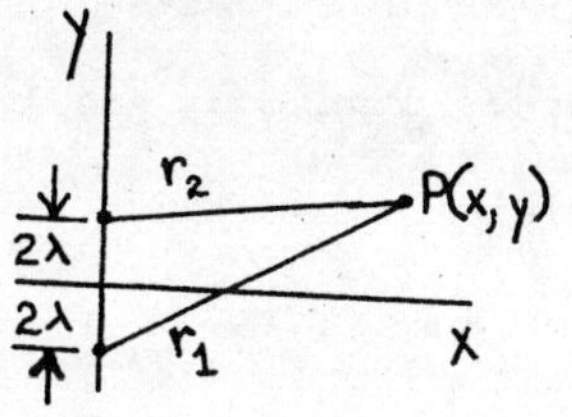

$$r_2^2 = x^2 + (y - 2\lambda)^2$$

$$\Delta r = \lambda = r_1 - r_2$$

$$r_1^2 = (\lambda + r_2)^2 = \lambda^2 + 2\lambda r_2 + r_2^2$$

$$x^2 + y^2 + 4\lambda y + 4\lambda^2 = \lambda^2 + 2\lambda\sqrt{x^2 + (y - 2\lambda)^2} + x^2 + y^2 - 4\lambda y + 4\lambda^2$$

$$8\lambda y = \lambda^2 + 2\lambda\sqrt{x^2 + (y - 2\lambda)^2}$$

$$8y - \lambda = 2\sqrt{x^2 + (y - 2\lambda)^2}$$

$$64y^2 - 16\lambda y + \lambda^2 = 4(x^2 + y^2 - 4\lambda y + 4\lambda^2)$$

$$60y^2 - 4x^2 = 15\lambda^2$$

This is an equation of a hyperbola.

(b) To find the angle of the asymptote,

$$\lim_{\substack{x\to\infty \\ y\to\infty}} \left(60\frac{y^2}{x^2}\right) = \lim_{x\to\infty}\left(\frac{15\lambda^2}{x^2} + 4\right)$$

$$60\frac{y^2}{x^2} = 4$$

$$\frac{y^2}{x^2} = \tan^2\theta = \frac{4}{60} = \frac{1}{15}$$

$$\theta = \tan^{-1}\frac{1}{\sqrt{15}} = 14.48°$$

$$\sin\theta = 0.25$$

From Eq. 20-15,

$$\sin\theta = \frac{n\lambda}{d} = \frac{\lambda}{d} = \frac{\lambda}{4\lambda} = 0.25$$

Alternately,

$$120y\,dy - 8x\,dx = 0$$

$$\frac{y}{x}\frac{dy}{dx} = \frac{1}{15}$$

For $y\to\infty$, $x\to\infty$, we know

$$\frac{y}{x} = \frac{1}{\sqrt{15}}$$

Thus,

$$\frac{dy}{dx} = \frac{\sqrt{15}}{15} = \frac{1}{\sqrt{15}}$$

20-19 (a) For an audience seated below the speaker, the angle of zero radiation intensity should be as close to zero as possible. Form Ex. 20-6,

$$\theta = \sin^{-1}\frac{\lambda}{w}$$

For w large, θ is small. Thus the long dimension of the speaker should be oriented vertically.

(b) $$\theta = \sin^{-1}\frac{\lambda}{w} = \sin^{-1}\frac{(331.5\ \text{m/s})}{(1000\ \text{Hz})(1.50\ \text{m})} = 12.8°$$

20-20 Using Eq. 20-16,

$$v_s = v(1 - \frac{\nu_o}{\nu}) = (331.5\ \text{m/s})(1 - \frac{440}{466.16}) = 18.6\ \text{m/s}$$

20-21 From Eq. 20-17,

$$\lambda = \frac{1}{\nu_o}(v - v_s \cos\alpha_s)$$

$$\lambda(0°) = \frac{1}{200}(331.5 - 15) = 1.58\ \text{m}$$

$$\nu(0°) = \frac{331.5\ \text{m/s}}{1.58\ \text{m}} = 210\ \text{Hz}$$

$$\lambda(45°) = \frac{1}{200}(331.5 - 15\cos 45°) = 1.60\ \text{m}$$

$$\nu(45°) = 207\ \text{Hz}$$

$$\lambda(90°) = \frac{1}{200}(331.5) = 1.66\ \text{m}$$

$$\nu(90°) = 200\ \text{Hz}$$

$$\lambda(135°) = \frac{1}{200}(331.5 - 15\cos 135°) = 1.71\ \text{m}$$

$$\nu(135°) = 194\ \text{Hz}$$

$$\lambda(180°) = \frac{1}{200}(331.5 + 15) = 1.73\ \text{m}$$

$$\nu(180°) = 191\ \text{Hz}$$

20-22 $\alpha_s' - \alpha_s = 15°$

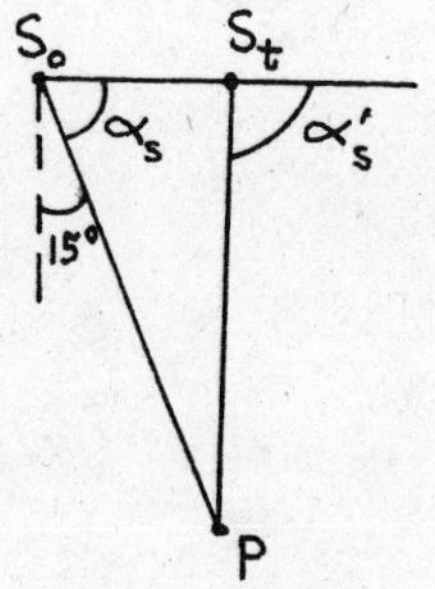

Using Eq. 20-19,

$$\sin(\alpha_s' - \alpha_s) = \frac{v_s}{v}\sin\alpha_s'$$

$$v_s = v\frac{\sin(\alpha_s' - \alpha_s)}{\sin\alpha_s'}$$

$$= v\frac{\sin 15°}{\sin 90°}$$

$$= (331.5 \text{ m/s})(0.259) = 85.8 \text{ m/s} = 309 \text{ km/h}$$

20-23 (a) Using Eq. 20-20,

$$\nu = (200 \text{ Hz}) \frac{331.5 + 18 \cos 75°}{331.5 - 15 \cos 75°} = 205.2 \text{ Hz}$$

(b) From Eq. 20-19,

$$\sin (\alpha_s' - \alpha_s) = \frac{v_s}{v} \sin \alpha_s'$$

$$\alpha_s' - \alpha_s = \sin^{-1} (\frac{15}{331.5} \sin 75°) = 2.5°$$

$$\alpha_s = 75° - 2.5° = 72.5°$$

$$\alpha_s' = 75° + 2.5° = 77.5°$$

(c) $$\nu = (200 \text{ Hz}) \frac{331.5 + 18 \cos 77.5°}{331.5 - 15 \cos 72.5°} = 205.1 \text{ Hz}$$

The correction makes a very small difference.

20-24 $$\beta = \sin^{-1} \frac{v}{v_s} = \sin^{-1} \frac{1}{1.38} = 46.4°$$

20-25 $$\frac{K_2}{K_1} = \frac{v_2^2}{v_1^2} = \frac{\sin^2 \beta_1}{\sin^2 \beta_2} = 1.50$$

The kinetic energy increased by 50%.

20-26 Since intensity is inversely proportional to the square of the distance,

$$I_4 = \frac{1}{100} I_{0.4}$$

The difference in sound intensity level is

$$\beta = 10 \log (\frac{I_{4 \text{ km}}}{I_{0.4 \text{ km}}}) = 10(-2) = -20$$

At 0.4 km,

$$\beta_{0.4} = 10 \log(\frac{10 \text{ W/m}^2}{10^{-12} \text{ W/m}^2}) = 130 \text{ dB}$$

$$\beta_4 = \beta_{0.4} + \beta = (130 - 20)\text{dB} = 110 \text{ dB}$$

Allowing for absorption of the wave over the distance travelled,

$$\beta_4' = \beta_4 - (7 \text{ dB/km})(3.6 \text{ km}) = 84.8 \text{ dB}$$

This is equivalent to the sound intensity level of very heavy traffic.

20-27 (a) $$\delta p_{mo} = \sqrt{2}\ \delta p_{rms} = A_o k \beta_a$$

$$A_o = \frac{\sqrt{2}\ \delta p_{rms}}{k\beta} = \frac{\sqrt{2}(\delta p_{rms}) v}{2\pi\nu(1.40\ p_o)}$$

$$= \frac{\sqrt{2}(2.07 \times 10^{-5}\ \text{Pa})(331.5\ \text{m/s})}{2\pi(1000\ \text{s}^{-1})(1.40 \times 1.013 \times 10^5\ \text{Pa})} = 1.1 \times 10^{-11}\ \text{m}$$

(b) Since the intensity has increased by 120 dB, it is 10^{12} times greater than at 0 dB. Since $A \propto I^{\frac{1}{2}}$,

$$A_{120} = 10^6\ A_o = 1.1 \times 10^{-5}\ \text{m}$$

20-28 (a) From Eq. 20-23,

$$I = \frac{1}{2} \frac{(0.5\ \text{Pa})^2}{(331.5\ \text{m/s})(1 + 10/273)^{\frac{1}{2}}(1.25\ \text{kg/m}^3)}$$

$$= 2.96 \times 10^{-4}\ \text{W/m}^2$$

From Eq. 20-24,

$$\beta = 20 \log \frac{\delta p_m}{\delta p_{rms}} = 20 \log \frac{\sqrt{2Iv\rho_o}}{\sqrt{I_o v\rho_o}}$$

$$= 20 \log \sqrt{\frac{2I}{I_o}} = 20 \log \sqrt{\frac{2(2.96 \times 10^{-4}\ \text{W/m}^2)}{10^{-12}\ \text{W/m}^2}}$$

$$= 87.7\ \text{dBA}$$

(b) The intensity in dB is given by

$$\beta = 10 \log \frac{I}{I_o} = 10 \log(2.96 \times 10^8) = 16.9\ \text{dB}$$

Using Figure 20-21, the loudness level is about 22 phons.

(c) $$A_{ear} = \frac{1}{3} A = \frac{1}{3} \frac{\delta p_m}{k\beta_a} = \frac{1}{3} \frac{(\delta p_m)v}{2\pi\nu\ \beta_a}$$

$$= \frac{1}{3} \frac{(0.5\ \text{Pa})(331.5\ \text{m/s})\sqrt{1 + 10/273}}{2\pi(3000\ \text{s}^{-1})(1.40 \times 1.013 \times 10^5\ \text{Pa})}$$

$$= 2.1 \times 10^{-8}\ \text{m}$$

(d) From Figure 20-22, the maximum strain would develop at about 12 mm from the oval window.

20-29 (a) Using Figure 20-22, the frequency is just under 2000 Hz.

(b) 16 mm from the oval window

(c) 400 Hz, 26 mm from the oval window

(d) $$\frac{\nu_1}{\nu_2} = \frac{2000}{400} = 5$$

$$\frac{d_1}{d_2} = \frac{35 - 16}{35 - 26} = 2.1$$

20-30 $\nu_c : \nu_E : \nu_G = 4:5:6$ (1)

$\nu_F : \nu_A : \nu_{C'} = 4:5:6$ (2)

$\nu_G : \nu_B : \nu_{D'} = 4:5:6$ (3)

For $\nu_C = 1$, ratio (1) gives

$$\nu_E = 5/4$$

$$\nu_G = 3/2$$

From ratio (3),

$$\nu_B = \frac{5}{4}\nu_G = \frac{5}{4}\left(\frac{3}{2}\right) = 15/8$$

$$\nu_{D'} = \frac{6}{5}\nu_B = \frac{6}{5}\left(\frac{15}{8}\right) = 9/4$$

Since

$$\nu_{D'} : \nu_D = 2:1,$$

$$\nu_D = \frac{1}{2}\nu_{D'} = 9/8$$

Since

$$\nu_{C'} : \nu_C = 2:1$$

$$\nu_{C'} = 2\nu_C = 2$$

From ratio (2),

$$\nu_A = \frac{5}{6}\nu_C = 5/3$$

$$\nu_F = \frac{4}{5}\nu_A = 4/3$$

20-31 (a)

m	n	ν, Hz	m	n	ν, Hz
0	±1	300	4	±2	200
0	±2	600	-1	±1	100
1	0	200	-1	±2	400
1	±1	500	-1	±3	700
1	±2	800	-2	±2	200
2	0	400	-2	±3	500
2	±1	100, 700	-2	±4	800
3	0	600	-3	±3	300
3	±1	300	-3	±4	600
4	0	800	-4	±3	100
4	±1	500	-4	±4	400

The frequencies heard are integer multiples of 100 Hz.

$= n(100\ \text{Hz}) \qquad n \quad 8$

(b)

m	n	ν, Hz	m	n	ν, Hz
0	±1	282	4	±2	236
0	±2	564	-1	±1	82
1	0	200	-1	±2	364
1	±1	482	-1	±3	646
1	±2	764	-2	±2	164
2	0	400	-2	±3	446
2	±1	682, 82	-2	±4	728
3	0	600	-3	±3	246
3	±1	318	-3	±4	528
4	0	800	-4	±3	46
4	±1	518	-4	±4	328

These frequencies do not follow a simple pattern; the frequencies produced when two frequencies related by a simple whole number are combined are much easier to define.

20-32 $v = v_o\sqrt{1 + T/273} \cong v_o\left(1 + \frac{T}{2(273)}\right)$

$$\frac{\Delta v}{v_o} \cong \frac{\Delta T}{2(273)} = \frac{34-11}{2(273)} = 0.042 = 4.2\%$$

20-33 $\nu_o = \frac{v}{2\pi}\sqrt{\frac{\pi r^2}{V\ell}} + \frac{v}{2\pi}\sqrt{\frac{\pi r^2}{(\pi r^2\ell)\ell}} = \frac{v}{2\pi}\sqrt{\frac{4}{L^2}} = \frac{v}{\pi L}$

$$\nu_{pipe} = \frac{v}{\lambda} = \frac{v}{4L}$$

The expression for ν_o used above assumes λ_o large compared to the dimensions of the pipe. In fact, this would not necessarily be the case; this could account for the discrepancy between the two expressions.

20-34 (a) $\nu_o = \frac{1}{2\pi}\sqrt{\frac{\kappa}{m}} = \frac{1}{2\pi}\sqrt{\frac{2000\ \text{N/m}}{0.010\ \text{kg}}} = 71.2\ \text{Hz}$

(b) From Ex. 20-3,

$$\kappa = v^2\rho_o\pi^2r^4/V = \frac{(331.5\ \text{m/s})^2(1.29\ \text{kg/m}^3)\pi^2(0.10\ \text{m})^4}{(0.5\ \text{m})(0.5\ \text{m})(0.2\ \text{m})}$$

$$= 2798\ \text{N/m}$$

(c) $\kappa' = (2000 + 2798)\ \text{N/m} = 4798\ \text{N/m}$

$$\nu_0' = \frac{1}{2\pi}\sqrt{\frac{4798\ \text{N/m}}{0.010\ \text{kg}}} = 110.2\ \text{Hz}$$

20-35 Assuming a Helmholtz resonator with a mass of excited air with effective length of $L = 8r/3\pi$,

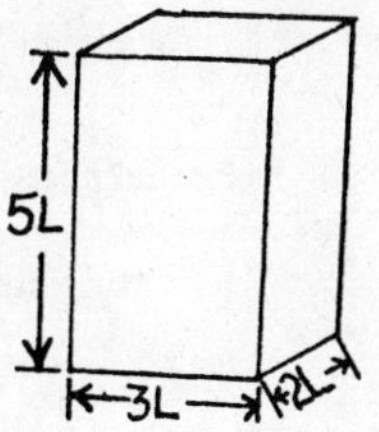

$$\nu_0 = \frac{v}{2\pi}\sqrt{\frac{3\pi^2 r}{8V}} = \frac{v}{2\pi}\sqrt{\frac{\pi^2 r}{80L^3}}$$

$$L^3 = \frac{v^2}{4\nu_0^2}\frac{r}{80} = \frac{(331.5\ \text{m/s})^2}{4(45\ \text{s}^{-1})^2}\frac{0.08\ \text{m}}{80} = 0.0136\ \text{m}$$

$$L = 0.239\ \text{m}$$

The dimensions of the cabinet are

$$48 \times 72 \times 120\ \text{cm}^3$$

20-36 $$\frac{\nu_1}{\nu_2} = \frac{v_1}{v_2}$$

$$\nu_1 = \frac{v_1}{v_2}\nu_2 = \frac{259}{331.5}(440\ \text{Hz}) = 343.8\ \text{Hz}$$

The note sounded would be close to F natural.

20-37 $$\frac{\nu_1}{\nu_2} = \frac{v_1}{v_2} = \sqrt{\frac{1 + T_1/273}{1 + T_2/273}}$$

For $T_2 = 0$,

$$\sqrt{1 + T_1/273} = \frac{277.18}{261.63}$$

$$T_1 = \left(\frac{277.18}{261.63}\right)^2 - 1$$

$$= 273\left[\left(\frac{277.18}{261.63}\right)^2 - 1\right] = 33.4°\ \text{C}$$

20-38 $$L_{effective} = \left(1 + \frac{1}{10}\right)(30.5\ \text{cm}) = 33.6\ \text{cm}$$

$$\lambda = 2L_{effective} = 67.2\ \text{cm}$$

$$\nu = \frac{v}{\lambda} = \frac{331.5\ \text{m/s}}{0.672\ \text{m}} = 493.3\ \text{Hz}$$

The note sounded is close to B natural.

20-39 For maxima,

$$d \sin\theta = \left(n + \frac{1}{2}\lambda\right)$$

For minima,

$$d \sin\theta = n\lambda$$

20-40 This problem is equivalent to that of an observer walking toward the source of a sound with the source moving at a velocity equal to the wind velocity given in the problem. The transformed observer velocity is

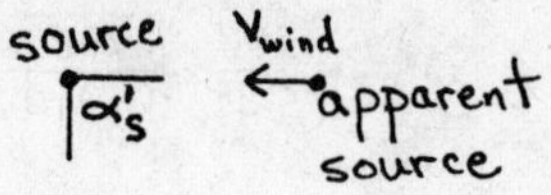

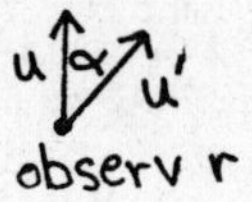

$$u' = \frac{u}{\cos \alpha}$$

Requiring the observer and the apparent source to arrive at the true source at the same time,

$$v_{wind}t = ut$$

Thus,

$$\tan \alpha = \frac{v_{wind}}{u}$$

Applying Eq. 20-20 to the equivalent problem,

$$\nu = \nu_0 \frac{v + u' \cos \alpha}{v - v_{wind} \cos \alpha_s}$$

$$= \nu_0 \frac{v + u}{v - v_{wind}(\cos 90°)}$$

$$= \nu_0 (1 + \frac{u}{v})$$

$$= (250 \text{ Hz})(1 + \frac{10}{331.5}) = 257.5 \text{ Hz}$$

Note that the solution is independent of wind velocity.

20-41 $\sin \beta = \frac{v}{v_s} = \frac{1}{N_M}$

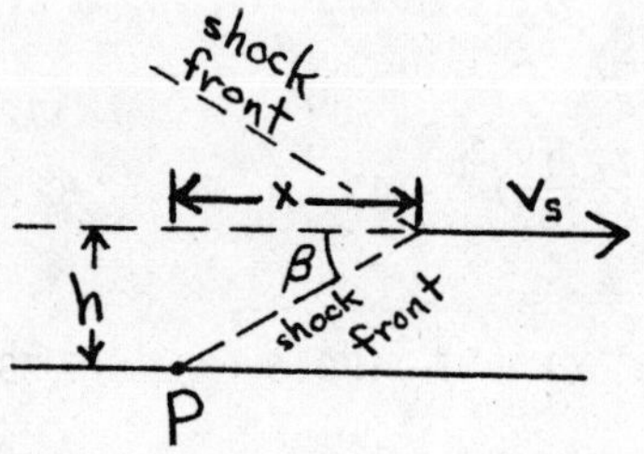

$h = v(12.8 \text{ s})$

$x = v_s(10 \text{ s})$

$\tan \beta = \frac{h}{x} = 1.28 \frac{v}{v_s} = \frac{1.28}{N_M}$

$\cos \beta = \frac{\sin \beta}{\tan \beta} = \frac{1}{1.28}$

$\beta = 38.6°$

$N_M = \frac{1}{\sin \beta} = 1.60$

20-42 For $u = 0$,

$$\nu = \frac{\nu_0}{1-(v_s/v)\cos \alpha_s}$$

$$\Delta\nu = \frac{-\nu_0}{[1-(v_s/v)\cos \alpha_s]^2}(\frac{v_s}{v} \sin \alpha_s)\Delta\alpha_s \cong -\nu_0(\frac{v_s}{v} \sin \alpha_s)\Delta\alpha_s$$

For v_s/v small,

$$\frac{v_s}{v}\sin \alpha'_s = \sin(\alpha'_s - \alpha_s) \cong \alpha'_s - \alpha_s$$

Also,

$$\Delta\alpha_s = \alpha'_s - \alpha_s$$

Thus,

$$\Delta\nu = -\nu_o(\alpha'_s - \alpha_s)^2$$

20-43 (a) $I = \frac{\overline{P}}{4\pi r^2} = \frac{10 \text{ W}}{4\pi(20 \text{ m})^2} = 0.00199 \text{ W/m}^2$

$$\beta = 10 \log\left(\frac{1.99 \times 10^{-3}}{10^{-12}}\right) = 93.0 \text{ dB}$$

(b) Using Eq. 20-23,

$$\delta p_m = \sqrt{2I v \rho_o}$$

$$= \sqrt{2(1.99 \times 10^{-3} \text{ W/m}^2)(331.5 \text{ m/s})(1.29 \text{ kg/m}^3)}$$

$$= 1.30 \text{ N/m}^2$$

(c) $\beta = 10 \log \frac{I}{I_o} = -3.0 \text{ dB}$

$$\frac{I}{I_o} = \left(\frac{R_o}{R}\right)^2 = 10^{-3/10} = 0.501$$

$$R = \frac{R_o}{\sqrt{0.501}} = \frac{20 \text{ m}}{\sqrt{0.501}} = 28.3 \text{ m}$$

21-1 The railway car frame is called S'. Observer in center of the car is 0'. The station frame is S. The observer in center of station is 0. The station rest frame is assumed to be the ether rest frame.

If the lightning strikes are simultaneous in S frame, then the observer 0' will see the flash from the front of the car before he sees the flash from the rear (assuming motion as given in the text). Now, since the observer 0' knows that he is not in the ether rest frame, he will say that signal from the front travelled faster than the signal from the rear of train.

By compensating for this difference in the speed of the light signals, 0' will see that both signals actually left their respective origins at the same time and therefore the events were simultaneous for him as it was for observer 0.

The postulates of special relativity are invoked to preserve the symmetries between different inertial frames.

21-2 Proper length L_o = 200 m, v = 275 km/h, ℓ_o = length of car in station frame. Ground observer "0" judges the flashes AA', BB' to be simultaneous.

Let S' be the frame of the railway car. Let S be the frame of the station.

$$\beta = \frac{v}{c} = (275\text{ km/h})\left(\frac{1\text{ h}}{3600\text{s}}\right)\left(\frac{1}{3 \times 10^8\text{ m/s}}\right) = 2.55 \times 10^{-7}$$

Therefore,

$$\gamma \cong 1$$

(a) Event A: $t_A = 0,\ x = \frac{\ell_o}{2}$ Event A': $x'_{A'} = \frac{L_o}{2}$

Event B: $t_B = 0,\ x = -\frac{\ell_o}{2}$ Event B': $x'_{B'} = -\frac{L_o}{2}$

$$ct' = \gamma[ct - \beta x]$$

$$c(t'_{B'} - t'_{A'}) = \gamma(c[t_B - t_A] - \beta[-\frac{\ell o}{2} - \frac{\ell o}{2}])$$

For $\Delta t = 0$,

$$c\Delta t' = \gamma\beta\ell_o, \quad \Delta t' = \gamma\frac{\beta}{c}\ell_o, \quad \ell_o = \frac{L_o}{\gamma} \cong L_o$$

$$\Delta t' = \frac{(2.55 \times 10^{-7})(200\text{ m})}{3 \times 10^8\text{ m/s}} = 1.7 \times 10^{-13}\text{s}$$

(b) 0' determines that clocks in S run more slowly than his own clocks.

$$\Delta t_o = \frac{\Delta t'}{\gamma} = \frac{\beta\ell_o}{c} = \frac{(2.55 \times 10^{-7})(200\text{ m})}{3 \times 10^8\text{ m/s}} = 1.7 \times 10^{-13}\text{ s}$$

21-3 (a) $$\Delta t' = \frac{\gamma\beta\ell_o}{c} = \frac{\beta L_o}{c}$$

$$= \frac{(0\cdot 6)(200\text{ m})}{3 \times 10^8\text{ m/s}} = 4 \times 10^{-7}\text{ s}$$

(b) $\Delta t_o = \frac{\Delta t'}{\gamma}$

$$\gamma = \frac{1}{\sqrt{1 - \beta^2}} = \frac{1}{0.8} = \frac{5}{4}$$

So,

$$\Delta t_o = \frac{4(4 \times 10^{-7} s)}{5} = 3.2 \times 10^{-7} \text{ s}$$

21-4 Two events in frame S: $(x_1, 0, 0, t_1)$, $(x_2, 0, 0, t_2)$

(a) Using Eqs. 21-8a,

$$x_1' - x_2' = \gamma([x_1 - x_2] - \beta c[t_1 - t_2])$$

Coincidence in S' means $(x_1' - x_2') = 0$

Therefore $(x_1 - x_2) = \beta c(t_1 - t_2)$

$$\beta c = u = \frac{(x_1 - x_2)}{(t_1 - t_2)}$$

(b) $u = \frac{x_1 - x_2}{t_1 - t_2}$, $t' = \gamma(t - \frac{\beta}{c}x)$

$$t_1' = \gamma\left(t_1 - \frac{1}{c^2}\left[\frac{x_1 - x_2}{t_1 - t_2}\right]x_1\right) \qquad t_2' = \gamma\left(t_2 - \frac{1}{c^2}\left[\frac{x_1 - x_2}{t_1 - t_2}\right]x_2\right)$$

$$t_1' - t_2' = \frac{\gamma}{(t_1 - t_2)}([t_1 - t_2]^2 - \frac{1}{c^2}[x_1 - x_2]^2)$$

$$\gamma = \frac{1}{(1 - \frac{u^2}{c^2})^{\frac{1}{2}}} = \frac{1}{\left(1 - \frac{1}{c^2}\frac{[x_1 - x_2]^2}{[t_1 - t_2]^2}\right)^{\frac{1}{2}}} = \frac{t_1 - t_2}{([t_1 - t_2]^2 - \frac{1}{c^2}[x_1 - x_2]^2)^{\frac{1}{2}}}$$

Plug γ in previous equation to get

$$t_1' - t_2' = ([t_1 - t_2]^2 - \frac{1}{c^2}[x_1 - x_2]^2)^{\frac{1}{2}}$$

(c) In order for two events to be spatially coincident in S',

$$u = \frac{x_1 - x_2}{t_1 - t_2}$$

Since $u < c$,

$$\frac{x_1 - x_2}{t_1 - t_2} < c$$

Thus, $(x_1 - x_2) < c(t_1 - t_2)$ is the new condition for coincidence in S'.

(d) $t_1' - t_2' = 0 = \gamma([t_1 - t_2] - \frac{\beta}{c}[x_1 - x_2])$

$$\frac{\beta}{c} = \frac{t_1 - t_2}{(x_1 - x_2)}\,; \quad u = \frac{c^2(t_1 - t_2)}{(x_1 - x_2)}$$

21-5 (a) From Example 21-5,

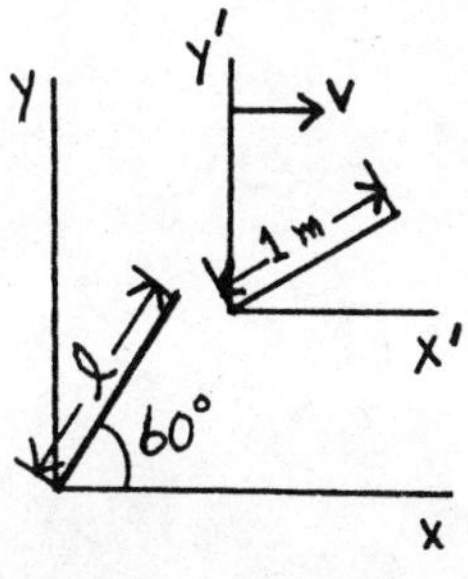

$$\tan\theta = \gamma \tan\theta_o$$

$$\gamma = \frac{\tan 60°}{\tan 30°} = 3$$

$$1 - \beta^2 = 1/\gamma^2 = 1/9$$

$$u = \beta c = c\sqrt{8/9} = 2.83 \times 10^8 \text{ m/s}$$

(b) $\ell = \ell_0[1 - \beta^2 \cos^2\theta_0]^{1/2}$

$= (1.0 \text{ m})(1 - \frac{8}{9} \times \frac{3}{4})^{1/2} = 0.577 \text{ m}$

21-6 $\beta = \frac{24}{25}$, distance = 10 L.Y.

Time to travel 10 L.Y. is,

$$t = \frac{d}{v} = \frac{10 \text{ L.Y.}}{(\frac{24}{25})c} = 10.4 \text{ y}$$

(a) According to astronaut, duration is

$$T' = \frac{2t}{\gamma} + 1 = (20 \text{ y})(1 - \frac{(24)^2}{(25)^2})^{1/2} + (1 \text{ y}) = (5.82 \text{ y}) + (1 \text{ y}) = 6.82 \text{ y}$$

(b) According to Earth Clock, duration of trip is

$$T = 2t + (1 \text{ y}) = 2(10.4 \text{ y}) + (1 \text{ y}) = 21.8 \text{ y}$$

(c) $T - T' = (21.82 - 6.82) \text{ y} = 15 \text{ y}$

The astronaut aged 15 y less.

21-7 Satellite at twice Earth's radius.

$$\frac{GMm_s}{(2R_E)^2} = \frac{m_s v^2}{(2R_E)}$$

$$\frac{GM}{R_E^2} = g$$

$$\left[\frac{g(2R_E)}{4}\right]^{1/2} = v = 5600 \text{ m/s}$$

$$\beta = \frac{v}{c} = 1.86 \times 10^{-5}, \quad \beta^2 = 3.48 \times 10^{-10}$$

$$v = \omega(2R_E) = \frac{(2\pi)(2R_E)}{T}$$

$$T = \frac{4\pi R_E}{5600 \text{ m/s}} = 1.43 \times 10^4 \text{ s}$$

T = time for satellite to go around the earth once, as measured on the earth

T' = time for satellite to go around the earth once, as measured by satelite

$$T' = \frac{T}{\gamma}, \quad \frac{1}{\gamma} = (1 - \beta^2)^{\frac{1}{2}} \approx 1 - \frac{1}{2}\beta^2$$

$$\Delta t = T(1 - \frac{1}{\gamma}) = \frac{T}{2}\beta^2 = \frac{(1.43 \times 10^4 \text{ s})}{2}(3.48 \times 10^{-10})$$

$$= 2.5 \times 10^{-6} \text{ sec}$$

The earth clock leads satellite clock by 2.5 ticks per revolution.

21-8 Show $x^2 - c^2t^2 = x'^2 - c^2t'^2$

$$x' = \gamma(x - \beta ct)$$

$$x'^2 = \gamma^2(x^2 + \beta^2c^2t^2 - 2x\beta ct)$$

$$ct' = \gamma(ct - \beta x)$$

$$c^2t'^2 = \gamma^2(ct^2 + \beta^2x^2 - 2\beta xct)$$

$$x'^2 - c^2t'^2 = \gamma^2(x^2[1-\beta^2] - c^2t^2[1 - \beta^2])$$

$$\gamma^2 = \frac{1}{[1-\beta^2]}$$

Therefore

$$x'^2 - c^2t'^2 = x^2 - c^2t^2$$

21-9 $u' = \dfrac{u + v}{1 + \frac{uv}{c^2}}$ In lab frame particles move in opposite directions at 0.6 c. If we are placed on one of the particles, then the other moves at

$$u' = \frac{0.6c + 0.6c}{1 + \frac{(0.6c)(0.6c)}{c^2}} = \frac{6/5}{1 + 9/25} = 0.88c$$

There is no violation of relativity when lab observer says that relative velocities of the two particles is 6/5 c.

21-10 $p = \gamma m_0 v = m_0 c$

$$\gamma v = (\frac{1}{v^2} - \frac{1}{c^2})^{-\frac{1}{2}}$$

So: $(\frac{1}{v^2} - \frac{1}{c^2})^{-\frac{1}{2}} = c$

$$\frac{1}{c^2} = \frac{1}{v^2} - \frac{1}{c^2}$$

$$\frac{1}{v^2} = \frac{2}{c^2}$$

$$v = \frac{c}{\sqrt{2}}$$

Given $E = K + m_0c^2 = \gamma m_0c^2$

for $v = \dfrac{c}{\sqrt{2}}$, $\gamma = (1 - \frac{v^2}{c^2})^{-\frac{1}{2}} = (1 - \frac{1}{2})^{-\frac{1}{2}} = \sqrt{2}$

$$K = m_0c^2(\gamma - 1) = m_0c^2(\sqrt{2} - 1) = 0.414\, m_0c^2$$

$$E = \sqrt{2}\, m_o c^2$$

21-11 Given $T = \frac{1}{2} m_o v^2$, $K = (\gamma - 1) m_o c^2$

We want v such that

$$\frac{K - T}{K} < 0.01, \qquad 1 - \frac{\frac{1}{2} m_o v^2}{(\gamma-1) m_o c^2} < 0.01$$

$$\frac{1}{2}\frac{v^2}{(\gamma-1)c^2} = \frac{1}{2}\frac{\beta^2}{(\gamma-1)} = \frac{1}{2}\left[\frac{1}{\gamma^2} + \frac{1}{\gamma}\right]$$

So,

$$0.99 - \frac{1}{2}\left[\frac{1}{\gamma^2} + \frac{1}{\gamma}\right] = 0$$

$$\frac{1}{\gamma^2} + \frac{1}{\gamma} - 1.98 = 0, \text{ but } \alpha = \frac{1}{\gamma}$$

$$\alpha^2 + \alpha - 1.98 = 0$$

$$\alpha_\pm = \frac{-1 \pm \sqrt{1+(1.98)(4)}}{2}.$$ Since $\gamma \geq 1$, choose α_+.

$$\alpha_+ = \frac{1}{\gamma} = 0.993$$

$$\beta^2 = 1 - \frac{1}{\gamma^2} = 1.39 \times 10^{-2}, \quad \beta = 1.18 \times 10^{-1}$$

$$v = 3.54 \times 10^7 \text{ m/s}$$

The velocity should be less than 3.54×10^7 m/s in order for the required condition to be satisfied.

21-12 Show velocity of composite particle is

$$V = \frac{\gamma v'}{(\gamma+1)} \qquad \gamma = \left(1 - \frac{v'^2}{c^2}\right)^{-\frac{1}{2}}$$

$$V_{CM} = \frac{\gamma_a m_o^a v' + m_o^b(0)}{\gamma_a m_o^a + m_o^b}$$

$$V_{CM} = \frac{\gamma_a v'}{\gamma_a + 1} \qquad \gamma_a = \left(1 - \frac{v^2}{c^2}\right)^{-\frac{1}{2}}$$

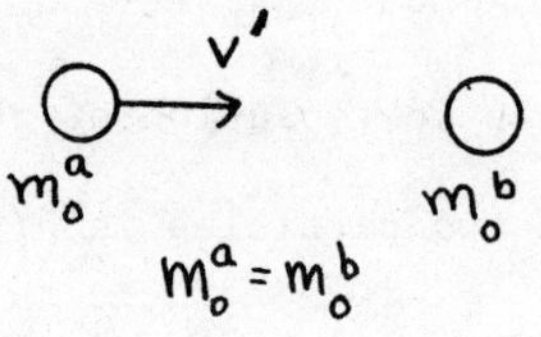

21-13 (a) Earth-Sun distance $= 1.5 \times 10^{11}$ m

Total area in spherical shell of above radius is

$$A = 4\pi(1.5 \times 10^{11} \text{ m})^2 = 2.83 \times 10^{23} \text{ m}^2$$

Total power radiated by sun is

$$P_R = (1.37 \frac{\text{kW}}{\text{m}^2})(2.83 \times 10^{23} \text{ m}^2) = 3.87 \times 10^{23} \text{ kW}$$

(b) $$P_R = c^2 \frac{dM_s}{dt}$$

$$\frac{dM_s}{dt} = \frac{3.87 \times 10^{26} \text{ W}}{9 \times 10^{16} \text{ m}^2/\text{s}^2} = 4.30 \times 10^9 \text{ kg/s}$$

$$\frac{\Delta M_s \text{(one year)}}{M_s} = \frac{(4.30 \times 10^9 \text{ kg/s})(3.16 \times 10^7 \text{ s})}{1.989 \times 10^{30} \text{ kg}} = 6.81 \times 10^{-14}$$

(c) $t = \frac{0.1\,M_s}{dM_s/dt} = \frac{0.1(1.989 \times 10^{30}\text{ kg})}{4.30 \times 10^{9}\text{ kg/s}}$

$= 1.47 \times 10^{12}$ y

21-14 $U + n \rightarrow Kr + Ba + 2n$

m(U) = 235.04392 u　　m(Kr) = 91.92569 u

m(n) = 1.00867 u　　m(Ba) = 141.91647 u

m(U) + m(n) = 236.05259u

m(Kr) + m(Ba) + 2m(n) = 235.85950u

$\Delta m = [m(u)+m(n)]-[(m(Kr)+m(Ba)+2\,m(n)] = (0.19309u)\left(\frac{1.66 \times 10^{-27}\text{ kg}}{1u}\right)$

$= 3.205 \times 10^{-28}$ kg

(a) Energy per event $= (3.205 \times 10^{-28}\text{ kg})(9 \times 10^{18}\ \frac{m^2}{s^2})$

$= 2.88 \times 10^{-9}$ J

(b) $(1\text{ kg})\left(\frac{6.024 \times 10^{26}\text{ u}}{1\text{ kg}}\right)\left(\frac{1}{235.04392\text{ u}}\right)$ = # of uranium atoms in 1 kg

$= 2.563 \times 10^{24}$ atoms

Total energy in 1 kg of uranium is

$E = (2.563 \times 10^{24})(2.88 \times 10^{-9}\text{ J})$

$= 7.38 \times 10^{15}$ J

(c) 2.5×10^{12} kW-h $= (2.5 \times 10^{15}\ \frac{J}{s})(3600\text{ s}) = 9 \times 10^{18}$ J

$9 \times 10^{18}\text{ J} = (0.32)(7.38 \times 10^{15}\text{ J/kg})(x\text{ kg})$

$x\text{ kg} = \frac{9 \times 10^{18}\text{ J}}{(0.32)(7.38 \times 10^{15}\text{J/kg}} = 3810.9$ kg

Therefore 3810.9 kg of uranium is needed to supply U.S. needs.

21-15 (a) Given $K = 8.0 \times 10^{-10}$ J

$\gamma - 1 = \frac{K}{m_o c^2} = \frac{8 \times 10^{-10}\text{ J}}{(9 \times 10^{-31}\text{ kg})(9 \times 10^{16}\text{ m}^2/\text{s}^2)} \simeq 10^4$

(b) $\gamma \sim 10^4$

$\beta^2 = 1 - \gamma^{-2} = 1 - 10^{-8}$

$\beta = 1 - \frac{1 \times 10^{-8}}{2}$, v = 299,999,998.5 m/s

(c) Earth-Moon distance $= 3.9 \times 10^{8}$ m = d

ct = d

$\Delta t = d\left(\frac{1}{c} - \frac{1}{v}\right) = (3.9 \times 10^{8}\text{ m})\left[\frac{1}{3 \times 10^{8}\text{ m/s}} - \frac{1}{299{,}999{,}998.5\text{ m/s}}\right]$

$c\Delta t = 1.99$ m = distance by which light wins!

(d) $\frac{m_s}{m_o} \simeq 10^4$

21-16 (a) $K = 8.0 \times 10^{-10}$ J

$$\gamma - 1 = \frac{K}{m_o c^2} = \frac{8 \times 10^{-10}\ \text{J}}{(1.67 \times 10^{-27}\ \text{kg})(9 \times 10^{16}\ \text{m}^2/\text{s}^2)} = 5.32$$

(b) $\gamma = 6.32$

$$\beta^2 = 1 - \frac{1}{\gamma^2} = 1 - \frac{1}{(6.32)^2} = 0.975, \quad \beta = 0.987$$

$$v = 296{,}224{,}021\ \text{m/s}$$

(c) $$\Delta t = d\left(\frac{1}{c} - \frac{1}{v}\right) = (3.9 \times 10^8\ \text{m})\left[\frac{1}{3 \times 10^8\ \text{m/s}} - \frac{1}{296{,}224{,}021\ \text{m/s}}\right]$$

$$c\Delta t = 4.97 \times 10^6\ \text{m}$$

(d) $E = \gamma m_o c^2 \qquad E_{REST} = m_o c^2$

$$\frac{E}{E_{REST}} = \frac{\gamma m_o c^2}{m_o c^2} = \gamma = 6.32$$

21-17 From Equation 21-19, after rearrangement,

$$t\left(\nu'\gamma + \frac{\nu' u}{c}\gamma \cos\theta' - \nu\right) + x\left(\frac{\nu}{c}\cos\theta - \frac{\gamma\nu' u}{c^2} - \frac{\nu'\gamma\cos\theta'}{c}\right)$$

$$+ y\left(\frac{\nu}{c}\sin\theta - \frac{\nu'}{c}\sin\theta'\right) = 0$$

Coefficient of "t" gives

i) $\nu'\gamma(1 + \beta\cos\theta') = \nu$

Coefficient of "x" gives

ii) $\nu\cos\theta = \gamma\nu'(\beta + \cos\theta')$

Plug i) into ii) to get

$$\cos\theta\, \nu'\gamma(1 + \beta\cos\theta') = \gamma\nu'(\beta + \cos\theta')$$

$$\cos\theta = \frac{\beta + \cos\theta'}{1 + \beta\cos\theta'}$$

$$\cos\theta' = \frac{\cos\theta - \beta}{1 - \beta\cos\theta}$$

Coefficient of "y" gives:

iii) $\nu'\sin\theta' = \nu\sin\theta$

Plug i) into iii) to get

$$\sin\theta' = \gamma(1 + \beta\cos\theta')\sin\theta = \gamma\left[1 + \beta\left(\frac{\cos\theta - \beta}{1 - \beta\cos\theta}\right)\right]\sin\theta$$

$$= \frac{\gamma\sin\theta}{1 - \beta\cos\theta}(1 - \beta^2) = \frac{\sin\theta}{\gamma(1 - \beta\cos\theta)}$$

21-18 Using Eq. 21-20,

$$\nu = \nu'\gamma(1 + \beta \cos \theta')$$

For $\cos \theta' = \dfrac{\cos \theta - \beta}{1 - \beta \cos \theta}$

$$\nu = \nu'\gamma\left(1 + \beta\frac{[\cos \theta - \beta]}{[1 - \beta \cos \theta]}\right) = \frac{\nu'\gamma}{1 - \beta \cos \theta}[1 - \beta \cos \theta + \beta \cos \theta - \beta^2]$$

$$= \frac{\nu'\gamma}{(1 - \beta \cos \theta)}(1 - \beta^2) = \frac{\nu'\gamma}{(1 - \beta \cos \theta)}\frac{1}{\gamma^2} = \frac{\nu'}{\gamma(1 - \beta \cos \theta)}$$

21-19 $\nu = \dfrac{\nu'}{\gamma(1 - \beta \cos 0)} = \nu'\sqrt{\dfrac{1 + \beta}{1 - \beta}}, \quad \theta = 0°$

$$\lambda' = \lambda\sqrt{\frac{1 + \beta}{1 - \beta}} \qquad \lambda'\sqrt{\frac{1 - \beta}{1 + \beta}} = \lambda$$

$$\Delta\lambda = \lambda - \lambda' = \lambda'\left(-1 + \sqrt{\frac{1 - \beta}{1 + \beta}}\right)$$

(a) $v = 50$ km/s

Therefore $\gamma \sim 1$

$$\sqrt{\frac{1 - \beta}{1 + \beta}} = (\sqrt{1 - \beta})(1 + \beta)^{-\frac{1}{2}} \cong (1 - \tfrac{1}{2}\beta)(1 - \tfrac{1}{2}\beta) \cong 1 - \beta$$

$$\Delta\lambda = \lambda[-1 + (1 - \beta)] = -\lambda\beta = -(6353\ \text{Å})\left(\frac{50 \text{ km/s}}{3 \times 10^8 \text{ m/s}}\right) = -1.05\ \text{Å}$$

(b) Given $\Delta\lambda = 0.085$ Å

$$\frac{\Delta\lambda}{2} = \lambda\beta, \quad \beta = \frac{\Delta\lambda}{2\lambda} = \frac{0.085\ \text{Å}}{2(6563\ \text{Å})} = 6.48 \times 10^{-6}$$

$v = 1942$ m/s

$R_s = 6.96 \times 10^8$ m

$$v = \omega R_s = \frac{2\pi}{T}R_s, \quad T = \frac{2\pi R_s}{v} = \frac{(2\pi)(6.96 \times 10^8 \text{ m})}{1942 \text{ m/s}}$$

$T = 2.25 \times 10^6$ s $= 25.8$ days.

21-20 $\lambda' = 0.10$ m since car acts as a new moving source

$\nu_1 = \nu'(1 + \beta) \rightarrow$ Frequency detected by car

$\nu_2 = \nu_1(1 + \beta) \rightarrow$ Detected by patrol radar operator

$$\nu_{beat} = \nu_2 - \nu' = \nu'[(1 + \beta)^2 - 1] \cong 2\beta\nu' = 2\frac{(100 \times 10^3)}{3600} \times \frac{1}{0.10} = 556 \text{ s}^{-1}$$

21-21 Transverse Doppler shift: $\nu = \dfrac{\nu'}{\gamma}, \quad \nu' =$ proper frequency

$\lambda = \lambda'\gamma$

$\Delta\lambda = \lambda' - \lambda = \lambda'(1 - \gamma)$

$-\dfrac{\Delta\lambda}{\lambda'} + 1 = \gamma, \quad \lambda' = 3968$ Å

$$\gamma = 1 + \frac{0.29}{3968} = 1.000073$$

$$v = c\left(1 - \frac{1}{\gamma^2}\right)^{1/2} = 3.62 \times 10^6 \text{ m/s}$$

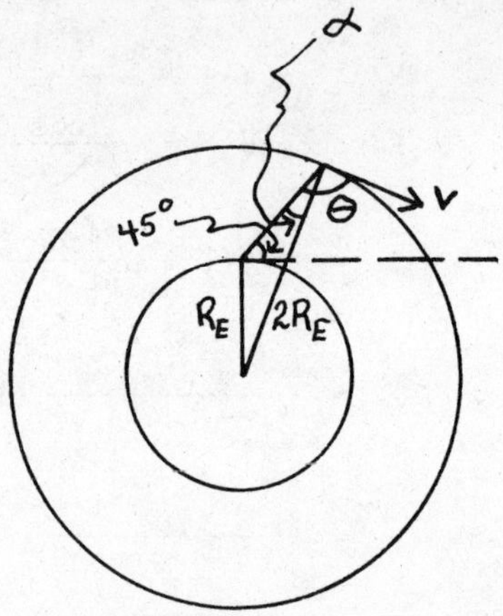

21-22 Satellite in orbit of two earth radii.

From Problem (21-7), $v = 5600$ m/s, $\gamma \sim 1$

From law of sines:

$$\frac{\sin\alpha}{R_E} = \frac{\sin(90° + 45°)}{2\,R_E}$$

$$\sin\alpha = \frac{1}{2}\frac{\sqrt{2}}{2}, \quad \alpha = 20.7°$$

$$\theta = 90° + 20.7° = 110.7°$$

$$\Delta\nu = \nu'\left[\frac{1}{\gamma(1 - \beta\cos\theta)} - 1\right] = (2 \times 10^7 \text{ s}^{-1})\left(\frac{1}{[1 - (1.8 \times 10^{-5})(\cos 110.7°)]} - 1\right)$$

$$= -127 \text{ Hz}$$

21-23 The velocity of the equatorial clock is

$$v = \omega r = \frac{2\pi}{T_E} R_E = (2\pi)\left(\frac{1 \text{ hr}}{(24 \text{ hr})(3600 \text{ sec})}\right)(6400 \times 10^3 \text{ m})$$

$$= 465 \text{ m/s}$$

$$\beta = \frac{v}{c} = 1.55 \times 10^{-6}$$

$$\gamma = (1 - \beta^2)^{-1/2} \sim 1 + \frac{1}{2}\beta^2$$

$$t_{polar} - t_{eq} = +\left(\frac{1}{2}\beta^2\right)(100 \text{ y})\left(\frac{365 \text{ d}}{1 \text{ y}}\right)\left(\frac{24 \text{ hr}}{1 \text{ d}}\right)\left(\frac{3600 \text{ s}}{1 \text{ hr}}\right)$$

$$= +\left(\frac{1}{2}\right)(1.55 \times 10^{-6})^2(3.15 \times 10^9 \text{s})$$

$$\Delta t = +0.0037 \text{ sec}$$

So when the equatorial clock reads 100 years, the polar clock reads 100 years + 0.0037 s.

21-24 Call the frame in which the meter stick is at rest the S' frame.

S' moves to right with respect to S.

$$x' = \gamma(x - \beta ct) \qquad ct' = \gamma(ct - \beta x)$$

In S frame, $\Delta t = \frac{\ell}{u}$

$\Delta t' = \gamma\Delta t - \frac{\gamma\beta}{c}\Delta x$, $\Delta x = 0$ since events occur in same place in S frame.

Therefore $\Delta t' = \gamma\Delta t$

Also $\Delta x' = \gamma\Delta x - \gamma\beta c\Delta t = -\gamma\beta c\Delta t$

Notice $\Delta x' = \ell'$, $\Delta x = 0$

$\ell' = \gamma\beta c\Delta t$, $\Delta t = \frac{\ell'}{\gamma\beta c}$

Combine Equations for Δt to get $\frac{\ell'}{\gamma\beta c} = \frac{\ell}{u}$, $\beta c = u$

So $\frac{\ell'}{\gamma} = \ell$

21-25

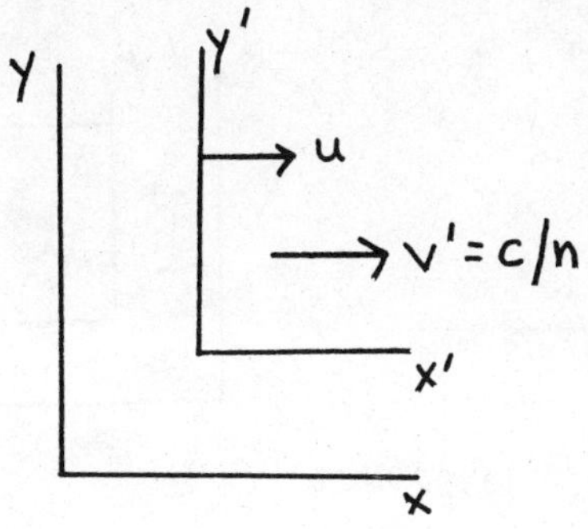

$$v_x = \frac{v'_x + u}{1 + \frac{v'_x u}{c^2}}$$

$$= \frac{c/n + u}{1 + (\frac{c}{n})(\frac{u}{c^2})} = (\frac{c}{n} + u)(1 + \frac{u}{nc})^{-1}$$

$$\simeq (\frac{c}{n} + u)(1 - \frac{u}{nc}) \cong \frac{c}{n} - \frac{u}{n^2} + u$$

$$v = \frac{c}{n} + u(1 - \frac{1}{n^2})$$

21-26

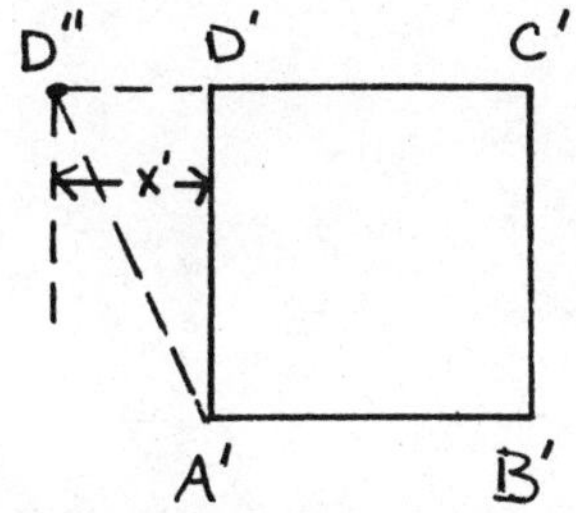

In order to arrive simultaneously at the camera, light from D' had to have left earlier than A' by amount $t' = \frac{\ell}{c}$. Since cube is moving with velocity u, the distance $x' = \overline{D'D''} = ut' = \frac{u}{c}\ell = \beta\ell$. Therefore, on the film D is to the left of A by an amount $\beta\ell$. The perpendicular distance $\overline{A'B'} = \ell$ is contracted in S to a length $\frac{\ell}{\gamma}$ or B appears on the film a distance $\frac{\ell}{\gamma}$ to the right of A. In the z-direction, the edge through A' appears on the film with length ℓ. All this is consistent with a cube rotated about the edge parallel to the z' axis through A', by an angle θ such that $\cos\theta = 1/\gamma$.

21-27 Show $cp = (K^2 + 2Km_oc^2)^{\frac{1}{2}}$

$E^2 = p^2c^2 + m_o^2c^4$, $\quad pc = (E^2 - m_o^2c^4)^{\frac{1}{2}}$

$E = K + m_oc^2$, $\quad E^2 = K^2 + m_o^2c^4 + 2Km_oc^2$

$pc = (K^2 + m_o^2c^4 + 2Km_oc^2 - m_o^2c^4) = (K^2 + 2Km_oc^2)^{\frac{1}{2}}$

For $K << m_oc^2$, $\quad pc = \sqrt{K}\,[K + 2\,m_oc^2]^{\frac{1}{2}} \cong \sqrt{K}\,(2\,m_oc^2)^{\frac{1}{2}}$

So $p = \sqrt{2\,Km_o}$ in the classical limit

$cp = \sqrt{2Km_oc^2}\quad[1 + \frac{K}{4m_oc^2}]$

$K/m_oc^2 = 0.020 \qquad cp = \sqrt{51}\,K$

$K/m_oc^2 = 0.20 \qquad cp = \sqrt{6}\,K$

$K/m_oc^2 = 2.0 \qquad cp = \sqrt{3/2}\,K$

$K/m_oc^2 = 20.0 \qquad cp = \sqrt{1.05}\,K$

21-28 If muons are produced at an altitude of 9 km above sea level and one-half reach sea level, what is the ratio of kinetic energy to rest-mass energy?

$\tau_\mu = 1.52\ \mu s$

$K = (\gamma - 1)m_o c^2, \quad E_{rest} = m_o c^2$

$\frac{K}{E_{rest}} = \gamma - 1, \quad dt = \gamma d\tau$

$vdt = \gamma v d\tau = 9 \times 10^3\ m$

$v = \beta c = c\sqrt{1 - \frac{1}{\gamma^2}}, \quad \gamma c\sqrt{1 - \frac{1}{\gamma^2}} = \sqrt{\frac{9 \times 10^3\ m}{d\tau}}$

$\sqrt{\gamma^2 - 1} = \frac{9 \times 10^3\ m}{(1.52 \times 10^{-6}\ s)(3 \times 10^8\ m/s)} = 1.97 \times 10^1$

$\gamma = 19.7$

$\frac{K}{E_{rest}} = \gamma - 1 = 19.7 - 1 = 18.7$

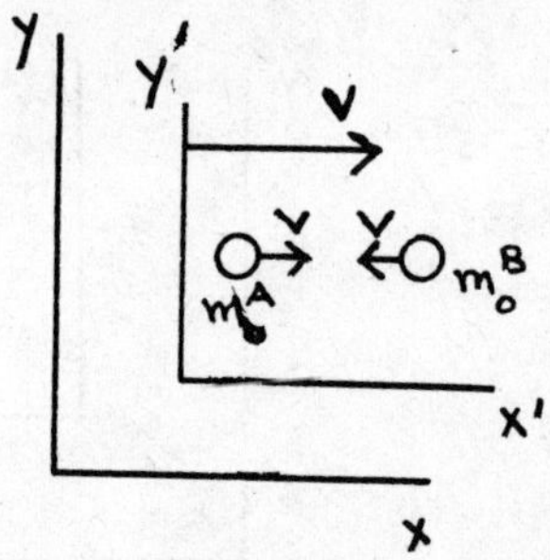

21-29

(a) $v = \frac{v' + u}{1 + \frac{v'u}{c^2}}$

for "A", $v_A = \frac{2v}{1 + \frac{v^2}{c^2}}$

for "B", $v_B = 0$

$\gamma = [1 - \frac{v^2}{c^2}]^{-\frac{1}{2}} = [1 - \frac{4v^2}{[1 + v^2/c^2]^2 c^2}]^{-\frac{1}{2}} = \frac{1 + v^2/c^2}{1 - v^2/c^2}$

$K = (\gamma - 1)m_o c^2 = \frac{1}{1 - v^2/c^2}[1 + v^2/c^2 - 1 + v^2/c^2]m_o c^2$

$= 2\gamma^2 v^2 m_o$

(b) In Newtonian limit, $\gamma \to 1$

$K = 2m_o v^2$

(c) single beam acceleration: $K_1 = (\gamma - 1)m_o c^2$

colliding beam: $K_2 = 2\gamma^2\beta^2 m_o c^2$

$\frac{K_2}{K_1} = \frac{2\gamma^2\beta^2}{(\gamma - 1)}$

$\beta = 4/5$ so $\gamma = 5/3$

Therefore $\frac{K_2}{K_1} = \frac{2 \times 16/9}{2/3} = \frac{16}{3}$

22-1 Convert 39.6° C to Fahrenheit. $t_F = \frac{9}{5} t_c + 32.0 = \frac{9}{5}(39.6) + 32.0 = 103.3°F$

Your fever is 103.3°F - 98.6°F = 4.7°F

22-2 $t_C = \frac{5}{9}(t_F - 32.0)$

highest temperature: $t_C = \frac{5}{9}(136 - 32.0) = 57.8°C$

lowest temperature: $t_C = \frac{5}{9}(-127 - 32.0) = -88.3°C$

22-3 $T = (273.16\text{ K})\frac{P}{P_o} = (273.16\text{ k})\frac{4.4\text{ torr}}{86.3\text{ torr}} = 13.93\text{ K}$

$t_C = T - 273.15 = 13.93 - 283.15 = -259.22°C$

22-4 The wire is 35 m when $t_C = -20°C$

$$\Delta\ell = \ell_o\bar{\alpha}(t - t_o)$$

$\bar{\alpha} \cong \alpha(20°C) = 1.67 \times 10^{-5}\text{ deg}^{-1}$ for Cu.

$$\Delta\ell = (35\text{ m})(1.67 \times 10^{-5}\text{ deg}^{-1})(35°C - (-20°C)) = +3.2\text{ cm}$$

22-5 $\bar{\alpha} \cong \alpha(20°C) = 1.05 \times 10^{-5}\text{ deg}^{-1}$ for steel.

$$\Delta\ell = (518\text{ m})(1.05 \times 10^{-5}\text{ deg}^{-1})(35°C - (-20°C)) = 0.299\text{ m}$$

22-6 The period of the pendulum is $T = 2\pi\sqrt{\frac{\ell}{g}}$. The period changes when the pendulum length changes.

$$\frac{dT}{d\ell} = \frac{2\pi}{\sqrt{g}}\left(\frac{1}{2}\frac{1}{\sqrt{\ell}}\right) \longrightarrow \frac{\Delta T}{T} = \frac{2\pi}{\sqrt{g}}\left(\frac{1}{2}\frac{1}{\sqrt{\ell}}\right)\frac{\Delta\ell}{T} = \frac{1}{2}\frac{\Delta\ell}{\ell}$$

The fractional change in the period, $\frac{\Delta T}{T}$, is also the fractional change in the time kept by the clock. As the pendulum length increases so does the period, so that the clock runs slowly. The discrepancy in one day is

$$\Delta T = \frac{1}{2}\frac{\Delta\ell}{\ell}(1\text{ day}) = \frac{1}{2}\frac{\bar{\alpha}\ell\Delta t}{\ell}(86400\text{ s})$$

$$= \frac{1}{2}(1.90 \times 10^{-5}\text{ deg}^{-1})(27°C - 18°C)(86400\text{ s}) = 7.39\text{ s}$$

22-7 $\frac{\Delta r}{r} = \alpha\,\Delta T$

$= (1.05 \times 10^{-5}\text{ deg}^{-1})(150°C - 20°C)$

$= 1.36 \times 10^{-3} = 0.136\%$

22-8 At 20°C the circular hole (in brass) has a diameter of 2.60 cm and the plug (steel) has a diameter of (2.60 cm − 0.010 mm). The hole diameter changes by the same law as the solid brass.

$$\Delta d_{hole}(T) = d_{o\,brass}\,\alpha_{brass}\,\Delta T$$

$$\Delta d_{plug}(T) = d_{o\,steel}\,\alpha_{steel}\,\Delta T$$

The diameters will be equal when

$$d_{hole} = d_{o\,brass} + \Delta d_{hole} = d_{plug} = d_{o\,steel} + \Delta d_{plug}$$

$$d_{o\,brass} - d_{o\,steel} = 2.60\text{ cm} - 2.599\text{ cm} = \Delta d_{plug} - \Delta d_{hole}.$$

Therefore,

$$0.001\text{ cm} = [2.599\text{ cm}(1.05\times 10^{-5}\text{ deg}^{-1}) - 2.6\text{ cm}(1.9\times 10^{-5}\text{ deg}^{-1})] \times (T - 20°C)$$

and $T = -25.2°C$.

22-9 By definition, $\beta = \frac{1}{V_o}\left.\frac{dV}{dT}\right|_{T_o}$

For a sphere, $V = \frac{4}{3}\pi R^3$. The radius changes with temperature as $R(T) = R_o\,[1 + \alpha(T - T_o)]$.

$$\left.\frac{dV}{dT}\right|_{T_o} = 4\pi R_o^2 \left.\frac{dR}{dT}\right|_{T_o}, \quad \left.\frac{dR}{dT}\right|_{T_o} = R_o\alpha$$

Therefore, $\beta = \frac{1}{\frac{4}{3}\pi R_o^3}(4\pi R_o^3\alpha) = 3\alpha$

22-10 From fluid statics, $p = p_o + \rho g h$, where p is the pressure a distance h below the surface, p_o is the surface pressure and ρ is the fluid density. At the level of Hg in the dish,

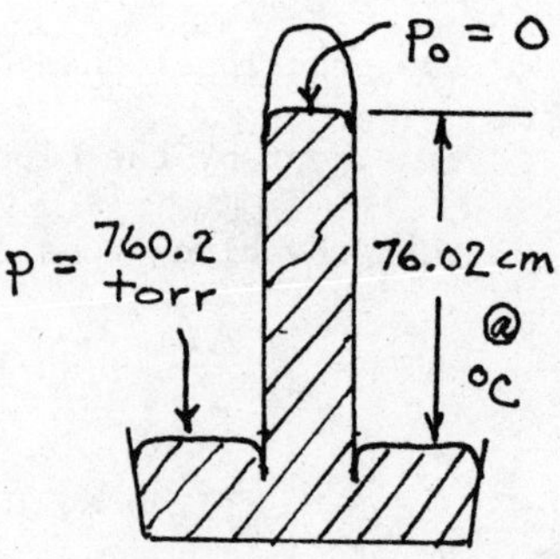

Torricelli barometer

$$760.2\text{ torr} = p_o + \rho_o g(76.02\text{ cm}) = \rho_o g(76.02\text{ cm}).$$

ρ_o is the Hg density at 0°C. When the temperature changes, the Hg density changes because of thermal expansion. If h is the new height of the Hg column at 30°C,

$$760.2\text{ torr} = \rho_{30}\,gh.$$

Dividing equations,

$$h = \frac{\rho_o}{\rho_{30}}(76.02\text{ cm})$$

The total mass of Hg is constant, so that

$$\frac{\rho_0}{\rho_{30}} = \frac{V_{30}}{V_0} = 1 + \bar{\beta}\,\Delta T$$

and h $= (1 + 0.182 \times 10^{-3}\ \text{deg}^{-1}(30°\text{C} - 0°\text{C}))(76.02\ \text{cm}) = 76.44\ \text{cm}$

The indicated pressure at 30°C is then 764.4 torr.

22-11 The change in the distance between the wire clamp points on the aluminum bar is

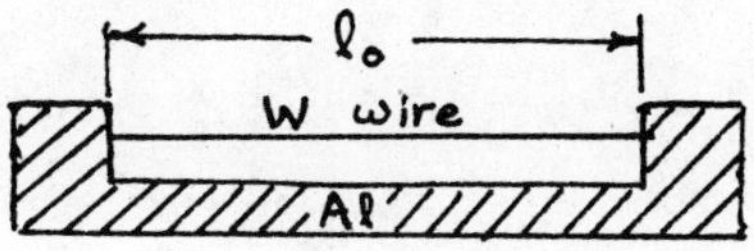

$$\Delta \ell_{Al} = \ell_0 \alpha_{Al} \Delta T$$

If the tungsten wire was also free to expand, its change in length would be

$\Delta \ell_w = \ell_0 \alpha_w \Delta T$. The wire is stretched

$$\Delta \ell = \Delta \ell_{Al} - \Delta \ell_w = \ell_0(\alpha_{Al} - \alpha_w)\Delta T$$

Young's modulus is defined as

$$Y = (\text{tensile stress}/\text{tensile strain})$$

$$= (\text{tension}/\text{cross-sectional area})/(\Delta \ell / \ell_0)$$

$$\text{tension} = Y(\Delta \ell / \ell_0)(\text{cross-sectional area})$$

$$= Y(\alpha_{Al} - \alpha_w)\Delta T(\pi r^2_{wire})$$

$$= (34 \times 10^{10}\ \frac{\text{N}}{\text{m}^2})(2.3 \times 10^{-5}\ \text{deg}^{-1} - 0.45 \times 10^{-5}\ \text{deg}^{-1})$$

$$\times (150°\text{C} - 20°\text{C})\pi\left(\frac{0.2\text{mm}}{2}\right)^2 = 25.7\ \text{N}$$

$$\text{total tension} = 50\ \text{N} + 25.7\ \text{N} = 75.7\ \text{N}$$

22-12 $Q = \left(\sum_i m_i \bar{c}_i\right)\Delta T$

$$Q = [2.4\ \text{kg} \times 0.216\ \frac{\text{Cal}}{\text{kg-deg}} + 1.6\ \text{kg} \times 0.0917\ \frac{\text{Cal}}{\text{kg-deg}}$$

$$+ 0.8\ \text{kg} \times 0.0931\ \frac{\text{Cal}}{\text{kg-deg}}]\ (80°\text{C} - 20°\text{C}) = 44.4\ \text{Cal}$$

22-13 The heat given up by the iron is equal to the heat gained by the water.

$$Q_{iron} = m_{iron}\,\bar{c}_{iron}\,\Delta T = (\rho_{iron} V_{iron})\,\bar{c}_{iron}\,\Delta T$$

$$= (3.93 \times 10^4\ \text{g})(0.145\ \frac{\text{Cal}}{\text{kg-deg}})(600°\text{C} - 250°\text{C})$$

$$= 1.994 \times 10^3\ \text{Cal}$$

$$Q_{water} = Q_{\text{heat to }100°\text{C}} + Q_{\text{water to steam}}$$

$$= m_{H_2O}\,\bar{c}_{H_2O}\,(100°\text{C} - 25°\text{C}) + m_{H_2O}\,L_v$$

$$= m_{H_2O}\,(1.00\ \frac{\text{Cal}}{\text{kg-deg}} \times 75°\text{C} + 539.12\ \frac{\text{Cal}}{\text{kg}})$$

Setting $Q_{iron} = Q_{water}$ gives $m_{H_2O} = 3.25$ kg

22-14 The friction brake turns all of the rotational kinetic energy into heat.

(a) $E_{rot} = \frac{1}{2} I\omega^2 = \frac{1}{2} M k^2\omega^2$, where k is the radius of gyration.

$= \frac{1}{2} (10 \text{ kg})(0.4 \text{ m})^2 (2\pi \times 2000 \frac{\text{rad}}{\text{min}} \frac{1 \text{ min}}{60 \text{ s}})^2$

$= 3.51 \times 10^4 \text{ J} = \underline{8.38 \text{ Cal}}$

(b) $E_{rot} = 8.38 \text{ Cal} = Q_{brake} = m\bar{c}\Delta T$

$8.38 \text{ Cal} = (6.0 \text{ kg})(0.15 \frac{\text{Cal}}{\text{kg deg}})(T - 60°\text{C})$

$T = 69.3°\text{C}$

22-15 The heat lost by the copper is equal to the heat gained by the water and vessel.

$Q_{Cu} = m_{Cu} \bar{c}_{Cu} \Delta T_{Cu} = (0.450 \text{ kg}) \bar{c}_{Cu} (87.2°\text{C} - 18.12°\text{C})$

$Q_{H_2O} + Q_{vessel} = m_{H_2O} c_{H_2O} \Delta T_{H_2O} + m_{vessel} c_{vessel} \Delta T_{vessel}$

$= (0.35 \text{ kg})(1 \frac{\text{Cal}}{\text{kg-deg}})(18.12°\text{C} - 10.00°\text{C}) + (0.0302 \text{ kg})(0.216 \frac{\text{Cal}}{\text{kg-deg}})$

$\times (18.12°\text{C} - 10.00°\text{C})$

assuming the vessel also starts at 10.00°C.

Setting $Q_{cu} = Q_{H_2O} + Q_{vessel}$ gives $\bar{c}_{Cu} = 0.0931 \frac{\text{Cal}}{\text{kg-deg}}$

22-16 The heat lost by the tin is used to melt the ice.

$Q_{tin} = (300 \text{ g})(0.0556 \frac{\text{Cal}}{\text{kg-deg}})(T - 0°\text{C})$

$Q_{ice} = m_{ice} L_f = (18.2 \text{ g})(79.71 \frac{\text{Cal}}{\text{kg}})$

Setting $Q_{tin} = Q_{ice}$ gives $T = 87.0°\text{C}$

22-17 The heat lost by the iron is equal to the heat gained by the water and calorimeter.

$Q_{iron} = (0.10 \text{ kg})(0.122 \frac{\text{Cal}}{\text{kg-deg}})(T - 45.0°\text{C})$

$Q_{water} + Q_{calorimeter} = [0.2 \text{ kg } (1 \frac{\text{Cal}}{\text{kg-deg}}) + 0.015 \frac{\text{Cal}}{\text{deg}}](45.0°\text{C} - 20.0°\text{C})$

Setting $Q_{iron} = Q_{water} + Q_{calorimeter}$ gives $T = 485.6°\text{C}$

22-18 The heat of combustion is equal to the heat gained by the water and calorimeter.

$Q_{lignite} = m_{lignite} L_c = (5.40 \text{ g})L_c$

$Q_{water} + Q_{calorimeter} = [0.900 \text{ kg}(1 \frac{\text{Cal}}{\text{kg-deg}}) + 0.23 \frac{\text{Cal}}{\text{deg}}]$

$\times (36.1°\text{C} - 20.2°\text{C})$

Setting $Q_{lignite} = Q_{water} + Q_{calorimeter}$ gives $L_c = 3327 \frac{Cal}{kg}$

22-19 Under conditions of steady water flow and steady burning rate, the rate of heat generation q is

$$q = \bar{c}(T_2 - T_1)r$$

The equilibrium output water temperature is $T_2 = 82°C$

The equilibrium input water temperature is $T_1 = 10°C$

The average specific heat of water over the temperature interval is $\bar{c} = 1.00 \frac{Cal}{kg\text{-}deg.}$

The mass flow rate of water is $r = 70 \frac{\ell}{min} (1 \frac{kg}{\ell})$.

Calculating, $q = 5040 \frac{Cal}{min}$

This is 32% of the heat production rate for the gas, so that $(0.32)q_{gas} = 5040 \frac{Cal}{min}$

Also,

$$q_{gas} = (\text{mass rate of gas})L_c$$

$$= (\text{volume rate of gas}) \rho_{gas} L_c$$

Therefore,

$$(0.32)(\text{volume rate})(1.12 \frac{kg}{m^3})(8400 \frac{Cal}{kg}) = 5040 \frac{Cal}{min}$$

and

$$\text{volume rate of natural gas} = 1.674 \frac{m^3}{min} = 100.4 \frac{m^3}{h}$$

22-20 The heating rate of the gasoline is

$$q_{gasoline} = (\text{mass rate of gasoline})L_c(\text{gasoline})$$

$$\text{mass rate} = (5.2 \frac{gal}{h})(3.79 \frac{\ell}{gal})(740 \frac{kg}{m^3})(\frac{1\ m^3}{10^3\ \ell})(\frac{1\ h}{3600\ s})$$

Therefore,

$$q_{gasoline} = 46.6 \frac{Cal}{s}$$

efficiency = energy rate out/energy rate in

$$= (60\ hp \times 0.178 \frac{Cal}{s}/hp)/46.6 \frac{Cal}{s} = 23\%$$

22-21 (a)

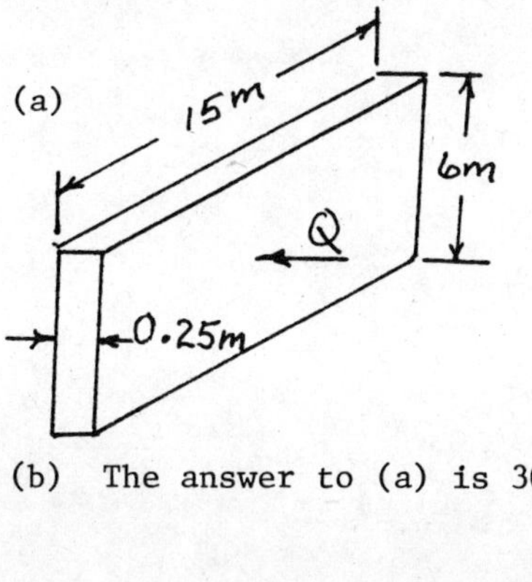

$$\frac{dQ}{dt} = \kappa \Delta T \frac{\text{area cross section}}{\text{width of wall}}$$

$$\frac{dQ}{dt} = (1.5 \times 10^{-4} \frac{Cal}{s\text{-}m\text{-}deg})[24°C - (-7.0°C)]\frac{6m \times 15m}{0.25m}$$

$$= 1.67 \frac{Cal}{s} = 6.03 \times 10^3 \frac{Cal}{h}$$

(b) The answer to (a) is 30% of the heating rate for the crude oil.

$$0.30\, q_{oil} = 6.03 \times 10^3 \frac{\text{Cal}}{\text{h}}$$

$$q_{oil} = L_c(\text{oil}) r_{oil} = \left(11500 \frac{\text{Cal}}{\text{kg}}\right) r_{oil}$$

Therefore,

$$r_{oil} = \text{mass rate of oil}$$

$$= \frac{6.03 \times 10^3 \frac{\text{Cal}}{\text{h}}}{0.3\left(11500 \frac{\text{Cal}}{\text{kg}}\right)} = 1.75 \frac{\text{kg}}{\text{h}}$$

In 8 hours the required amount is

$$(8\text{h}) r_{oil} = (8\text{h})\left(1.75 \frac{\text{kg}}{\text{h}}\right) = 14.0 \text{ kg}$$

$$= 14.0 \text{ kg}\left(\frac{1 \text{ m}^3}{919 \text{ kg}}\right)\left(\frac{1 \ell}{10^{-3} \text{ m}^3}\right)\left(\frac{1 \text{ gal}}{3.79 \ell}\right) = 4.0 \text{ gal}$$

22-22

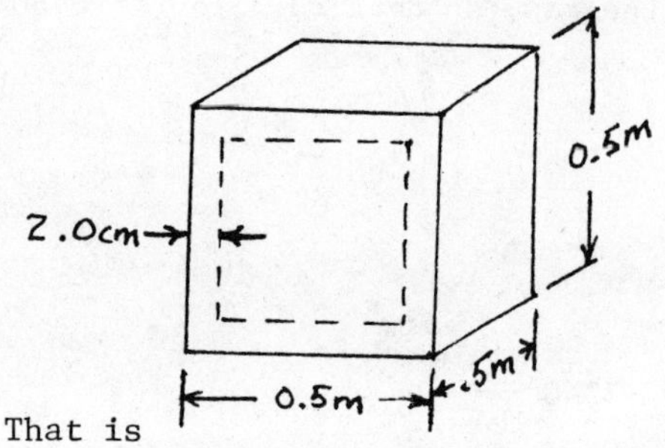

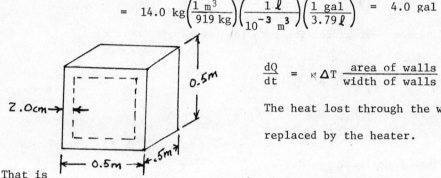

$$\frac{dQ}{dt} = \kappa \Delta T \frac{\text{area of walls}}{\text{width of walls}}$$

The heat lost through the walls is just replaced by the heater.

That is

$$110 \text{ W} = \kappa(85.0°\text{C} - 20.0°\text{C}) \frac{6 \times 0.5\text{m} \times 0.5\text{m}}{0.02\text{m}}$$

$$\kappa = \frac{110 \frac{\text{J}}{\text{s}}(1 \text{ Cal}/4.186 \text{ J})}{65 \text{ deg} \times 75\text{m}} = 5.39 \times 10^{-3} \frac{\text{Cal}}{\text{s-deg-m}}$$

22-23 (a) The rate of heat flow must be the same through ℓ_1 and ℓ_2.

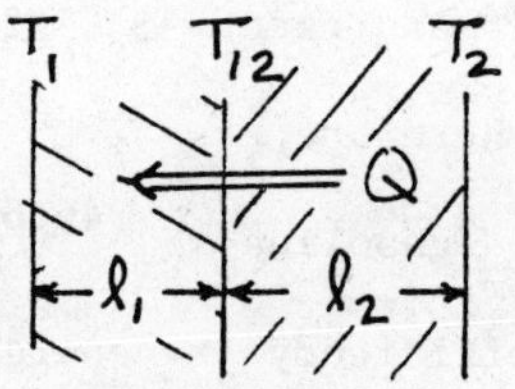

Thus, $$\frac{dQ}{dt} = \kappa_1 (T_{12} - T_1)\frac{A}{\ell_1} = \frac{T_{12} - T_1}{R_1}$$

$$= \kappa_2 (T_2 - T_{12})\frac{A}{\ell_2} = \frac{T_2 - T_{12}}{R_2}$$

where $R_1 = \frac{\ell_1}{\kappa_1 A}$ and $R_2 = \frac{\ell_2}{\kappa_2 A}$.

(1) $R_1 \frac{dQ}{dt} = T_{12} - T_1$

(2) $R_2 \frac{dQ}{dt} = T_2 - T_{12}$

Adding (1) and (2): $\frac{dQ}{dt} = \frac{T_2 - T_1}{R_1 + R_2} = \frac{T_2 - T_1}{R}$

(b) Dividing (1) by (2): $\frac{R_1}{R_2} = \frac{T_{12} - T_1}{T_2 - T_{12}}$

Thus,

$$T_{12} = \frac{R_1 T_2 + R_2 T_1}{R_1 + R_2} = T_1 + (T_2 - T_1)\frac{R_1}{R_1 + R_2}$$

22-24 Use the result of Problem 28-23 to find the rate of heat loss with the insulation in place.

$$\left.\frac{dQ}{dt}\right|_{\text{insulated}} = \frac{T_2 - T_1}{R_1 + R_2} = \frac{(T_2 - T_1)A}{\frac{\ell_2}{\kappa_1} + \frac{\ell_2}{\kappa_2}}$$

Without insulation,

$$\left.\frac{dQ}{dt}\right|_{\text{without}} = \frac{T_2 - T_1}{R_1} = \frac{(T_2 - T_1)A}{\frac{\ell_1}{\kappa_1}}$$

Dividing these equations,

$$\frac{\left.\frac{dQ}{dt}\right|_{\text{insulated}}}{\left.\frac{dQ}{dt}\right|_{\text{without}}} = \frac{R_1}{R_1 + R_2} = \frac{1}{1 + \frac{\ell_2}{\ell_1}\frac{\kappa_1}{\kappa_2}}$$

$$= \left[1 + \left(\frac{3.0\ \text{cm}}{25\ \text{cm}}\right)\left(\frac{1.5 \times 10^{-4}}{0.1 \times 10^{-4}}\right)\right]^{-1} = 0.36$$

22-25 At equilibrium, $\frac{1}{4}$ hp is the rate at which energy leaves the calorimeter.

That is, $$\frac{1}{4}\ \text{hp} = \kappa\,\Delta T\,\frac{A}{\ell} = \frac{1}{4}\ \text{hp}\left(0.1782\ \frac{\text{Cal}}{\text{s}}\Big/\text{hp}\right)$$

$$\Delta T = \frac{\ell}{A\kappa}\left(0.04455\ \frac{\text{Cal}}{\text{s}}\right) = 0.0145\ \text{deg}$$

22-26 The end of the rod in boiling water is at 100°C. The other end is at 0°C.

$$\frac{dQ}{dt} = \kappa_{Al}\,\Delta T\,\frac{A}{\ell} = \left(0.050\ \frac{\text{Cal}}{\text{s-m-deg}}\right)(100°\text{C})\ \frac{\pi\left(\frac{1.0\ \text{cm}}{2}\right)^2}{35\ \text{cm}}$$

$$= 1.122 \times 10^{-3}\ \frac{\text{Cal}}{\text{s}}$$

This heat is used to melt ice.

$$Q_{\text{melt ice}} = L_f(\text{ice})m_{\text{ice}} \qquad \frac{dQ}{dt} = L_f(\text{ice})\frac{dm}{dt}\text{ice},$$

$$\frac{dm}{dt}\text{ice} = \frac{1.122 \times 10^{-3}\ \frac{\text{Cal}}{\text{s}}}{79.71\ \frac{\text{Cal}}{\text{kg}}} = 0.0507\ \frac{\text{kg}}{\text{h}}$$

22-27 (a) If $t_C = t_F = \frac{9}{5}t_C + 32.00$

then, $t_C = -\frac{5}{4}(32.00)°\text{C} = -40°\text{C} = -40°\text{F}$

On the absolute scale this is $T = -40°\text{C} + 273.15\ \text{deg} = 233.15\ \text{K}$

(b) $t_F = 2t_C = \frac{9}{5} t_C + 32.00$

$t_C = 5(32.00)°C = 160°C = 433.15 K$

22-28 T(triple point) = 273.16 K

$t_R = \frac{9}{5} T = \frac{9}{5} (273.16)°R = 491.688°R$

and $t_F = t_R - 459.68 = 32.008 °F$

22-29 Any linear dimension changes according to the same law.

$$\Delta R_{belt} = \alpha R_o \Delta T = \alpha R_{equator} \Delta T$$

$$= \alpha(\text{steel})(6.38 \times 10^6 \text{ m})(1 \text{ deg})$$

$$= (1.05 \times 10^{-5} \text{ deg}^{-1})(1 \text{ deg})(6.38 \times 10^6 \text{m}) = 67.0 \text{ m}$$

22-30 (a) $\alpha \ell = \frac{d\ell}{dT}$ implies $\int_{\ell_0}^{\ell} d\ell = \int_{T_0}^{T} \alpha \ \ell dT$

Since $\ell \cong \ell_o$, $\ell - \ell_o \cong \ell_o \int_{T_0}^{T} \alpha(T)dT$

or $\ell = \ell_o[1 + \int_{T_0}^{T} \alpha(T)dT]$

(b) Substitute $\alpha(T) = 2.25 \times 10^{-5} \text{ deg}^{-1} + (2.3 \times 10^{-8} \text{ deg}^{-2})T$

and $\ell_o = 1$m into the result from part (a).

$$\Delta \ell = \ell - \ell_o = (1 \text{ m})\int_{0°C}^{100°C} [2.25 \times 10^{-5} \text{ deg}^{-1} + (2.3 \times 10^{-8} \text{ deg}^{-2})T]dT$$

$$= (1 \text{ m})[2.25 \times 10^{-5} \times 10^2 + \frac{1}{2} \times 2.3 \times 10^{-8} \times 10^4] = 2.365 \text{ mm}$$

(c) $\alpha(T = 20°C) = 2.25 \times 10^{-5} \text{ deg}^{-1} + (2.3 \times 10^{-8} \text{ deg}^{-2})(20 \text{ deg})$

$$= 2.296 \times 10^{-5} \text{ deg}^{-1}$$

which when rounded off is the tabulated result. The average value of α in the given temperature interval is by definition

$$\bar{\alpha} = \frac{\Delta \ell}{\ell_o \Delta T} = \frac{2.365 \text{ mm}}{(1 \text{ m})(100°C)} = \underline{2.365 \times 10^{-5} \text{ deg}^{-1}}$$

22-31 (a) Assuming that the steel rails aren't free to expand, the suppressed thermal expansion produces a stress given by

$$s = Y\frac{\Delta \ell}{\ell} = Y \alpha \Delta T = (20.5 \times 10^{10} \text{ Pa})(1.05 \times 10^{-5} \text{ deg}^{-1})(25°)$$

$$= \underline{5.38 \times 10^7 \text{ Pa}}$$

(b) Compared to the yield strength this is
$(5.38 \times 10^7 \text{ Pa})/(52.2 \times 10^7 \text{ Pa}) = \underline{10.3\%}$

22-32 (a) $\Delta R = \alpha_{Ag} R_o (T - T_o)$

$= (1.90 \times 10^{-5} \text{ deg}^{-1})(5.00 \text{ cm})(30°C - 10°C) = 1.9 \times 10^{-3} \text{ cm}$

(b) $\Delta S = 2\alpha_{Ag} S_o (T - T_o)$, $S_o = 4\pi R_o^2$

$= 2(1.90 \times 10^{-5} \text{ deg}^{-1})(4\pi \times 25.00 \text{ cm}^2)(30°C - 10°C) = 0.239 \text{ cm}^2$

(c) $\Delta V = 3\alpha_{Ag} V_o (T - T_o)$, $V_o = \frac{4}{3}\pi R_o^3$

$= 3(1.90 \times 10^{-5} \text{ deg}^{-1})(\frac{4}{3}\pi \times 125.0 \text{ cm}^3)(30°C - 10°C) = 0.597 \text{ cm}^3$

22-33 (a) The final volume of the liquid is

$$V_\ell = V + \Delta V_\ell = V + \bar{\beta}_\ell V \Delta T$$

The final volume contained within the glass is

$$V_g = V + \Delta V_g = V + \bar{\beta}_g V \Delta T$$

The difference, $V_\ell - V_g$, is the volume of liquid that overflows.

$$\Delta V = V_\ell - V_g = \Delta V_\ell - \Delta V_g = V(\bar{\beta}_\ell - \bar{\beta}_g)\Delta T$$

(b) Using the result from (a),

$$6.8 \text{ cm}^3 = 1 \ell(\bar{\beta}_\ell - 3.2 \times 10^{-5} \text{ deg}^{-1})(25°C - 5°C)$$

$$\bar{\beta}_\ell = 0.372 \times 10^{-3} \text{ deg}^{-1}$$

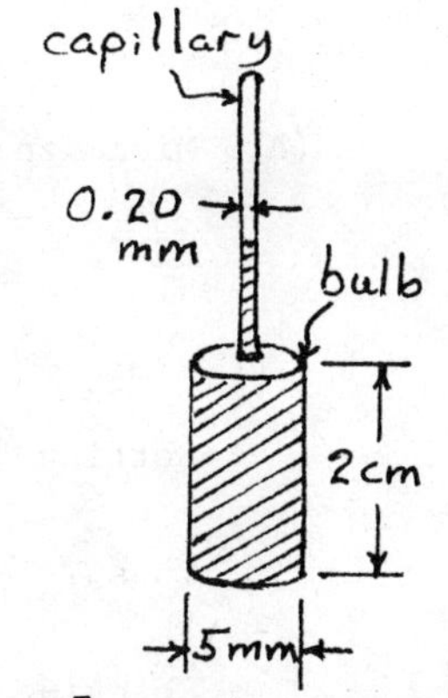

22-34 The overflow from the bulb produces the change in the mercury height in the capillary tube. As in Problem 22-33 the overflow is

$$\Delta V = V_{bulb}(\bar{\beta}_{Hg} - \bar{\beta}_{pyrex})\Delta T$$

$$= \pi\left(\frac{5 \text{ mm}}{2}\right)^2(2 \text{ cm})\ [0.182 \times 10^{-3} \text{ deg}^{-1} - 3(0.32 \times 10^{-5} \text{ deg}^{-1})](100 \text{ deg})$$

$$= 6.77 \times 10^{-3} \text{ cm}^3$$

The volume is displaced into the capillary tube and represents a height given by

$$\Delta V = 6.77 \times 10^{-3} \text{ cm}^3 = h\pi\left(\frac{0.02 \text{ cm}}{2}\right)^2$$

$$h = 21.55 \text{ cm}$$

22-35 (a) The bulk modulus is defined as

$$B_\ell = \frac{\Delta p}{(\Delta V/V)}$$

where ΔV is the change in volume due to compression. As $B_{shell} \gg B_{liquid}$, it is the liquid that is compressed. The thermal expansion of the liquid is greater than the thermal expansion of the Cu, so that the compression of the liquid is

$$\Delta V = V_{liquid} - V_{shell}$$

Thus,

$$\Delta p = \frac{\Delta V}{V} B_\ell = B_\ell (\bar{\beta}_\ell - 3\bar{\alpha}_s) \Delta T$$

(b) $\Delta p = (2.05 \times 10^9 \text{ Pa})[0.207 \times 10^{-3} \text{ deg}^{-1} - 3 \times 1.67 \times 10^{-5} \text{ deg}^{-1}]$

$$\times (100°C - 10°C) = 2.895 \times 10^7 \text{ Pa}$$

is the pressure produced by the compression. The fraction of the yield strength is $(2.90 \times 10^7 \text{ Pa})/(34.0 \times 10^7 \text{ Pa}) = 8.53\%$

22-36 (a) $c(T) = \frac{1}{m}\frac{dQ}{dT}$ $\qquad dQ = m\,c(T)dT$

Therefore, $Q = m\int_{T_0}^{T} c(T)dT$

or, $Q = 1\text{kg}[1.00680(100°C) - \frac{1}{2} \times 5.509 \times 10^{-4} (100°C)^2$

$$+ \frac{1}{3} \times 9.277 \times 10^{-6} (100°C)^3 - \frac{1}{4} \times 3.768 \times 10^{-8} (100°C)^4]$$

$$= \underline{100.076 \text{ Cal}}$$

(b) By definition, $Q(0°C \text{ to } 100°C) = m\bar{c}\,\Delta T$

$$\bar{c} = \frac{Q}{m\Delta T} = \frac{100.076 \text{ Cal}}{(1 \text{ kg})(100 \text{ deg})} = 1.00076 \frac{\text{Cal}}{\text{kg-deg}}$$

(c) Minimize the function c(T) with respect to T.

$$\frac{dc}{dT} = 0 - 5.509 \times 10^{-4} + 2 \times 9.277 \times 10^{-6} T$$
$$- 3 \times 3.768 \times 10^{-8} T^2$$

Setting $\frac{dc}{dT} = 0$ determines the extrema of the function.

Thus, $T_{extrema} = 38.92°C, 125.22°C$

Checking that $\left.\frac{d^2c}{dT^2}\right|_{38.92°C} > 0$, shows that

38.92°C is the temperature at which c(T) is a minimum. This agrees with the actual graph of c(T).

22-37 The heat lost by the aluminum is equal to the heat gained by the calorimeter and ice water.

$$Q_{Al} = m_{Al}\,c_{Al}\,\Delta T_{Al} = 0.55 \text{ kg}\left(0.216 \frac{\text{Cal}}{\text{kg-deg}}\right)(100°C - T)$$

where T is the final equilibrium temperature. The heat transferred from the Al must first melt the ice in the ice water at 0°C.

$$Q_{melt\ ice} = m_{ice}L_f(\text{ice}) = 0.08 \text{ kg}(79.71 \frac{\text{Cal}}{\text{kg}}) = 6.38 \text{ Cal}$$

At this point there are 380 g of water at 0°C which must be raised to the final equilibrium temperature.

$$Q_{water} = m_w c_w \Delta T = 0.38\ kg\left(1.00\ \frac{Cal}{kg\text{-}deg}\right)(T - 0°C)$$

$$Q_{calorimeter} = H\,\Delta T = 0.0193\ \frac{Cal}{deg}\ (T - 0°C)$$

Setting $Q_{Al} = Q_{ice} + Q_{water} + Q_{calorimeter}$ and solving for T, T = 10.6°C

22-38 The heat capacity of the whiskey is given by

$$H_{whiskey} = \sum_i m_i \bar{c}_i = m_{ethyl}\,\bar{c}_{ethyl} + m_{water}\,\bar{c}_{water}$$

$$= \frac{1}{2} m_{whiskey}\,\bar{c}_{ethyl} + \frac{1}{2} m_{whiskey}\,\bar{c}_{water}$$

$$= \frac{1}{2}(\text{volume whiskey})\,\rho_{whiskey}(\bar{c}_{ethyl} + \bar{c}_{water})$$

$$= \frac{1}{2}\left(4\ oz \times 29.58\ \frac{cm^3}{oz}\right)\left(0.919\ \frac{g}{cm^3}\right)(0.572 + 0.999)\ \frac{cal}{g\text{-}deg}$$

$$= 85.40\ \frac{cal}{deg}$$

The heat lost by the ice is gained by the whiskey

$$Q_{ice} = Q_{to\ melt\ ice} + Q_{of\ melted\ ice}$$

$$= m_{ice}\,L_f(ice) + m_{ice}\,\bar{c}_{water}\,\Delta T$$

$$= 20\ g\left[79.71\ \frac{cal}{g} + 1.00\ \frac{cal}{g\text{-}deg}\ (T - 0°C)\right]$$

where T is the final equilibrium temperature.

$$Q_{whiskey} = H_{whiskey}\,\Delta T = 85.40\ \frac{cal}{deg}\ (25.0°C - T)$$

Setting $Q_{ice} = Q_{whiskey}$ gives T = 5.13°C

22-39 For a mass of water, m, falling from 400 m the initial gravitational potential energy is mgh. If this energy is converted into heat which is absorbed by the water, then

$$mgh = Q_{water} = m\bar{c}_{water}\,\Delta T_{water}$$

$$\Delta T_{water} = \frac{gh}{c} = \left(9.8\ \frac{m}{s^2}\right)(400\ m)\Big/\left(4186\ \frac{J}{kg\text{-}deg}\right)$$

$$= 0.94°C$$

22-40 Ninety percent of the kinetic energy of the bullet is converted to heat energy which is absorbed by the bullet.

$$K_{bullet} = \frac{1}{2} mv^2$$

$$Q_{bullet} = \frac{9}{10} K_{bullet} = m\bar{c}_{bullet} \Delta T_{bullet}$$

$$\frac{9}{10} (\frac{1}{2} mv^2) = m\bar{c}\Delta T$$

$$\Delta T_{bullet} = 0.45 \frac{v^2}{c} = 0.45 \frac{(250 \frac{m}{s})^2}{136 \frac{J}{kg\text{-}deg}} = 206.8°C$$

22-41 The mechanical energy input after n rotations of the movable paddles is

$$E = \tau\Delta\theta = (2 \text{ mg R})(2\pi n) = 4\pi n \text{ mg R}$$

$$= 4\pi(470.0 \text{ turns})(2.50 \text{ kg})(9.8 \frac{m}{s^2})(0.08 \text{ m})$$

$$= 11576 \text{ J}$$

If all of this energy is turned into heat within the calorimeter,

$$E = Q_{water} + Q_{calorimeter} = m_{H_2O} \bar{c}_{H_2O} \Delta T + H_{calorimeter} \Delta T$$

$$= [1.200 \text{ kg}(1.000 \frac{Cal}{kg\text{-}deg}) + 0.072 \frac{Cal}{deg}](16.21°C - 14.00°C)$$

$$= 2.8111 \text{ Cal}$$

Thus, 1 Cal = 4118 J

The result of this experiment might come out too small on the heat side as some heat escapes through the sides of the calorimeter.

22-42 The answer is required in watts/meter of wire, so look at a 1-m length of wire

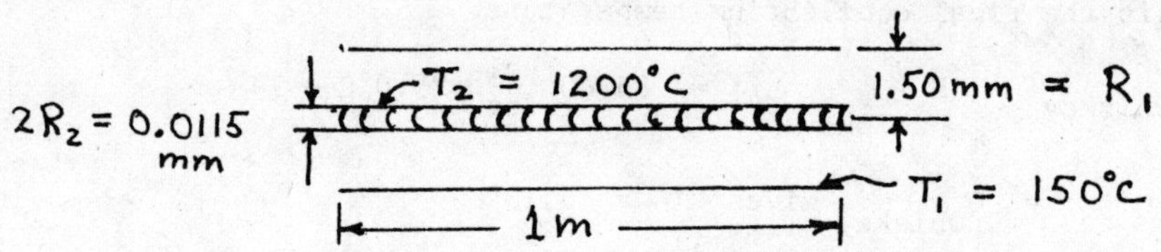

For a system (the porcelain sheath) with cylindrical symmetry, the heat conduction formula is

$$\frac{dQ}{dt} = \frac{2\pi L\kappa(T_2 - T_1)}{\ln(R_1/R_2)}$$

L is the length of the system, and 1 and 2 denote the outside and inside cylindrical surfaces between which conduction takes place.

$$\left.\frac{dQ}{dt}\right|_{1\,m} = \frac{2\pi(2.5 \times 10^{-4} \frac{Cal}{s\text{-}m\text{-}deg})(1200°C - 150°C)}{\ln(1.50 \text{ mm} / 0.00575 \text{ mm})}$$

$$= \left(0.2964 \frac{Cal}{s\text{-}m}\right)\frac{4186 \text{ J}}{Cal} = 1241 \frac{W}{m}$$

22-43 When thermal conduction is perpendicular to a compound slab, the total thermal resistance is the sum of the resistances for each section.

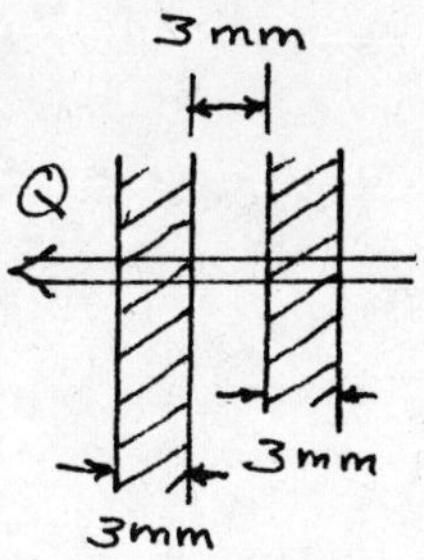

$$\frac{dQ}{dt}(\text{double pane}) = \frac{\Delta T}{R}$$

$$R = R_{\text{pane 1}} + R_{\text{airgap}} + R_{\text{pane 2}}$$

$$= \frac{\ell_{\text{pane 1}}}{\kappa_{\text{glass}} A} + \frac{\ell_{\text{air gap}}}{\kappa_{\text{air}} A} + \frac{\ell_{\text{pane 2}}}{\kappa_{\text{glass}} A}$$

$$= \frac{2\ell_{\text{glass}}}{\kappa_{\text{glass}} A} + \frac{\ell_{\text{air}}}{\kappa_{\text{air}} A}$$

For a single pane of glass,

$$\frac{dQ}{dt}(\text{single pane}) = \frac{\Delta T}{R_{\text{single}}}$$

$$R_{\text{single}} = \frac{\ell_{\text{glass}}}{\kappa_{\text{glass}} A}$$

Thus,

$$\frac{\frac{dQ}{dt}(\text{double pane})}{\frac{dQ}{dt}(\text{single pane})} = \frac{\frac{\Delta T}{R}}{\frac{\Delta T}{R_{\text{single}}}} = \frac{\frac{\ell_{\text{glass}}}{\kappa_{\text{glass}}}}{\frac{2\ell_{\text{glass}}}{\kappa_{\text{glass}}} + \frac{\ell_{\text{air}}}{\kappa_{\text{air}}}}$$

$$= \left[2 + \frac{\ell_{\text{air}}}{\ell_{\text{glass}}} \frac{\kappa_{\text{glass}}}{\kappa_{\text{air}}}\right]^{-1} = \left[2 + \frac{2.5 \times 10^{-4}}{6.2 \times 10^{-6}}\right]^{-1} = 2.36\%$$

Double pane windows improve the insulation by a factor of about 40 over single pane windows.

22-44 Compound layers in cylindrical symmetry follow the additive thermal resistance rule.

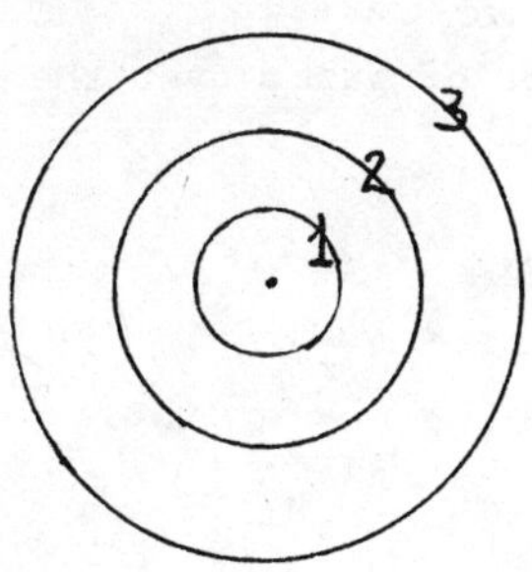

$$\left.\frac{dQ}{dt}\right|_{1 \text{ to } 2} = \frac{T_1 - T_2}{R_{12}}, \quad R_{12} = \frac{\ln(R_2/R_1)}{2\pi L \kappa_{12}}$$

$$\left.\frac{dQ}{dt}\right|_{2 \text{ to } 3} = \frac{T_2 - T_3}{R_{23}}, \quad R_{23} = \frac{\ln(R_3/R_2)}{2\pi L \kappa_{23}}$$

$$(1) \quad R_{12}\frac{dQ}{dt} = T_1 - T_2, \quad (2) \quad R_{23}\frac{dQ}{dt} = T_2 - T_3$$

Adding (1) and (2): $\quad (3) \quad \frac{dQ}{dt} = \frac{T_1 - T_3}{R_{12} + R_{23}}$

Dividing (1) by (2): $\quad (4) \quad T_2 = T_1 + (T_3 - T_1)\frac{R_{12}}{R_{23} + R_{12}}$

(a) From (3), $\frac{dQ}{dt} = \frac{\Delta T}{R_{asbestos} + R_{styrofoam}}$

$$R_{asbestos} = \frac{\ln(8.50\text{ cm}/2.50\text{ cm})}{2\pi(1\text{ m})\left(0.19\times10^{-4}\frac{\text{Cal}}{\text{s-m-deg}}\right)} = 1.025\times10^{4}\ \frac{\text{s-deg}}{\text{Cal}}$$

$$R_{styrofoam} = \frac{\ln(10.50\text{ cm}/8.50\text{ cm})}{2\pi(1\text{ m})\left(0.024\times10^{-4}\frac{\text{Cal}}{\text{s-m-deg}}\right)} = 1.401\times10^{4}\ \frac{\text{s-deg}}{\text{Cal}}$$

Therefore,

$$\frac{dQ}{dt} = \frac{80\text{ deg}}{1.025\times10^{4}\ \frac{\text{s-deg}}{\text{Cal}} + 1.401\times10^{4}\ \frac{\text{s-deg}}{\text{Cal}}} = 11.87\ \frac{\text{Cal}}{\text{h}}$$

(b) From (4), T between asbestos and styrofoam is

$$T_2 = T_1 + \Delta T\frac{R_{12}}{R_{23}+R_{12}}$$

$$= 100°\text{C} = 80°\text{C}\ \frac{1.025\times10^{4}\ \frac{\text{s-deg}}{\text{Cal}}}{2.426\times10^{4}\ \frac{\text{s-deg}}{\text{Cal}}} = 66.2°\text{C}$$

22-45 In the first case the cabin air effectively loses heat by the addition of colder air. The heat lost by the cabin air is equal to the heat gained by the cold air.

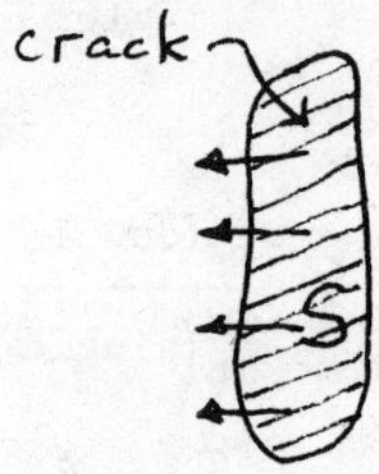

$$Q_{cabin\ air} = Q_{cold\ air} = m_{cold\ air}\,\bar{c}_{cold\ air}\,\Delta T$$

$$\frac{dQ}{dt}(\text{cabin air}) = \bar{c}_{cold\ air}\,\Delta T\,\frac{dm}{dt}(\text{cold air})$$

As the average speed of the cold air is 2.5 $\frac{m}{s}$, in one second a volume of air 2.5 m × S will enter the cabin, where S is the area of the crack.

Then, $\frac{dm}{dt}(\text{cold air}) = \rho_{cold\ air}\frac{dV}{dt} = \rho_{cold\ air}\,vS$

where v is the average speed of the cold air. With a sheet of paper over the crack the heat loss is by conduction through the paper:

$$\frac{dQ}{dt}(\text{paper}) = \kappa_{paper}\,\Delta T\,\frac{S}{\delta}$$

The paper thickness is $\delta = 0.18$ mm

The ratio of the heat loss in the two cases is

$$\frac{\frac{dQ}{dt}(\text{cabin air})}{\frac{dQ}{dt}(\text{paper})} = \frac{\bar{c}_{air}\,\Delta T\,\rho_{air}\,v\,S}{\kappa_{paper}\,\Delta T\,\frac{S}{\delta}} = \frac{\bar{c}_{air}\,v\,\rho_{air}\,\delta}{\kappa_{paper}}$$

$$= \frac{\left(0.496 \frac{\text{Cal}}{\text{kg-deg}}\right)\left(2.5 \frac{\text{m}}{\text{s}}\right)\left(1.3 \frac{\text{kg}}{\text{m}^3}\right)\left(0.18 \times 10^{-3} \text{ m}\right)}{0.70 \times 10^{-4} \frac{\text{Cal}}{\text{s-m-deg}}} = 4.15$$

22-46

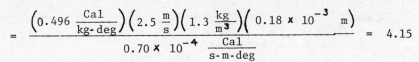

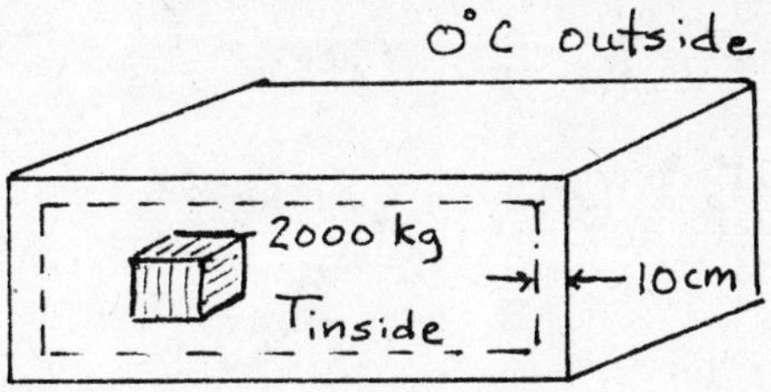

300 m² of surface area through which conduction takes place

If the furnace is turned off when the inside temperature is uniformly 20°C, the heat conducted through the walls begins to lower the temperature.

$$\frac{dQ}{dt}(\text{out}) = \bar{\kappa}_{\text{walls}} \Delta T_{\text{inside to outside}} \frac{A}{\ell}$$

$$= \bar{\kappa}_{\text{walls}} \frac{A}{\ell} (T_{\text{inside}} - 0°\text{C})$$

Also, the change in the inside temperature is related to the total amount of heat conducted out. The heat lost by the 2000 kg is conducted to the outside.

$$Q_{\text{out}} = Q_{\text{of 2000 kg}} = (m\bar{c})_{2000\text{ kg}} \Delta T_{\text{total change inside}}$$

$$= (m\bar{c})_{2000\text{ kg}} (20°\text{C} - T_{\text{inside}})$$

$$\frac{dQ}{dt}(\text{out}) = -(m\bar{c})_{2000\text{ kg}} \times \frac{dT}{dt}(\text{inside})$$

Equating the two expressions for $\frac{dQ}{dt}(\text{out})$,

$$\frac{dT}{dt}(\text{inside}) = \frac{\bar{\kappa}_{\text{walls}} A}{(m\bar{c})_{2000\text{ kg}} \ell} T_{\text{inside}}$$

$$= -3.75 \times 10^{-5} \text{ s}^{-1} T_{\text{inside}}$$

The solution of this differential equation which satisfies the conditions,

$T_{\text{inside}}(t = 0) = 20°\text{C}$ and $T_{\text{inside}}(t = \infty) = 0°\text{C}$, is

$$T_{\text{inside}} = (20°\text{C})\, e^{-(3.75 \times 10^{-5} \text{ s}^{-1})t}$$

(a) After 6 hours = 2.16×10^4 s,

$$T_{\text{inside}} = (20°\text{C})\, e^{-(3.75 \times 10^{-5} \text{ s}^{-1})(2.16 \times 10^4 \text{ s})} = 8.90°\text{C}$$

(b) The heat required to raise the temperature back to 20°C inside must be equal to the heat lost in lowering the temperature to 8.90°C.

$$Q_{\text{out}}(\text{after 6 hours}) = Q_{\text{inside}} = (m\bar{c})_{2000\text{ kg}} (20°\text{C} - T_{6h})$$

$$= (2000 \text{ kg})\left(0.20 \frac{\text{Cal}}{\text{kg-deg}}\right)(20°\text{C} - 8.90°\text{C}) = 4440 \text{ Cal}$$

(c) In order to maintain the inside temperature at 20°C, the furnace must supply heat as quickly as it is lost.

$$\frac{dQ}{dt}(\text{out}) = \bar{\kappa}_{\text{walls}} \frac{A}{\ell} (T_{\text{inside}} - 0°\text{C}) = \bar{\kappa}_{\text{walls}} \frac{A}{\ell} (20 \text{ deg})$$

$$= \left(5.0 \times 10^{-6} \frac{\text{Cal}}{\text{s-m-deg}}\right)\left(\frac{300 \text{ m}^2}{0.10 \text{ m}}\right)(20 \text{ deg}) = 0.30 \frac{\text{Cal}}{\text{s}}$$

Over a period of six hours,

$$Q = \int_0^{6h} \frac{dQ}{dt}\, dt = \left(0.30 \frac{\text{Cal}}{\text{s}}\right)(6\text{h}) = 6480 \text{ Cal}$$

(d) The amount of heat saved over the six hours by shutting off the furnace was

6480 Cal − 4440 Cal = 2040 Cal

22-47

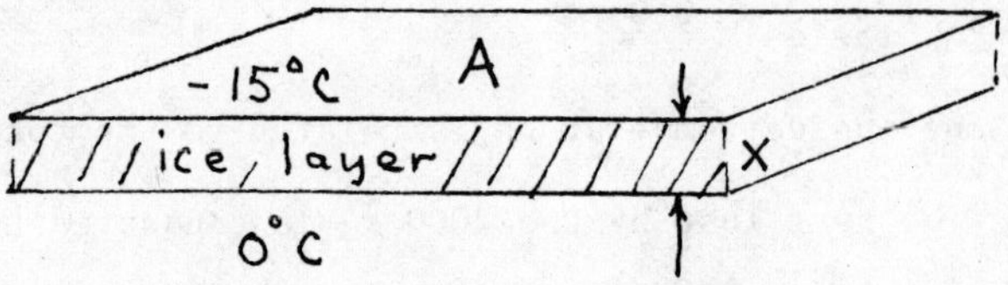

(a) The temperature at the ice-water junction is 0°C. The conduction equation for heat loss through a section of the ice layer of surface area A is

$$\frac{dQ}{dt} = \kappa_{\text{ice}} \Delta T \frac{A}{x}$$

where x is the ice thickness at time t. The heat loss turns the water just below the ice layer of surface area A into ice according to

$$Q_{\text{out}} = Q_{\text{water}} = m_{\text{water}} L_f$$

$$m_{\text{water}} = V\rho = A\rho\, dx$$

dx is the width of the thin layer of water below the ice which is freezing.

Therefore,

$$Q_{\text{out}} = L_f A \rho_{H_2O\ @\ 0°\text{C}}\, dx$$

$$\frac{dQ}{dt}(\text{out}) = L_f A \rho \frac{dx}{dt} = \kappa_{\text{ice}} \Delta T \frac{A}{x}$$

$$\frac{dx}{dt} = \frac{\kappa_{\text{ice}} \Delta T}{L_f \rho} \frac{1}{x}$$

(b) Solving the differential equation,

$$x\, dx = \frac{\kappa \Delta T}{L_f} dt \qquad \text{with } x(t = 0) = 0$$

$$x(t) = \left[\frac{2\kappa \Delta T}{L_f} t\right]^{\frac{1}{2}} = \left[\frac{2 \times 4.0 \times 10^{-4} \frac{\text{Cal}}{\text{s-m-deg}} \times 15 \text{ deg}}{79.71 \frac{\text{Cal}}{\text{kg}} \times 1.00 \times 10^3 \frac{\text{kg}}{\text{m}}} t\right]^{\frac{1}{2}}$$

$$x(2h) = [(1.505 \times 10^{-7} \frac{m^2}{s})(7200s)]^{\frac{1}{2}} = 3.29 \text{ cm}$$

22-48 The temperature at the outside surface of the brass (at R_2) is 20°C. 296 W is the rate of heat generation within the capsule. Noting that isothermal surfaces in the brass will be concentric spheres for uniform heat conduction, look at the conduction through a thin shell of thickenss dr at a radius r from the center.

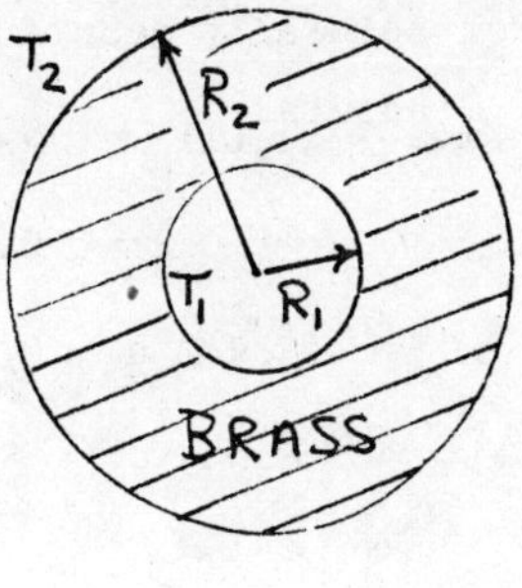

$$\frac{dQ}{dt} = -\kappa \frac{S_\perp}{d\ell_{\parallel}} dT$$

$S_\perp$, the shell area perpendicular to conduction, is $4\pi r^2$. $d\ell_{\parallel}$, the width of the shell through which conduction takes place, is dr.

$$\frac{dQ}{dt} = -\kappa \frac{dT}{dr} 4\pi r^2$$

$$r^2 \frac{dT}{dr} = -\frac{1}{4\pi\kappa}\frac{dQ}{dt} = A = \text{constant}$$

$$\frac{dT}{dr} = \frac{A}{r^2} \quad \text{implies} \quad T(r) = -\frac{A}{r} + B$$

$$\text{But } T(R_2) = T_2 = -\frac{A}{R_2} + B$$

$$T(R_1) = T_1 = -\frac{A}{R_1} + B$$

$$T_1 - T_2 = A\left(\frac{1}{R_2} - \frac{1}{R_1}\right)$$

$$T_1 = T_2 - \frac{1}{4\pi K}\frac{dQ}{dt}\left(\frac{1}{R_2} - \frac{1}{R_1}\right)$$

Once steady temperature conditions are achieved, the rate at which heat is produced in the capsule must be equal to the conduction rate throught the brass layer.

Therefore,

$$T_1 = 20°C - \frac{1}{4\pi\left(0.026 \frac{Cal}{s\text{-}m\text{-}deg}\right)}\left(296 \frac{J}{s}\right)\left(\frac{1}{10.0cm} - \frac{1}{1.0cm}\right)$$

$$= 20°C + 19.48°C = 39.48°C$$

Chapter 23

23-1 mass of ^1H = $1.0007825 \, u(1.66057 \times 10^{-27} \frac{kg}{u})$

$$= 1.6736 \times 10^{-27} \text{ kg}$$

23-2 $m(^7Li) = 7.016004\ u$ 29.58%

$m(^6Li) = ? = x$ 7.42%

For naturally occuring lithium,

$$m(^{natural}Li) = 6.941\ u$$

$$= 0.9258\ (7.016004\ u) + 0.0742\ x$$

$$x = (6.941\ u - 0.9258 \times 7.016004\ u)/0.0742$$

$$= 6.005\ u$$

23-3 $m(^{35}Cl) = 34.96885\ u$ x%

$m(^{37}Cl) = 36.96590\ u$ (100-x)%

$$m(^{natural}Cl) = 35.453\ u$$

$$= x(34.96885\ u) + (1 - x)(36.96590\ u)$$

$$1.99705\ x = 1.5129\ u$$

Therefore,

$$x = \%\ ^{35}Cl = 75.757\%$$

$$100 - x = \%\ ^{37}Cl = 24.243\%$$

23-4 $m(H) = 1.00797\ u$

$m(C) = 12.01115\ u$

$m(0) = 15.9994\ u$

$$m(C_2H_5OH) = 2(12.01115\ u) + 6(1.00797\ u) + 15.9994\ u$$

$$= 46.06952\ u$$

Thus, C_2H_5OH is $\frac{6 \times m(H)}{m(C_2H_5OH)} \times 100\%$ hydrogen by mass

In 1 kg of ethyl alcohol there are

$$\frac{6(1.00797\ u)}{46.06952\ u}(1000\ g) = 131.276\ g \text{ of hydrogen}$$

23-5 The molecular mass of C_4H_9OH is

$$4(12.01115\ u) + 10(1.00797\ u) + 15.9994\ u = 74.1237\ u$$

There are 74.1237 g of C_4H_9OH per mol

$$\frac{\text{molecules}}{cm^3} = \left(\frac{\text{molecules}}{\text{mol}}\right)\left(\frac{\text{mol}}{g}\right)\left(\frac{g}{cm^3}\right)$$

$$= \left(6.02205 \times 10^{23}\ \frac{\text{molecules}}{\text{mol}}\right)\left(\frac{1\ \text{mol}}{74.1237\ g}\right)\left(0.80567\ \frac{g}{cm^3}\right)$$

$$= 6.54552 \times 10^{21}\ \frac{\text{molecules}}{cm^3}$$

23-6 The volume occupied by the spread out oil droplet is

$d_{molecule}\ A$, so that

$$\rho = \frac{m}{V} \text{ gives } d_{molecule} = \frac{m}{\rho A}$$

$$d_{molecule} = (8 \times 10^{-4}\ g)/\left(0.90\frac{g}{cm^3} \times 0.55m^2 \times 10^4 \frac{cm^2}{m^2}\right)$$

$$= 1.62 \times 10^{-7}\ cm = 16.2\ \mathring{A}$$

23-7 Let d be the size (diameter) of a water molecule. Each molecule occupies a volume d^3.

Therefore,

$$\frac{\text{molecules}}{\text{unit volume}} = \frac{1 \text{ molecule}}{d^3}$$

$$= \left(\frac{\text{molecules}}{\text{mol}}\right)\left(\frac{\text{mol}}{g}\right)\left(\frac{g}{\text{unit volume}}\right)$$

$$= \frac{N_o}{m(H_2O)\text{in grams}}$$

$$d = \left[\frac{(2 \times 1.00797 + 15.9994)\frac{g}{mol}}{6.02205 \times 10^{23}\frac{\text{molecules}}{mol} \times 1\frac{g}{cm^3}}\right]^{1/3} = 3.1043\ \mathring{A}$$

23-8 Use the ideal gas law $pV = nRT$.

Initially, $p_oV_o = n_oRT_o$

$$(10.0\ atm)V_o = n_oR(15 + 273.15)K$$

Finally, $p_fV_f = n_fRT_f$

or, $p_fV_o = \frac{1}{2}n_oR(65 + 273.15)K$

Dividing these equations,

$$\frac{p_f}{10.0\ atm} = \frac{1}{2}\left(\frac{338.15\ K}{288.15\ K}\right)$$

$p_f = 5.87$ atm

23-9 (a) Initially, $P_oV_o = n_oRT_o$

$$(1\ atm)V_o = n_oR(10 + 273.15)K$$

Finally, $P_fV_f = n_fRT_f$

$$P_f(0.28\ V_o) = n_oR(40 + 273.15)K$$

Dividing these equations,

$$\frac{0.28 \times P_f}{1\ atm} = \frac{313.15\ K}{283.15\ K} \text{ giving } P_f = 3.95\ atm$$

$$= 2.95 \text{ atm gauge} = 4.002 \times 10^5 \text{ Pa(abs.)} = 43.362\ lb/in^2 \text{ (gauge)}$$

(b) After being driven,

$$P_d(1.02)(0.28V_o) = n_oR(85 + 273.15)K$$

Dividing equations,

$$\frac{(1.02)p_d}{p_f} = \frac{358.15\ K}{313.15\ K}$$

$$p_d = 1.1213 \times p_f = 4.48 \times 10^5\ Pa(abs.) = 50.405\ lb/in^2(gauge)$$

23-10 (a) Initially the air in the bell satisfies $p_o V_o = n_o R T_o$

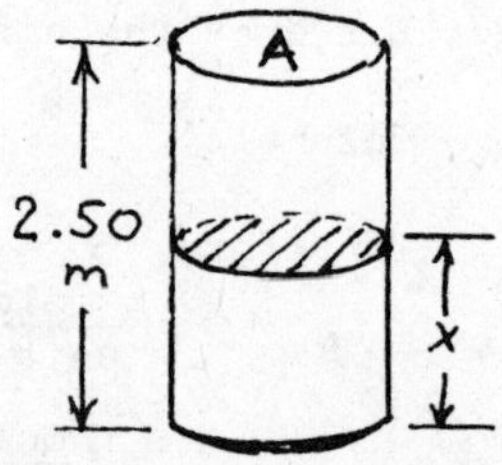

$$p_{surface} V_{bell} = nRT_o$$

or, (1) $p_{surface}(2.5m \times A) = nRT_o$

When the bell is lowered, the air in the bell satisfies

(2) $p_{bell}(2.5m - x)A = nRT_f$

where x is the height the water rises in the bell. Also, the pressure in the bell, once it is lowered, is equal to the seawater pressure at the depth of the water level in the bell.

(3) $$p_{bell} = p_{surface} + \rho g(82.3m - x)$$
$$\cong p_{surface} + \rho_{sea}\, g(82.3m)$$

The approximation is good, as $x < 2.50$ m. Substituting (3) into (2) and substituting nR from (1) into (2),

$$[p_{surface} + \rho_{sea}\, g(82.3m)](2.5m - x)A = p_{surface} V_{bell} \frac{T_f}{T_o}$$

Using $p_{surface} = 1\ atm = 1.01325 \times 10^5$ Pa

and $\rho_{sea} = 1.035 \times 10^3 \frac{kg}{m^3}$

$$x = 2.5m \left[1 - \frac{T_f}{T_o}\left(1 + \frac{\rho_{sea}\, g \times 82.3m}{p_{surface}}\right)^{-1}\right]$$

$$= 2.5m \left[1 - \frac{277.15\ K}{293.15\ K}\left(1 + \frac{1.035 \times 10^3 \frac{kg}{m^3} \times 9.8 \frac{m}{s^2} \times 82.3m}{1.01325 \times 10^5\ Pa}\right)^{-1}\right] = 2.24m$$

(b) If the water in the bell is to be expelled, the air pressure in the bell must be raised to the water pressure at the bottom of the bell.

That is,

$$p_{bell} = p_{surface} + \rho_{sea}\, g(82.3m)$$

$$= 1.01325 \times 10^5\ Pa + (1.035 \times 10^3 \frac{kg}{m^3})(9.8 \frac{m}{s^2})(82.3m)$$

$$= 9.3609 \times 10^5\ Pa = 9.2385\ atm$$

23-11 The volume of the gas is constant at 2.0 ℓ, and the number of moles is constant.

$$n = \text{(grams)}/\text{(grams/mol)}$$

$$= 1.50\ \text{g}/(4.0026\ \text{g/mol}) = 0.3748\ \text{mol}$$

Therefore,

$$p = \frac{nRT}{V} = \frac{(0.3748\ \text{mol})(8.3144\ \frac{\text{J}}{\text{mol-K}})(-150 + 273)\text{K}}{2 \times 10^{-3}\ \text{m}^3}$$

$$= 1.9165 \times 10^{5}\ \text{Pa} = 1.8914\ \text{atm}$$

23-12 Initially the pressure in the tube is zero (it is evacuated).

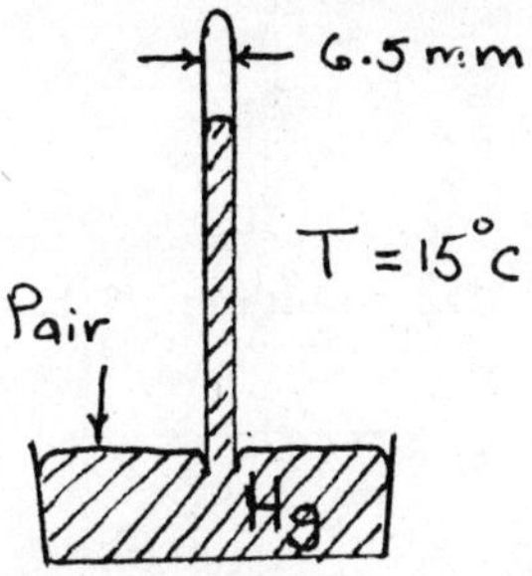

$$p_{air} = \rho_{Hg} gh$$
$$= \rho_{Hg} g(0.759\ \text{m})$$

When the nitrogen is injected,

$$p_{air} = \rho_{Hg} gh' + p_{N_2}$$
$$= \rho_{Hg} g(0.622\text{m}) + p_{N_2}$$

$$p_{N_2} = \rho gh - \rho gh' = (1.36 \times 10^4\ \frac{\text{kg}}{\text{m}^3})(9.8\ \frac{\text{m}}{\text{s}^2})(0.759\ \text{m} - 0.622\ \text{m})$$

$$= 1.8259 \times 10^4\ \text{Pa}$$

$$p_{N_2} V_{N_2} = n_{N_2} RT_{N_2} \text{ gives } n_{N_2} = \frac{p_{N_2} V_{N_2}}{RT_{N_2}}$$

$$V_{N_2} = (0.08\text{m} + 0.759\text{m} - 0.622\text{m})\pi \left(\frac{6.5\ \text{mm}}{2}\right)^2$$

$$n_{N_2} = \frac{(1.8259 \times 10^4\ \text{Pa})(7.201 \times 10^{-6}\ \text{m}^3)}{(8.3144\ \frac{\text{J}}{\text{mol-deg-K}})(15 + 273.15)\text{K}}$$

$$= 5.488 \times 10^{-5}\ \text{mol of nitrogen}$$

Mass of injected N_2 $= (5.488 \times 10^{-5}\ \text{mol})(28.01\ \frac{\text{g}}{\text{mol}}) = 1.54\ \text{mg}$

23-13 (a) For the mixture,

$$(1)\quad p_{mixture} V = (n_{He} + n_{O_2})RT$$

$$m_{He} = n_{He}\ (4.0026\ \frac{\text{g}}{\text{mol}})$$

$$m_{O_2} = n_{O_2}\ (31.9988\ \frac{\text{g}}{\text{mol}})$$

$$\rho_{mixture} = \frac{m_{mixture}}{V} = \frac{m_{He} + m_{O_2}}{V} = 0.980\ \frac{\text{kg}}{\text{m}^3}$$

$$\text{or (2)}\quad 0.980\ \frac{\text{kg}}{\text{m}^3} = \frac{1}{V}\left[n_{He}(4.0026\ \frac{\text{g}}{\text{mol}}) + n_{O_2}\ (31.9988\ \frac{\text{g}}{\text{mol}})\right]$$

Eliminating the unknown V from (1) and (2),

$$p_{mix}\left[n_{He}(4.0026\ \frac{\text{g}}{\text{mol}}) + n_{O_2}\ (31.9988\ \frac{\text{g}}{\text{mol}})\right] = 0.980\ \frac{\text{kg}}{\text{m}^3}\ (n_{He} + n_{O_2})RT$$

$$\frac{n_{O_2}}{n_{He}} = 2.1716$$

The gases are in the same volume at the same temperature, so that

$$\frac{p_{He}}{p_{mix}} = \frac{n_{He}}{n_{O_2} + n_{He}} = \frac{1}{\frac{n_{O_2}}{n_{He}} + 1} = \frac{1}{2.1716 + 1} = 0.3153$$

Therefore,

$$p_{He} = 0.3153(1 \text{ atm}) = 0.3153 \text{ atm}$$

$$p_{O_2} = p_{mix} - p_{He} = 0.6847 \text{ atm}$$

(b) The fraction of the mixture that is O_2 by mass is

$$\frac{m_{O_2}}{m_{He} + m_{O_2}} = \left(1 + \frac{m_{He}}{m_{O_2}}\right)^{-1} = \left(1 + \frac{n_{He} \times 4.0026 \frac{g}{mol}}{n_{O_2} \times 31.9988 \frac{g}{mol}}\right)^{-1}$$

$= 0.9455 = 94.55\%$ by mass is O_2

23-14 The air and water vapor can be treated separately. Assuming that there is essentially no water vapor in the air when the tank is open, the pressure is due only to the air. The same amount of air is in the tank before and after it is sealed, and it is in the same volume.

For the air in the tank,

$$p_oV = nRT_o \qquad p_fV = nRT_f$$

$$p_f = p_o \frac{T_f}{T_o} = 1 \text{ atm} \frac{(120 + 273.15)\text{K}}{(20 + 273.15)\text{K}} = 1019.25 \text{ torr}$$

Once the tank is closed and the temperature is raised, water evaporates until the air is saturated. The partial pressure at 120°C for the vapor is

$1488.9 \text{ torr} = p_v$ (tabulated value)

p_{total}(absolute) $= p$ (air) $+ p_v$

$= 1019.25 \text{ torr} + 1488.9 \text{ torr} = 2508.15 \text{ torr}$

$= 3.300 \text{ atm} = 2.300 \text{ atm gauge} = 1748.1 \text{ torr gauge}$

23-15 Atmospheric pressure on the top of Mt Whitney, California is

$$p = p_o e^{-\alpha y} = (1 \text{ atm}) e^{-0.125 \text{ km}^{-1} (4.42 \text{ km})}$$

$= 0.5755 \text{ atm} = 437.4 \text{ torr}$

Water boils when the vapor pressure exceeds atmospheric pressure.

At 80°C, $p_v = 355.1$ torr

At 90°C, $p_v = 525.8$ torr

Linear interpolation gives $T_{boiling} = 84.8°C$

23-16 $M(C_6H_{12}O_6) = (6 \times 12.01115 + 12 \times 1.00797 + 6 \times 15.9994) \frac{g}{mol} = 180.1589 \frac{g}{mol}$

Therefore, $250\ g = 250\ g/(180.1589 \frac{g}{mol})$

$= 1.3877$ mol

Also, as $\rho_{water} = 1.0 \times 10^3 \frac{kg}{m^3}$

3 kg of water $= 3\ kg\left(\frac{1\ m^3}{1.0 \times 10^3\ kg}\right) = 3 \times 10^{-3}\ m^3 = 3\ \ell$

The molarity of the solution is

$$\frac{1.3877\ mol}{3\ \ell} = 0.4626\ molal$$

Then, $T_{boil} = 100°C + 0.512 \frac{deg}{molal}(0.4626\ molal)$

$= 100.24°C$

23-17

$$\text{relative humidity @ 25°C} = \frac{\text{actual partial pressure (25°C)}}{\text{saturated vapor pressure (25°C)}}$$

The air in question becomes saturated at 15°C (the dew point). This means that the actual partial pressure at 25°C is equal to the saturated vapor pressure at 15°C (the actual partial pressure does not change during the cooling to produce dew). Using tabulated values for the saturated vapor pressures,

$$\text{relative humidity @ 25°C} = \frac{12.79\ \text{torr}}{23.76\ \text{torr}} = 53.8\%$$

23-18

$$\text{rel humidity @ 30°C} = \frac{\text{actual partial p (30°C)}}{\text{saturated vapor p (30°C)}} = 70\%$$

or, $p_{actual} = 0.7\ p_v(30°C)$

Water vapor must be removed from the air to decrease the relative humidity to 40%.

$$p'_{actual} = 0.4\ p_v(30°C)$$

Also,

$$p_{actual} = \frac{nRT}{V}, \quad p'_{actual} = \frac{n'RT}{V}$$

$$n - n' = \frac{V}{RT}(p_{actual} - p'_{actual}) = \frac{V}{RT}(0.7 - 0.4)p_v(30°C)$$

$$= \frac{1800\ m^3\ (0.3)(31.71\ torr)}{8.3144\ \frac{J}{mol\text{-}deg\text{-}K}\ (30 + 273.15)\ K} \left(\frac{1.01325 \times 10^5\ Pa}{760\ torr} \right)$$

$$= 9.057 \times 10^2\ mol = 9.057 \times 10^2\ mol\ (18.015\ \frac{g}{mol})$$

$= 16.32$ kg of water removed from the air.

23-19 When the three roots of Van der Waal's equation are equal, we may expand:

(1) $(V - V_c)^3 = V^3 - 3V^2 V_c + 3VV_c^2 - V_c^3 = 0$

where V is a molar volume.

Also, $$(p + \frac{a}{V^2})(V - b) = RT$$

$$pV^3 - (pb + RT)V^2 + aV - ab = 0$$

or, (2) $$V^3 - (b + \frac{RT}{p})V^2 + \frac{a}{p} V - \frac{ab}{p} = 0$$

At the critical point (1) and (2) must be identical, so that

(3) $3V_c = b + \frac{RT_c}{p_c}$ from V^2 coefficients

(4) $3V_c^2 = \frac{a}{p_c}$ from V coefficients

(5) $V_c^3 = \frac{ab}{p_c}$ from constant terms

From (4) and (5)

$$V_c^3 = V_c V_c^2 = V_c(\frac{1}{3}\frac{a}{p_c}) = \frac{ab}{p_c} \quad \text{giving} \quad V_c = 3b$$

Substituting this result into (4)

$$3(3b)^2 = \frac{a}{p_c} \quad \text{gives} \quad p_c = \frac{a}{27b^2}$$

Substituting for V_c and p_c in (3),

$$3(3b) = b + \frac{R}{\left(\frac{a}{27b^2}\right)} T_c$$

$$T_c = \frac{8b\left(\frac{a}{27b^2}\right)}{R} \quad \text{or} \quad T_c = \frac{8a}{27bR}$$

Also, putting $V_c = 3b$ into (3) gives

$$3(3b) = b + \frac{RT_c}{p_c} \quad \text{so that } b = \frac{RT_c}{8p_c}$$

Substituting this value of b into $T_c = \frac{8a}{27bR}$,

$$a = \frac{27}{8}\left(\frac{RT_c}{8p_c}\right) RT_c = \frac{27}{64}\ \frac{R^2\ T_c^2}{p_c}$$

23-20 As b represents the volume occupied by a mole of the gas,

$$b = N_o(\text{volume of a single molecule})$$

$$\cong (6.022 \times 10^{23} \frac{\text{molecules}}{\text{mol}})(2 \times \frac{4}{3} \pi r^3_{\text{atom}}).$$

The touching hard-sphere approximation is made for diatomic molecules.

In Problem 23-19 it was shown that $b = \frac{1}{3} V_c$.

For N_2 , $b = \frac{1}{3} (90.0 \frac{\text{cm}^3}{\text{mol}})$

For O_2 , $b = \frac{1}{3} (78.0 \frac{\text{cm}^3}{\text{mol}})$

Therefore,

$$\frac{1}{3} (90.0 \frac{\text{cm}^3}{\text{mol}}) = (6.022 \times 10^{23} \frac{\text{molecules}}{\text{mol}})(2 \times \frac{4}{3} \pi r^3_N)$$

and $\frac{1}{3} (78.0 \frac{\text{cm}^3}{\text{mol}}) = (6.022 \times 10^{23} \frac{\text{moleucules}}{\text{mol}})(2 \times \frac{4}{3} \pi r^3_O)$

$$r_N = 1.812 \times 10^{-8} \text{ cm} = 1.812 \text{ Å}$$

$$r_O = 1.727 \times 10^{-8} \text{ cm} = 1.727 \text{ Å}$$

23-21 (a) Van der Waal's equation is

$$(p + \frac{a}{V^2})(V - b) = RT$$

Solving for p

$$p = \frac{RT}{V - b} - \frac{a}{V^2}$$

$$pV = RT(1 - \frac{b}{V})^{-1} - \frac{a}{V}$$

Expanding $(1 - \frac{b}{V})^{-1}$

$$pV = RT(1 + \frac{b}{V} + \frac{b^2}{V^2} + \frac{b^3}{V^3} + \cdots) - \frac{a}{V}$$

$$= RT + (RTb - a)/V + RTb^2/V^2 + \cdots$$

In the expansion,

$$pV = RT + A_1(T)/V + A_2(T)/V^2 + \cdots$$

then, $A_1(T) = RTb - a$, $A_2(T) = RTb^2$, $A_3(T) = RTb^3$

and so forth.

(b) As $V = \frac{RT}{P}$ to lowest order, substitute this into the expansion for pV .

$$pV \cong RT + A_1(T)/(\frac{RT}{p}) + \cdots$$

$$= RT + \frac{A_1(T)}{RT} p + \cdots$$

Comparing this with the expansion

$$pV = RT + B_1(T)\,p + B_2(T)p^2 + \cdots$$

shows that

$$B_1(T) = \frac{A_1(T)}{RT} = \frac{RTb - a}{RT} = b - \frac{a}{RT}$$

23-22 Initially

$$p_o V_o = nRT_o$$

$$(1\text{ atm})V_o = nR(20 + 273.15)\text{K}$$

Finally,

$$p_f V_v = nRT_f$$

$$(22\text{ atm})\left(\frac{1}{15}V_o\right) = nRT_f$$

Dividing equations,

$$\left(\frac{22\text{ atm}}{1\text{ atm}}\right)\left(\frac{\frac{1}{15}V_o}{V_o}\right) = \frac{T_f}{293.15\text{ K}} \quad \text{which gives } T_f = 429.95\text{ K} = 156.8°\text{C}$$

23-23 At the lake bottom the bubble of marsh gas satisfies (1) $p_o V_o = nR\,T_o$

At the bottom of the lake the pressure inside the bubble must be equal to the water pressure there, so

(2) $p_o = p_{atm} + \rho gh$

h is the lake depth and p_{atm} the surface pressure. Once the bubble is at the surface,

(3) $p_f V_f = nR\,T_f$

and (4) $p_f = p_{atm}$

Dividing (3) by (1),

$$\frac{p_f V_f}{p_o V_o} = \frac{T_f}{T_o} \qquad \frac{V_f}{V_o} = \frac{p_o}{p_f}\frac{T_f}{T_o}$$

Substituting from (2) and (4),

$$\frac{V_f}{V_o} = \frac{p_{atm} + \rho gh}{p_{atm}}\frac{T_f}{T_o} = \left(1 + \frac{\rho gh}{p_{atm}}\right)\frac{T_f}{T_o}$$

$$\frac{\frac{4}{3}\pi r_f^3}{\frac{4}{3}\pi r_o^3} = \left[1 + \frac{(1.0\times 10^3\ \frac{\text{kg}}{\text{m}^3})(9.8\ \frac{\text{m}}{\text{s}^2})(4.2\text{ m})}{1.01325\times 10^5\text{ Pa}}\right]\frac{(12 + 273.15)\text{K}}{(5 + 273.15)\text{K}}$$

$$\frac{r_f^3}{r_o^3} = 1.4416 \text{ gives } \frac{r_f}{r_o} = 1.130$$

23-24 Substituting into $pV = nRT$,

$$n = \frac{pV}{RT} = \frac{(1.01325 \times 10^5 \text{ Pa})(8 \text{ m} \times 6\text{m} \times 3.5 \text{ m})}{(8.3144 \frac{\text{J}}{\text{mol-deg}})(18 + 273.15)\text{K}}$$

$$= 7.032 \times 10^3 \text{ mol}$$

Therefore, the mass of air is given by

$$m_{air} = n\,M(\text{air}) = (7.032 \times 10^3 \text{ mol})(28.96 \frac{\text{g}}{\text{mol}}) = 203.6 \text{ kg}$$

23-25 Volume per mole of air $= \frac{V}{n} = \frac{RT}{p}$

$$= \frac{(8.3144 \frac{\text{J}}{\text{mol-K}})(25 + 273.15)\text{K}}{1.01325 \times 10^5 \text{ Pa}} = 2.447 \times 10^{-2} \frac{\text{m}^3}{\text{mol}}$$

Also, Volume per mole of air $= N_o d^3$

where d is the average distance between molecules.

$$d = \left(2.447 \times 10^{-2} \frac{\text{m}^3}{\text{mol}} \Big/ 6.022 \times 10^{23} \frac{\text{molecules}}{\text{mol}}\right)^{1/3}$$

or, $d = 34.4 \times 10^{-10} \text{ m} = \underline{34.4 \text{ Å}}$

23-26 $p_{Ar} = 350 \text{ torr} = \frac{350}{760} \text{ atm} = \frac{n_{Ar} RT}{V}$

$p_{CO_2} = 850 \text{ torr} = \frac{850}{760} \text{ atm} = \frac{n_{CO_2} RT}{V}$

(a) $n_{Ar} = \frac{V}{RT} p_{Ar}$

$$= \frac{2.0 \times 10^{-3} \text{ m}^3}{(8.3144 \frac{\text{J}}{\text{mol-K}})(15 + 273.15)\text{K}} (350 \text{ torr})\left(\frac{1.01325 \times 10^5 \text{ Pa}}{760 \text{ torr}}\right)$$

$$= (1.113 \times 10^{-4} \text{ mol})350 = 3.895 \times 10^{-2} \text{ mol}$$

$m_{Ar} = n_{Ar} M(\text{Ar}) = (3.895 \times 10^{-2} \text{ mol})(39.95 \frac{\text{g}}{\text{mol}}) = 1.556 \text{ g}$

$n_{CO_2} = (1.113 \times 10^{-4} \text{ mol})850 = 9.460 \times 10^{-2} \text{ mol}$

$m_{CO_2} = n_{CO_2} M(CO_2) = (9.460 \times 10^{-2} \text{ mol})(44.01 \frac{\text{g}}{\text{mol}}) = 4.163 \text{ g}$

(b) $\rho = \frac{m}{V} = \frac{m_{Ar} + m_{CO_2}}{V} = \frac{1.556 \text{ g} + 4.163 \text{ g}}{2.0 \times 10^{-3} \text{ m}^3} = 2.86 \frac{\text{kg}}{\text{m}^3}$

23-27 (a) Given $\gamma = -\frac{dT}{dy}$, where γ is constant, the solution to this differential equation is

$T(y) = -\gamma y + C$

Setting $T(y = 0) = T_o$ gives $C = T_o$

Therefore, $T(y) = -\gamma y + T_o$

(b) Treating the atmosphere as a liquid ocean of nonuniform density, Pascal's law does not apply. However, over an infinitesimal layer,

$$\frac{dp}{dy} = -\rho g$$

where ρ is the almost constant density of the layer. Also, for the layer at height y,

$$\rho = \frac{m}{V} = \frac{nM_{air}}{V} = \frac{p}{RT}M_{air} \quad \text{with T in deg K}$$

Using the result for T from part (a),

$$\rho(y) = \frac{p(y)}{R(T_o - \gamma y)}M_{air} \quad \text{giving} \quad \frac{dp}{dy} = -\frac{p(y)}{R(T_o - \gamma y)}M_{air}\, g$$

$$\frac{dp}{p} = -\frac{M_{air}\,g}{R}\,\frac{dy}{T_o - \gamma y}$$

$$\int_{p(0)}^{p(y)} \frac{dp}{p} = -\frac{M_{air}\,g}{R}\int_0^y \frac{dy}{T_o - \gamma y}$$

$$\ln\frac{p(y)}{p(0)} = \frac{M_{air}\,g}{\gamma R}\ln\left(\frac{T_o - \gamma y}{T_o}\right)$$

$$p(y) = p_o\left(\frac{T_o - \gamma y}{T_o}\right)^{M_{air}\,g/R\gamma} \qquad \text{where } p(y = 0) = p_o$$

(c) $p = p_o\left(1 - \frac{\gamma y}{T_o}\right)^{Mg/R\gamma}$

Let $x = -\frac{\gamma y}{T_o}$ giving $\gamma = -\frac{xT_o}{y}$

so that $\lim_{\gamma \to 0} x = 0$

Then, $p = p_o(1 + x)^{-Mgy/RT_o x}$

$$= p_o\left[(1 + x)^{\frac{1}{x}}\right]^{-Mgy/RT_o}$$

and

$$\lim_{\gamma \to 0} p(\gamma) = \lim_{x \to 0} p(x) = p_o\left[\lim_{x \to 0}(1 + x)^{\frac{1}{x}}\right]^{-\frac{Mgy}{RT_o}}$$

$$= p_o\, e^{-\frac{Mgy}{RT_o}}$$

(d) At $y = 8882$ m, with $T_o = 15°C$ and $\gamma = \frac{0.6°C}{100m}$,

$$T(8882\text{ m}) = 15°C - \frac{0.6°C}{100\text{ m}}(8882\text{ m})$$

$$= -\underline{38.3°C} = 234.9\text{ K}$$

$$\frac{M_{air}\,g}{R\gamma} = \frac{(0.02896\,\frac{kg}{mol})(9.8\,\frac{m}{s^2})}{(8.3144\,\frac{J}{mol\cdot K})(6 \times 10^{-3}\,\frac{°C}{m})} = 5.689$$

$$p = p_o\left(\frac{T}{T_o}\right)^{\frac{Mg}{R\gamma}} = (1\text{ atm})\left(\frac{234.9\text{ K}}{288.15\text{ K}}\right)^{5.689} = 0.312\text{ atm}$$

The actual values for T and p are lower than these (temperature drops more quickly than linearly)

23-28 h = 759 mm (see Problem 23-12).

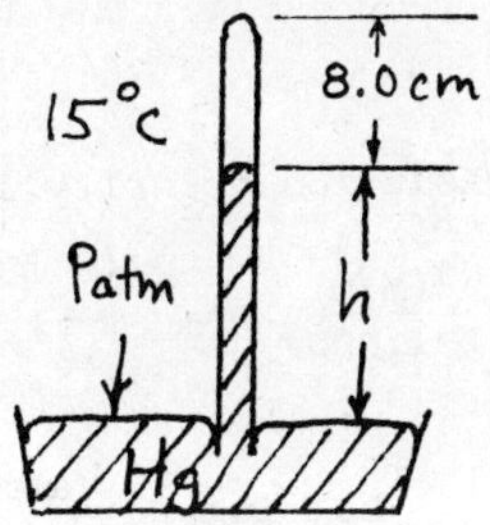

The mercury vapor pressure is

$p_v(\text{Hg}) = 7.8 \times 10^{-4}$ torr

(a) The pressure at the top of the mercury column is $p_v(\text{Hg})$.

Therefore, $p_{atm} = \rho_{Hg} gh + p_v(\text{Hg})$ is exact.

The approximation is that $p_v(\text{Hg}) \cong 0$, so that

$$p'_{atm} = \rho_{Hg} gh$$

The percentage error of the approximation is

$$\frac{p_{atm} - p'_{atm}}{p'_{atm}} = \frac{p_v(\text{Hg})}{759 \text{ mm-Hg}} = \frac{7.8 \times 10^{-4} \text{ torr}}{759 \text{ torr}}$$

$$= 1.0 \times 10^{-4} \ \%$$

(b) $n_{Hg} = \dfrac{pV}{RT}$

$$V \text{ above Hg column} = \pi r^2 (8.0 \text{ cm}) = \pi \left(\frac{6.5 \text{ mm}}{2}\right)^2 (8.0 \text{ cm})$$

Therefore,
$$n_{Hg} = \frac{(7.8 \times 10^{-4} \text{ torr})\left(\dfrac{1.013 \times 10^5 \text{ Pa}}{760 \text{ torr}}\right)(2.65 \times 10^{-6} \text{ m}^3)}{(8.3144 \frac{\text{J}}{\text{mol-K}})(15 + 273.15)\text{K}}$$

$$= 1.15 \times 10^{-10} \text{ mol}$$

$$m_{Hg} = n_{Hg} M(\text{Hg}) = (1.15 \times 10^{-10} \text{ mol})(200.59 \tfrac{\text{g}}{\text{mol}}) = 0.023 \ \mu\text{g}$$

$$= 2.3 \times 10^{-8} \text{ g}$$

23-29 The absolute steam pressure is p_{gauge} + 1 atm = 2.50 atm + 1.00 atm = 3.50 atm = 2660 torr

If the pressure is due entirely to steam, the vapor pressure is 2660 torr at the saturation point. The tabulated values are

t_c = 130°C: p_v = 2025.6 torr

t_c = 140°C: p_v = 2709.5 torr

Linear interpolation gives

$$t_c(p_v = 2660 \text{ torr}) = 130°\text{C} + \left(\frac{2660 - 2025.6}{2709.5 - 2025.6}\right) 10°\text{C} = 139.3°\text{C}$$

23-30 $M(C_{12}H_{22}O_{11}) = (12 \times 12.01115 + 22 \times 1.00797 + 11 \times 15.9994)\ \frac{g}{mol} =$

$$0.3423\ \frac{kg}{mol}$$

The solution molarity is given in mol/liter:

$$M = \frac{(m/M)}{1.00\ \ell} = \frac{(0.50\ kg)\left(\frac{1\ mol}{0.3423\ kg}\right)}{1.00\ \ell} = 1.461\ \text{molal}$$

(a) $T_{boiling}\ (@\ 1\ atm) = 100°C + \left(0.52\ \frac{deg}{molal}\right)(1.461\ molal)$

$= \underline{100.76°C}$

(b) $T_{freezing}\ (@\ 1\ atm) = 0°C - \left(1.86\ \frac{deg}{molal}\right)(1.461\ molal)$

$= -2.72°C$

23-31

$$\text{rel humidity @ } 20°C = \frac{\text{actual partial pressure (20°C)}}{\text{saturated vapor pressure (20°C)}}$$

rel humidity = 20%

actual partial pressure = $p = 0.20\ p_v(20°C)$

or $p = 0.20\ (17.51\ torr) = 3.502\ torr$

After water is vaporized into the room air the relative humidity is raised to 40%.

$p' = 0.40(17.51) = 7.004\ torr$

Since the temperature and volume remain the same the ideal gas law implies that

$$p' - p = (7.004 - 3.502)\,torr = \frac{RT}{V}(n' - n)$$

The number of moles of water vaporized into the air is

$$n' - n = \frac{6m \times 5m \times 3.5m}{(8.3144\ \frac{J}{mol\text{-}K})(20 + 273.15)K}(7.004 - 3.502)\,torr$$

$= 20.11\ mol$

mass H_2O vaporized $= (n' - n)M(H_2O) = (20.11\ mol)(18.015\ \frac{g}{mol}) = \underline{0.362\ kg}$

Assuming $\rho_{H_2O} = 1.0 \times 10^3\ \frac{kg}{m^3}$

$$\text{volume } H_2O \text{ vaporized} = \frac{m}{\rho} = \frac{(0.362\ kg)}{(1.0 \times 10^3\ \frac{kg}{m^3})} \times \left(10^3\ \frac{\ell}{m^3}\right)$$

$= 0.362\ \ell$

23-32 $$\text{rel humidity @ } 20°\text{C} = \frac{\text{actual partial pressure } (20°\text{C})}{\text{saturated vapor pressure } (20°\text{C})}$$

rel humidity = 45%

actual partial pressure = p = 0.45 $p_v(20°C)$

or, p = 0.45(17.51 torr) = 7.880 torr

$$n_{\text{water vapor}} = \frac{pV}{RT} = \frac{(7.880\text{ torr})\left(\frac{1.013 \times 10^5\text{ Pa}}{760\text{ torr}}\right)(8\text{m} \times 6\text{m} \times 2.8\text{m})}{(8.3144 \frac{\text{J}}{\text{mol}\cdot\text{K}})(20 + 273.15)\text{K}}$$

$$= 57.93\text{ mol}$$

$$m_{\text{water vapor}} = n_{\text{water vapor}} M(\text{water vapor})$$

$$= (57.93\text{ mol})(18.015 \frac{\text{g}}{\text{mol}}) = 1.044\text{ kg}$$

23-33 $$\text{rel humidity @ } 40°\text{C} = \frac{\text{actual partial pressure } (40°\text{C})}{\text{saturated vapor pressure } (40°\text{C})}$$

$$\text{actual partial pressure } (40°\text{C}) = p = \frac{n_{H_2O\text{ vapor}} RT}{V} \qquad (1)$$

At 10°C the air is saturated with water vapor and in addition some water has been removed from the air.

actual partial pressure (10°C) = p' = $p_v(10°C)$ = 9.21 torr

$$9.21\text{ torr} = \frac{n'_{H_2O\text{ vapor}} RT'}{V'} \qquad (2)$$

$$\text{Therefore, } n'_{H_2O\text{ vapor}} = 9.21\text{ torr}\frac{V'}{RT'}$$

$$n' = (9.21\text{ torr})\left(\frac{1.013 \times 10^5\text{ Pa}}{760\text{ torr}}\right)\frac{(2.5\text{m}^3 - 35\text{ cm}^3)}{(8.3144 \frac{\text{J}}{\text{mol K}})(10 + 273.15)\text{K}}$$

$$= 1.304\text{ mol}$$

Since the volume of water removed from the air was 35 cm^3, the number of moles of water vapor removed was

$$(35\text{ cm}^3)\left(1 \frac{\text{g}}{\text{cm}^3}\right)\left(\frac{1\text{ mol}}{18.015\text{ g}}\right) = 1.943\text{ mol}$$

Therefore n = n' + 1.943 mol = 3.247 mol

Substituting this into (1),

$$p = \frac{nRT}{V} = \frac{(3.247\text{ mol})(8.3144\text{ J/mol}\cdot\text{K})(40 + 273.15)\text{K}}{2.5\text{ m}^3}$$

$$= 3.381 \times 10^3\text{ Pa} = 25.36\text{ torr}$$

$$\text{rel humidity @ } 40°C = \frac{p(40°C)}{p_v(40°C)} = \frac{25.36 \text{ torr}}{55.13 \text{ torr}} = 46.0\%$$

23-34 Initially for the water vapor,

$$\text{rel humidity @}25°C = \frac{\text{actual partial pressure } (25°C)}{\text{saturated vapor pressure } (25°C)} = 60\%$$

$$\text{actual partial pressure} = p = 0.60(23.76 \text{ torr})$$

$$= 14.26 \text{ torr}$$

Once the bucket is placed in the room, water will evaporate until the saturated vapor pressure is reached (rel humidity = 100%)

$$p' = 1.00\ p_v(25°C) = 23.76 \text{ torr}$$

The number of moles of water evaporated from the bucket is then

$$n' - n = \frac{V}{RT}(p' - p)$$

$$= \frac{200 \text{ m}^3}{(8.3144 \frac{\text{J}}{\text{mol}\cdot\text{K}})(25 + 273.15)\text{K}}(23.76 \text{ torr} - 14.26 \text{ torr})$$

$$= 1.022 \times 10^2 \text{ mol}$$

$$m_{\text{water evaporated}} = (n' - n)M(H_2O)$$

$$V_{\text{evaporated}} = \frac{m}{\rho_{H_2O}} = (1.022 \times 10^2 \text{ mol})(18.015 \frac{\text{g}}{\text{mol}}) / (1 \frac{\text{g}}{\text{cm}^3})$$

$$= 1.842 \times 10^3 \text{ cm}^3 = 1.842\ \ell$$

23-35

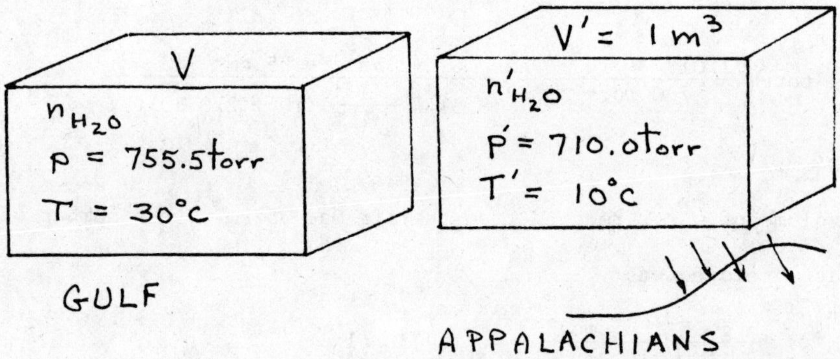

Look at 1m^3 of the air mass over the mountains and assume that all of the rain falls over the mountains under the given conditions. Let

n''_{H_2O} = moles of rain from 1 m^3 of air over the mountains

n'_{H_2O} = moles of water vapor in 1 m^3 of mountain air

n_{H_2O} = moles of water vapor in the correponding volume of air over the Gulf of Mexico.

Thus, $n_{H_2O} = n'_{H_2O} + n''_{H_2O}$

To find V, the volume of air over the Gulf that corresponds to $1 m^3$ over the mountains, note that the amounts of dry air in the corresponding volumes are equal.

Therefore, $n_{dry} = n'_{dry} \qquad \frac{p_{dry} V}{RT} = \frac{p'_{dry} V'}{RT'}$

The air is saturated with water vapor at both locations.

$$p_{H_2O} = p_v(30°C) = 31.71 \text{ torr}$$

$$p'_{H_2O} = p_v(10°C) = 9.21 \text{ torr}$$

$$p_{dry} = p - p_{H_2O} = 755.5 \text{ torr} - 31.71 \text{ torr} = 723.79 \text{ torr}$$

$$p'_{dry} = p' - p'_{H_2O} = 710.0 \text{ torr} - 9.21 \text{ torr} = 700.79 \text{ torr}$$

Therefore,

$$V = \left(\frac{p'_{dry}}{p_{dry}}\right)\left(\frac{T}{T'}\right) V' = \left(\frac{700.79 \text{ torr}}{723.79 \text{ torr}}\right)\left(\frac{30 + 273.15}{10 + 273.15}\right)(1 \text{ m})^3$$

$$= 1.0366 \text{ m}^3$$

Then

$$n_{H_2O} = \frac{p_{H_2O} V}{RT} = \frac{(31.71 \text{ torr})\left(\frac{1.013 \times 10^5 \text{ Pa}}{760 \text{ torr}}\right)(1.0366 \text{ m}^3)}{(8.3144 \frac{J}{mol\cdot K})(30 + 273.15)K}$$

$$= 1.739 \text{ mol}$$

$$n'_{H_2O} = \frac{p'_{H_2O} V'}{RT'} = \frac{(9.21 \text{ torr})\left(\frac{1.013 \times 10^5 \text{ Pa}}{760 \text{ torr}}\right)(1.0366 \text{ m}^3)}{\left(8.3144 \frac{J}{mol\cdot K}\right)(10 + 273.15)K}$$

$$n'_{H_2O} = 0.541 \text{ mol}$$

$$n''_{H_2O} = n_{H_2O} - n'_{H_2O} = 1.198$$

is the amount of rain from the $1 m^3$ of air.

Converting to mass,

$$m'' = n''_{H_2O} M(H_2O) = (1.198 \frac{mol}{m^3})(18.015 \frac{g}{mol}) = 21.6 \frac{g}{m^3}$$

23-36 Given $\left(p + \frac{a'}{T V^2}\right)(V - b') = RT$

with $a' = \frac{16}{3} p_c V_c^2 T_c \qquad (1)$

$b' = \frac{1}{4} V_c \qquad (2)$

$p_c V_c = \frac{9}{32} RT_c \qquad (3)$

(a) From (3), $V_c = \frac{9}{32}\frac{RT_c}{p_c}$ (4)

Substituting into (1),

$$a' = \frac{16}{3}p_c\left(\frac{9}{32}\frac{RT_c}{p_c}\right)^2 T_c = \frac{27}{64}\frac{R^2T_c^3}{p_c}$$

Combining (2) and (4),

$$b' = \frac{1}{4}V_c = \frac{1}{4}\left(\frac{9}{32}\frac{RT_c}{p_c}\right) = \frac{9}{128}\frac{RT_c}{p_c}$$

(b) Rearranging Berthelot's equation,

$$p + \frac{a'}{TV^2} = \frac{RT}{V - b'},$$

$$pV = RT(1 - \frac{b'}{V})^{-1} - \frac{a'}{TV}$$

$$= RT(1 + \frac{b'}{V} + \frac{b'^2}{V^2} + \cdots) - \frac{a'}{TV}$$

Using the lowest order, $V = RT/p$, on the right-hand side, there results

$$pV = RT + (b' - \frac{a'}{RT^2})\,p + \cdots$$

(c) At Boyle's temperature the first order coefficient vanishes.

$$b' - \frac{a'}{RT_B^2} = 0 \text{ gives } b' = \frac{a'}{RT_B^2}$$

Using the results of part (a),

$$b' = \frac{9}{128}\frac{RT_c}{p_c} = \frac{a'}{RT_B^2} = \left(\frac{27}{64}\frac{R^2T_c^3}{p_c}\right)\frac{1}{RT_B^2}$$

$$T_B^2 = 6T_c^2 \longrightarrow T_B = \sqrt{6}\,T_c$$

(d) Tables give $T_c(N_2) = 126.2$ K

Van der Waal's equation has $T_B = \frac{27}{8}T_c = 425.9$ K

Berthelot's equation has $T_B = \sqrt{6}\,T_c = 308.6$ K

The observed value is $T_B = 327.2$ K

(e) For Van der Waal's equation,

$$pV = RT + (b - \frac{a}{RT})p + \cdots$$

with $a = \frac{27}{64}\frac{R^2T_c^2}{p_c}$ and $b = \frac{RT_c}{8p_c}$ (see Problem 23-19)

$$\frac{pV}{RT} = 1 + \left(\frac{1}{8} - \frac{27}{64}\frac{T_c}{T}\right)\frac{T_c}{T}\frac{p}{p_c} + \cdots$$

For Berthelot's equation,

$$pV = RT + (b' - \frac{a'}{RT^2})p + \cdots$$

with $a' = \frac{27}{64}\frac{R^2 T_c^3}{p_c}$ and $b' = \frac{9}{128}\frac{RT_c}{p_c}$

$$\frac{pV}{RT} = 1 + \left(\frac{9}{128} - \frac{27}{64}\frac{T_c}{T}\right)\frac{T_c}{T}\frac{p}{p_c} + \cdots$$

At T = 300 K and p = 70 atm with $T_c(H_2)$ = 33.2K and p_c = 12.8 atm,

$$\left.\frac{pV}{RT}\right|_{\text{Berthelot's}} = 1 + 0.0394 = 1.039$$

$$\left.\frac{pV}{RT}\right|_{\text{Van der Waal's}} = 1 + 0.047 = 1.047$$

The observed value is $\frac{pV}{RT} = 1.042$

CHAPTER 24

24-1 (a)

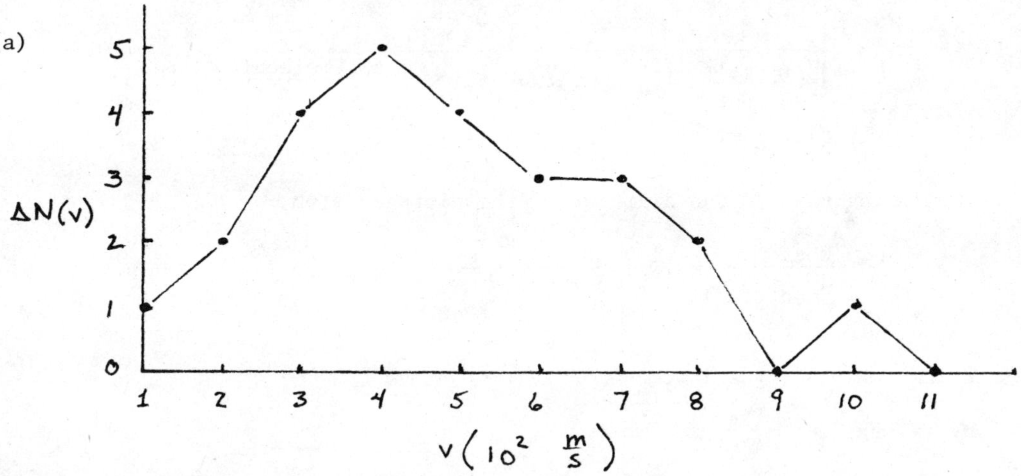

(b) $$\bar{v} = \frac{\sum_i n_i v_i}{\sum_i n_i}$$

$$= \frac{\begin{pmatrix}1 \times 1 + 2 \times 2 + 4 \times 3 + 5 \times 4 + 4 \times 5 \\ + 3 \times 6 + 3 \times 7 + 2 \times 8 + 1 \times 10\end{pmatrix} \times 10^2 \frac{m}{s}}{1 + 2 + 4 + 5 + 4 + 3 + 3 + 2 + 1}$$

$$= \frac{122 \times 10^2 \frac{m}{s}}{25} = 488 \frac{m}{s}$$

(c) $$v_{rms} = \sqrt{\overline{v^2}} = \left(\frac{\sum_i n_i v_i^2}{\sum_i n_i}\right)^{1/2}$$

$$= \left[\begin{pmatrix}1 \times 1^2 + 2 \times 2^2 + 4 \times 3^2 + 5 \times 4^2 + 4 \times 5^2 \\ + 3 \times 6^2 + 3 \times 7^2 + 2 \times 8^2 + 1 \times 10^2\end{pmatrix} \times \frac{10^4 \frac{m^2}{s^2}}{25}\right]^{1/2}$$

$$v_{rms} = 532 \frac{m}{s}$$

(d) $$\frac{v_{rms}}{\bar{v}} = \frac{532\frac{m}{s}}{488\frac{m}{s}} = 1.090$$

For an ideal gas, $\frac{v_{rms}}{\bar{v}} = \sqrt{\frac{3\pi}{8}} = 1.085$

The given distribution follows the Maxwell-Boltzmann distribution in this respect.

24-2

(a) $$v_{rms} = \left(\frac{3p}{\rho}\right)^{1/2} = \left(\frac{3pV}{m}\right)^{1/2}$$

$$V = \frac{m}{3p} v_{rms}^2$$

$$= \frac{(5.0 \times 10^{-6}\ \text{kg})(300\ \frac{\text{m}}{\text{s}})^2}{(3 \times 0.10\ \text{atm})(1.013 \times 10^5\ \frac{\text{Pa}}{\text{atm}})} = 14.8\ \text{cm}^3$$

(b) If d is the average spacing between molecular centers, $n N_o d^3 = V$

$$d = \left(\frac{V}{nN_o}\right)^{1/3}$$

$$n(N_2) = \frac{m_{N_2}}{M(N_2)} = \frac{5.0 \times 10^{-3}\ \text{g}}{28.01\ \frac{\text{g}}{\text{mol}}} = 1.785 \times 10^{-4}\ \text{mol}$$

$$\text{and } d = \left[\frac{1.48 \times 10^{-5}\ \text{m}^3}{(1.785 \times 10^{-4}\ \text{mol})(6.022 \times 10^{23}\ \frac{\text{molecules}}{\text{mol}})}\right]^{1/3}$$

$$= 5.16 \times 10^{-9}\ \text{m}$$

Comparing this to the diameter of the nitrogen atom,

$$\frac{d_{N\ atom}}{d_{centers}} = \frac{1.06 \times 10^{-10}\ \text{m}}{5.16 \times 10^{-9}\ \text{m}} = 2.1\%$$

24-3 (a) Using the ideal gas law,

$$pV = nRT = \frac{m_{Xe}}{M(Xe)} RT$$

$$M(Xe) = \frac{mRT}{pV} = \frac{(5.551\ \text{g})(8.3144\ \frac{\text{J}}{\text{mol}\cdot\text{K}})(15 + 273.15)\text{K}}{(1.01325 \times 10^5\ \text{Pa})(10^{-3}\ \text{m}^3)}$$

$$= 1.3125 \times 10^2\ \frac{\text{g}}{\text{mol}}$$

Therefore, $m(Xe) = 131.25\ \text{u}$

(b)

$$v_{rms} = \left(\frac{3p}{\rho}\right)^{1/2} = \left(\frac{3pV}{m}\right)^{1/2}$$

$$= \left[\frac{(3 \times 1.01325 \times 10^5\ \text{Pa}) \times (10^{-3}\ \text{m}^3)}{5.551 \times 10^{-3}\ \text{kg}}\right]^{1/2} = 234.0\ \frac{\text{m}}{\text{s}}$$

(c) $$\Delta U_{grav} = U_{grav\ top} - U_{grav\ bottom}$$

$$= m_{Xe\ atom}\, gh$$

where h is the cube height.

$$\Delta U_{grav} = (131.25\ \text{u})(1.66057 \times 10^{-27}\ \frac{\text{kg}}{\text{u}})(9.8\ \frac{\text{m}}{\text{s}^2})(0.10\ \text{m})$$

$$= 2.136 \times 10^{-25}\ \text{J}$$

$$\overline{K} = \frac{1}{2}(m_{Xe\ atom})\overline{v^2} = \frac{1}{2}\mu v_{rms}^2$$

$$= \frac{1}{2}(131.25\ u)(1.66057 \times 10^{-27}\ \frac{kg}{u})(234.0\ \frac{m}{s})^2$$

$$= 5.967 \times 10^{-21}\ J$$

The gravitational energy is negligible when compared to the kinetic energy of the atoms.

$$\overline{K}/\Delta U_{grav} = 2.79 \times 10^4$$

24-4 (a) $v_{rms} = \sqrt{\frac{3kT}{\mu}}$ where μ is the molecular mass.

At 20°C,

$$\sqrt{3kT} = [(3 \times 1.38066 \times 10^{-23}\ \frac{J}{K})(20 + 273.15)K]$$

$$= 1.1019 \times 10^{-10}\ kg^{+1/2}\ \frac{m}{s}$$

$$\mu_{H_2} = 2 \times 1.00797u \times 1.66057 \times 10^{-27}\ \frac{kg}{u} = 3.348 \times 10^{-27}\ kg$$

$$\mu_{He} = 4.0026u \times 1.66057 \times 10^{-27}\ \frac{kg}{u} = 6.647 \times 10^{-27}\ kg$$

$$\mu_{CO_2} = 44.01u \times 1.66057 \times 10^{-27}\ \frac{kg}{u} = 7.308 \times 10^{-26}\ kg$$

Therefore,

$$v_{rms}(H_2) = (1.1019 \times 10^{-10}\ kg^{1/2}\ \frac{m}{s})(3.348 \times 10^{-27}\ kg)^{-1/2} = 1904.5\ \frac{m}{s}$$

$$v_{rms}(He) = (1.1019 \times 10^{-10}\ kg^{1/2}\ \frac{m}{s})(6.647 \times 10^{-27}\ kg)^{-1/2} = 1351.6\ \frac{m}{s}$$

$$v_{rms}(CO_2) = (1.1019 \times 10^{-10}\ kg^{1/2}\ \frac{m}{s})(7.308 \times 10^{-26}\ kg)^{-1/2} = 407.6\ \frac{m}{s}$$

(b) At 100°C,

$$\sqrt{3kT} = [(3 \times 1.38066 \times 10^{-23}\ \frac{J}{K})(100 + 273.15)K]$$

$$= 1.2432 \times 10^{-10}\ kg^{1/2}\ \frac{m}{s}$$

$$\mu_{O_2} = 31.9988u \times 1.66057 \times 10^{-27}\ \frac{kg}{u} = 5.3136 \times 10^{-26}\ kg$$

$$\mu_{Ar} = 39.948u \times 1.66057 \times 10^{-27}\ \frac{kg}{u} = 6.6336 \times 10^{-26}\ kg$$

$$\mu_{NH_3} = (14.0067u + 3 \times 1.00797u) \times 1.66057 \times 10^{-27}\ \frac{kg}{u}$$

$$= 2.828 \times 10^{-26}\ kg$$

Therefore,

$$v_{rms}(O_2) = (1.2432 \times 10^{-10}\ kg^{1/2}\ \frac{m}{s})(5.3136 \times 10^{-26}\ kg)^{-1/2} = 539.3\ \frac{m}{s}$$

$$v_{rms}(Ar) = (1.2432 \times 10^{-10}\ kg^{1/2}\ \frac{m}{s})(6.6336 \times 10^{-26}\ kg)^{-1/2} = 482.7\ \frac{m}{s}$$

$$v_{rms}(NH_3) = (1.2432 \times 10^{-10}\ kg^{1/2}\ \frac{m}{s})(2.828 \times 10^{-26}\ kg)^{-1/2} = 739.3\ \frac{m}{s}$$

24-5 At 15°C assume that only the translational and rotational modes are present.

$$\nu = \text{degrees of freedom} = 3_{trans} + 3_{rot} \text{(triatomic)}$$

$$= 6$$

Then, $C_v = \frac{1}{2}\nu R = \frac{1}{2}(6)(8.3144 \frac{J}{mol \cdot K}) = 24.94 \frac{J}{mol \cdot K}$ is the molar specific heat.

$$M(H_2S) = (2 \times 1.00797 + 32.06)\frac{g}{mol}$$

$$\text{Therefore, } C_v = (24.94 \frac{J}{mol \cdot K})(\frac{1 \text{ mol}}{0.03408 \text{ kg}})(\frac{1 \text{ Cal}}{4186 \text{ J}})$$

$$= 0.175 \frac{\text{Cal}}{\text{kg} \cdot \text{deg}}$$

24-6 $T = \frac{1}{4} T_D$

T_D(diamond)	= 1860 K	$\longrightarrow$	$T_{diamond}$	= 465 K
T_D(Al)	= 398 K	$\longrightarrow$	T_{Al}	= 99.5 K
T_D(Cd)	= 168 K	$\longrightarrow$	T_{Cd}	= 42 K
T_D(Pb)	= 88 K	$\longrightarrow$	T_{Pb}	= 22 K

24-7 $C_v(FeS_2 \text{ at } T = 45 \text{ K}) = 0.653 \frac{J}{mol\text{-}K}$

The Debye theory predicts that at low enough temperature ($\frac{T}{T_D} \lesssim 0.15$),

$$C_v = \left(1945 \frac{J}{mol\text{-}K}\right)\left(\frac{45 \text{ K}}{T_D}\right)^3$$

$$T_D = 45 \text{ K}\left(\frac{1945}{0.653}\right)^{1/3} = 647.5 \text{ K}$$

$$\text{Checking, } \frac{T}{T_D} = \frac{45K}{647.5 \text{ K}} = 0.07 < 0.15$$

24-8 (a) $C_e = \gamma T$

$$C_e(\text{Ni at } 40 \text{ K}) = \left(7.3 \times 10^{-3} \frac{J}{mol\text{-}K^2}\right)(40 \text{ K}) = 0.292 \frac{J}{mol\text{-}K}$$

$$C_v(\text{Debye at } 40 \text{ K}) = \left(1945 \frac{J}{mol\text{-}K}\right)\left(\frac{40 \text{ K}}{413 \text{ K}}\right)^3 = 1.767 \frac{J}{mol \cdot K}$$

Therefore, $\frac{C_e}{C_{v(Debye)}}$ at 40K = 16.5%

(b) $\Delta U_{\text{internal energy per mole}} = \int C_v(T)dT$

where $C_v(T)$ is the molar specific heat.

Using $C_v(\text{Debye}) = \left(1945 \frac{J}{mol \cdot K}\right)(\frac{T}{T_D})^3$ and $C_e = \gamma T$

$$\Delta U = \int_{0 K}^{40 K} (C_v + C_e)dT$$

$$= \left(1945 \frac{J}{mol \cdot K}\right)\int_{0 K}^{40 K}\left(\frac{T}{413 \text{ K}}\right)^3 dT + \left(7.3 \times 10^{-3} \frac{J}{mol\text{-}K^2}\right)\int_{0 K}^{40 K} T \, dT$$

$$= \left(1945 \frac{J}{mol\text{-}K}\right)\frac{1}{4}\ \frac{(40\ K)^4}{(413\ K)^3} + \left(7.13 \times 10^{-3}\ \frac{J}{mol\text{-}K^2}\right)\frac{1}{2}\ (40\ K)^2$$

$$= 17.67 \frac{J}{mol} + 5.70 \frac{J}{mol} = 23.37 \frac{J}{mol}$$

24-9 The Maxwell-Boltzmann distribution is

$$\frac{dN(v)}{dv} = 4\pi N \left(\frac{\mu}{2\pi kT}\right)^{3/2} v^2\, e^{-\frac{1}{2}\mu v^2/kT} = f(v)$$

This function is a maximum when $\frac{df(v)}{dv} = 0$

or,

$$4\pi N \left(\frac{\mu}{2\pi kT}\right)^{3/2} \left(2v - v^2\ \frac{\mu v}{kT}\right) e^{-\frac{1}{2}\mu v^2/kT} = 0$$

$$2v - \frac{\mu}{kT} v^3 = 0 \quad \text{gives} \qquad v(2 - \frac{\mu}{kT} v^2) = 0$$

The maximum of the distribution is then at

$$v = \sqrt{\frac{2kT}{\mu}}$$

24-10 Setting $\xi = \frac{v}{v_{rms}} = \sqrt{\frac{\mu}{3kT}}(v)$, the Maxwell-Boltzmann distribution can be rewritten as

$$dN(v) = \frac{N}{v_{rms}}\ 6\sqrt{\frac{3}{2\pi}}\ \xi^2\ e^{-\frac{3}{2}\xi^2}\ dv$$

For "small enough" velocity intervals the approximation can be made that

$$\Delta N(v) = \frac{N}{v_{rms}}\ 6\sqrt{\frac{3}{2\pi}}\ \xi^2\ e^{-\frac{3}{2}\xi^2}\ \Delta v$$

where v is taken to be in the middle of the interval (to improve the approximation).

In this case

$$v_{rms} = \sqrt{\frac{3kT}{\mu}} = \left[\frac{3 \times 1.3807 \times 10^{-23}\ \frac{J}{K} \times 250\ K}{4.0026\ u \times 1.6606 \times 10^{-27}\ \frac{kg}{u}}\right]^{1/2} = 1248.2\ \frac{m}{s}$$

With intervals $\Delta v = 250\ \frac{m}{s}$ and $N = 1000$ molecules,

$$\Delta N(v) = \Delta N(v_{middle}) = \left(\frac{1000\ \text{molecules}}{1248.2\ \frac{m}{s}}\right) 6\sqrt{\frac{3}{2\pi}}\ \xi^2 e^{-\frac{3}{2}\xi^2} (250\ \frac{m}{s})$$

$$= 830.4\ \xi^2\ e^{-\frac{3}{2}\xi^2} \quad \text{molecules}$$

where ξ is evaluated at v_{middle} $\left(\xi_{middle} = \frac{v_{middle}}{v_{rms}}\right)$.

v interval $\left(\frac{m}{s}\right)$	$v_{middle}\left(\frac{m}{s}\right)$	ξ_{middle}	$\Delta N(\xi_{middle})$
0 - 250	125	0.10015	13.17
250 - 500	375	0.30044	65.46

500 - 750	625	0.50073	142.94
750 - 1000	875	0.70103	195.26
1000 - 1250	1125	0.90132	199.45
1250 - 1500	1375	1.10161	163.23
1500 - 1750	1625	1.30191	110.73
1750 - 2000	1875	1.50220	63.49
2000 - 2250	2125	1.70249	31.14
2250 - 2500	2375	1.90279	13.17

$\sum \Delta N = 998$

24-11 (a) The upward acceleration of the elevator is equivalent to a downward acceleration of the molecules.

g' = effective gravitational acceleration

$= g + a = g + 3g = 4g$

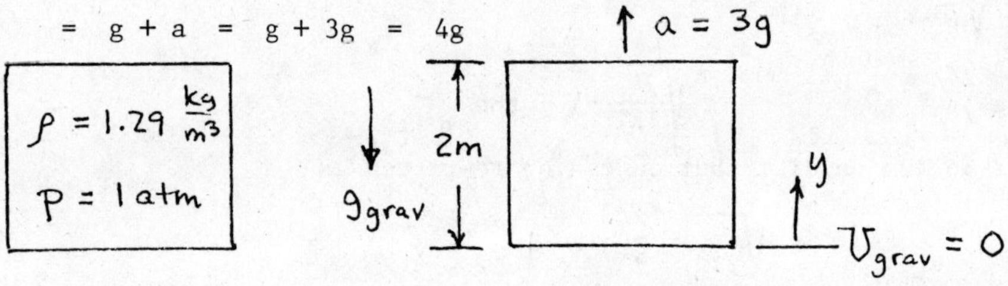

At Rest   Accelerating

The Boltzmann statistical factor implies that

$$N(y) = N(0)e^{-\mu g' y/kT}$$

where N(y) is the number of molecules per unit volume at a height y in the elevator. Then, $N(y) = N(0)e^{-\alpha y}$

where $\alpha = \frac{\mu g'}{kT} = \frac{N(0)\mu g'}{N(0)kT} = \frac{\rho_0 g'}{p_0} = \frac{4\rho_0 g}{p_0}$

as $\rho_0 = N(0)\mu$ and $p_0 = N(0)kT$

$$\text{Now, } y_{C.M.} = \frac{\int_0^h yN(y)dy}{\int_0^h N(y)dy} = \frac{\int_0^h yN(0)e^{-\alpha y}dy}{\int_0^h N(0)e^{-\alpha y}dy}$$

where h is the elevator height.

Integrating by parts,

$$y_{C.M.} = \frac{-y\frac{e^{-\alpha y}}{\alpha}\Big|_0^h + \int_0^h \frac{e^{-\alpha y}}{\alpha}dy}{-\frac{e^{-\alpha y}}{\alpha}\Big|_0^h}$$

$$= \frac{1}{\alpha} + h\left(1 - e^{\alpha h}\right)^{-1} = y$$

(b) $\Delta y = \frac{1}{2}h - y = \frac{1}{2}h - \left(\frac{1}{\alpha} + h\,(1 - e^{\alpha h})^{-1}\right)$

$$= \frac{1}{2}h - \frac{1}{\alpha} - \frac{h}{1 - e^{\alpha h}}$$

Expanding $e^{\alpha h}$,

$$\Delta y = \frac{1}{2}h - \frac{1}{\alpha} - \frac{h}{1 - \left(1 + \alpha h + \frac{1}{2}(\alpha h)^2 + \cdots\right)}$$

$$= \frac{1}{2}h - \frac{1}{\alpha} + \frac{1}{\alpha}\,\frac{1}{\left[1 + \frac{1}{2}(\alpha h) + \frac{1}{6}(\alpha h)^2 + \cdots\right]}$$

Since $\alpha h \ll 1$, let $x = \frac{1}{2}(\alpha h) + \frac{1}{6}(\alpha h)^2 + \cdots$

Then, as $(1 + x)^{-1} = 1 - x + x^2 - \cdots$

$$\Delta y = \frac{1}{2}h - \frac{1}{\alpha} + \frac{1}{\alpha}\left[1 - \left(\frac{1}{2}(\alpha h) + \frac{1}{6}(\alpha h)^2 + \cdots\right) + \left(\frac{1}{2}(\alpha h) + \frac{1}{6}(\alpha h)^2 + \cdots\right)^2 + \cdots\right]$$

$$= \frac{1}{2}h - \frac{1}{\alpha}\left(\frac{1}{2}(\alpha h) + \frac{1}{6}(\alpha h)^2 + \cdots\right) + \frac{1}{\alpha}\left(\frac{1}{2}(\alpha h) + \frac{1}{6}(\alpha h)^2 + \cdots\right)^2 + \cdots$$

$$= \frac{1}{12}\alpha h^2 + \text{terms of order } h(\alpha h)^2$$

$$\Delta y = \frac{1}{12}\alpha h^2 \quad \text{for } \alpha h \ll 1$$

(c) The given values of pressure and density are calculated using the entire volume of the elevator, V.

$$\rho_{rest} = \frac{m_{air}}{V} = 1.29\ \frac{kg}{m^3}\,, \quad p_{rest} = \frac{n_{air}RT}{V} = 1\ \text{atm}$$

$$\frac{\rho_{rest}}{p_{rest}} = \frac{1.29\ \frac{kg}{m^3}}{1.01325 \times 10^5\ \text{Pa}} = 1.273 \times 10^{-5}\ m^{-2}\ s^2$$

$$= \frac{m_{air}}{n_{air}RT} = \frac{M(air)}{RT}$$

Once the elevator is accelerated,

$\rho(y) = \frac{m(y)}{(y)}$ is computed in a slab of thickness dy at height y.

$p(y) = \frac{n(y)RT}{V(y)}$, T remains constant.

Therefore, $\frac{\rho(y)}{p(y)} = \frac{m(y)}{n(y)RT} = \frac{M(air)}{RT} = \frac{\rho_{rest}}{p_{rest}} = \frac{\rho_o}{p_o}$

and $\alpha = g'\frac{\rho_o}{p_o} = 4(9.8\ \frac{m}{s^2})(1.273 \times 10^{-5}\ m^{-2}\ s^2)$

$$= 4.991 \times 10^{-4}\ m^{-1}$$

Then, $\Delta y = \frac{1}{12}\alpha h^2 = \frac{1}{12}(4.991 \times 10^{-4}\ m^{-1})(2\ m)^2 = 0.166$ mm

Note that $\alpha h \cong 10^{-3} \ll 1$ as required.

24-12 For an ideal gas, $pV = nRT = nN_o kT = NkT$

$$\mathcal{N} = \frac{N}{V} = \frac{p}{kT}$$

σ = total scattering cross section for hard spheres = πd^2

Therefore,
$$\lambda = \frac{1}{\sqrt{2}\,\sigma \mathcal{N}} = \frac{1}{\sqrt{2}} \frac{1}{\pi d^2} \frac{kT}{p}$$

24-13

Gas	$\rho\left(\frac{kg}{m^3}\right)$	$\eta\,(\times 10^{-5}$ Pa - s)	$D(\times 10^{-5}\ \frac{m^2}{s})$	$\frac{\eta}{\rho}(\times 10^{-5}\ \frac{m^2}{s})$
Argon	1.604	2.229	1.76	1.39
Hydrogen	0.0809	0.887	14.3	10.96
Krypton	3.366	2.496	0.90	0.74
Methane	0.645	1.098	2.08	1.70
Neon	0.810	3.138	4.73	3.87
Nitrogen	1.125	1.757	2.00	1.56
Xenon	5.303	2.274	0.44	0.43

24-14 D(Xenon at 1 atm, T = 20°C) = $0.44 \times 10^{-5}\ \frac{m^2}{s}$

$$D = \frac{2}{3\sqrt{\pi}\,\sigma p}\sqrt{\frac{k^3 T^3}{\mu}} \quad \text{gives} \quad \frac{D}{D_o} = \left(\frac{T}{T_o}\right)^{3/2} \frac{p_o}{p}$$

$$D\left(\frac{1}{10}\text{ atm, } T = 100°C\right) = \left(0.44 \times 10^{-5}\ \frac{m^2}{s}\right)\left(\frac{100 + 273.15}{20 + 273.15}\right)^{3/2} \frac{1\text{ atm}}{\frac{1}{10}\text{atm}}$$

$$= 6.32 \times 10^{-5}\ \frac{m^2}{s}$$

24-15 $\sigma = \pi d^2$, where d is the hard sphere diameter.

$$D = \frac{2}{3\sqrt{\pi}\,\sigma p}\sqrt{\frac{k^3 T^3}{\mu}} \quad \text{gives} \quad d = \left[\frac{2}{3Dp\sqrt{\mu}}\left(\frac{kT}{\pi}\right)^{3/2}\right]^{1/2}$$

Using D(krypton at 1 atm, T = 20°C) = $0.90 \times 10^{-5}\ \frac{m^2}{s}$,

$d = 3.01 \times 10^{-10}$ m = <u>3.01 Å</u>

$$\eta = \frac{2}{3\sqrt{\pi}\,\sigma}\sqrt{\mu kT} \quad \text{gives} \quad d = \left[\frac{2}{3\,\eta\,\pi^{3/2}}\sqrt{\mu kT}\right]^{1/2} = 3.37\ \text{Å}$$

for η(krypton at 1 atm, T = 20°C) = 2.496×10^{-5} Pa-s

24-16 At T = 20°C, p = 1 atm,

$\eta(\text{He}) = 1.961 \times 10^{-5}$ Pa-s

$\eta(\text{Xe}) = 2.274 \times 10^{-5}$ Pa-s

As $\eta = \dfrac{2}{3\sqrt{\pi}\,\sigma} \cdot \sqrt{\mu k T}$ where $\sigma = \pi d^2$,

$$\frac{\eta(\text{He})}{\eta(\text{Xe})} = \frac{\sqrt{\mu_{\text{He}}}}{\sigma_{\text{He}}} \cdot \frac{\sigma_{\text{Xe}}}{\sqrt{\mu_{\text{Xe}}}} = \sqrt{\frac{\mu_{\text{He}}}{\mu_{\text{Xe}}}} \left(\frac{d_{\text{Xe}}}{d_{\text{He}}}\right)^2$$

Therefore, $\dfrac{d_{\text{Xe}}}{d_{\text{He}}}$ (T = 20°C, p = 1 atm) $= \left(\dfrac{\mu_{\text{Xe}}}{\mu_{\text{He}}}\right)^{1/4} \left(\dfrac{\eta(\text{He})}{\eta(\text{Xe})}\right)^{1/2}$

$$= \left(\frac{131.30\ u}{4.0026\ u}\right)^{1/4} \left(\frac{1.961 \times 10^{-5}\ \text{Pa-s}}{2.274 \times 10^{-5}\ \text{Pa-s}}\right)^{1/2} = 2.222$$

$$\text{ratio of atomic volumes} = \frac{\frac{4}{3}\pi\left(\frac{d_{\text{Xe}}}{2}\right)^3}{\frac{4}{3}\pi\left(\frac{d_{\text{He}}}{2}\right)^3} = \left(\frac{d_{\text{Xe}}}{d_{\text{He}}}\right)^3 = 10.98$$

$$\text{ratio of atomic masses} = \frac{\mu_{\text{Xe}}}{\mu_{\text{He}}} = \frac{131.30\ \text{u}}{4.0026\ \text{u}} = \underline{32.80}$$

The larger atom (Xe) has the larger mass although the ratios differ by a factor of about 3.

24-17 To calculate the specific heat of air at constant volume in $\dfrac{\text{Cal}}{\text{kg-deg}}$, look at 1 kg of air. It contains 0.76 kg of N_2 and 0.24 kg of O_2. The number of moles of each is

$$n_{N_2} = \frac{m_{N_2}}{M(N_2)} = \frac{0.76\ \text{kg}}{28.01\ \frac{\text{g}}{\text{mol}}} = 27.133 \text{ moles of } N_2$$

$$n_{O_2} = \frac{m_{O_2}}{M(O_2)} = \frac{0.24\ \text{kg}}{31.9988\ \frac{\text{g}}{\text{mol}}} = 7.5003 \text{ moles of } O_2$$

The total specific heat of the kg of air is the sum of the specific heats of the N_2 and O_2 components.

$$C_v(\text{kg air}) = n_{N_2} C_v(\text{per mole } N_2) + n_{O_2}\, C_v(\text{per mole } O_2)$$

Diatomic molecules with no vibrational modes (rigid) have the number of degrees of freedom, $\nu = 5$. For such an ideal gas C_v(molar specific heat) $= \frac{1}{2}\nu R$

$$C_v(\text{molar specific heat}) = \frac{1}{2}(5)(8.3144\ \frac{\text{J}}{\text{mol-K}}) = 20.79\ \frac{\text{J}}{\text{mol-K}}$$

for both O_2 and N_2.

Then,

$$c_v(\text{kg air}) = (27.133 \text{ mol} + 7.500 \text{ mol})(20.79 \frac{J}{\text{mol-deg}})$$

$$c_v = 719.9 \frac{J}{\text{kg-deg}} = 0.172 \frac{\text{Cal}}{\text{kg-deg}}$$

24-18 $$C_v\left(\frac{J}{\text{mol K}}\right) = \left(1945 \frac{J}{\text{mol K}}\right)\left(\frac{T}{T_D}\right)^3 + \gamma T$$

$$\frac{C_v}{T} = \frac{1945 \frac{J}{\text{mol}\cdot\text{K}}}{T_D^3} T^2 + \gamma$$

Therefore, plotting $\frac{C_v}{T}$ as a function of T^2 should produce a straight line of slope $= \left(1945 \frac{J}{\text{mol K}}\right)/T_D^3$ and intercept $= \gamma$.

T(K)	T^2 (K^2)	$C_v\left(10^{-2} \frac{J}{\text{mol K}}\right)$	$\frac{C_v}{T}\left(10^{-2} \frac{J}{\text{mol K}^2}\right)$
4	16	0.78	0.1950
6	36	1.45	0.2417
8	64	2.70	0.3375
10	100	4.47	0.4470

A least squares fit determines a line with

$$\text{slope} = 3.0565 \times 10^{-5} \frac{J}{\text{mol K}^4}$$

$$\text{intercept} = 1.4024 \times 10^{-3} \frac{J}{\text{mol K}^2}$$

$$T_D = \left(1945 \frac{J}{\text{mol K}} \Big/ \text{slope}\right)^{1/3} = 399.2 \text{ K}$$

$$\gamma = \text{intercept} = 1.402 \times 10^{-3} \frac{J}{\text{mol-K}^2}$$

24-19 $P(x) = e^{-x/\lambda}$ = probability that a given molecule will travel a distance x before suffering a collision. Then the average distance traveled by the molecule is

$$\overline{x} = \frac{\int_0^\infty x\, P(x)dx}{\int_0^\infty P(x)dx} = \frac{\int_0^\infty x\, e^{-x/\lambda} dx}{\int_0^\infty e^{-x/\lambda} dx}$$

or,
$$\overline{x} = \frac{-\lambda x e^{-x/\lambda}\Big|_0^\infty + \lambda\int_0^\infty e^{-x/\lambda} dx}{\int_0^\infty e^{-x/\lambda} dx} = 0 + \lambda = \lambda$$

24-20 (a) $\tau = \frac{\lambda}{\overline{v}}$, $\lambda = \frac{kT}{\sqrt{2}\pi p d^2}$, $\overline{v} = \sqrt{\frac{8kT}{\pi\mu}}$

So, $$\tau = \sqrt{\frac{\mu k}{\pi}} \cdot \frac{T^{1/2}}{4d^2 p}$$

From tabulated values,

$$\tau(N_2 \text{ at 1 atm, } T = 20°C) = \frac{\lambda}{\bar{v}} = \frac{887 \times 10^{-9} \text{ m}}{471 \frac{m}{s}} = 1.883 \times 10^{-9} \text{ s}$$

As $\dfrac{\tau}{\tau_o} = \left(\dfrac{T}{T_o}\right)^{1/2} \dfrac{p_o}{p}$,

$$\tau(T,p) = \left(1.883 \times 10^{-9} \text{ s}\right)\left(\frac{T}{293.15 \text{ K}}\right)^{1/2}\left(\frac{1 \text{ atm}}{p}\right)$$

(b) $$\tau(100°C, \tfrac{1}{20} \text{ atm}) = \left(1.883 \times 10^{-9} \text{ s}\right)\left(\frac{373.15 \text{ K}}{293.15 \text{ K}}\right)^{1/2}\left(\frac{1 \text{ atm}}{\frac{1}{20} \text{ atm}}\right) = 4.249 \times 10^{-8} \text{ s}$$

24-21 $\mu(H_2) = 2.016\,u$, $d(H_2) = 2.35 \times 10^{-10}$ m

$$\rho = \frac{\text{mass}}{\text{volume}} = \frac{n_{H_2} M(H_2)}{V} = \frac{p}{RT} M(H_2)$$

$$\rho(20°C, 1 \text{ atm}) = \frac{1.01325 \times 10^5 \text{ Pa}}{(8.3144 \frac{J}{mol\cdot K})(20 + 273.15)K}\left(2.016 \times 10^{-3} \frac{kg}{mol}\right)$$

$$= 8.381 \times 10^{-2} \frac{kg}{m^3}$$

$$D = \frac{2}{3\sqrt{\pi}\,\sigma\, p}\sqrt{\frac{k^3 T^3}{\mu}} \quad \text{with} \quad \sigma = \pi d^2$$

$$D(20°C, 1 \text{ atm}) = 9.5235 \times 10^{-5} \frac{m^2}{s}$$

$$\eta(20°C, 1 \text{ atm}) = \rho D = 7.982 \times 10^{-6} \text{ Pa-s}$$

$$\bar{v} = \sqrt{\frac{8kT}{\pi\mu}} = \left(\frac{8 \times 1.38066 \times 10^{-23} \frac{J}{K} \times 293.15K}{\pi \times 2.016 \text{ u} \times 1.66 \times 10^{-27} \text{ kg/u}}\right)^{1/2} = 1755 \frac{m}{s}$$

As $D = \frac{1}{3}\bar{v}\lambda$,

$$\lambda = \frac{3D}{\bar{v}} = \frac{3 \times 9.5235 \times 10^{-5} \frac{m^2}{s}}{1755 \frac{m}{s}} = 1.628 \times 10^{-7} \text{ m}$$

CHAPTER 25

25-1 $dW = p\,dV \qquad \int_a^b dW = W_{ab} = \int_a^b p\,dV$

During an isobaric process p = constant.

Therefore, $W_{ab} = p\int_a^b dV = p(V_b - V_a)$

25-2 $W_{ab} = \int_a^b p\,dV$

The work done by the gas is just the area under the curve $p = \alpha V^2$ between V_a and V_b.

$$W_{ab} = \int_a^b \alpha V^2 \, dV = \frac{1}{3} \alpha (V_b^3 - V_a^3)$$

$$V_b = 2V_a = 2(1 \text{ m}^3) = 2 \text{ m}^3$$

$$W_{ab} = \frac{1}{3}(5.0 \frac{\text{atm}}{\text{m}^6} \times 1.01325 \times 10^5 \frac{\text{Pa}}{\text{atm}})[(2 \text{ m}^3)^3 - (1 \text{ m}^3)^3]$$

$$= 1.182 \times 10^6 \text{ J}$$

25-3 (a) For an isothermal expansion $W_{ab} = nRT \ \ell n \frac{V_b}{V_a}$.

$$n_{Ar} = \frac{m_{Ar}}{M(Ar)} = \frac{5.0 \text{ g}}{39.94 \frac{\text{g}}{\text{mol}}} = 0.1252 \text{ mol}$$

Then, as $\frac{V_b}{V_a} = 2$,

$$W_{ab} = (0.1252 \text{ mol})(8.3144 \frac{\text{J}}{\text{mol} \cdot \text{K}})(20 + 273.15)\text{K}(\ell n \ 2)$$

$$= 211.5 \text{ J}$$

(b) $p_a = 1$ atm

$pV = nRT =$ constant for isothermal process

$$p_a V_a = p_b V_b \qquad p_b = p_a \frac{V_a}{V_b} = (1 \text{ atm})(\frac{1}{2}) = 0.50 \text{ atm}$$

25-4

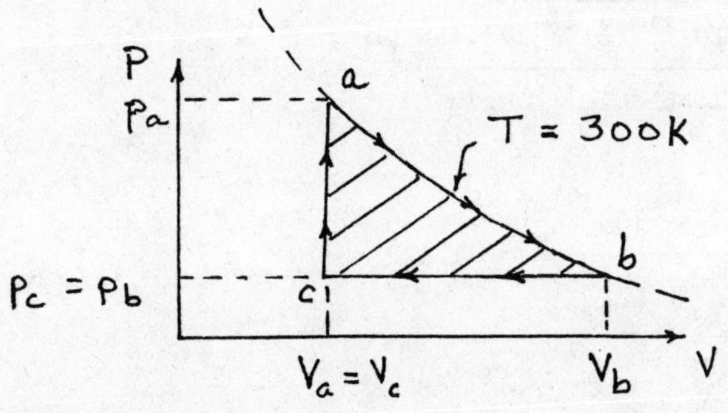

$$W_{cycle} = W_{ab} + W_{bc} + W_{ca}$$

From c to a the volume doesn't change so that no work is done on this leg of the cycle ($dV = 0$)

$$W_{ab}(\text{isothermal}) = nRT \ \ell n \frac{V_b}{V_a}$$

$$W_{bc}(\text{isobaric}) = p_{bc}(V_c - V_b)$$

$$p_a V_a = nRT \qquad V_a = \frac{nRT}{p_a}$$

$$\text{So } V_a = \frac{(1\text{ mol})(8.3144 \frac{J}{mol\cdot K})(300\text{ K})}{(5\text{ atm})(1.01325 \times 10^5 \frac{Pa}{atm})} = 4.923 \times 10^{-3}\ m^3 = V_c$$

$$p_bV_b = nRT \qquad \frac{p_bV_b}{p_aV_a} = \frac{T_b}{T_a} = 1 \qquad \frac{V_b}{V_a} = \frac{p_a}{p_b} = \frac{5\text{ atm}}{1\text{ atm}}$$

$$V_b = 5V_a = 2.462 \times 10^{-2}\ m^3$$

Then

$$W_{cycle} = W_{ab} + W_{bc} = nRT\ \ln 5 + p_{bc}(V_a - 5V_a)$$

$$= (1\text{ mol})(8.3144 \frac{J}{mol\cdot K})(300\text{ K})\ \ln 5 - 4(4.923 \times 10^{-3}\ m^3)(1.013 \times 10^5\ Pa)$$

$$= 2019\ J$$

25-5 compressibility $= -\frac{1}{V}\frac{dV}{dp}$

$$pV = nRT \qquad V = \frac{nRT}{p} \qquad \frac{dV}{dp} = -\frac{nRT}{p^2}$$

$$\text{Therefore } -\frac{1}{V}\frac{dV}{dp} = -\frac{1}{V}(-\frac{nRT}{p^2}) = -\frac{p}{nRT}(-\frac{nRT}{p^2}) = \frac{1}{p}$$

25-6 (a) The pressure against which the mercury must do work in order to expand is atmospheric pressure.

$$W_{expansion} = \int_a^b p\ dV = 1\text{ atm }(V_b - V_a) = (1\text{ atm})\ \Delta V$$

Using $\bar{\beta}_{Hg} = 0.182 \times 10^{-3}\ deg^{-1}$

$$\Delta V = \bar{\beta}\ V_a\ \Delta T = (0.182 \times 10^{-3}\ deg^{-1})(1.2 \times 10^{-3}\ m^3)(100°C - 20°C)$$

Then $W_{expansion} = (1.01325 \times 10^5\ Pa)\Delta V = 1.77\ J$

(b) For the constant pressure expansion

$$Q = mc_p\ \Delta T$$

$$m_{Hg} = \rho_{Hg}\ V = (1.36 \times 10^4 \frac{kg}{m^3})(1.2 \times 10^{-3}\ m^3) = 16.32\ kg$$

Then, with $c_p = 138 \frac{J}{kg\cdot deg}$

$$Q = (16.32\ kg)(138 \frac{J}{kg\cdot deg})(80\ deg) = 1.80 \times 10^5\ J$$

(c) The first law of thermodynamics gives

$$U = Q - W = 1.80 \times 10^5\ J - 1.77\ J \cong Q = 1.80 \times 10^5\ J$$

(d)
$$c_p = \frac{1}{m}\frac{\Delta U}{\Delta T}\Big|_{p\ constant} = \frac{1}{m}\frac{Q - W}{\Delta T}\Big|_{p\ constant}$$

$$\cong \frac{1}{m}\frac{Q}{\Delta T}\Big|_{p\ constant} = \frac{1}{m}\frac{Q}{\Delta T}\Big|_{v\ constant} = c_v$$

25-7 If the walls are diathermic, heat is free to pass through them, so that the temperature of the gas within the cylinder remains at 300 K.

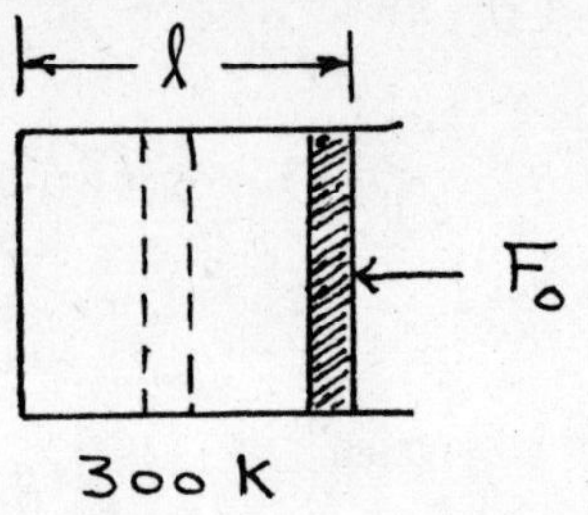

(a) $$W_{\text{done by external force}} = 10\,F_o \int_0^{h/2} dr = 5F_o h$$

where h is the height of the cylinder.

$$p_o = \frac{F_o}{A}$$

$$V_o = Ah$$

So $P_o V_o = F_o H$

Thus,

$$W_{\text{done by external force}} = 5p_o V_o = 5nRT$$

$$= 5(4\text{ mol})(8.314\text{ J/mol·K})(300\text{ K})$$

$$= 4.99 \times 10^4\text{ J}$$

(b) Since the temperature remains constant, U = 0.

The work done by the gas is given by

$$W = -W_{\text{done by external force}} = Q - U = Q$$

$$Q = -4.99 \times 10^4\text{ J}$$

The energy from the external force is given off by the cylinder in the form of heat.

25-8

a		b	
n =	1 mole	n =	1 mole
p =	1 atm	p =	1 atm
T =	300 K	T =	400 K

The gas is heated isobarically so that the pressure is constant.

(a) $$W_{ab} = \int_a^b p\,dV = 1\text{ atm}(V_b - V_a)$$

Since $pV = nR'T$, $V = \frac{nR'T}{p}$ with $R' = 0.99967\text{ R}$

$$V_b - V_a = \frac{nR'}{p}(T_b - T_a).$$

$$W_{ab} = p(V_b - V_a) = nR'(T_b - T_a)$$

$$= (1\text{ mole})(0.99967\text{ R})(400\text{ K} - 300\text{ K}) = (99.97\ \text{ R})\text{J}$$

(b) $$Q_{ab} = nc_p\,\Delta T \quad \text{as the pressure is constant}$$

$$= (1\text{ mole})(2.5043\text{ R})(400\text{ K} - 300\text{ K}) = (250.43\text{ R})\text{J}$$

(c) $U_{ab} = Q_{ab} - W_{ab} = (250.43 - 99.97)R$

$= (150.46\ R)J$

(d) According to kinetic theory

$$U_{ab} = \frac{1}{2} n\nu R(T_b - T_a)$$

ν , for monatomic argon in this temperature regime, is 3.

Therefore $U_{ab} = \frac{1}{2}(1 \text{ mole})(3\ R)(100\ K) = (150\ R)J$

which is very near to the result from (c).

25-9 For an adiabatic process

$$p = (\text{constant})V^{-\gamma} \quad \text{and} \quad p_a V_a^\gamma = p_b V_b^\gamma = \text{constant}$$

Then

$$W_{ab} = \int_a^b p\, dV = p_a V_a^\gamma \int_a^b V^{-\gamma}\, dV$$

$$W_{ab} = p_a V_a^\gamma \left. \frac{V^{-\gamma+1}}{1-\gamma} \right|_a^b = p_a V_a^\gamma \frac{V_b^{1-\gamma} - V_a^{1-\gamma}}{1 - \gamma}$$

$$= \frac{p_a V_a - p_a V_a^\gamma V_b^{1-\gamma}}{\gamma - 1} = \frac{p_a V_a - p_b V_b}{\gamma - 1}$$

25-10 For argon at 1 atm and $T = 15°C$

$$\frac{pV}{T} = 208.0 \frac{J}{kg \cdot deg}, \quad c_p = 0.1253 \frac{Cal}{kg \cdot deg} \quad \text{and} \quad \gamma = 1.668.$$

The value for $\frac{pV}{T}$ is for a 1-kg sample. Then, treating argon as an ideal gas,

$$\frac{pV}{T} = nR$$

where n is the number of moles in 1 kg.

Also $C_p - C_v = R$ and $\frac{C_p}{C_v} = \gamma$

so that $C_p(1 - \frac{1}{\gamma}) = R$ where C_p is the molar specific heat at constant pressure.

Then

$$\left.\frac{pV}{T}\right|_{1\ kg} = n_{1\ kg}\ C_p(\text{molar})(1 - \frac{1}{\gamma}) = c_p(1 - \frac{1}{\gamma})$$

since (n moles in 1 kg)C_p(molar) $= c_p$(per kg).

$$208.0 \frac{J}{kg \cdot deg} = 0.1253 \frac{Cal}{kg \cdot deg} (1 - \frac{1}{1.668})$$

$$1\ Cal = 4145\ J$$

25-11 compressibility $= k = -\frac{1}{V}\frac{dV}{dp}$

To calculate the adiabatic compressibility of an ideal gas

$$pV^\gamma = \text{constant} = A \quad \text{so} \quad V = (\frac{A}{p})^{1/\gamma}$$

$$\frac{dV}{dp} = \frac{1}{\gamma}\left(\frac{A}{p}\right)^{-1+1/\gamma} \left(-\frac{A}{p^2}\right)$$

Therefore,

$$k_{adiabatic} = -\left(\frac{A}{p}\right) \times \left(\frac{1}{\gamma}\right)\left(\frac{A}{p}\right)^{-1+1/\gamma}\left(-\frac{A}{p^2}\right) = \frac{1}{\gamma p}$$

25-12 $v = \sqrt{\frac{\gamma p}{\rho}}$ is the speed of sound in a gas.

Assuming air to be an ideal gas,

$$pV = nRT \quad \text{so} \quad \frac{n}{V} = \frac{p}{RT}$$

Then $\rho = \frac{m}{V} = \frac{nM}{V} = M_{air}\frac{p}{RT}$

$$v = \sqrt{\frac{\gamma RT}{M}} = \frac{(1.403)\ \left(8.3144\ \frac{J}{mol\cdot K}\right)\ (288.15)K}{\left(28.96\ \frac{g}{mol}\right)\ \left(\frac{1\ kg}{1000\ g}\right)}$$

$$= 340.7\ \frac{m}{s}$$

25-13 $Q_{N_2} = +25.0$ Cal $\qquad \Delta U_{N_2} = +8.0$ Cal

$m_{N_2} = 1.00$ kg $\qquad p$ = constant (isobaric)

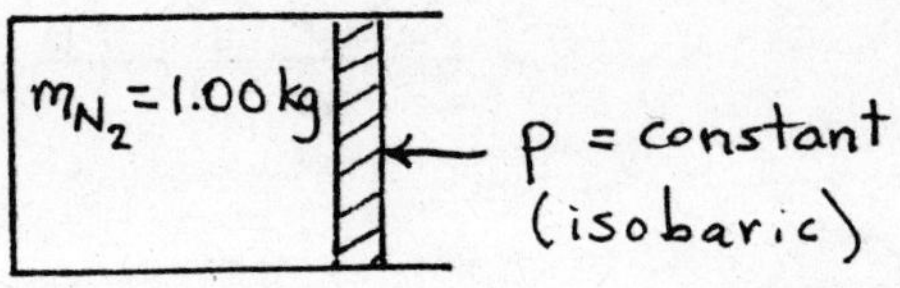

(a) The first law of thermodynamics is

$Q = W + \Delta U$ between equilibrium states.

W = work done by gas $= Q - \Delta U$

$= 25.0\ \text{Cal} - 8.0\ \text{Cal} = 17.0\ \text{Cal} = 7.116 \times 10^4\ \text{J}$

(b) $W = p\,\Delta V$ for an isobaric process.

Therefore

$$\Delta V = \frac{W}{p} = \frac{7.116 \times 10^4\ J}{1.01325 \times 10^5\ Pa} = 0.702\ m^3$$

(c) $\Delta U = nC_p\,\Delta T$

n = moles of N_2 = $1.00\ kg/28.0134\ \frac{g}{mol}$

C_p = molar specific heat of N_2 at constant pressure

$= 29.05\ \frac{J}{mol\ K}$

Therefore

$$\Delta T = \frac{\Delta U}{nC_p} = \frac{(8.0\ Cal)\ \left(4186\ \frac{J}{Cal}\right)}{(35.70\ mol)\left(29.05\ \frac{J}{mol\cdot K}\right)} = 32.3\ K$$

25-14 (a) W = work done by gas per cycle

= area contained within the cycle path

$= (p_a - p_d)(V_b - V_a)$

$= (3.2 - 1.0)\text{atm} \times 1.013 \times 10^5 \frac{\text{Pa}}{\text{atm}} \times (5.0 - 1.0) \times 10^{-2}\ \text{m}^3$

$= 8.92 \times 10^3\ \text{J} = 892\ \text{J}$

(b) Since the gas returns to its initial conditions at the completion of the cycle, $\Delta T_{cycle} = 0$ and $\Delta U_{cycle} = 0$

Therefore $Q = W$ = heat supplied to the gas

$= 892\ \text{J} = 0.213\ \text{Cal}$

25-15 $W_{ab} = \int_a^b p\, dV$

Solving Van der Waal's equation for p,

$$(p + \frac{a}{\mathcal{V}^2})(\mathcal{V} - b) = RT \text{ gives } p = \frac{RT}{\mathcal{V} - b} - \frac{a}{\mathcal{V}^2}$$

where $\mathcal{V} = \frac{V}{n}$ is the molar volume.

To calculate the work done per mole in an isothermal expansion, let $\mathcal{V} = V(n = 1 \text{ mole})$

Therefore

$$W_{per\ mole} = \int_a^b (\frac{RT}{V - b} - \frac{a}{V^2})\, dV$$

$$= RT\ \ell n\left(\frac{V_b - b}{V_a - b}\right) + a\left(\frac{1}{V_b} - \frac{1}{V_a}\right)$$

25-16 $C_p - C_v = R(1 + \frac{9}{4}\frac{\mathcal{V}_c}{\mathcal{V}}\frac{T_c}{T})$

For argon, $T_c = 150.7$ K and $\mathcal{V}_c = 75.3\ \text{cm}^3/\text{mol}$ are the critical temperature and critical molar volume respectively.

The correction term is $\frac{9}{4} R \frac{\mathcal{V}_c}{\mathcal{V}} \frac{T_c}{T}$

At $T = 0°\text{C}$ and $p = 1$ atm, $\mathcal{V} = 22.41 \frac{\ell}{\text{mol}}$

Therefore

$$(C_p - C_v) - R = \frac{9}{4} \frac{75.3 \frac{\text{cm}^3}{\text{mol}}}{22.41 \times 10^3 \frac{\text{cm}^3}{\text{mol}}} \frac{150.7\ \text{K}}{273.15\ \text{K}} R$$

$$= 4.17 \times 10^{-3}\ R$$

25-17 (a) $\frac{pV}{n} = RT + B_1(T)p$

$p = nRT/(V - nB_1(T))$

Therefore

$$W_{ab} = \int_a^b p\, dV = nRT \int_a^b \frac{dV}{V - nB_1(T)}$$

for an isothermal process. Thus,

$$W_{ab} = nRT \ \ell n \frac{V_b - nB_1(T)}{V_a - nB_1(T)}$$

(b) For air at 0°C with $B_1(T) = -13.5 \frac{cm^3}{mol}$,

$V_a = 250\ cm^3$, $V_b = 1000\ cm^3$ and $n = 1$ mol,

$$W_{ab} = (1\ mol)(8.3144 \frac{J}{mol \cdot K})(273.15\ K)\ \ell n \frac{1000\ cm^3 + 13.5\ cm^3}{250\ cm^3 + 13.5\ cm^3}$$

$$= 3059\ J$$

(c) If air is considered to be an ideal gas

$$W_{ab} = nRT\ \ell n \frac{V_b}{V_a} = (1\ mol)(8.3144 \frac{J}{mol\ k})(273.15\ k)\ \ell n \frac{1000\ cm^3}{250\ cm^3}$$

$$= 3148\ J$$

(d) Comparing the results of (b) and (c)

$$\frac{W_{ideal} - W_{virial}}{W_{ideal}} = \frac{3148\ J - 3059\ J}{3148\ J} = 3\%$$

If the pressure is 89.6 atm the correction term is

$$B_1(T = 0°C)p = -13.5 \frac{cm^3}{mol}(89.6\ atm)$$

$$= -123 \frac{J}{mol}$$

whereas $RT = 2271 \frac{J}{mol}$.

Since the correction is proportional to pressure it will be insignificant near $p = 1$ atm.

25-18 (a)

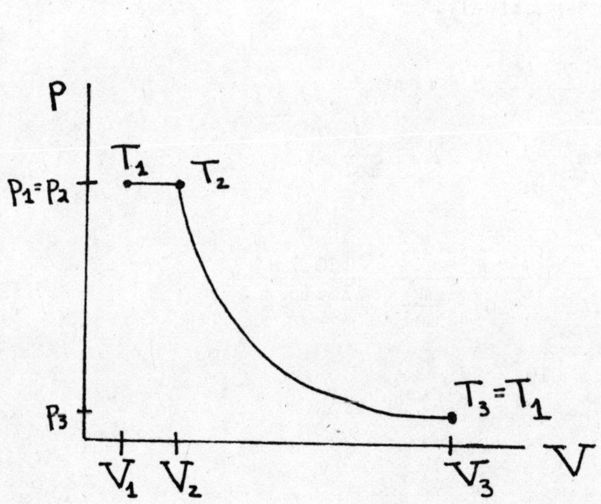

(b) Since the second phase of the expansion is adiabatic, no heat is exchanged in that phase. For the first part of the expansion,

$$Q = nC_p(T_2 - T_1) = nC_pT_1\left(\frac{V_2}{V_1} - 1\right)$$

$$= (3\ \text{mol})(20.80\ \text{J/mol}\cdot\text{K})(293.15\ \text{K})\left(\frac{25}{10} - 1\right)$$

$$= 2.74 \times 10^4\ \text{J}$$

(c) Since $T_1 = T_3$, there is no difference in temperature between the initial and final states. Thus, the change in internal energy, U, is equal to zero.

(d) $W = Q - U = Q = 2.74 \times 10^4\ \text{J}$

(e) $p_2V_2^\gamma = p_3V_3^\gamma = p_3V_3V_3^{\gamma-1} = p_1V_1V_3^{\gamma-1}$

$$V_3 = \left(\frac{p_2V_2^\gamma}{p_1V_1}\right)^{1/(\gamma-1)} = \left(\frac{V_2^\gamma}{V_1}\right)^{1/(\gamma-1)}$$

For argon, $\gamma = 1.67$. Thus,

$$V_3 = (25^{1.67}/10)^{1/0.67} = 98.15\ \text{m}^3$$

Also,

$$p_3 = p_1\frac{V_1}{V_3} = 0.10p_1$$

25-19 (a) At any point in the expansion

force of spring = force due to gas pressure

$\kappa x_b = p_b A, \qquad \kappa x_a = p_a A, \qquad x_b = \ell_b - \ell_o$

Therefore $p_b = \frac{\kappa}{A}(\ell_b - \ell_o) = \frac{\kappa}{A}[\ell_b - (\ell_a - x_a)]$

Also

$V_a = \ell_a A$ and $V_b = \ell_b A$

so that

$$p_b = \frac{\kappa}{A}\left[\frac{V_b}{A} - \left(\frac{V_a}{A} - \frac{p_a A}{\kappa}\right)\right]$$

$$\text{or } p_b = p_a + \frac{\kappa}{A^2}(V_b - V_a)$$

Equivalently, $p = p_a + \frac{\kappa}{A^2}(V - V_a)$

(b) $W_{ab} = \int_a^b p\, dV = \int_a^b p_a + \frac{\kappa}{A^2}(V - V_a)\, dV$

Therefore,

$$W_{ab} = p_a(V_b - V_a) + \frac{\kappa}{A^2}[\tfrac{1}{2}(V_b^2 - V_a^2) - V_a(V_b - V_a)]$$

$$= p_a(V_b - V_a) + \frac{\kappa}{A^2}\frac{1}{2}(V_b - V_a)^2$$

$$= (V_b - V_a)[p_a + \frac{1}{2}\frac{\kappa}{A^2}(V_b - V_a)]$$

$$= \frac{1}{2}(V_b - V_a)(p_b + p_a)$$

(c) Since $V_b = \ell_b A$, $V_a = \ell_a A$

$$V_b - V_a = (\ell_b - \ell_a)A$$

Also $p_b = \frac{\kappa}{A}x_b$, $p_a = \frac{\kappa}{A}x_a$

Therefore

$$W_{ab} = \frac{1}{2}(V_b - V_a)(p_b + p_a)$$

$$= \frac{1}{2}(\ell_b - \ell_a)A(x_b + x_a)\frac{\kappa}{A}$$

Then $\ell_b - \ell_a = x_b - x_a$ implies that

$$W_{ab} = \frac{1}{2}\kappa(x_b^2 - x_a^2)$$

(d) $\frac{1}{2}\kappa(x_b^2 - x_a^2) = \frac{1}{2}(7500\ \frac{N}{m})(x_b^2 - x_a^2)$

$$x_a = \frac{A}{\kappa}p_a = \frac{75\ cm^2}{7500\ \frac{N}{m}}(10^5\ \frac{N}{m^2}) = 0.10\ m$$

$$x_b = x_a + (\ell_b - \ell_a) = x_a + (0.10\ \text{m}) = 0.20\ \text{m}$$

Therefore,

$$W_{ab} = \tfrac{1}{2}\kappa(x_b^2 - x_a^2) = \tfrac{1}{2}(7500\ \tfrac{\text{N}}{\text{m}})[(0.20\ \text{m})^2 - (0.10\ \text{m})^2] = 112.5\ \text{J}$$

$$V_b = A\ell_b = (75 \times 10^{-4}\ \text{m}^2)(0.40\ \text{m}) = 3 \times 10^{-3}\ \text{m}^3$$

$$V_a = A\ell_a = (75 \times 10^{-4}\ \text{m}^2)(0.30\ \text{m}) = 2.25 \times 10^{-3}\ \text{m}^3$$

$$p_b = \frac{\kappa}{A} \times b = \frac{7500\ \text{N/m}}{75 \times 10^{-4}\ \text{m}^2}(0.20\ \text{m}) = 2 \times 10^5\ \frac{\text{N}}{\text{m}^2}$$

Therefore

$$W_{ab} = \tfrac{1}{2}(V_b - V_a)(p_b + p_a) = \tfrac{1}{2}(3 - 2.25) \times 10^{-3}\ \text{m}^3 (2 + 1) \times 10^5\ \frac{\text{N}}{\text{m}^2}$$
$$= 112.5\ \text{J}$$

(e) $U_{ab} = \tfrac{1}{2} n\nu R\,(T_b - T_a)$

$\nu = 3$ for a monatomic ideal gas.

Therefore

$$U_{ab} = \frac{3}{2} nR\left(\frac{p_b V_b}{nR} - \frac{p_a V_a}{nR}\right) = \frac{3}{2}(p_b V_b - p_a V_a)$$

$$= \frac{3}{2}(2 \times 3 - 1 \times 2.25) \times 10^5\ \frac{\text{N}}{\text{m}^2} \times 10^{-3}\ \text{m}^3 = 562.5\ \text{J}$$

(f) $Q_{ab} = U_{ab} + W_{ab} = 562.5\ \text{J} + 112.5\ \text{J} = 675\ \text{J}$

25-20

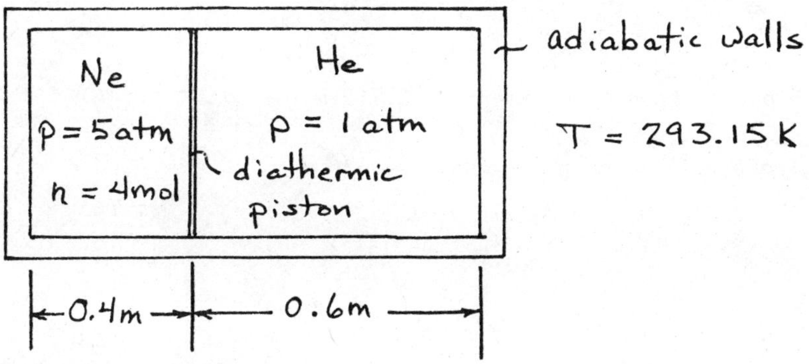

Once the piston is released and the final equilibrium is reached

(1) $p'_{Ne} = p'_{He}$. If x is the distance that the piston moves to the right to reach the final equilibrium

(2) $V'_{Ne} = (0.4\ \text{m} + x)A \qquad V'_{He} = (0.6\ \text{m} - x)A$

where A is the piston's cross-sectional area.

Considering both gases together as a system, the total volume doesn't change.

Therefore

$$W_{\text{whole system}} = 0$$

Since the walls are adiabatic $Q_{whole\ system} = 0$

$U_{system} = 0$ and (3) $T'_{final} = T_{initial} = 293.15$ K

This implies that

(4) $p_{Ne} V_{Ne} = p'_{Ne} V'_{Ne}$

(5) $p_{He} V_{He} = p'_{He} V'_{He}$

(6) $\dfrac{p_{Ne} V_{Ne}}{n_{Ne}} = \dfrac{p_{He} V_{He}}{n_{He}}$

The last equation follows from the fact that the piston is diathermic, so that $T_{Ne} = T_{He}$. Dividing (4) by (5)

$$\frac{V'_{Nc} p'_{Ne}}{V'_{He} p'_{He}} = \frac{V_{Ne} p_{Ne}}{V_{He} p_{He}} \qquad \frac{V_{Ne'}}{V_{He'}} = \frac{V_{Ne} p_{Ne}}{V_{He} p_{He}}$$

From (6), $\dfrac{p_{Ne}}{p_{He}} = \dfrac{V_{He} n_{Ne}}{V_{Ne} n_{He}} \qquad \dfrac{V_{Ne'}}{V_{He'}} = \dfrac{n_{Ne}}{n_{He}}$

Also from (6), $\dfrac{n_{He}}{n_{Ne}} = \dfrac{p_{He} V_{He}}{p_{Ne} V_{Ne}} = \dfrac{(1\text{ atm})A(0.6\text{ m})}{(5\text{ atm})A(0.4\text{ m})} = \dfrac{3}{10}$

Then from (2),

$$\frac{V_{Ne'}}{V_{He'}} = \frac{0.4\text{ m} + x}{0.6\text{ m} - x} = \frac{10}{3} \qquad x = 0.369\text{ m}$$

The final length of the neon compartment is 0.769 m.

25-21 (a) Initially, before ignition, $pV = nRT$

$$n_{air} = \frac{pV}{RT} = \frac{(30\text{ atm})(0.4\ \ell)}{(8.3144\ \frac{J}{mol\cdot K})(293.15\text{ K})} = 0.499\text{ mol}$$

(b) After ignition and before the expansion

$$Q = 1.2\text{ Cal} = n_{air} C_V \Delta T$$

(ignoring the small heat capacity of the oxidation products)

$$C_V(\text{air}) = 20.68\ \frac{J}{mol\cdot K}\ \text{(tabulated value)}$$

Therefore

$$\Delta T = \frac{Q}{nC_V} = \frac{1.2\text{ Cal}}{(0.499\text{ mol})(20.68\ \frac{J}{mol\ K})} = 486.9\text{ deg}$$

and $T_{final} = 20°\text{ C} + 486.9°\text{ C} = 506.9°\text{ C}$

The pressure and temperature before and after the constant volume explosion are related.

$$\frac{p_{after}}{T_{after}} = \frac{p_{before}}{T_{before}} \qquad p_{before}\frac{T_{after}}{T_{before}} = p_{after}$$

Therefore,

$$p_{after} = 30\text{ atm}\ \frac{780.05\text{ K}}{293.15\text{ K}} = 79.8\text{ atm}$$

(c) The volume before the adiabatic expansion is 0.40 ℓ and the volume after the expansion is 2.0 ℓ. If the subscript b represents quantities before and a represents quantities after the expansion,

$$p_b V_b^{\gamma} = p_a V_a^{\gamma}$$

$$p_b V_b = nRT_b \quad , \quad p_a V_a = nRT_a$$

$$T_a = \frac{p_a V_a}{p_b V_b} T_b = \left(\frac{V_b}{V_a}\right)^{\gamma} \frac{V_a}{V_b} T_b = \left(\frac{V_b}{V_a}\right)^{\gamma-1} T_b$$

$$\gamma_{air} = 1.403$$

$$\rightarrow\ T_a = \left(\frac{0.40}{2.0}\right)^{1.403-1} (780.05\text{ K}) = 407.8\text{ K}$$

(d) $$W_{explosion} = 0 \quad \text{(no expansion)}$$

$$W_{adiabatic} = \frac{p_b V_b - p_a V_a}{\gamma - 1} \quad , \quad p_a = p_b\left(\frac{V_b}{V_a}\right)^{\gamma}$$

$$W_{total} = W_{adiabatic} = p_b[V_b - (V_b/V_a)^{\gamma} V_a]/(\gamma-1)$$

$$= \frac{(79.8\text{ atm})(1.013\times 10^5\text{ Pa/atm})[0.4 - 2(0.4/2)^{1.403}]\times 10^{-3}\text{ m}^3}{1.403 - 1}$$

$$= 3830\text{ J}$$

(e) $$\text{efficiency} = \frac{W}{\Delta Q} = \frac{3830\text{ J}}{1.2\text{ Cal}\ (4186\ \frac{\text{J}}{\text{Cal}})} = 76.2\%$$

'25-22 (a) Let p_o be the atmospheric pressure. The compression of the gas is adiabatic, giving

$$p_o V_{initial}^{\gamma} = p_{final}\, V_{final}^{\gamma}$$

At equilibrium, the pressure of the piston is equal to the pressure of the gas.

$$\rho g y + p_o = p_{final} = p_o\left(\frac{V_{initial}}{V_{final}}\right)^{\gamma} = p\left(\frac{h}{h-y}\right)^{\gamma}$$

$$\frac{h}{h-y} = \left(1 + \frac{\rho g y}{p_o}\right)^{1/\gamma}$$

$$h[1 - (1 + \rho g y/p_o)^{1/\gamma}] = -y(1 + \rho g y/p_o)^{1/\gamma}$$

$$h = \frac{y}{1 - (1 + \rho g y/p_o)^{-1/\gamma}}$$

(b) For an isothermal process $\gamma = 1.$

$$h = \frac{y}{1-[p_o/(p + \rho gy)]} = \frac{y(p_o + \rho gy)}{\rho gy}$$

$$= \frac{p_o}{\rho g} + y$$

(c) $$W_{ab} = \int_{V_{initial}}^{V_{final}} p\, dV = nRT \int_{V_{initial}}^{V_{final}} \frac{dV}{V} = p_o V_{initial} \ln \frac{V_{final}}{V_{initial}}$$

$$= p_o V_{initial} \ln \frac{h-y}{h}$$

$U_b - U_A = \frac{\nu}{2} n R (T_b - T_a)$ 25-6/P.696

$c_p - c_v = R$ #25-12/P.699 & $C_v = \frac{\nu}{2} \cdot R$ # 25-13/P.699

ν = D.o.F.

$\gamma = \frac{C_P}{C_V} =$ { 1.3 (APPROX) POLYATOMIC GASES AT 1 ATM & T=15°C P.700) ; $\frac{5}{3}$ FOR MONOATOMIC GASES - P.700 WITH ν=3.) ; 1.40 FOR DIATOMIC IDEAL GASES P.701) ν = 5 RIGID, ν = 7 NON RIGID P.699 }

pV^γ = CONSTANT (ADIABATIC PATH)

FOR TRIATOMIC (NON COLLINEAR) ν = 6 RIGID, ν = 12 NON RIGID } P699

DEG. CELSIUS + 273.15 = KELVINS

R = 8.3144 J/mol-K = (P.647)

CAL × 4186 J/CAL = # JOULES (P.625)

CHAPTER 26

26-1 $$e_{carnot} = 1 - \frac{T_2}{T_1} = 1 - \frac{303.15\text{ K}}{373.15\text{ K}} = 0.1876$$

$$e'_{carnot} = 1 - \frac{T'_2}{T'_1} = 1 - \frac{303.15\text{ K}}{853.15\text{ K}} = 0.6447$$ ✓

$$\frac{e'}{e} = 3.44$$ ✓

26-2 (a) $e_{carnot} = 1 - \dfrac{T_2}{T_1} \qquad T_2 = (1 - e)T_1$

Therefore,

$$T_2 = (1 - 0.40)(250 + 273.15)\text{K} = 313.9 \text{ K}$$

$$= 40.7° \text{ C}$$

(b) $e_{carnot} = 1 - \dfrac{Q_2}{Q_2} \qquad Q_2 = (1 - e)Q_1$

Therefore

$$Q_2 = (1 - 0.40)(5 \text{ Cal}) = 3 \text{ Cal per cycle}$$

(c) $W = Q_1 - Q_2 = 5 \text{ Cal} - 3 \text{ Cal} = 2 \text{ Cal per cycle}$

(d) For the isothermal expansion ($\Rightarrow \Delta U = 0 \Rightarrow \Delta W_{in} = \Delta Q_{in}$)

$\Delta U = 0 \Rightarrow W_{in} =$ $$Q_1 = nRT \ \ell n \frac{V_b}{V_a}$$ #25-2/P.692 & P712

where $\dfrac{V_b}{V_a}$ is the volume expansion ratio.

Therefore,

$$5 \text{ Cal } (4186 \frac{\text{J}}{\text{Cal}}) = (0.8 \text{ mol})(8.3144 \frac{\text{J}}{\text{mol·K}})(523.15 \text{ K}) \ell n \frac{V_b}{V_a}$$

$$\frac{V_b}{V_a} = 409.5$$

(e) For the adiabatic expansion IDEA

IDEAL GAS ⇒ P700-701

$$p_b V_b^\gamma = p_c V_c^\gamma \quad , \quad p_b = \frac{nRT_b}{V_b} \quad , \quad p_c = \frac{nRT_c}{V_c}$$

$$\frac{V_c}{V_b} = \left(\frac{T_b}{T_c}\right)^{\frac{1}{\gamma - 1}}$$

where $\dfrac{V_c}{V_b}$ is the adiabatic volume expansion ratio

Thus $T_b = T_1$ and $T_c = T_2$

Then with $\gamma_{\text{ideal gas}} = \dfrac{5}{3}$

$$\frac{V_c}{V_b} = \left(\frac{523.15 \text{ K}}{313.89 \text{ K}}\right)^{3/2} = 2.15$$

26-3

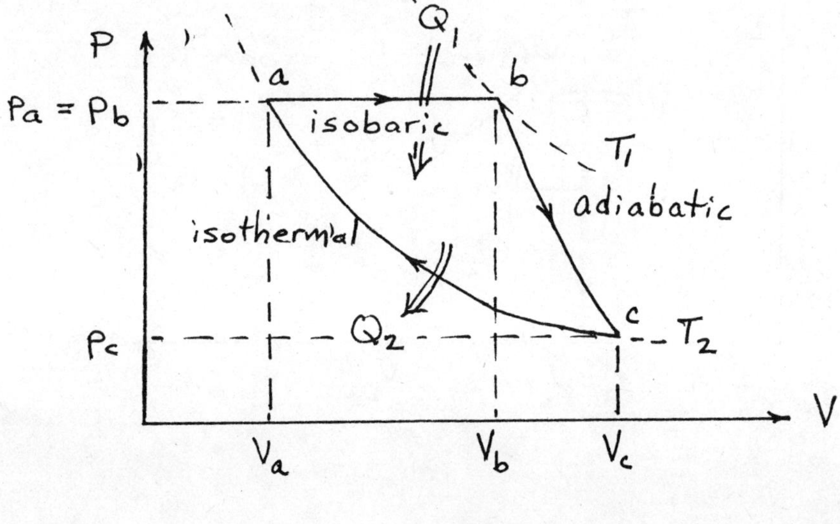

$$e = \frac{W}{Q_1} = \frac{Q_1 - Q_2}{Q_1} = 1 - \frac{Q_2}{Q_1}$$ # 26-2/p 712

$Q_1 = nC_p(T_1 - T_2)$ for isobaric expansion

$Q_2 = nRT_2 \ \ln \frac{V_c}{V_a}$ for isothermal compression P.712 BOTTOM

For isobaric expansion $\frac{T}{V}$ = constant

$$\frac{V_b}{V_a} = \frac{T_b}{T_a} = \frac{T_1}{T_2}$$

For adiabatic expansion pV^{γ} = constant

$pVV^{\gamma-1}$ = constant so $TV^{\gamma-1}$ = constant

$$\frac{V_c}{V_b} = \left(\frac{T_b}{T_c}\right)^{\frac{1}{\gamma-1}} = \left(\frac{T_1}{T_2}\right)^{\frac{1}{\gamma-1}}$$

Therefore, $\frac{V_c}{V_a} = \frac{V_c}{V_b} \times \frac{V_b}{V_a} = \frac{T_1}{T_2}\left(\frac{T_1}{T_2}\right)^{\frac{-1}{1-\gamma}} = \frac{T_1}{T_2}^{\frac{\gamma}{\gamma-1}}$

Then,

$$\text{efficiency} = 1 - \frac{nRT_2 \ \ln \left(\frac{T_1}{T_2}\right)^{\frac{\gamma}{\gamma-1}}}{nC_p(T_1 - T_2)}$$

or

$$e = 1 - \frac{RT_2 \frac{\gamma}{\gamma-1} \ln \frac{T_1}{T_2}}{C_p(T_1 - T_2)}$$

$\gamma = \frac{C_p}{C_v}$ gives $\frac{\gamma}{\gamma-1} = \frac{C_p}{C_p - C_v} = \frac{C_p}{R}$

so that $e = 1 - \frac{T_2}{T_1 - T_2} \ln \frac{T_1}{T_2}$

26-4

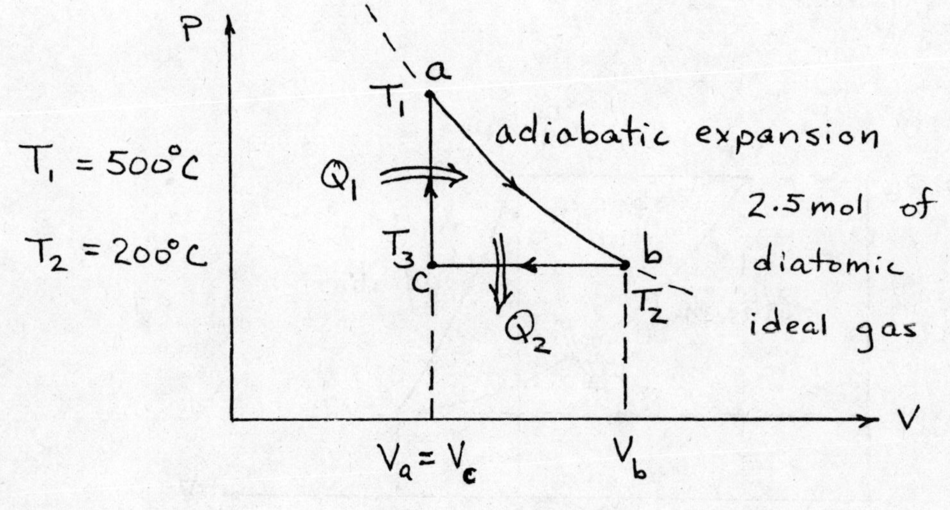

(a) $e = 1 - \frac{Q_2}{Q_1}$

$Q_1 = nC_v(T_1 - T_3)$ - constant volume

$Q_2 = nC_p(T_2 - T_3)$ - isobaric process

Therefore, $\frac{Q_2}{Q_1} = \frac{C_p(T_2 - T_3)}{C_v(T_1 - T_3)} = \gamma \frac{T_2 - T_3}{T_1 - T_3}$

and $e = 1 - \gamma \frac{(T_2 - T_3)}{(T_1 - T_3)}$

(b) For adiabatic expansion,

$pV^{\gamma} = \text{constant}$ or $TV^{\gamma-1} = \text{constant}$

Therefore, $T_1 V_a^{\gamma-1} = T_2 V_b^{\gamma-1}$

For an isobaric process $\frac{T}{V} = \text{constant}$

$$\frac{T_3}{T_2} = \frac{V_c}{V_b} = \frac{V_a}{V_b} = \left(\frac{T_2}{T_1}\right)^{\frac{1}{\gamma-1}}$$

or $T_3 = \left(\frac{T_2{}^{\gamma}}{T_1}\right)^{\frac{1}{\gamma-1}}$

$$\gamma = \frac{C_p}{C_v} = \frac{C_v + R}{C_v} \quad , \quad C_v = \tfrac{1}{2}\,\nu R$$

For a diatomic ideal gas $\nu = 5$, so that $\gamma = 1.4$

Then

$$T_3 = (473.15\text{ K})\left(\frac{473.15\text{ K}}{773.15\text{ K}}\right)^{\frac{1}{1.4-1}} = 138.6\text{ K}$$

(c) $e = 1 - 1.4\,\frac{473.15\text{ K} - 138.6\text{ K}}{773.15\text{ K} - 138.6\text{ K}} = 0.262$

(d) $W = Q_1 - Q_2 = eQ_1$

$Q_1 = nC_v(T_1 - T_3)$

$= (2.5\text{ mol})(\frac{5}{2} \times 8.3144\ \frac{\text{J}}{\text{mol K}})(773.15\text{ K} - 138.6\text{ K})$

$W = eQ_1 = (0.262)(3.297 \times 10^4\text{ J}) = 8639\text{ J}$

26-5

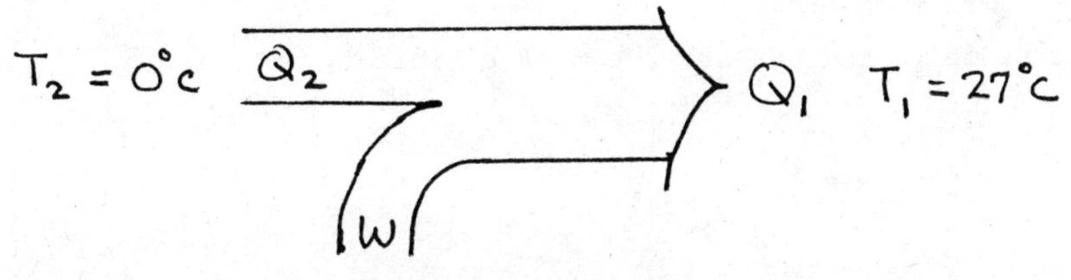

(a) Q_2 = heat removed from the ice

$$= mL_f = 50\ \text{kg}(3.337 \times 10^5\ \frac{\text{J}}{\text{kg}}) = 1.669 \times 10^7\ \text{J}$$

For a Carnot refrigerator,

$$\frac{Q_2}{Q_1} = \frac{T_2}{T_1} \qquad Q_1 = Q_2\frac{T_1}{T_2}$$

Therefore,

$$Q_1 = (1.669 \times 10^7\ \text{J})\frac{(27 + 273.15)\text{K}}{273.15\ \text{K}} = 1.833 \times 10^7\ \text{J}$$

(b) $W = Q_1 - Q_2 = 1.645 \times 10^6\ \text{J}$

is the energy that must be supplied to operate the refrigerator.

$$P = \frac{W}{t} \text{ so } t = \frac{W}{P} = \frac{1.645 \times 10^6\ \text{J}}{10^3\ \frac{\text{J}}{\text{s}}} = 1645\ \text{s}$$

$$= 27.42\ \text{min}$$

26-6 Given two adiabatic paths that intersect at c, construct an isothermal path that intersects the two adiabatic paths. There is only one path, ab, along which heat can enter the system. Therefore $W = Q$, which violates the second law of thermodynamics.

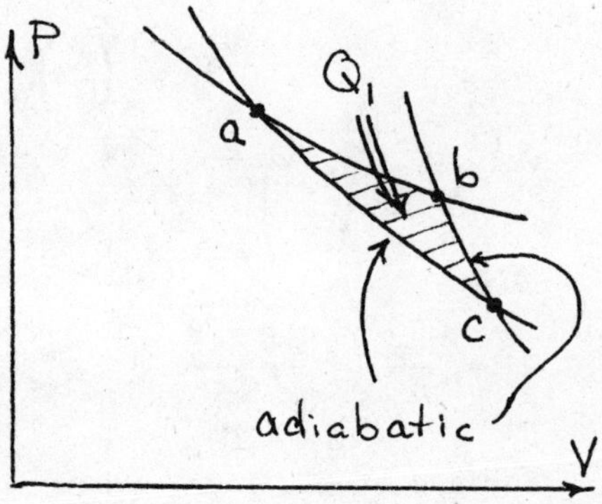

26-7 $$e_{actual} = \frac{W}{Q_1} = \frac{W}{W + Q_2}$$

$$= \frac{1500\ \text{J}}{1500\ \text{J} + 4.8\ \text{Cal}(4186\ \frac{\text{J}}{\text{Cal}})} = 6.95\%$$

$$e_{reversible} = e_{carnot} = \frac{T_1 - T_2}{T_1} = 1 - \frac{(100 + 273.15)\text{K}}{(180 + 273.15)\text{K}}$$

$$= \underline{17.65\%}$$

$$\frac{e_{actual}}{e_{reversible}} = \frac{6.95}{17.65} = 0.394$$

26-8 $W_{minimum} = W_{Carnot}$

because Carnot (reversible) engines are the most efficient.

$Q_2 = 0.10$ Cal

$$Q_1 = \frac{T_1}{T_2}Q_2 = \frac{(20 + 273.15)\text{K}}{(0 + 273.15)\text{K}}(0.10\ \text{Cal})$$

Therefore $W_{minimum} = Q_1 - Q_2 = 0.10732\ \text{Cal} - 0.10\ \text{Cal}$

$$= 0.00732 \text{ Cal} = 30.6 \text{ J}$$

26-9 A Carnot refrigerator delivers 0.50 Cal to a hot reservoir at $\theta_1 = 5.21$ K while doing 1750 J of work.

$$\frac{\theta_1}{\theta_2} = \frac{Q_1}{Q_2} \text{ for Carnot engines}$$

$$\theta_2 = \theta_1 \frac{Q_2}{Q_1} = \theta_1 \frac{Q_1 - W}{Q_1} = \theta_1 (1 - \frac{W}{Q_1})$$

Therefore

$$\theta_2 = 5.21 \text{ K} \left[1 - \frac{1750 \text{ J}}{(0.50 \text{ Cal})(4186 \frac{\text{J}}{\text{Cal}})}\right] = 0.854 \text{ K}$$

26-10

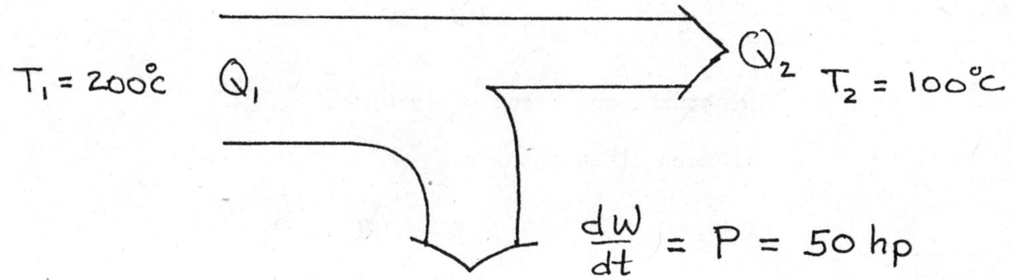

$$e_{actual} = 0.4 \, e_{reversible}$$

$$e_{reversible} = 1 - \frac{T_2}{T_1} = 1 - \frac{373.15 \text{ K}}{473.15 \text{ K}} = 0.2113$$

$$e_{actual} = 0.4 \, (0.2113) = 0.0845$$

As $e_{actual} = \frac{W}{Q_1}$, $Q_1 = \frac{1}{e_{actual}} W$

$$\frac{dQ_1}{dt} = \frac{1}{e_{actual}} \frac{dW}{dt} = \frac{1}{0.0845} (50 \text{ hp})(745.7 \frac{\text{W}}{\text{hp}})$$

$$= 4.41 \times 10^5 \text{ W}$$

or Q per cycle $= \frac{Q_1}{\text{sec}} \div \frac{\text{cycles}}{\text{sec}}$

$$= \frac{4.41 \times 10^5 \text{ W}}{(500 \text{ rpm})(\frac{1 \text{ min}}{60 \text{ s}})} = \frac{4.41 \times 10^5 \text{ W}}{8.333 \frac{\text{cycle}}{\text{s}}}$$

$$= 5.29 \times 10^4 \frac{\text{J}}{\text{cycle}}$$

26-11

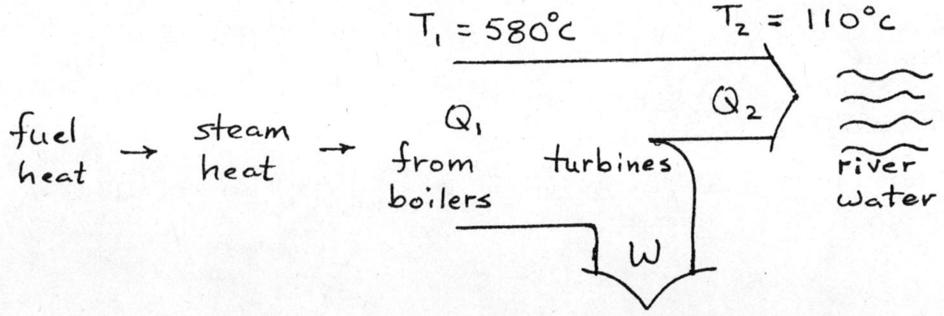

26-11 $e_{boilers} = 0.80$, $e_{turbines} = 0.40$, $e_{generators} = 0.95$

river flow rate $= \frac{dV}{dt} = 1800 \frac{m^3}{s}$

$\frac{1}{10}$ of the flow is used for cooling the turbine output

fuel heat of combustion $= L_c = 4.40 \times 10^7 \frac{J}{kg}$

(a) $e_{generator} = 0.95 = \frac{\text{power out of generators}}{\text{power into generators}}$

$= \frac{\text{power out of generators}}{\text{power out of turbines}}$

Therefore, $P_{turbines} = \frac{P_{generators}}{0.95} = \frac{1250 \text{ MW}}{0.95}$

$= 1315.8 \text{ MW}$

(b) $W_{turbines} = Q_{into\ turbines} - Q_{out\ of\ turbines} = Q_1 - Q_2$

$= e_{turbines} Q_{into\ turbines} = eQ_1$

$Q_1 = \frac{W}{e}$, $Q_2 = Q_1 - W = \frac{W}{e} - W = W(\frac{1-e}{e})$

Therefore

$$\frac{dQ_1}{dt} = \frac{1}{e}\frac{dW}{dt} = \frac{1}{0.40}(1315.8 \text{ MW}) = 3289.5 \text{ MW}$$

$$\frac{dQ_2}{dt} = \frac{dW}{dt}\frac{1-e}{e} = (1315.8 \text{ MW})\frac{0.6}{0.4} = 1973.7 \text{ MW}$$

(c) $e_{reversible} = e_{Carnot} = 1 - \frac{T_2}{T_1} = 1 - \frac{(110 + 273.15)\text{K}}{(580 + 273.15)\text{K}}$

$= 0.551$

This is the ideal turbine efficiency.

$$e_{overall} = \frac{W_{out\ gen}}{Q_{into\ boiler}} = \frac{W_{out\ gen}}{Q_{into\ gen}} \frac{Q_{into\ gen}}{Q_{into\ boiler}}$$

$$= e_{gen} \frac{W_{out\ turb}}{Q_{into\ turb}} \frac{Q_{into\ turb}}{Q_{into\ boiler}}$$

$$= e_{gen}\, e_{turb} \frac{W_{out\ boiler}}{Q_{into\ boiler}} = e_{gen}\, e_{turb}\, e_{boiler}$$

$$= (0.95)(0.40)(0.80) = 0.304$$

$$\frac{e_{overall}}{e_{reversible}} = \frac{0.304}{0.551} = 0.552$$

(d) As $e_{overall} = \frac{W_{out\ gen}}{Q_{into\ boiler}}$

$$\frac{dQ_{into\ boiler}}{dt} = \frac{1}{e_{overall}}(1250 \text{ MW}) = \frac{1}{0.304}(1250 \text{ MW}) = 4112 \text{ MW}$$

$$= \frac{dQ_{from\ oil}}{dt} = \frac{d}{dt}(m_{oil} L_c) = L_c \frac{dm_{oil}}{dt}$$

Now $\frac{\text{barrels}}{\text{hr}} = \frac{\text{kg}}{\text{s}} \frac{\text{s}}{\text{hr}} \frac{\text{barrel}}{\text{kg}}$

$$\frac{\text{kg}}{\text{barrel}} = (0.159 \frac{\text{m}^3}{\text{barrel}})(920 \frac{\text{kg}}{\text{m}^3}) = 146.3 \text{ kg/barrel}$$

$$\frac{\text{barrels}}{\text{hr}} = \frac{dm_{oil}}{dt}(146.3 \frac{\text{kg}}{\text{barrel}})^{-1}$$

$$= \frac{1}{L_c}\frac{dQ_{oil}}{dt}(146.3 \frac{\text{kg}}{\text{barrel}})^{-1}$$

$$= \frac{1}{4.40 \times 10^7 \frac{\text{J}}{\text{kg}}}(4112 \text{ MW})(3600 \frac{\text{s}}{\text{hr}}) \quad (146.3 \frac{\text{kg}}{\text{barrel}})^{-1}$$

$$= 2300 \frac{\text{barrels}}{\text{hr}}$$

(e) The rate at which heat is expelled from the turbines is

$$\frac{dQ_2}{dt} = 1973.7 \text{ MW}$$

$$Q_2 = m_{river}\, c_{river}\, \Delta T \qquad \frac{dQ_2}{dt} = \frac{dm_{river}}{dt} c_{river}\, \Delta T$$

where m_{river} is from $\frac{1}{10}$ of the river's flow.

Therefore,

$$\Delta T_{cooling\ water} = \frac{(1973.7 \times 10^6 \text{ W})}{(1.00 \frac{\text{Cal}}{\text{kg deg}})(\frac{1}{10} \times 1800 \frac{\text{m}^3}{\text{s}} \times 10^3 \frac{\text{kg}}{\text{m}^3})(4186 \frac{\text{J}}{\text{Cal}})}$$

$$= 2.62 \text{ deg}$$

Only half of this heat goes into heating the river water, the other half being dissipated into the air. This water is now remixed with the river.

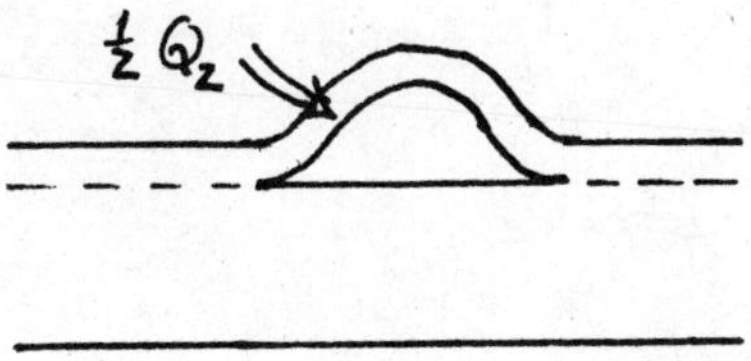

$\frac{1}{10} M_{river}$ is at $T_{river} + \frac{1}{2}\Delta T$

$\frac{9}{10} M_{river}$ is at T_{river}

Heat is exchanged between $\frac{1}{10} M_{river}$ and $\frac{9}{10} M_{river}$

$$\frac{1}{10} M_{river}\, c(T_{river} + \tfrac{1}{2}\Delta T - T_f) = \frac{9}{10} M_{river} c(T_f - T_{river})$$

$$10\, T_f = 10\, T_{river} + \tfrac{1}{2}\Delta T$$

so that

$$T_f - T_{river} = \Delta T_{river} = \frac{1}{20}\Delta T = 0.13 \text{ deg}$$

26-12 $e_{\text{ideal Otto cycle}} = 1 - \dfrac{1}{r^{\gamma-1}}$

With a compression ratio, $r = 6.8$

and $\gamma_{air} = 1.403$

$$e_{ideal} = 1 - (6.8)^{1-1.403} = 0.538$$

$$e_{actual} = (0.45)e_{ideal} = 0.242$$

$$e_{actual} = \frac{W}{Q_1} \text{ so } \frac{dQ_1}{dt} = \frac{1}{e}\frac{dW}{dt}$$

$$\frac{dQ_1}{dt} = \frac{1}{0.242}(30 \text{ hp} \times 747.7 \frac{W}{hp}) = 9.27 \times 10^4 \text{ W}$$

But $Q_1 = Q_{combustion} = m_{gasoline} L_c$

Therefore

$$\frac{dm}{dt} = \frac{1}{L_c}\frac{dQ_1}{dt} = \frac{9.27 \times 10^4 \text{ W}}{4.81 \times 10^7 \frac{J}{kg}} = 1.93 \times 10^{-3} \frac{kg}{s}$$

$$\frac{\ell}{h} = \frac{kg}{s} \times \frac{s}{h} \times \frac{\ell}{m^3} \times \left(\frac{kg}{m^3}\right)^{-1}$$

$$= (1.93 \times 10^{-3} \frac{kg}{s})(3600 \frac{s}{h})(10^3 \frac{\ell}{m^3}) \quad (740 \frac{kg}{m^3})^{-1} = 9.39 \frac{\ell}{h}$$

$$\frac{km}{\ell} = \frac{km}{h}\Big/\frac{\ell}{h} = \frac{95 \frac{km}{h}}{9.39 \frac{\ell}{h}} = 10.1 \frac{km}{\ell} \quad (23.8 \text{ mi/gal})$$

26-13

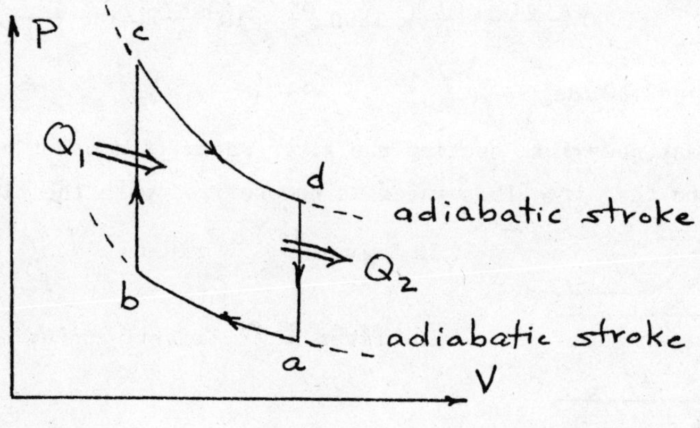

For path cd,

$$p_c V_c^\gamma = p_d V_d^\gamma \qquad p_c V_c V_c^{\gamma-1} = p_d V_d V_d^{\gamma-1}$$

$$T_c V_c^{\gamma-1} = T_d V_d^{\gamma-1}$$

$$\frac{T_d}{T_c} = \left(\frac{V_c}{V_d}\right)^{\gamma-1}$$

Similarly, for path ab,

$$\frac{T_a}{T_b} = \left(\frac{V_b}{V_a}\right)^{\gamma-1}$$

Then, as $V_b = V_c$ and $V_a = V_d$

$$\frac{T_d}{T_c} = \left(\frac{V_c}{V_d}\right)^{\gamma-1} = \left(\frac{V_b}{V_a}\right)^{\gamma-1} = \frac{T_a}{T_b} \quad \text{so} \quad \frac{T_d}{T_a} = \frac{T_c}{T_b}$$

Therefore,

$$\frac{1}{r^{\gamma-1}} = \frac{1}{\left(\frac{V_a}{V_b}\right)^{\gamma-1}} = \left(\frac{V_b}{V_a}\right)^{\gamma-1} = \frac{T_a}{T_b}$$

and

$$\frac{T_d - T_a}{T_c - T_b} = \frac{T_a\left(\frac{T_d}{T_a} - 1\right)}{T_b\left(\frac{T_c}{T_b} - 1\right)} = \frac{T_a}{T_b} = \frac{1}{r^{\gamma-1}}$$

26-14

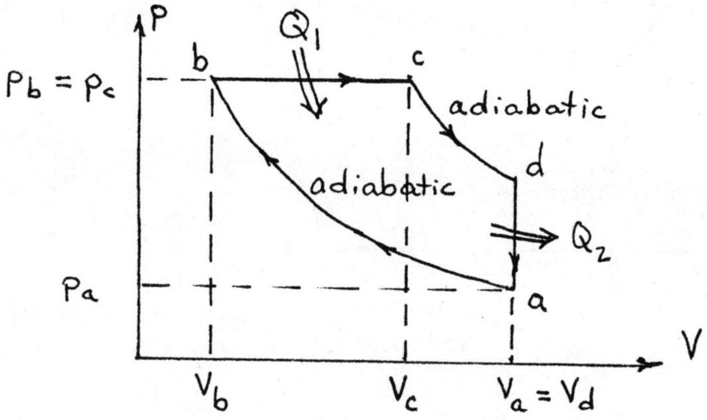

Along cd $\quad p_cV_c^\gamma = p_dV_d^\gamma \qquad T_cV_c^{\gamma-1} = T_dV_d^{\gamma-1} \quad (1)$

Along ab $\quad p_aV_a^\gamma = p_bV_b^\gamma \qquad T_aV_a^{\gamma-1} = T_bV_b^{\gamma-1} \quad (2)$

Along bc $\quad \frac{V_b}{T_b} = \frac{V_c}{T_c} \qquad T_c = T_b\frac{V_c}{V_b} \quad (3)$

Therefore all of the temperatures can be expressed in terms of T_b and the volumes.

From (1) $\quad T_d = T_c\left(\frac{V_c}{V_d}\right)^{\gamma-1} = T_b\frac{V_c}{V_b}\left(\frac{V_c}{V_d}\right)^{\gamma-1}$

From (2) $\quad T_a = T_b\left(\frac{V_b}{V_a}\right)^{\gamma-1}$

Then $e = 1 - \frac{1}{\gamma}\frac{T_d - T_a}{T_c - T_b}$

$$= 1 - \frac{1}{\gamma}\,\frac{T_b\left(\frac{V_c}{V_b}\right)\left(\frac{V_c}{V_d}\right)^{\gamma-1} - T_b\left(\frac{V_b}{V_a}\right)^{\gamma-1}}{T_b\left(\frac{V_c}{V_b}\right) - T_b}$$

Since $V_d = V_a$

$$e = 1 - \frac{1}{\gamma}\,\frac{\left(\frac{V_c}{V_d}\right)^{\gamma} - \left(\frac{V_b}{V_a}\right)^{\gamma}}{\frac{V_c}{V_d} - \frac{V_b}{V_a}} = 1 - \frac{1}{\gamma}\,\frac{\left(\frac{1}{r_E}\right)^{\gamma} - \left(\frac{1}{r_C}\right)^{\gamma}}{\frac{1}{r_E} - \frac{1}{r_C}}$$

26-15

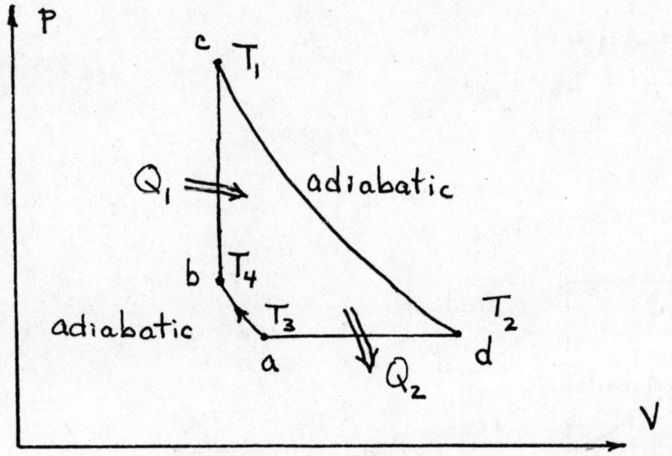

$$e = \frac{W}{Q_1} = \frac{Q_1 - Q_2}{Q_1} = 1 - \frac{Q_2}{Q_1}$$

Along bc V is constant.

Therefore, $Q_1 = nC_v\Delta T = nC_v(T_c - T_b)$

Along da p is constant

Therefore, $Q_2 = nC_p\Delta T = nC_p(T_d - T_a)$

Then

$$e = 1 - \frac{nC_p(T_d - T_a)}{nC_v(T_c - T_b)} = 1 - \gamma\,\frac{T_2 - T_3}{T_1 - T_4}$$

26-16

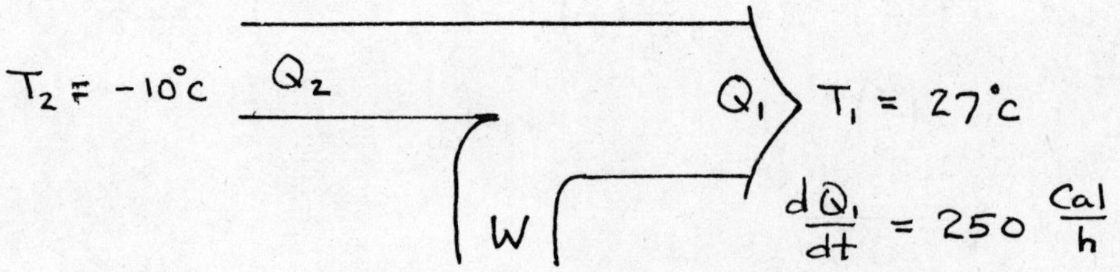

$$K_r(\text{Carnot}) = \frac{T_2}{T_1 - T_2} = \frac{263.15\ \text{K}}{300.15\ \text{K} - 263.15\ \text{K}} = 7.112$$

$$K_r(\text{actual}) = (0.55)K_r(\text{Carnot}) = 3.912 = K_r$$

By definition

$$K_r = \frac{Q_2}{W} = \frac{Q_2}{Q_1 - Q_2} = \frac{Q_1 - W}{W} \qquad Q_1 = (1 + K_r)W$$

$$\text{Therefore } \frac{dW}{dt} = \frac{1}{1 + K_r}\frac{dQ_1}{dt} = \frac{1}{1 + 3.912}\ 250\ \frac{\text{Cal}}{\text{h}}$$

$$= 50.9\ \frac{\text{Cal}}{\text{h}} = 0.0141\ \frac{\text{Cal}}{\text{h}} = 59.2\ \text{W}$$

26-17

$$\Delta S = \int_{T_1}^{T_2} \frac{dQ}{T}$$

$$Q = m_{H_2O}\, c_{H_2O}\, \Delta T \qquad dQ = m_{H_2O}\, c_{H_2O}\, dT$$

$$\Delta S = (mc)_{H_2O} \int_{T_1}^{T_2} \frac{dT}{T} = (mc)_{H_2O}\, \ell n\, \frac{T_2}{T_1}$$

In the given temperature range $c_{H_2O} \cong 1.00\ \frac{\text{Cal}}{\text{kg}\cdot\text{deg}}$

Therefore,

$$\Delta S = (1\ \text{kg})(1.00\ \frac{\text{Cal}}{\text{kg}\cdot\text{deg}})\,\ell n(\frac{373.15\ \text{K}}{273.15\ \text{K}}) = 0.312\ \frac{\text{Cal}}{\text{deg}}$$

26-18 For an ideal monatomic gas (He)

$$S(T,V) = nR(\frac{3}{2}\ \ell n\ T + \ell n\ V) + S_o$$

so that

$$\Delta S = S_2 - S_1 = nR(\frac{3}{2}\ \ell n\ \frac{T_2}{T_1} + \ell n\ \frac{V_2}{V_1})$$

Since $p_1V_1 = nRT_1$ and $p_2V_2 = nRT_2$

$$\frac{V_2}{V_1} = \frac{T_2}{T_1}\frac{p_1}{p_2} = \frac{373.15\ \text{K}}{273.15\ \text{K}}\ \frac{1\ \text{atm}}{10\ \text{atm}} = 0.1366$$

Therefore

$$\Delta S = (3\ \text{mol})(8.3144\ \frac{\text{J}}{\text{mol}\cdot\text{K}})[\frac{3}{2}\ \ell n\ \frac{373.15\ \text{K}}{273.15\ \text{K}} + \ell n(0.1366)]$$

$$= -38.0\ \text{J/K}$$

26-19 Since the system is thermally isolated heat lost by Al = heat gained by H_2O.

Therefore,

$$m_{Al}\, c_{Al}(T_{Al} - T) = m_{H_2O}\, c_{H_2O}(T - T_{H_2O})$$

where T is the final equilibrium temperature.

$$(0.30\ \text{kg})(0.216\ \frac{\text{Cal}}{\text{kg deg}})(85°\text{C} - T) = (0.50\ \text{kg})(1.00\ \frac{\text{Cal}}{\text{kg}\cdot\text{deg}})(T - 25°\text{C})$$

$$T = 31.9°\text{C}$$

Then

$$\Delta S_{system} = \Delta S_{Al} + \Delta S_{H_2O} = \int_{T_{Al}}^{T} \frac{dQ_{Al}}{T} + \int_{T_{H_2O}}^{T} \frac{dQ_{H_2O}}{T}$$

$$\text{or } \Delta S = m_{Al}\, c_{Al}\, \ell n \frac{T}{T_{Al}} + m_{H_2O}\, c_{H_2O}\, \ell n \frac{T}{T_{H_2O}}$$

$$\Delta S = (0.30 \text{ kg})(0.216 \frac{\text{Cal}}{\text{kg deg}}) \ell n \frac{(31.9 + 273.15)\text{K}}{(85 + 273.15)\text{K}}$$

$$+ (0.50 \text{ kg})(1.00 \frac{\text{Cal}}{\text{kg deg}}) \ell n \frac{(31.9 + 273.15)\text{K}}{(25 + 273.15)\text{K}} = 1.00 \text{ Cal/K}$$

26-20 i = state with ice cube and water

f = equilibrium state

P_i = probability that state i occurs $\propto \Omega_i$

P_f = probability that state f occurs $\propto \Omega_f$

where $S_i = k\ell n\Omega_i$ and $S_f = \ell n\Omega_f$

$\Omega_i = e^{S_i/k}$, $\Omega_f = e^{S_f/k}$

Then $$\frac{P_i}{P_f} = \frac{\Omega_i}{\Omega_f} = e^{(S_i - S_f)/k} = e^{-\Delta S/k} = 10^{-6}$$

$$\Delta S = -k\, \ell n(10^{-6})$$

$$= f \times \Delta S_{example} = f(0.100 \text{ J/K})$$

Therefore $$f = \frac{\Delta S}{\Delta S_{example}} = \frac{-k\ell n(10^{-6})}{0.100 \text{ J/K}} = \frac{6\ k\ \ell n(10)}{0.100 \text{ J/K}}$$

$$= \frac{6 \times 1.38 \times 10^{-23} \text{J/K} \times \ell n\ 10}{0.100 \text{ J/K}} = 1.91 \times 10^{-21}$$

f is the factor by which the masses in the example must be reduced to make $\frac{P_i}{P_f} = 10^{-6}$.

In the example m_{ice} = 10 g.

The reduced mass is f × 10 g = 1.91×10^{-20} g.

molecules in ice cube = (numbers of moles)N_o

$$= \frac{1.91 \times 10^{-20} \text{ g}}{18.015 \frac{\text{g}}{\text{mol}}} (6.022 \times 10^{23} \frac{\text{molecules}}{\text{mol}}) = 637 \text{ molecules}$$

The return to the state with the ice cube is still quite improbable

$(10^{-6} \times P_{equilibrium})$ even when only 637 molecules of ice are to be formed.

26-21 The temperature is constant during the phase change

$$\Delta S = \int \frac{dQ}{T} = \frac{Q}{T_{vapor}}$$

For one mole

$$Q = \frac{mass}{mol} L_v = ML_v$$

$$\Delta S_{H_2O} = \frac{(18.015 \frac{g}{mol})(2.2567 \times 10^6 \frac{J}{kg})}{(100 + 273.15)K} = 1.089 \times 10^2 \text{ J/K}$$

$$\Delta S_{Ag} = \frac{(107.87 \frac{g}{mol})(2.323 \times 10^6 \frac{J}{kg})}{(2.63 + 273.15)K} = 1.029 \times 10^2 \text{ J/K}$$

$$\Delta S_{Hg} = \frac{(200.59 \frac{g}{mol})(2.956 \times 10^5 \frac{J}{kg})}{(356.58 + 273.15)K} = 0.942 \times 10^2 \text{ J/K}$$

$$\Delta S_{N} = \frac{(28.013 \frac{g}{mol})(1.994 \times 10^5 \frac{J}{kg})}{(-195.8 + 273.15)K} = 0.722 \times 10^2 \text{ J/K}$$

The disorder introduced into a system with a fixed number of molecules by changing from the liquid to the gaseous state is approximately independent of the type of molecule.

26-22 (a) During the isothermal expansion heat is absorbed from the hot reservoir at $T_1 = 450$ K

$$Q_1 = nRT_1 \ \ell n \frac{V_b}{V_a}$$

$$= (2 \text{ mol})(8.3144 \frac{J}{mol \cdot K})(450 \text{ K}) \ell n(\frac{3.5\ell}{0.8\ell}) = 1.10 \times 10^4 \text{ J}$$

(b) For a Carnot engine $\frac{Q_2}{Q_1} = \frac{T_2}{T_1}$

$$Q_2 = \frac{T_2}{T_1} Q_1 = \frac{300 \text{ K}}{450 \text{ K}} (1.10 \times 10^4 \text{ J}) = 7.36 \times 10^3 \text{ J}$$

(c) $W = Q_1 - Q_2 = 3.68 \times 10^3$ J

26-23 $$Q = \frac{T_2}{T_1} \frac{T_1 - T_3}{T_2 - T_3} Q_1 \qquad T_1 = (200 + 273.15)K$$

$$T_2 = (20 + 273.15)K, \quad T_3 = 273.15 \text{ K}$$

$$Q = \frac{293.15 \text{ K}}{473.15 \text{ K}} \frac{200 \text{ deg}}{20 \text{ deg}} Q_1 = 6.2 Q_1$$

26-24 Cost per mile driving $= \frac{cost}{gal} \frac{gal}{mi}$

$$\cong \frac{\$1.25}{gal} \frac{1 \text{ gal}}{12 \text{ mi}} \cong \frac{10¢}{mile}$$

$$\text{cost per mile running} = \left(\frac{\text{cost}}{\text{cheeseburger}}\right)\left(\frac{\text{cheeseburger}}{\text{mi}}\right)$$

$$\cong \frac{\$1.00}{\text{burger}} \frac{90\ \frac{\text{Cal}}{\text{mi}}}{350\ \frac{\text{Cal}}{\text{burger}}} \cong \frac{26¢}{\text{mile}}$$

The driving is more cost effective (1980)

26-25

$$e_{max} = e_{Carnot} = 1 - \frac{T_2}{T_1} = 1 - \frac{(10 + 273)\text{K}}{(30 + 273)\text{K}} = 6.6\%$$

26-26

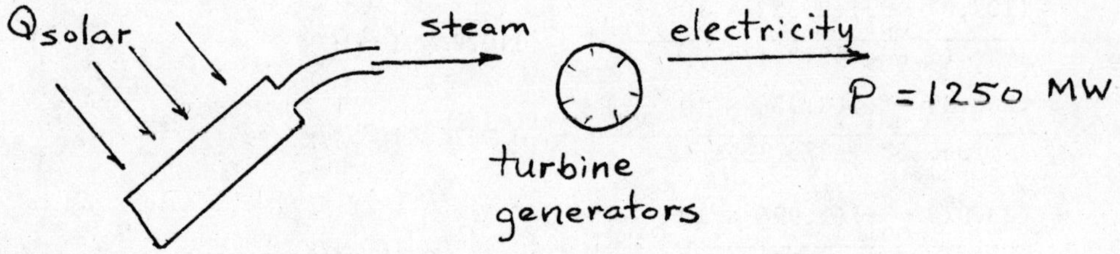

$$e_{overall} = (e_{solar})(e_{turbine}) = 0.30$$

$$\frac{dQ_{solar}}{dt}\Big/\text{m}^2 = 0.68\ \frac{\text{KW}}{\text{m}^2} \text{ in daylight}$$

$$e_{overall} = \frac{W_{out}}{Q_{solar}} \qquad \frac{dQ_{solar}}{dt} = \frac{1}{e}\frac{dW_{out}}{dt} = \frac{P}{e}$$

$$\frac{dQ_{solar}}{dt} = \left(\frac{dQ_{solar}}{dt}\Big/\text{m}^2\right)\left(\text{Area}_{reflector}\right) = \frac{1250\ \text{MW}}{0.30}$$

$$A = \frac{1250 \times 10^6\ \text{W}}{0.30} \frac{1}{0.68\ \frac{\text{KW}}{\text{m}^2}} = 6.13 \times 10^6\ \text{m}^2$$

$$= 6.13\ \text{km}^2$$

26-27

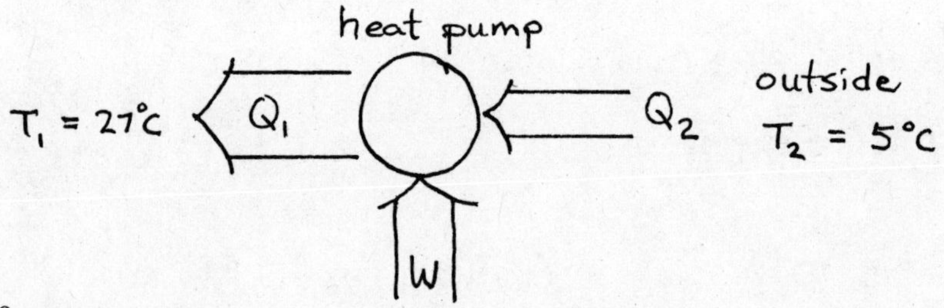

$$\frac{dQ_1}{dt} = 10^5\ \frac{\text{Cal}}{\text{h}}$$

$$K_h = (0.55)K_h(\text{Carnot}) = (0.55)\frac{T_1}{T_1 - T_2} = (0.55)\frac{300.15\ \text{K}}{22\ \text{K}} = 7.50$$

$$K_h = \frac{Q_1}{W} \text{ so } \frac{dW}{dt} = \frac{1}{K_h}\frac{dQ_1}{dt} = \frac{1}{7.50}\left(10^5\ \frac{\text{Cal}}{\text{h}}\right)$$

$$\text{or } \frac{dW}{dt} = 1.33 \times 10^4\ \frac{\text{Cal}}{\text{h}} = 15.5\ \text{kW}$$

26-28 $\Delta S_{universe} = \Delta S_{H_2O} + \Delta S_{rest\ of\ universe} \cong S_{H_2O}$

$$\Delta S_{H_2O} = \frac{\Delta Q}{T} = \frac{m_{H_2O}L_V}{T} = \frac{(1\ \text{kg})(585\ \frac{\text{Cal}}{\text{kg}})}{(20 + 273)\text{K}} = +\ 2.00\ \frac{\text{Cal}}{\text{deg}}$$

26-29 During the adiabatic steps of the Carnot cycle $dQ = 0$ so $dS = \frac{dQ}{T} = 0$; thus S is constant. As the temperature is constant during the isothermal steps, the Carnot cycle is a rectangle on a T - S diagram.

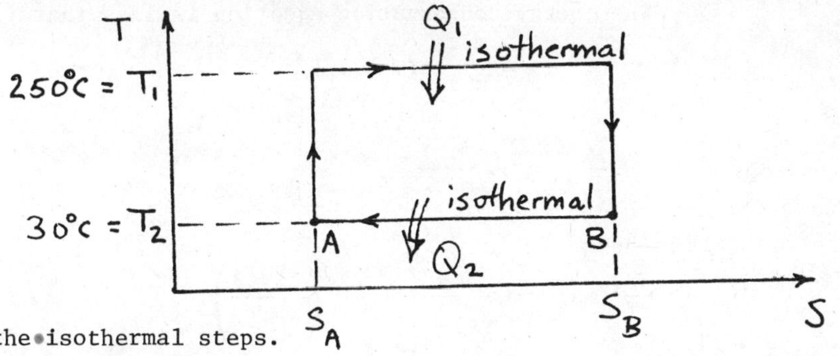

The work is done during the isothermal steps.

$$W = Q_1 - Q_2 = \int_1 dQ - \int_2 dQ$$

where Q_1 and Q_2 are positive.

$$\rightarrow W = \int_1 T_1\ dS - \int_2 T_2\ dS$$

$$= T_1(S_B - S_A) - T_2(S_B - S_A) = (T_1 - T_2)(S_B - S_A)$$

Therefore

$$W = 2000\ \text{J} = (T_1 - T_2)(S_B - S_A)$$

$$\rightarrow S_B - S_A = \frac{2000\ \text{J}}{250\ \text{K} - 30\ \text{K}} = 9.09\ \frac{\text{J}}{\text{K}}$$

26-30

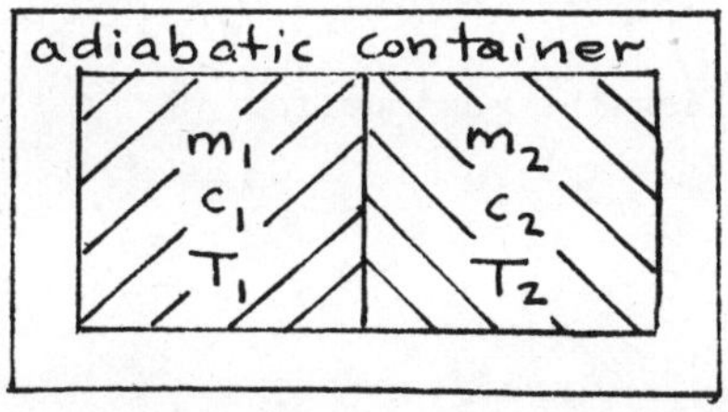

$T_2 < T_1$

$T_1 \rightarrow T$

$T_2 \rightarrow T'$

(a) Energy conservation requires that

$Q_1 + Q_2 = 0$ (no heat escapes the container)

$Q_1 = m_1c_1(T - T_1) \qquad Q_2 = m_2c_2(T' - T_2)$

$m_1c_1(T - T_1) = m_2c_2(T_2 - T')$

$$\Delta S = \Delta S_1 + \Delta S_2 = \int_{T_1}^{T} \frac{dQ_1}{T} + \int_{T_2}^{T'} \frac{dQ_2}{T}$$

$$= m_1c_1 \int_{T_1}^{T} \frac{dT}{T} + m_2c_2 \int_{T_2}^{T'} \frac{dT}{T}$$

$$= m_1c_1 \ln\frac{T}{T_1} + m_2c_2 \ln\frac{T'}{T_2}$$

(b) $$\frac{d}{dT}(\Delta S) = \frac{d}{dT}(m_1c_1\ln\frac{T}{T_1}) + \frac{d}{dT}(m_2c_2 \ln\frac{T'}{T_2})$$

$$= m_1c_1\frac{1}{T} + \frac{d}{dT'}(m_2c_2 \ln\frac{T'}{T_2})\frac{dT'}{dT}$$

$$= \frac{m_1c_1}{T} + \frac{m_2c_2}{T'}\frac{dT'}{dT}$$

The energy conservation equation implies that

$$T' = T_2 - \frac{m_1c_1}{m_2c_2}(T - T_1)$$

$$\frac{dT'}{dT} = -\frac{m_1c_1}{m_2c_2}$$

Then

$$\frac{d}{dT}(\Delta S) = \frac{m_1c_1}{T} + \frac{m_2c_2}{T'}\left(-\frac{m_1c_1}{m_2c_2}\right)$$

$$= m_1c_1(\frac{1}{T} - \frac{1}{T'})$$

Therefore $\frac{d}{dT}(\Delta S) = 0$ and $T = T'$

$$\frac{d^2}{dT^2}(\Delta S) = -\frac{m_1c_1}{T^2} - m_1c_1\frac{d}{dT}\left(\frac{1}{T'}\right)$$

$$= -m_1c_1[\frac{1}{T^2} - \frac{1}{T'^2}\frac{dT'}{dT}]$$

$$= -m_1c_1[\frac{1}{T^2} + \frac{m_1c_1}{m_2c_2}\frac{1}{T^2}]$$

Thus

$$\frac{d^2}{dT^2}(\Delta S) < 0$$

so that the equilibrium condition, $T = T'$ provides the maximum for S.

If i = initial state, f = final state

probability of state i = $P_i \propto \Omega_i = e^{S_i/K}$

probability of state f = $P_f \propto \Omega_f = e^{S_f/K}$

then $\frac{P_f}{P_i} = e^{(S_f - S_i)/K} = e^{\Delta S/K}$

The equilibrium condition has ΔS at its maximum, so that the equilibrium state (f) is the state of largest probability.